Intellectual Capital and Technological Innovation: Knowledge-Based Theory and Practice

Pedro López Sáez
Universidad Complutense de Madrid, Spain

Gregorio Martín de Castro
Universidad Complutense de Madrid, Spain

José Emilio Navas López
Universidad Complutense de Madrid, Spain

Miriam Delgado-Verde
Universidad Complutense de Madrid, Spain

Information Science REFERENCE

INFORMATION SCIENCE REFERENCE

Hershey · New York

Director of Editorial Content: Kristin Klinger
Director of Book Publications: Julia Mosemann
Acquisitions Editor: Lindsay Johnston
Development Editor: Joel Gamon
Typesetter: Gregory Snader
Production Editor: Jamie Snavely
Cover Design: Lisa Tosheff
Printed at: Yurchak Printing Inc.

Published in the United States of America by
Information Science Reference (an imprint of IGI Global)
701 E. Chocolate Avenue
Hershey PA 17033
Tel: 717-533-8845
Fax: 717-533-8661
E-mail: cust@igi-global.com
Web site: http://www.igi-global.com/reference

Library of Congress Cataloging-in-Publication Data

Intellectual capital and technological innovation : knowledge-based theory and practice / Pedro Lopez Saez ... [et al.], editors.
p. cm.
Includes bibliographical references and index.
Summary: "This book provides theoretical frameworks and empirical research findings to assist managers, academics and consultants gain a thorough understanding of intellectual capital and social capital as a base for improving their tasks"--Provided by publisher.
ISBN 978-1-61520-875-3 (hardcover) -- ISBN 978-1-61520-876-0 (ebook) 1. Intellectual capital. 2. Technological innovations--Social aspects. I. Lopez Saez, Pedro.
HD53.I5774 2010
338'.064--dc22

2010010166

British Cataloguing in Publication Data
A Cataloguing in Publication record for this book is available from the British Library.

Editorial Advisory Board

Table of Contents

Detailed Table of Contents

examination of these issues with an eye toward educational solutions at all levels. A pilot study was conducted of international college students at a small midwestern university, to serve as a model for conducting your own needs analysis. The analysis should help you to begin thinking differently about the educational needs for globalized classrooms and work environments while building the needed intellectual capital to survive and thrive under changing conditions.

This chapter examines the notion of national intellectual capital, which could bring new insights to the existing national science and technology policy thinking. This notion led the chapter proposes a framework to measure national intellectual capital, and the investigation based on the framework was applied in Panama. The results revealed that the Panamanian intellectual capital faces decreasing supply of human capital in S&T fields, service-dominant market capital, weak and less indigenous innovation capital, and a steady increase of process capital. Some intellectual capital policy implications are drawn for Panama and other developing countries.

This chapter studies the relations between the emergent concept of Organizational Identity and the concept of Intellectual Capital. Specifically, the chapter focuses on the idea that when new technology-based firms develop the social process of construction of their Organizational Identity, at the same time they develop the social interaction needed to define variables of their Organizational Capital. This proposition is grounded in a theoretical review of the concept of Organizational Identity and Organizational Capital in the Intellectus Model, and it is empirically tested in five case studies of new technology-based firms created at Madrid Science Park. Finally, the limitations and suggestions for future research as well as conclusions of the study are presented.

Section 2
Technological Innovation

In this chapter we focus on a problem-oriented view of managing innovation. In particular, we ask what it means to manage innovation well, so that innovation outcomes are enhanced. We start by briefly reflecting on the rise in value of intangibles as an asset class (notably knowledge), and by defining terms. We then briefly review selected literature on strategic management as an example of the typical manner in which innovation has been previously approached. Looking at selected macro- and micro-organizational theories we reflect on several shortcomings of extant literature. We then introduce the knowledge-based view and identify a particular sub-field, the problem finding/problem solving (PF/PS) perspective, as one promising way to address these issues. We identify several challenges in the present formulation of the perspective, and discuss further research opportunities. Though the PF/PS perspective presently lacks compelling empirical evidence, it shows promise for developing into a useful logic for managing and enhancing innovation.

Technological capital drives technological innovation if it is appropriately used; in reality, however, many kinds of management constraints hamper its effective use. This chapter elucidates changes in the use of technological capital from the technology spillover perspective. Tracking patent applications and analyzing the relationships among them have enabled us to conduct microscopic analyses of technology spillovers. We classify technology spillover structures by organization and technological field, forming a matrix of two-by-two arrays and applying it to Japan's R&D in compound semiconductor materials. In the 1980s, technology management was affected by the characteristics of an industrial society and spillover was therefore limited to individual firms and technological fields. In the 1990s, technology spillover shrank, reflecting organizational inertia in technology management. In the early 2000s, technology spillover broadened across a wider range of industries and technological fields, reflecting changes in science and technology policy and technology management that was ripe for open innovation.

The dynamic capabilities Perspective has emerged in the last decade to improve the explanatory power of the Resource-Based View of the firm with regard to obtaining sustainable competitive advantage by firms operating in rapid change environments. However, some aspects of its conceptualization, especially in relation to the factors that make up them, do not appear to be entirely clear. Accordingly, the aim of the present paper is to provide a conceptual synthesis of the construct, trying to identify the key factors involved in the development of dynamic capabilities by firms. Based on a sample of 177 companies around the world operating in the software industry, we derive an empirical study. Our findings suggest that firms that jointly develop absorptive capacity and innovative capacity achieve better results.

Section 3
Intellectual Capital and Technological Innovation

The growing importance of knowledge as a key productive factor in today's society requires a change in thinking about innovation in general terms as well as in specific terms, such as technological innovation, product innovation and organizational innovation (Nonaka, 1994). Also, companies are aware that knowledge is the most valuable and strategic resource to face the current environment (Chen et al., 2004). In this sense, intellectual capital is increasingly interesting to companies that get their benefits through innovation and knowledge. For this reason, it's important to study the relationship between intellectual capital and innovation (Edvinsson and Sullivan, 1996). In fact, it is widely accepted that an organization capability to innovate is closely tied to it is intellectual capital, or its ability to utilize its knowledge resources (Subramaniam and Youndt, 2005). So, this research proposes a theoretical model to identify and classify the different elements of intellectual capital (human capital, organizational capital, technological capital, relational capital and social capital) as a source of various types of technological innovation (product, process, radical and incremental innovation), analyzing the relationship between each of those components and the four types of innovation.

Organizational knowledge, as well as their different manifestations, may play a critical role for effective development of the firm processes and products (Kogut and Zander, 1996; Teece, 1998; Alegre and Lapiedra, 2005). The present work explores the influence that the different manifestations of organizational knowledge, or structural capital, show on the innovation capability of professional service firms. To achieve this purpose, in a first section an exploratory factor analysis is carried out. From it, the main blocks of organizational capital are obtained. In the next step, the basic hypotheses about the positive influence of these types of structural assets, on firm innovation are presented.

This chapter examines a comprehensive list of intellectual capital (IC)-related indicators for developing corporate R&D capabilities along the input-process-result (IPR) processes. Via factor analysis, 43 R&D related IC indicators were abstracted into 11 factors. Corporate R&D managers prioritized these IC factors by completing analytical hierarchy process (AHP) questionnaires. The results of AHP are as

Firm's environment is becoming more complex and dynamic, consequently continuous innovation is considered a key element to gain competitive advantage. According to the knowledge-based view, knowledge is considered the central element in the innovation process and its effective management is the key to achieve higher innovation performance. Since social networks involve knowledge and information embedded into inter-organizational network, social capital is considered the engine fuelling the innovation activities. Through an intensive review of the literature, the chapter seeks to uncover the effect of social capital on innovation through enhancing its fundamental knowledge-based activities. The social capital literature provided controversial approaches about its contribution to the innovation performance. The chapter contributes to the literature by mapping the different views, trying to clarify its contribution to the enhancement of knowledge-based activities, and revealing the effect of different social configurations on innovation outcomes. To do so, the chapter first analyzes the different views of social capital and explains its contribution within the organization studies. Further, the key role of the knowledge and its management to the innovation outcomes is discussed. Finally, the chapter clarifies the implications of social capital to the innovation performance and its typologies as radical or incremental ones, as well as designs a theoretical framework for recognizing both its beneficial and constraining effect. Further, the chapter proposes a balanced approach to the structural social capital reconciling two controversial camps in the literature.

In the last decades, the cluster approach to innovation has become very influential among policymakers. Clusters are recognized as the primary locus of social capital (SC) and civic engagement and are thus able to foster company innovativeness. This chapter investigates the effectiveness of top-down cluster policies in creating new SC and the ability of this SC to create new intellectual capital (IC), hence improving firms' innovativeness. For this purpose, two case studies about top-down clusters in Italy were developed. While in literature the SC-IC-innovation link has been largely investigated as well as the power of traditional clusters to promote SC, to the knowledge of the authors, the same power has not been investigated with reference to top down clusters. The purpose of this chapter is to investigate such links, thus contributing to fill a gap in the extant literature. Examination of the cases in this study revealed almost opposite results in terms of the links investigated.

Preface

INTELLECTUAL CAPITAL AND TECHNOLOGICAL INNOVATION: KNOWLEDGE-BASED THEORY AND PRACTICE. AN INTRODUCTION

This volume compiles a series of works devoted to the analysis of intellectual capital, innovation processes and outcomes, the link between the former two concepts, and also related to a phenomenon increasingly important for both intellectual capital and innovation, that is, social capital. This publication follows the way of the Intellectual Capital-Based View, considering that innovation is the best representation of competitive advantage in the present day. This first chapter introduces the four sections and fourteen contents chapters that built up the book, as well as the theoretical framework and concepts that lie at the heart of this project.

MAIN OBJECTIVE OF THE BOOK

This volume is titled 'Intellectual Capital and Technological Innovation: Knowledge-Based Theory and Practice'. In short, it compiles a series of works devoted to the analysis of intellectual capital, innovation processes and outcomes, the link between the former two concepts, and also related to a phenomenon that is quite close for both intellectual capital and innovation, that is, social capital.

The main idea behind a research paper compilation like this is quite simple in essence, but quite complex in practice, because it involves intangible and multifaceted theoretical constructs as intellectual capital, technological innovation, and social capital. The idea, trying to express it in a single sentence, is that knowledge stocks (or intellectual capital), is one of the main determinant factors (if not the most) for obtaining innovations, which in turn improve the initial knowledge base and allow the starting of a virtuous cycle, when the innovator and owner of that intellectual capital socially interacts with other agents, developing also what can be labelled as social capital.

Although in the following chapters, readers may find formal and fully-developed definitions based on literature reviews and academic conversations, we are outlining the three main concepts that the book deals with. Thus, in the following paragraphs, brief definitions for intellectual capital, innovation, and social capital are provided, as well as the main assumptions of the Resource-Based View and the Intellectual Capital-Based View.

Intellectual capital is usually considered as a synonym for intangible assets (Edvinsson and Sullivan, 1996; Bontis, 1998; CIC, 2003). Assets and capital are economic and accountability terms, but the intellectual adjective makes reference to a certain dimension that has been generally neglected by economists and accountants due to the inherent difficulties that intangibles show for an appropriate treatment and modelling. For this reason, intangible assets have also been named as hidden assets or invisible assets. And why has been intellectual capital neglected while land, labour, and capital have been so deeply studied? Because the intangible nature of intellectual capital makes especially difficult to measure, monitor, and in some cases enforce ownership for this kind of assets. Besides, the term intellectual capital includes a wide range of assets, each of them tied to a different kind of knowledge piece. Thus, we can find human, organizational, technological, or market knowledge that can be encapsulated in commitment and motivation, organizational culture and design, patents and routines, customer loyalty and reputation, and so on, showing different characteristics and implications in each case. For this reason, a deep study of intellectual capital is needed, trying to open the usually hidden or invisible black box in which they are generally contained for economists and accountants.

Innovation essentially involves the introduction of something new (Myers and Marquis, 1969; Damanpour and Evan, 1984; OECD, 2005). The Schumpeterian notion of innovation ties this novelty to inputs and outputs, regarding the introduction of new raw materials, productions processes, markets, supplies, or organization. As we can see, all the fields in which innovation can take place are tied to different kinds of intellectual capital or intangible assets. For this reason, it makes sense to think about intellectual capital as an antecedent for innovation. In general, academic literature has focused on radical versus incremental, and product versus process innovation. Nevertheless, in a rapid change environment as the present one, in which evolution and disruption coexist, and in economies becoming more service-oriented every day, this theoretical distinctions become increasingly blurred, and well defined and contextualized settings for analyzing innovation are needed. Thus, taking into account the most critical intangible assets needed for each kind of innovative outcomes in a certain setting becomes the only way for improving the study of the innovation phenomena.

Social capital represents the benefits that a certain agent can obtain from the network of relations in which it gets involved (Granovetter, 1985; Nahapiet and Ghoshal, 1998; Adler & Kwon, 2002). This concept is gaining and outstanding interest in recent years for several disciplines. Social capital has been related to intellectual capital because it shows an intangible nature itself, and because it can allow to develop and strengthen intellectual capital from external sources thanks to network ties. Social capital has also been considered as a critical antecedent or contextual trigger for innovation, as it can be followed from the notion of inter-firm collaborative agreements, innovation networks and systems, or clusters and industrial districts. Nevertheless, if the intangible nature and heterogeneous contents of intellectual capital require contextualized analysis for each kind of innovation according to firm, industry, or geographical variables, social capital shows an even more fuzzy nature that makes it an extremely specific asset. Thus, even two agents taking part in the same network of relations will not be at the same positions within that network. This determines the range of benefits that each of them could reach. Besides, the own intellectual capital available also determines the value of each relation because it locks in the possible combinations in order to create novel or improved applications, services, products or business processes. Social capital is an understudied phenomenon that, for this volume deserves great attention, because as it has been mentioned, can be critical for both intellectual capital and technological innovation.

The Resource-Based View (Wernerfelt, 1984; Barney, 1991, 2001; Peteraf, 1993) is the most influential paradigm for analyzing competitive advantage in the present. Although the notions of the Resource-Based View can be applied to different levels (i.e. business, corporation, network, region, nation…) explaining competitive advantages among agents, this theoretical stream was born within the field of strategic management. Its basic claim is that sustained differences in performance come mainly from the differences in assets endowments among competitors. This way, two axioms are at the heart of the Resource-Based View: First, agents show heterogeneity about their asset endowments, and second, that asset endowments can not be transformed at will by decision-makers, because some assets show imperfect transferability and barriers for imitation. In an intuitive way, it makes sense to think that the firms which a better understanding about their customers, products, technologies, markets and organizational relations, that is, firms with a wide and strong endowment of intellectual capital will perform excellently. Therefore, according to the Resource-Based View, knowledge and intellectual capital should be strategically considered as the most important resources, and the capability to acquire, integrate, store, share and to apply them, will be the most valuable organizational capability to get and sustain competitive advantage (Grant, 1996).

The Intellectual Capital-Based View (Subramaniam and Yound, 2005; Reed, Lubatkin, and Srinivasan, 2006) is a refinement and middle range theory that aims to make operative the main assumptions of the Resource-Based View through the measurement indicators provided by the intellectual capital literature. The thoughtful Resource-Based thinking has been considered as excessively theoretical, assuming (but not explaining) the relationships between resources (or assets) and corporate value (or performance) (Tseng and Goo, 2005). Critics of the Resource-Based View (Priem and Butler, 2001; Foss and Knudsen, 2003) have raised questions about its legitimacy as theory, precisely because it is extremely difficult to parameterize and test empirically its main axioms without measuring intangible assets in some detail. Reed, Lubatkin, and Srinivasan (2006) propose a pragmatic, though partial, resolution from a mid-range theory that they call 'an intellectual capital-based view of the firm'. Their main contribution is that, as a mid-range theory, the intellectual capital perspective enables a better hypotheses development and empirical testing for the Resource-Based View, adopting a partial rather than generalized view in this research.

This publication follows the way of the Intellectual Capital-Based View, considering that innovation is the best representation of competitive advantage in the present day. Thus, according to the point of view shown in this volume, Schumpeterian rents, which can be captured by firms, innovation networks and systems, regions or nations, come necessarily from their intellectual capital endowments, and especially from their ability to create and improve those endowments.

STRUCTURE AND CONTENTS

Along these pages, readers will find mainly academic papers dealing with the previously mentioned concepts and theoretical arguments. Most of the chapters are written by scholars and aimed for scholars, showing thoughtful literature reviews, conceptual models, and empirical research with findings supporting theory. However, some of them also are oriented toward debate and opinions, trying to provide advice for the management practice, and suggesting novel directions for managerial and academic thinking.

The intended audience for this publication are people interested in cutting-edge research in the field of intellectual capital and technological innovation, and especially interested in the link between these two concepts. Besides, this research interest in complemented with one of the most recent and promising notions in supporting both intellectual capital building and shared innovation processes, that is, social capital, which adds network and social perspectives to the mentioned analysis. Both academics and decision-makers interested in the implications and new directions for study suggested by the mentioned concepts and theoretical streams may find useful recommendations in the chapters of this volume.

The book is structured according to four sections. The first of them introduces the concept of intellectual capital, which constitutes the base for all the theoretical and empirical work gathered within this volume. The second addresses, in general, innovation processes, and technological innovation in particular. Then, the third section explores the causal relations between intellectual capital and innovation, what place it at the heart of the entire work because it represents the essential idea that triggered editing this cutting edge research publication. The fourth and last section is focused on social capital, bearing in mind that in the present day this has become an essential condition both for obtaining intellectual capital and strengthen the firm knowledge base, and for fostering innovation at a firm, industry, region, or national scale.

In the following pages the different chapters are briefly described, providing a comprehensive guide of contents for the readers. For each chapter, the main topics discussed, the theoretical streams and essential notions presented, as well as the theoretical or empirical character of the chapter are commented. Besides, for empirical papers, a research resume with the main research techniques and sample characteristics is also provided.

The first section of this volume addresses the intellectual capital topic. Within it, readers may find four papers. The first of them is entirely theoretical, with a review of the concept, components, and evolution of this research stream. The following two chapters show how intellectual capital represents an interesting phenomenon not only for firms, but also for regions or countries. In this vein, they provide interesting insights about how intellectual capital can be measured, analyzed, and developed. The fourth chapter included in this section shows the findings of a qualitative research with new technology firm, focusing on how certain components of intellectual capital, namely organizational capital, are developed through the process of organizational identity building.

The section devoted to intellectual capital starts with the theoretical review carried out by Sohrabi, Raeesi, and Khanlari, which is titled 'Intellectual Capital Components, Measurement and Management: A Literature Survey of Concepts and Measures'. This chapter deals with the definitions, proposed components, and the advances in reporting initiatives for intellectual capital. Although it can be considered as an introductory chapter, this work provides a very useful tool for managers in charge of intellectual capital, because in its metrics section, it includes a comprehensive list of the most widely accepted indicators for measuring and managing intangible assets in practice. Besides, readers can find the method and utility, as well as a framework for preparing and using intellectual capital statements.

The second chapter, also within the intellectual capital section takes a national or institutional perspective for intellectual capital. In 'Globalization of Instruction: Developing Intellectual Capital', Holland and Holland address the problems that educational institutions have to face when trying to provide intellectual capital for individuals and educational systems that compete in an increasingly globalized context. As required from the Intellectual Capital-Based View, in this chapter authors focus the analysis on a certain kind of intangible assets related to human capital, trying to frame how educational institutions should change in order to provide that human capital with skills for global work environments. The chapter provokes a particular theoretical discussion, highlighting also the role of social capital and technological capital in the process of global adaptation. Besides, empirical evidence is also provided from a pilot study with college students at a small North-American University.

Another chapter dealing with intellectual capital at a regional scale, specifically at a national scale, is Chapter 3: 'Measuring Science & Technology in Panama: Towards a National Intellectual Capital Framework'. In this work, Chang and Kan develop a framework for measuring and reporting the intellectual capital endowments of an entire country. This way, national competitive advantages (and disadvantages) are also supposed to be based on intellectual capital, and it is considered as a platform for innovation and economic development, which is especially interesting for developing countries. The perspective of the presented framework is closely related to other systems of national accountability and science and technology measurement like those employed by the OECD or the WEF. In addition to the methodological guidance for policy-makers that can be found in the chapter, readers will also find on it an application of the proposed framework for Panama, measuring different kinds of intangible assets as human capital, innovation capital, market capital and process capital.

The last chapter included in the intellectual capital section is titled 'A Study on the Relations Between Organizational Identity and Intellectual Capital: Empirical Evidence in New Technology Based Firms at Madrid Science Park'. On the contrary to the previous chapters, the perspective for analyzing intellectual capital in this case is firm level. Besides, the focus of the work of Bueno, Longo, and Salmador is on understanding the connections of the process of framing organizational identity with the development of different aspects of a specific type of intellectual capital, this is organizational capital. Besides, qualitative research has been conducted for explaining this phenomenon. This way, the cases of five new technology-based firms created at Madrid Science Park, characterized by a highly reduced size and a rapid growth process are analyzed, using both primary data obtained from semi-structured interviews, and secondary data from direct observation and secondary sources. Findings of this research reveal that the social process of building an organizational identity also entails the development of some forms of organizational capital as organizational culture, organizational design, learning environments, and channels oriented towards internal and external customers.

The second section is devoted to innovation. Three heterogeneous chapters are included in it. The first one is methodological, introducing the problem-find-solving as a tool for improving innovation performance. The second is an empirical research about spillover in the case of technological capital, relating this phenomenon to the evolution and innovation results of the nonferrous metal industry in Japan. The section is closed with a theoretical-empirical chapter in which, innovation capacity, along with absorptive capacity, are taken as the main antecedents of firm dynamic capabilities. This framework is tested on an international sample of high-innovative firms from the software development industry.

Entering the innovation section, readers will find the chapter titled 'Problem Finding and Solving: A Knowledge-Based View of Managing Innovation' by Heiman & Hurmelinna-Laukkanen. This work is basically theoretical, but with the particularity that it introduces a novel perspective and methodology for innovation analysis: the problem finding/problem solving (PF/PS) perspective. This chapter also follows the theoretical background of this publication, arguing that knowledge-based processes and managerial decisions regarding these processes play a critical role in determining innovation outcomes, and ultimately firm success. Thus, organizations need a better understanding about what problems must they find and solve in order to improve their performance. This requires taking problems as unit of analysis for attaining a high performance innovation, and authors of this chapter support their claim with different contributions form the innovation management literature, as well as from the strategic management and micro-organizational behaviour fields. Analyzing problems involves identification of different kinds of biases (individual, team, and environmental), and solving problems is mainly a task related to overcoming these biases, which leads to superior innovation outcomes.

Nakagawa and Watanabe also address technological innovation, although they provide an empirical piece of research with 'Innovation Dynamics of Materials Technology: An Empirical Analysis and Recommendations'. They carry out a longitudinal study with patent data focusing on one specific kind of intellectual capital, namely technological capital, and its relevance for innovation dynamics at the industry level. For conducting their empirical research regarding changes in the use of technological capital from the technology spillover perspective, which is quite related to the notion of social capital and its evolution, Nakagawa and Watanabe focused on the nonferrous metal industry in Japan over the past quarter of a century, tracking patent applications and analyzing the relationships among them have enabled us to conduct microscopic analyses of technology spillovers, classifying them according to organization and technological field. Their results show that the management of technology spillover influences firm performance and profitability. Thus, changes from intra-intra spillover, to intra-inter and inter-inter spillover for firm technological capital are used to explain firm performance and the evolution towards the open innovation and social capital based perspectives that rule in the present day, which authors consider a post-information society.

The last chapter devoted especially to innovation is titled 'Absorptive Capacity, R&D Intensity and Innovation Results'. The nature of this work is twofold, because it attempts to make both theoretical and empirical contributions. Along its theoretical part, Cruz & Delgado provide a literature review for the concept and components of dynamic capabilities, ending up with a definition following the seminal work of Teece, Pisano, and Shuen (1997) and framing two related concepts, namely absorptive capacity and innovation capacity, as the key components for developing dynamic capabilities. Innovation, according to this chapter, and in line with the research stream followed by this publication, appears as a way for capitalizing firm intellectual capital, capturing value, and improving performance. An interesting perspective of this chapter is that chances for obtaining superior returns from innovation come directly from the ability of the firm to identify and internalize the new external knowledge through its absorptive capacity. As a complement for the proposed relations between the notions of dynamic capabilities, absorptive capacity and innovation capacity, authors carry out an empirical study with an international set of firms from the software development industry. Their results support the claim that those companies that jointly develop their absorptive and innovative capacities, showing the theoretical balance proposed as dynamic capabilities, reach a higher performance than those that only develop absorptive capacity, which show a negative impact on performance when acting alone, or innovative capacity, that only provides a limited impact on firm results when appear individually.

The third and central section of the book introduces the key question that has guided the process of compiling this volume: the relationship between intellectual capital and technological innovation. Three chapters are included within this section. The first one presents the 'intellectual-capital based view', that as has been commented previously, is a refinement of the Resource Based View, and provides some propositions for its empirical development on the field of technological innovation. The second chapter of this section follows this vein, but with an empirical research focused on only one intellectual capital component, namely structural capital, analyzing its impact on firm innovation capabilities. The last chapter of this section is also an empirical research about how choosing intellectual capital components and indicators can be critical for developing firm innovation capabilities along the input, process, and result oriented stages.

In the chapter titled 'An Intellectual Capital-Based View of Technological Innovation', Delgado and Cruz address the main topic of this publication, deepening in the idea of Subramaniam and Youndt (2005) that organizational innovation capabilities are closely related to intellectual capital. This theoretical chapter provides a framework for the identification and classification of different kinds of intellectual capital, namely human, organizational, technological, relational, and social capitals. Besides, this theoretical model takes the different elements of intellectual capital as sources of different types of technological innovation, which are considered according to the classical notions of radical vs. incremental, and product vs. process innovation. The main contribution of this work is explicitly considering the relationship of each component of intellectual capital with the different kinds of innovation, trying to provide a deeper focus for the main thesis of the Intellectual Capital-Based View. Although it has a strictly theoretical nature, this chapter provides a guideline for all readers interested in exploring the relations between intellectual capital and innovation that may be attracted by the topic of this book. This way, some of the types of intellectual capital considered in other chapters, as technological, organizational, or social capital, can be easily framed within this model in order to gain insights about their connections with innovation and even with performance.

Taking into account the framework presented in the previous chapter, a perfect complementary empirical analysis of its logic can be found on 'Structural Capital and Innovation Capabilities. Theory and Empirical Evidence from Spain'. In this chapter, Alama provides empirical evidence in order to test the main claim of the Intellectual Capital based View, focusing the analysis on one kind of intellectual capital: structural capital, which includes those pieces of knowledge that provide coherence and guidance for the whole organization (Edvinsson and Malone, 1997), and is quite close to the definition of organizational capital provided in other chapters. As dependent variable for testing the mentioned research approach, competitive advantage has been considered in the shape of innovation outcomes, whereas it is understood as the successful development of new projects, products, services and ideas. Taking a sample of Spanish professional service firms, factor analysis reveals five components of structural capital, as well as only one kind of innovation for this type of knowledge-intensive industry. Regression analysis results support the idea that structural capital is one of the main sources of firm innovation. Thus, this chapter contributes with empirical evidence for moving from 'Resource Based View of the innovation process' (Pike, Roos, and Marr, 2005) towards an 'Intellectual Capital based View of Innovation'.

Another empirical chapter for testing the relationship between intellectual capital and innovation is 'Prioritizing Corporate R&D Capabilities: The Intellectual Capital Perspective'. In this research piece, Chang, Yu, and Chi apply the conception that managing intellectual capital directly impacts business performance, but for making it operative, they focus their study in searching critical components and indicators of intellectual capital in order to develop firm innovation capabilities, distinguishing three stages: input, process, and result. With 43 indicators collected from the intellectual capital literature on IC research, R&D management and innovation management, authors surveyed Taiwanese corporate R&D managers of high-tech manufacturing companies, in order to asses the relevance of the different intellectual capital components on each of the mentioned R&D stages. The results of factor analysis obtained 11 intellectual capital components according to the perspective of the R&D managers. From them, analytical hierarchy process (AHP) methodology allowed to find the most critical elements of intellectual capital for input, process, and result phases. Findings show that the relevance of intellectual capital dimensions and indicators is phase specific. According to the results of this research, the most influential factors of intellectual capital for innovation are organizational and human capital, in the input phase, and relational and process capital in the result phase. R&D managers may find advice in this chapter about what pieces of intellectual capital must be monitored and enforced for success and performance improvement when managing firm innovation.

The fourth and last section of the book introduces social capital as an increasingly relevant complementary topic for the fields of intellectual capital and technological innovation. This section includes four chapters. The first three deal with the relevance of this topic for innovation and the development of intellectual capital, presenting the concept and its dimensions from a theoretical perspective, as well as its different configurations in order to improve the innovation performance. Then, the last chapter, that also closes this volume, provides empirical evidence from case studies in order to demonstrate the interaction among social capital, intellectual capital, and technological capital in a cluster scale.

The first chapter in the section devoted to social capital keeps contact with the innovation section and also with the intellectual capital-innovation one, because it suggest an intangible organizational asset (the organizational sweet spot) and describes its usefulness for firm innovation. In 'Co-Evolving Relationships and Innovation Dynamics', Charles Ehin raises the need of social systems to be fully recognized as organizational 'hidden assets', in order to address their management and capture the value that they can provide for firms. The reflection of the author is oriented towards the self-organization principles that rule social interactions among agents, trying to emphasize how developing constructive social contexts can support the dynamics that allow people to establish meaningful relationships and organizations to expand their sweet spot. This chapter claims for the development of pragmatic approaches from multi-disciplinary research from fields like evolutionary psychology and social neuroscience, in order to gain a better understanding of social capital and its benefits, explicitly remarking that higher emphasis must be placed on the organizational shared-access ecology success factors such as individual autonomy for responsible experimentation, a culture of sharing information, and fostering situational leadership.

After raising the relevance of social perspectives in analyzing innovation and firm performance and claiming for multi-disciplinary approaches for this concept, in 'Social Capital: Definition, Dimensions and Measurement' a theoretical review of the concept and dimensions of this construct is provided. Blasco devotes his work on this chapter to propose a consensus definition of social capital and to identify the ways in which extant literature has tried to make this topic operative. In spite of the fact that in other chapters social capital is considered as a component or dimension of intellectual capital, this work argues that it shows characteristics that allow us to include it within any of the dimensions of intellectual capital, as human, structural, or relational capital. After illustrating that social capital has acquired an important relevance in a wide range of disciplines, and the increasing interest of academics in explaining how the social context influences firm behavior and results, the chapter adopts the definition proposed by Nahapiet & Ghoshal (1998) and explains two dimensions for this construct which also provide two different easy methods for capturing or measuring it. Thus, on one hand the structural dimension of social capital implies the analysis of the essential features of a network structure that connects diverse groups in order to obtain and exchange information and transfer knowledge, and how this can improve intellectual capital boosting technological innovation. On the other hand, the relational dimension of social capital is shaped by the characteristics and processes of relating each other between agents. This chapter provides a theoretically sound base for designing future research about social capital, especially when considering its connections with intellectual capital and innovation, as is the case of this book.

In particular, when relating social capital to intellectual capital and innovation, the chapter titled 'Social Capital, Knowledge and Technological Innovation' analyzes the different views of social capital and explains its contribution within the organization studies, discussing the implications of this construct for innovation performance, distinguishing between radical and incremental innovation. In this theoretical work, Alguezaui and Fillieri propose a theoretical framework for analyzing the effect of different configurations of the structural dimension of social capital on the innovation performance. For doing so, cohesive and sparse network configurations are presented, as well as its main benefits and constraints for innovation. Cohesive networks are based on strong and highly interconnected ties and this kind of social capital provides trust, norms, and commitment. Whereas, sparse networks provide the focal firm the ability to access novel information, benefiting firm's innovation performance through a wider number of loose ties. Nevertheless, both kind of social capital can pose problems. On one side, cohesive networks may turn over-embedded and lead to lock-in situations, with a high cost of maintaining strong ties, and an inability for exploration and experimentation beyond their existing network. On the other side, sparse networks can lead to the inability to a fully development of the innovation. With a thoughtful explanation of both types of social capital networks, this chapter proposes that cohesive network configurations are more suitable for radical and product innovation, and that sparse networks are more appropriate for incremental and process innovation.

The last chapter of the book was written by Massa and Testa, and it provides empirical support for the arguments presented in the rest of chapters included in the social capital section, taking a cluster based perspective and using the case study method. Thus, 'Do Top-Down Cluster Policies Succeed in Fostering Social Capital and Innovation? Some insights from Italian cases' gather all the topics presented in this publication: intellectual capital, technological innovation, and social capital at once. The scheme presented in this chapter propose that the effectiveness of top-down cluster policies is critical for creating new social capital at this regional level, which in turn lead to the development of new intellectual capital, hence improving the innovation capabilities of the firms located at the cluster. For illustrating this proposed scheme, two case studies about deliberate cluster promoted programs in Italy are included in the chapter. These cases present different drivers for cluster development: demand side based and supply side based. Results argue the superiority of supply-side driven clusters, because although they may experience slower return on investments on the short term, their strong export-orientation has a higher probability of success in the long-term. Thus, this chapter explains to policy makers and managers of cluster based firms how clusters may become a privileged locus of social capital that enhances dynamics of learning and knowledge creation based on socially embedded vertical and horizontal linkages of co-locating firms and on their interaction with education and research and development agents. This way, social capital provides access to an improved base of intellectual capital that fosters firm innovation.

Pedro López Sáez
Universidad Complutense de Madrid, Spain

Gregorio Martín de Castro
Universidad Complutense de Madrid, Spain

José Emilio Navas López
Universidad Complutense de Madrid, Spain

Miriam Delgado-Verde
Universidad Complutense de Madrid, Spain

REFERENCES

Adler, P.S., & Kwon, S.W. (2002). Social capital: Prospects for a new concept. Academy of Management Review, 27(1), 17-40.

Barney, J.B. (1991). Firm Resources and Sustained Competitive Advantage. Journal of Management, 17(1), 99-120.

Barney, J. (2001). Is the Resource-Based View a Useful Perspective for Strategic Management Research? Yes. Academy of Management Review, 26, 41-57.

Bontis, N. (1998). Intellectual Capital: an Exploratory Study that Develops Measures and Models. Management Decision, 36, 63-76.

CIC (2003). Modelo Intellectus: Medición y Gestión del Capital Intelectual (Serie Documentos Intellectus No. 5). Madrid: Centro de Investigación sobre la Sociedad del Conocimiento (CIC).

Damanpour, F. & Evan, W.M. (1984). Organizational Innovation and Performance: The Problem of 'Organizational Lag'. Administrative Science Quarterly, 29, 392-409.

Edvinsson, L. & Malone, M. (1997) Intellectual Capital. Realizing your Company's True Value by Findings its Hidden Brainpower. New York: Harper Collins Publishers, Inc.

Edvinsson, L. & Sullivan, P. (1996). Developing a Model for Managing Intellectual Capital. European Management Journal, 14, 356-364.

Foss, N.J. & Knudsen, T. (2003). The Resource-Based Tangle: Towards a Sustainable Explanation of Competitive Advantage. Managerial and Decision Economics, 24, 291-307.

Granovetter, M.S. (1985). Economic action and social structure: The problem of embeddedness. American Journal of Sociology, 91, 481-510.

Grant, R.M. (1996). Prospering in Dynamically Competitive Environments. Organization Science, 7(4), 375-387.

Myers, S. & Marquis, D. G. (1969). Successful Industrial Innovation. Washington, DC: National Science Foundation.

Nahapiet, J., & Ghoshal, S. (1998). Social capital, intellectual capital and the organizational advantage. Academy of Management Review, 23(2), 242-266.

OECD (2005). Oslo Manual. Guidelines for Collecting and Interpreting Innovation Data, OECD-Eurostat, Luxembourg.

Peteraf, M. (1993). The Cornerstones of Competitive Advantage: a Resource Based View. Strategic Management Journal, 14, 179-191.

Pike, S., Roos, G., & Marr, B. (2005). Strategic Management of Intangible Assets and Value Drivers in R&D Organizations. R&D Management, 35, 111-124.

Priem, R. & Butler, J. (2001). Is the Resource-Based View a Useful Perspective for Strategic Management Research?. Academy of Management Review, 26, 22-40.

Reed, K., Lubatkin, M., & Srinivasan, N. (2006). Proposing and Testing an Intellectual Capital-Based View of the Firm. Journal of Management Studies, 43, 867-893.

Subramaniam, M. & Youndt, M. A. (2005). The Influence of Intellectual Capital on the Types of Innovative Capabilities. Academy of Management Journal, 48, 450-463.

Teece, D.J. & Pisano, G. & Shuen, A. (1997). dynamic capabilities and Strategic Management. Strategic Management Journal, 18(7), 509-533.

Tseng, C. & Goo, Y.J. (2005). Intellectual Capital and Corporate Value in an Emerging Economy: Empirical Study of Taiwanese Manufacturers. R&D Management, 35, 187-201.

Wernerfelt, B. (1984). A Resource-Based View of the Firm. Strategic Management Journal, 5, 171-180.

KEY TERMS AND DEFINITIONS

Intellectual Capital: Intangible assets or knowledge stocks.

Innovation: Introduction of something new, whatever it will be a new thing (product, raw material, supply source…) or way to do things (production or management process).

Social Capital: Benefits that a certain agent can obtain from the network of relations in which it gets involved.

Resource-Based View (RBV): Theory of competitive advantage which argues that sustained differences in performance among competitors come mainly from the differences in asset endowments. Its explanation of competitive advantage is based on asset endowments heterogeneity among competitors, and on the imperfect transferability and barriers for imitation regarding those asset endowments.

Intellectual Capital-Based View: Refinement and middle range theory that aims to make operative the main assumptions of the Resource-Based View through the measurement indicators provided by the intellectual capital literature. This perspective enables a better hypotheses development and empirical testing for the Resource-Based View, adopting a partial rather than generalized view in this research.

Acknowledgment

The authors wish to express their acknowledgement to all the contributors to this edited book, as well as to all the members of the editorial advisory board and 'ad hoc' reviewers. Authors would like to highlight their gratitude to Fundación Rafael del Pino (Spain), and to its Director, Professor Amadeo Petitbò Juan, for trusting in our work and funding our research. To all of them, thank you very much.

Pedro López Sáez
Gregorio Martín de Castro
José Emilio Navas López
Miriam Delgado-Verde
Editors

Section 1
Intellectual Capital

Chapter 1
Intellectual Capital Components, Measurement and Management:
A Literature Survey of Concepts and Measures

Babak Sohrabi
University of Tehran, Iran

Iman Raeesi
University of Tehran, Iran

Amir Khanlari
University of Tehran, Iran

ABSTRACT

Intellectual Capital (IC) is neither a new phrase nor an old formally accepted term to measure the intangible assets of an organization. This concept has been used in different theories of organization and management. In the 1990s, the focus of attention dramatically changed from financial aspects and the techniques used for their measurement to the knowledge and intangible assets and the way businesses could manage them strategically in order to gain a competitive edge. By this way, IC has been recognized as one of the core components for managing and analyzing the strategic organizational intellectual capabilities, hence; to improve the quality of knowledge management (KM). The chapter aims to conceptually review and analyze the importance, definitions, components, and reporting mechanisms in the field of IC measurement and management.

INTRODUCTION

With the beginning of the new era of knowledge-based production and mentality development, the need for a comprehensive background on knowledge and the methods on its measurement gained importance. However, there is a lack of published reviews in the related field. A review of prior relevant literature is an essential feature of any academic and practical project, because an effective review cre-

DOI: 10.4018/978-1-61520-875-3.ch001

ates a firm foundation for advancing knowledge (Webster and Watson, 2002). The present chapter aims to cover the need for a better understanding of conceptual and practical contributions on various approaches of intellectual capital. Although there are many articles and working papers focusing on intellectual capital, no major attempt has been made on the conceptual and practical review of it for about a decade. As a matter of fact, we often need to draw on theories and views from a variety of fields. Since, the reviews are considered to be the foundation for research and development of theories, they tend to be critical in strengthening the related field of study. Analyzing the works done in the past, removes the threat of non-comprehensive and partial understanding on the subject and therefore helps in organizing and making a better foundation for preparing new theoretical and practical developments. In the present chapter, the scope of study on intellectual capital is in the following manner:

- *Background*

In this section, a review of the literature on intellectual capital and knowledge management is provided and some evidences on importance of intellectual capital and the reasons for growing attention to it, is presented.

- *Definitions, Components and Metrics of Intellectual Capital*

In the next three sections, a rather comprehensive set of definitions, a chronological review of classifications in the literature, and the trends of IC in the recent years have been reviewed. A pervasive list of various classifications of IC components is also presented. The metrics section provides a comprehensive list of the most accepted IC metrics (or indicators) across the literature for each of the IC components that are intended to be measured and managed. This section significantly enables the scholars and practitioners in establishing the initial IC measurement foundation. Also, the indicators can notably assist in configuring and implementing KM systems according to the specific organizational needs in order to provide a continuous and accurate assessment mechanism of measurement and reporting.

- *Intellectual Capital Measurement*

Through acquiring a rather comprehensive viewpoint on IC definitions and classifications, the measurement is the next step in which firms attempt to continuously evaluate the overall status of their intellectual capabilities. The IC measurement capability of the organization is one the main factors influencing the visibility of the managerial efforts on empowering the current IC status with regard to the overall strategies specified by top managers.

- *Intellectual Capital Statement and Reporting*

One of the managerial inputs for strategic decision making has been recently stated to be the IC statement, a new approach to embracing the importance of IC and evaluating intellectual capabilities of the firm. IC statement can play a significant role in initiating the process of identifying the difference between market and book values of the firm. This section provides a framework to guide managers in their initial efforts of creating IC statements.

- *Future Research Directions*

In this section, an integrative model relates the concepts of human, organizational, and relational capitals with R&D (or innovation capital). This model has been added to complement the literature on R&D which has recently related the importance of R&D concept to the components of intellectual capital. Although the research and development aspect has been repeatedly stated in the IC literature, few attempts have been made to explicitly relate the IC components and R&D concepts together and generate new ideas and practical outcomes.

- *Conclusion*

Finally, the chapter concludes with a roadmap containing explicit steps which help in using the IC potential toward a better knowledge-based management of organizations. At the end, Some specific suggestions have been provided which may assist the scholars and practitioners in the process of IC management.

BACKGROUND

One of the main reasons for a literature survey is that there are many new conceptual and practical articles and books, each focusing on a special topic without elaborating on the overall picture of the main concept. It must be kept in mind that materials which are specific to a domain, are not a good place for presenting a holistic view of what they attempt to concentrate on. Another reason for this review is that there has been no encompassing review of the IC literature for about a decade. It is important to notice that during this period, hundreds of working papers and books have been published but lacked collective attempts to bring them together and present an integrative view so as to make the way clear for the future of intellectual capital theorizing and practical management.

In this chapter, we are trying to integrate and clarify the conceptual foundations of IC. Intellectual Capital can be defined, categorized, measured, and through IC management efforts, it can be transformed into intellectual assets. These assets make the organization capable of attaining a competitive edge over it's competitors through improving the knowledge-based performance of the business processes. Before exploring the IC literature, it will be beneficial to generally differentiate the words "Asset" and "Capital" based on the non-financial approaches used in IC literature. As Litschka et al.(2006) stressed for a clear distinction between human capital (as one of the main components of IC as described later in the chapter) and human assets, the same will apply for intellectual capital and intellectual asset. As an example, the abilities like "book-keeping" and "speaking a foreign language" pertain to the human capital of an employee, even though they might be of minor relevance to his/her current job. However, these abilities are not to be seen as human assets in the context of the organization the person is working for. If the person is working for an organization, where such capabilities are being directly put into work, they belong to the human assets of the organization while belonging to the human capital of individual at the same time (Litschka et al., 2006). The same structure is applied to intellectual capital and intellectual asset. The organizational processes, IT infrastructure and applications, documented knowledge, relationships with partners and other industries and the organizational regulations are examples of different intellectual assets being employed for creating value. This is because they belong to the organization. The knowledge which is hidden in the minds of human resources or in the relationships of organization with the market can be thought of as intellectual capital possessed by the organization but not readily documented and formulated.

So far, numerous convincing arguments have been presented in support of the need to better understand the intellectual capital (e.g. Brooking, 1996; DATI, 1998; 1999; Petty and Guthrie, 2000b; SMAC, 1998; Sveiby, 1998). These arguments range from an intuitive understanding that IC matters and should be taken into consideration (Stewart, 1997) to the evidence that recognizing intellectual capital has the potential to improve the efficiency of both capital and labor markets (Bukh et al., 1999; OECD, 1999; Petty and Guthrie, 2000).

Wernerfelt's (1984) conceptual article entitled "A Resource-based view of the firm" was recently selected as one of the most influential papers regarding two important aspects of the firm. With the emergence of Resource-based and the Knowledge-based views, it was observed that success of companies was no longer dependent solely on access to material resources rather to intangible assets, and it was that access that could provide a competitive edge in the long term, specially for knowledge intensive organizations (Argyris, 1992; Argyris and Scho¨n, 1996; Barney, 1991, 2001; Bontis et al., 2002; Grant, 1991, 1996a, 1996b; Itami and Roehl, 1987; Nonaka and Takeuchi, 1995; Peteraf, 1993; Pedrini, 2007; Priem and Butler, 2001a, b; Senge, 1990; Wernerfelt, 1984, 1995), However, each of the views has contributed to the field in a different manner.

One aspect of Resource-Based View (RBV) relates to the products and the other regards the resources available to a firm (Priem and Butler, 2001a). RBV focuses on resources and their deployment in organizations, leading to the development of value creation (Cheng et al., 2008; Peppard and Rylander, 2001). Obviously, products are being provided through resources and a lack of resources might generate a greater gap in providing the final products to the market. This might result in a great profit loss. Therefore, resources are of supreme importance to the firms.

In the RBV, those resources that are valuable, rare, difficult to imitate and none-substitutable can yield sustained competitive advantage (Meyer, 1991, Priem and Butler, 2001a). The RBV attributes advantage in an industry to a firm's control over bundles of unique material, human, organizational and geographical resources and skills that enable unique value-creating strategies (Barney, 1991, Priem and Butler, 2001a). According to this view, a firm's resources – particularly intangible ones – are more likely to contribute to firms attaining and sustaining superior performance when they are combined or integrated (Barney, 1991, Reed et al., 2006).

These resources can be divided into two distinct parts. One relating to financial and tangible resources as directly represented in the financial statements of the firm, and the other one which includes the non-financial and intangible resources of a firm, mostly referred to as intellectual assets and capabilities. In this regard, a specific view on intangible resources of a firm has been recently created. Intellectual Capital View (ICV) deals solely with knowledge that is created by and stored in a firm's intangible capitals (Edvinsson and Malone, 1997; Wright et al., 2001, Reed et al., 2006). The IC view is most important in knowledge-based firms which are more reliant on their intellectual resources.

A knowledge-based perspective of the firm has also emerged in the knowledge management literature (Alavi and Leidner, 2001; Cole, 1998; Spender, 1996a, 1996b; Nonaka and Takeuchi, 1995). Knowledge is both a resource and a source of competence and capability. Adopting a knowledge-based approach, the relevance of a resource is related either to its embedded knowledge or to its role of stimulating and sustaining knowledge dynamics (Schiuma et. Al., 2008). ICV is complementary to Leonard-Barton's (1992) which is more widely understood as knowledge-base view (KBV). While both seek to explain the hidden knowledge-based dynamics that underlie a firm's value, and both are grounded in an RBV logic, they differ in focus. KBV is primarily interested in evaluating the effectiveness of a firm's use of knowledge-management concepts and tools as knowledge-generating mechanisms, such as its information technology and information management systems (Leonard-Barton, 1992; Nonaka et al., 2001, Reed et al., 2006). In contrast, ICV's focus is on the stocks and flows of knowledge capital embedded in an organization and is posited to have direct associations with its financial performance (Youndt et al., 2004, Reed et al., 2006). In other words, the KBV is more encompassing and attempts to efficiently manage the knowledge intensive processes and sources of new knowledge that are of significance to the firm while the ICV mainly concentrates on the embedded knowledge in business processes and tries to manage and direct the flow of knowledge toward more valuable outcomes for stakeholders. Since the knowledge-based capabilities are of significance in the information era as pointed out in various working papers and books in the last decade, it can be recognized that knowledge-based view is gaining a special position among scholars and industry practitioners. Since the KBV is a general viewpoint and the ICV is more specific as it attempts to relate the knowledge of the firm to values delivered to stakeholders, the ICV is directly related to the KBV and is a more practical term that is utilized in the context of knowledge-intensive organizations at which, efforts of intellectual capital management are being planed and implemented.

According to the latest study conducted by OECD (2006), intellectual capital (IC) has played a significant role, as much as tangible capital, in improving labor productivity in the USA from 1995 to 2003. The importance of IC is shown not only by the strong support of international organizations (e.g. OECD and European Union (EU) Commission) but also by the large numbers of articles since the 1990s (Kim and Kumar, 2009).There are some unique characteristics of intellectual capital that makes it capable of generating value-added for organizations. IC is viewed as an important strategic resource for gaining competitive advantage. This becomes more obvious when evaluating the organizations with knowledge intensive business processes – such as software development and hardware manufacturing companies in the ICT sector. Therefore, prior to deeper investigation of IC definitions and classifications, it would be of value to clarify the importance of IC concept as repeatedly mentioned by scholars of the field. The IC importance is due to it's significant and unique characteristics as follows:

1. **Scalability.** Physical assets are limited by their tangibility. This means a choice must be made or the best use of these assets following the criteria of cost-opportunity. Intangible assets on the other hand can be put to non-competitive use, since their use by one person does not exclude another. Their use is limited only by the size of the internal market (Romer, 1994, 1998).
2. **Growing returns.** Intangible resource benefits from a process of accumulation, which can produce an increasing scale of return. The growing return process is mainly created through the accumulation of documented knowledge. The accumulated knowledge helps the experts gain a firm base for further in-depth studies. Since there is no depreciation with knowledge, it is obvious that the knowledge creation process and rate of returns would be increasingly upward (Pedrini, 2007).
3. **Difficulty to imitate.** The generation and development of immaterial resources is difficult to imitate. As intellectual capital is embedded in organizational processes and minds of human resources, it would be difficult to obtain or imitate it (Pedrini, 2007).
4. **Network effect.** At the center of a network, an innovation protected by a patent is often found (Lev, 2001), consequently the intangible resources benefit from the main rule of networks which states that the advantages of inclusion grow with the growing number of people involved. The bigger the network, more advantages each individual will gain (Shapiro and Varian, 1999).

In general, the importance of IC can be emphasized in the revolution in information technology and the information society; the growing importance of knowledge and the knowledge-based economy; the changing patterns of interpersonal activities and the network society; and the emergence of innovation as the principal determinant of competitiveness (Petty and Guthrie, 2000). One of the practical methods of showing the importance of each concept to practitioners can be through the measurement of that concept with regard to the realities of the businesses. By providing the top executives with the positive results of measuring the IC in the firm, continuous management of intellectual capital can take place and there will be an ongoing managerial commitment and support as a prerequisite to the success of IC management efforts.

Intellectual capital is a term with various definitions in different theories of management and economics. Most of the literature dealing with intellectual capital is concerned with the middle or upper level management, professionals, and organizations (Petty and Guthrie, 2000). As literature suggests, a coherence in IC approaches and developments is being gradually achieved. More attention is recently being paid on how to get the theories and conceptual frameworks into action and produce practically impressive outcomes for upper managers and stakeholders. Going through different conferences on intellectual capital, various invaluable books, papers, and a large number of firms providing organizations with knowledge-based products and services (Petty and Guthrie, 2000), initial efforts have been typically focused on consciousness raising activities that try to communicate the importance of recognizing and understanding the potential of intellectual capital in creating and managing a sustainable competitive advantage (Petty and Guthrie, 2000) As a matter of fact, the development of brands, stakeholder relationships, reputation and the culture of organizations are viewed as providing sustainable sources of business advantage. The ability to build and leverage the value of these intangible assets constitute a core competency for organizations (Kannan and G.Aulbur, 2004).

IC is the major element used for measuring the performance of knowledge management. Here, it is worthy to concisely explore through the KM relevant literature. KM can be simply defined as leveraging knowledge in a strategic manner which comprises a range of practices used to identify, create, store, retrieve, represent, and share knowledge. Leveraging knowledge in a strategic manner means to become able to measure the knowledge management capabilities of a firm in line with the organizational strategies and goals. Measuring knowledge management through intellectual capital components helps the organizations align their intellectual capabilities with their overall strategies and the objectives pertaining to them.

As an overall definition for building the relation between knowledge and IC, Intellectual Capital can be defined as an object which is mainly used to measure the knowledge management performance and efficiency. Generally, Knowledge management is about the management of the intellectual capital controlled by a company. Further, the Knowledge management, as a function, describes the act of managing the object, intellectual capital (Petty and Guthrie, 2000; Guthrie et. Al., 1999) and intellectual capital, on the other hand, can be seen as an analytical object used for KM measurement.

Considering IC as a factor of measuring the knowledge management performance, intellectual capital helps us become capable of reporting on the overall KM performance of an organization. In other words, measuring intellectual capital makes the organization ready for distinguishing its strengths and weaknesses in terms of human, organizational, and relational capitals in order to remove the obstacles or take corrective actions for making the KM management mechanisms more effective. Nowadays, knowledge being the most important productive element, all enterprises should attach great importance to the significance and urgency of IC in their survival and development (Chen et al., 2004). Existing knowledge can be of two kinds (Berends et al., 2001; Cook and Brown,

1999; Fischer, 2001; Marsili, 2000; Nonaka and Takeuchi, 1995; Van Aken, 2001):

- **Explicit (codified or articulated):** refers to the knowledge which can be readily documented and can be codified and restored for later use.
- **Tacit (implicit or non-codified):** refers to the knowledge which is not written down or otherwise codified, and is contained within the minds of individuals (Carson et al., 2004).

Human Capitals are the owners of tacit knowledge as this kind of knowledge is mentally stored and become available to the organization. Converting the tacit knowledge into explicit knowledge occurs during the process of knowledge management. Human Capital does not directly belong to the firm unless it is transformed into human asset or possessed as Organizational Capital. Therefore, the practices of knowledge management encompasses and influences the management of intellectual capital. The concepts of KM and IC are required to be considered at the same time, as they are interrelated subjects. The view of knowledge as a capability suggests a knowledge management perspective centered on building core competencies, understanding the strategic advantage of know-how, and creating intellectual capital (Alavi and Leidner, 2001).

Intellectual capital management is a process of value creation; in contrast, intellectual capital is a process of discovery and development (Roos et al., 1997). Here, value does not only imply calculating, but also understanding the creation and development of value (Bukh, 2003; Hussi and Ahonen, 2002; Guthrie, 2001; Mouritsen et al., 2002; Mouritsen, 2004; Petty and Guthrie, 2000). Consequently, managing the intangibles of an organization effectively, means managing and leveraging the power of knowledge towards a competitive edge and therefore, managing the company toward a higher rate of return on investments and on assets. In other words, intellectual capital management can be seen as a prerequisite to knowledge management and an effective IC management strategy can lead to a more effective management of knowledge within an organization.

This chapter is centered around main IC concepts while taking other related concepts into account. In this way, an attempt has been made so as to gather and analyze as many conceptual and practical IC concepts as useful. Concepts mostly determine the organizing framework of a review (Webster and Watson, 2002). Over two hundred articles and books have been reviewed and analyzed according to the concepts mainly related to intellectual capital as illustrated in Table 1. This table shows a pervasive list of IC related concepts. Each concept can be an interesting topic for the future empirical and conceptual studies.

Synthesizing the articles, a trend has appeared that resulted in a comprehensive list of qualitative indicators that has been presented later in the intellectual capital components section. Getting familiar with the advantages, limitations, and the reasons for increasing attention on the importance of intellectual capital and it's relation to knowledge management, the next sections present the IC definitions and components, IC measurement indicators, IC reporting, and final views on IC management.

INTELLECTUAL CAPITAL DEFINITIONS

Before reviewing and analyzing the most important IC definitions, it is necessary to review a brief chronological description of contributions to the intellectual capital field, as shown in Table 2 (Petty and Guthrie, 2000), which particularly focuses on the works done prior to 2000. Most of the surveys and papers in this period concentrate on the development of working definitions of IC and also expanding the IC measurement concepts.

Table 1. List of categories and concepts

Central Category	Concepts
Fundamentals	Knowledge management (KM) Intellectual capital management IC components Human capital Innovation and process capitals Organizational capital Research and development Organizational culture Conceptual frameworks & models for IC Intellectual capital identification & importance Drivers of IC Organizational learning
Strategy	Metaphor Intellectual capital & strategy IC & competitive advantage IC & marketing Patents and ownership IC & career outcomes Intellectual capital implementation Balanced Score Card
Information systems	Intellectual capital reporting IT systems in IC Strategic information systems Knowledge management systems IC web IC commercialization Investment in IC
Analysis and Measurement	Value chain analysis Epistemological analysis Value network analysis IC evaluation IC measurement Credit risk analysis Performance

During 2000-2009, many published papers have directly concentrated on the IC measurement indicators for obtaining a better insight on the issues like human assets and the way they have been utilized, the potential of organizational capital and the processes behind it, the relational capital, and finally research and development. Generally, the works done after 2000 can be described as summarized in table 3.

IC has been defined from different perspectives, focusing on the analysis level (individual or organizational), the perception of the value of time (current value or future value), and objectivity (input or output) (Swart, 2006). John Kenneth Galbraith is credited for being the first to use the term "intellectual capital" in 1969 (Bontis, 1998), moreover, the term attracted the attention of scholars in the early 1990s when it was used by Stewart (1997) in Fortune Magazine. The concept of IC although has been prevalent for more than a decade, nevertheless there is a lack of consensus on its components and definitions (Marr, 2005). IC is a multi-disciplinary concept and the understanding of it varies across different business-related disciplines (Huang et al., 2007).

Various theoretical researches have attempted to define and classify IC (Brooking, 1996; Huang et

Table 2. Brief chronological review of literature on IC

Period	Theories / Concepts / Conferences
Early 1980s	General notion of intangible value (often broadly labeled as "goodwill")
Mid-1980s	The "information age" takes hold and the gap between book values and market value widens noticeably for many companies
Late 1980s	Early attempts by practitioner consultants to construct statements/accounts that measure intellectual capital (Sveiby, 1988)
Early 1990s	Kaplan and Norton introduce the concept of a balanced scorecard (1992). Initiatives to systematically measure and report on company stocks of intellectual capital to external parties (e.g. SCSI 1995)
Mid-1990s	Celemi's Tango simulation tool was launched in 1994, which is the first widely marketed product to enable executive education on the importance of intangibles. Also in 1994, a supplement to Skandia's annual report was produced, which focused on presenting an evaluation of the company's stock of intellectual capital. "Visualizing intellectual capital" generated a great deal of interest from other companies seeking to follow Skandia's lead (Edvinsson, 1997). Nonaka and Takeuchi (1995) present their highly influential work on "the knowledge creating company". Although the book concentrates on "knowledge", the distinction between knowledge and intellectual capital is sufficiently fine as to make the book relevant to those with a pure focus on intellectual capital. Another sensation caused in 1995 when Celemi used a "knowledge audit" to offer a detailed assessment of the state of its intellectual capital
Late 1990s	An increasing number of large-scale projects (e.g. the MERITUM project; Danish; Stockholm) which introduced some academic rigor into research on intellectual capital. In 1999, the OECD convened an international symposium in Amsterdam on intellectual capital.

Table 3. IC in the post-2000 period

Period	Theories / Concepts / Conferences
2000 – 2009	Works on metrics and questionnaires continued and expanded, and an increasing number of books and articles focused on classification of IC metrics and their implementations. These range from the papers merely using prior classifications for evaluating the companies in different countries to the papers updating the existing or proposing new models. Also, a broad range of authors and practitioners are working in the field of knowledge engineering, which directly focuses on capturing, storage, retrieval and application of knowledge. Knowledge engineering mainly focuses on the techniques and methods that can be applied to actually retrieve the tacit knowledge of human capital (e.g. see Lai, 2007). Alongside the conceptual papers, a number of empirical studies have been conducted to evaluate the validity of proposed models for IC measurement and management or utilizing them to evaluate the non-financial status of the firms. Recent articles have also been mainly focused on the implementation of the IC classification schemes attributable to a new stage of conceptual and practical development of IC. There are also new attempts on breaking the IC components into more refined and detailed sub-components. These attempts might lead to perceiving the interrelationships among IC component and subcomponents and as a result, a pervasive scientific foundation might appear which paves the way as an infrastructure for the future research and practical implementations efforts. (see for example, Bruggen et al., 2009; Castro and Lopez, 2008; Chan, 2009; Kim and Kumar, 2009; Palacios et al., 2007; Pearse, 2009; Tovstiga and Tulugurova, 2009)

al. 2007; Roos et al., 1998; Sveiby, 1997; Edvinsson and Malone, 1997). According to Andriessen (2006), the use of metaphors may be well suited for understanding IC. The author further suggests that IC, itself, is a metaphorical understanding of knowledge. This is consistent with the view of some IC scholars. From an epistemological perspective, IC is said to be knowledge about knowledge, and the understanding of IC appears to require an assessment of the language used in its definition and application (Chan, 2009; Jørgensen and Boje, 2006).

Stewart (1997, p.66) characterizes intellectual capital as "the sum of the knowledge, information, intellectual property and experience held by everybody in a company, in order to create a competitive edge and therefore to generate wealth". According to another definition, intellectual capital can be defined as intellectual resources that have been "formalized, captured and leveraged" to create assets of higher value (Prusak, 1998). Based on the given definitions, it can be implied that measuring intellectual materials or in other words, intangible assets is naturally difficult and requires an extensive amount of effort in addition to considering a number of organizational and individual variables (Kannan and G.Aulbur, 2004).

As a summarization for the final definition, the term intellectual capital refers to all immaterial resources that determine the value of the repository of intellectual properties possessed by an organization. In other words, IC refers to the Knowledge within organizational processes that can be converted into value. It can be concluded as main definition of this chapter that IC is a term which refers to the non-financial value creation capabilities of a firm that represents the difference between market and book values and can be further described by it's components. It would be of value to define IC according to the definition of each of it's components. The next section covers the most accepted set of IC components.

INTELLECTUAL CAPITAL COMPONENTS

It is widely agreed that IC is grouped into three areas – human, organizational, and relational capitals – on both organizational and individual levels (Allee, 1998, 1999; Bontis, 1999; Carson et al. 2004; Cascio, 1998; Cuganesan, 2006; Ferrier, 2001; Fischer, 2001; Huang, 1998; Kaplan and Norton, 1992; Kim and Kumar, 2009; Lynn, 2000; Miller et al., 1999; Mouritsen et al., 2001; Mouritsen et al., 2005; Nordhaug, 1998; Petty and Guthrie, 2000; Rastogi, 2000b; Romijn and Albu, 2001; Tseng and Goo, 2005Van Buren, 1999). There are slight variations in some of the models, however, most of them comprise these three interrelated components as indicated in Table 4.

As discussed earlier, There is a rather comprehensive consensus that the Human, Organizational, Relational Capitals are the main components of IC and they can cover most of IC important aspects. The other components can be sub-categorized within these three ones. An explanation of each component is provided as following:

- **"Human Capital" (HC)** is the capabilities of individuals who are the source of innovation and renewal within companies (Stewart, 1997). Similarly, Lynn (2000) defines it as "an inventory of the skill sets and knowledge of individuals within an organization". Although research and development has been considered as a process and as an artifact produced by human capital, it has gained a high priority in the literature of intellectual capital and has been stated separately in some of the articles and books as R&D or innovation capital. Here, it is regarded as a sub-component of organizational capital as discussed later in the chapter.
- **"Organizational Capital" (OC) or "Structural Capital" (SC)** is essentially

Table 4. IC classifications

Developed by	Framework / Country	Classification
Saint-Onge (1996)	Canadian Imperial Bank of Commerce (Canada)	Human capital Structural capital Relational capital
Edvinsson and Malone (1997)	Skandia Value Scheme (Sweden)	Human capital Structural capital
Stewart (1997)	USA	Human capital Structural capital Customer capital
Sveiby (1997)	Intangible Assets Monitor (Australia)	Employee competence Internal structure External structure
Bontis (1998)	Canada	Human capital Structural capital Customer capital
Roos et al. (1998)	UK	Human capital Structural capital Relational capital
OECD (1999)	OECD	human capital organizational capital
Van Buren (1999)	American Society for Training and Development (USA)	Human capital Innovation capital Process capital Customer capital
O'Donnell and O'Regan (2000)	Ireland	People Internal structure External structure
Carson et al. (2004)	University of Australia	Human Capital Customer Competitor Relational Capital Personal Attributes Skills Structural Capital Fluid Crystallized
Chen et al. (2004)	Zhejiang University	Human Capital Structural Capital Customer Capital Innovation Capital
Huang et al.(2007)	England, Malaysia	Human Capital Employee Capabilities Employee Development and Retention Employee Behavior Structural Capital Development of products/Ideas Organization Infrastructure Customer Capital Market Perspectives Data on customers Customer service and relationship

continued on following page

Table 4. continued

Developed by	Framework / Country	Classification
Pedrini (2007)	Italy	Human Capital Organizational Capital Relational Capital
Cheng et al. (2008)	Taiwan	Human Capital Process Capital Customer Capital Innovation Capital
Bruggen et al. (2009)	Netherlands	Human Capital Structural Capital Relational Capital
Chan (2009)	Australia	Human Capital Structural Capital Physical Capital

"captured" human capital, and relies on some of the skills of human capital for its existence, such as the ability to communicate and the willingness to share information and allow it to be stored and maintained in Organizational Capital (Carson et al., 2004).It is conventionally used to refer to the processes and procedures that are in some ways recorded and hence accessible to the organization as how things are done in order to maintain effectiveness. It is formed by the intellectual input of the employees, but it "belongs" to the firm. In other words, Organizational Capital has been further subcategorized by some authors into process capital, intellectual property and innovation capital (Chatzkel, 2002; Chen et al., 2004). Edvinsson and Malone (1997) classified Organizational Capital into organizational, process and innovation capital. Here, organizational capital includes the organization philosophy and systems for leveraging its capability. Process capital includes the techniques, procedures, and programs that implement and enhance the delivery of goods and services. Innovation capital includes intellectual properties and intangible assets. Therefore, an organization's policies and procedures, customized software applications, research and development programs, training courses and patents are examples of organizational (structural) capital (Kannan and G.Aulbur, 2004). Organizational Capital by itself, refers to things like proprietary software systems, distribution networks, and supply chains. It can be seen as the knowledge that does not go home at night and belongs to the organization and has the ability to be reproduced and shared (Stewart, 1997).

- **"Relational Capital" (RC) or "Customer Capital" (CC)** is one of the most important components for classifying and measuring the status of intellectual capital in a firm. This importance has been repeatedly recognized in the literature (as discussed later in the chapter). Relational Capital refers to the value of an organization's external relationships with other organizations, suppliers, people, and even competitors with which, it has communications and goes through partnerships, alliances, purchase and sell contracts, etc. Roos et al. (1997) define this type of external capital as the network with external stakeholders who have a key influence on the company. Therefore, Relational Capital refers to issues like trust and understanding and the strength and loyalty of external relations. Customer satisfaction, repetition in the business processes, financial well-being, and price sensitivity which may be used as indicators of Relational Capital (Kannan and G.Aulbur, 2004).

Since the mentioned components represent a holistic view of the important aspects of intellectual capital, it would be of value to study them in greater depth. As the importance of each component lies in the important elements within it, it can be implied that the study of these elements would provide a better insight of what they try to explore and clarify. The intellectual capital metrics section provides most of the accepted elements of each component. These elements have been accepted and considered as applicable metrics in many books and papers.

INTELLECTUAL CAPITAL METRICS

Metrics are needed to further convince management and stakeholders as to the value of knowledge management initiatives (Liebowitz and Suen, 2000). From a strategic point of view, Intellectual Capital is used to create and enhance firm value. The measurement angle of IC focuses on how new reporting mechanisms can be constructed. These mechanisms enable non-financial, qualitative items of intellectual capital to be measured alongside traditional, quantifiable, financial data (Johanson et al., 1999).

As discussed in the chapter, the debates on intellectual capital are by now more concentrated on comprehensive and integrative theories and models encompassing the IC measurement and management in practice. It has been taken for granted that concentrating on a widely accepted and supported set of measures helps the practitioners in pervasively evaluating the status of IC and extracting the potential pitfalls and strengths in the process of IC management. Grasenick and Low (2004), suggest that IC indicators should offer simple information that can be understood and communicated. These measures, sometimes, have been stated in divergent sets in the literature while at another period of time, in a convergent manner. In recent years, most of the authors have rather agreed upon the same measures. In other words, a trend of converging theories and sets of indicators can be recognized in the literature. Here, there is an attempt to go deeper into this broad literature and extract the existing convergence on the indicators among different authors and present them in an integrative view. All of the indicators in the following tables (Tables 5,6, and 7) have been directly or indirectly stated or applied in the articles and surveys. These indicators are mostly oriented toward non-financial and qualitative aspects, however, some of them might be transformed into quantitative ones according to the requirements of the business under study.

It has been impossible for R&D organizations to express their efforts and performances in their financial statement in a way that demonstrates the real value of the organization (Riegler and Hollerschmid, 2006). This gap between the financial statement and an organization's real value is encouraging R&D organizations to pay attention to developing IC indicators that more accurately measure their value (Kim and Kumar, 2009), specially the value-added of R&D in their overall organizational value. Here, IC is a key driver of innovation and core value in the new economy (Pike et al., 2005a; Kim and Kumar, 2009); In intense circumstances, R&D activity has been considered central to efforts to innovate, that is, applying and utilizing new opportunities to produce services or products (van de Ven, 1986). Romijn and Albaladejo (2002) designed an experimental innovation index and used it alongside conventional ones, such as patents granted or new products, to test relationships with indicators of internal sources of innovation capability. Their findings supported the importance of R&D (Cheng et al., 2008). The success of innovation mainly

Table 5. Human capital indicators

Index	Supporting Articles
Capabilities and skills	(Mouritsen et. al., 2002) - (Huang et. al., 2007) - (Cohen & Kaimenakis, 2007) - (Flostrand, 2006) - (Guthrie, 2001) - (Bozbura, 2004) - (Ordonez, 2002) - (Seleim and Ashour, 2004) - (Johanson et. al., 2006) - (Tovstiga and Tulugurova, 2007) - (Ordonez, 2003) - (Marr et. al., 2004) - (Pedrini, 2007) - (Litschka et. Al., 2006) - (Guthrie and Petty, 2000) - (Lim and Dallimore, 2004) - (Brennan, 2000) - (Skandia, 1994, 1995a, 1995b, 1996a, 1996b, 1997, 1998) - (Mayo, 2001) - (Bontis, 1998) - (Roos and Roos, 1997) - (Konrad Group, 1988) - Bruggen et al. (2009)
Loyalty and commitment	(Huang et. al., 2007) - (Cohen & Kaimenakis, 2007) -(Flostrand, 2006) - (Ordonez, 2002) - (Tovstiga and Tulugurova, 2007) - (Ordonez, 2003) - (Pedrini, 2007) - (Litschka et. Al., 2006) - (Guthrie and Petty, 2000) - (Brennan, 2000) - (Ordonez, 2005) - (Skandia, 1994, 1995a, 1995b, 1996a, 1996b, 1997, 1998) - (Mayo, 2001) - (Roos and Roos, 1997)
Satisfaction	(Mouritsen et. al., 2002) - (Huang et. al., 2007) - (Cohen & Kaimenakis, 2007) - (Bozbura, 2004) - (Litschka et. Al., 2006) - (Lim and Dallimore, 2004) - (Skandia, 1994, 1995a, 1995b, 1996a, 1996b, 1997, 1998) - (Mayo, 2001) - - (Bontis, 1998) - (Konrad Group, 1988)
Valuing the attitudes and believes	(Mouritsen et. al., 2002) - (Huang et. al., 2007) - (Bose, 2004) - (Darroch, 2003) - (Ordonez, 2002) - (Johanson et. al., 2006) - (Tovstiga and Tulugurova, 2007) - (Lim and Dallimore, 2004) - (Ordonez, 2005) - (Liebowitz and Suen, 2000) - (Roos et. al., 1998)
Quality of knowledge-oriented teams	(Mouritsen et. al., 2002) - (Bozbura, 2004) - (Pedrini, 2007) - Bruggen et al. (2009)
Cooperating with expert consultants	(Mouritsen et. al., 2002) - (Bozbura, 2004) - (Ordonez, 2002) - (Seleim and Ashour, 2004) - (Johanson et. al., 2006) - (Tovstiga and Tulugurova, 2007) - (Ordonez, 2003) - (Marr et. al., 2004) - (Pedrini, 2007) - (Lim and Dallimore, 2004)
Employee retention capabilities	(Huang et. al., 2007) - (Bose, 2004) - (Flostrand, 2006) - (Bozbura, 2004) - (Ordonez, 2003) - (Marr et. al., 2004) - (Ordonez, 2005) - (Liebowitz and Suen, 2000)
Promotion mechanisms based on individual capabilities	(Bose, 2004) - (Ordonez, 2002)
Attitudes toward female managers	(Bose, 2004) - (Ordonez, 2002) - (Ordonez, 2003) - (Roos et. al., 1998)
Attitudes toward higher levels of education	(Huang et. al., 2007) - (Bose, 2004) - (Flostrand, 2006) - (Bozbura, 2004) - (Johanson et. al., 2006) - (Ordonez, 2003) - (Marr et. al., 2004) - (Pedrini, 2007) - (Guthrie and Petty, 2000) - (Lim and Dallimore, 2004) - (Brennan, 2000) - (Roos et. al., 1998)
Quality of training programs	(Huang et. al., 2007) - (Bose, 2004) - (Bozbura, 2004) - (Ordonez, 2002) - (Seleim and Ashour, 2004) - (Marr et. al., 2004) - (Pedrini, 2007) - (Ordonez, 2005) - (Liebowitz and Suen, 2000) - (Roos et. al., 1998)
IT literacy	(Bose, 2004) - (Liebowitz and Suen, 2000) - (Roos et. al., 1998)
Managerial capabilities	(Bose, 2004) - (Bozbura, 2004) - (Seleim and Ashour, 2004) - (Guthrie and Petty, 2000) - (Brennan, 2000) - (Liebowitz and Suen, 2000) - (Roos et. al., 1998) - (Jacobsen and Hofman, 2005)
Organizational capability in creating motivation	(Huang et. al., 2007) - (Bose, 2004) - (Ordonez, 2003) - (Ordonez, 2005) - (Liebowitz and Suen, 2000) - (Roos et. al., 1998)
Acquaintance with the organizational strategy	(Huang et. al., 2007) - (Bose, 2004) - (Litschka et. Al., 2006) - (Liebowitz and Suen, 2000)

depends on a highly educated or experienced individual's capabilities in areas such as innovation capacity, skills, and know-how. Unfortunately, an individual's capabilities never cover the entire complex of multiple and complicated innovation activities because the individual must also interact with organizational level capabilities such as process, culture, and brand value (Subramaniam and Youndt, 2005).

Research and development (or in other words, innovation capital) has been considered as one of the most important factors in measuring the performance of intellectual capital management. In order to show the importance of this component in

Table 6. Organizational (structural) capital indicators

index	Supporting articles
Knowledge management capabilities	(Cohen & Kaimenakis, 2007) - (Guthrie, 2001) - -(Bozbura, 2004) - (Skandia, 1994, 1995a, 1995b, 1996a, 1996b, 1997, 1998) - (Mayo, 2001) - (Bontis, 1998) - (Roos and Roos, 1997) - (Konrad Group, 1988) - (Jacobsen and Hofman, 2005) - Bruggen et al. (2009)
Organizational knowledge	(Huang et. al., 2007) - (Cohen & Kaimenakis, 2007) -(Bose, 2004) - (Guthrie, 2001) -(Bozbura, 2004) - (Johanson et. al., 2006) - (Tovstiga and Tulugurova, 2007) - (Marr et. al., 2004) - (Litschka et. Al., 2006) - (Guthrie and Petty, 2000) - (Brennan, 2000) - (Skandia, 1994, 1995a, 1995b, 1996a, 1996b, 1997, 1998) - (Mayo, 2001) - (Bontis, 1998) - (Jacobsen and Hofman, 2005) - (Sanchez-Canizares et. Al., 2007)
Efficiency of organizational processes	(Cohen & Kaimenakis, 2007) - (Bozbura, 2004) - (Litschka et. Al., 2006) - (Guthrie and Petty, 2000) - (Skandia, 1994, 1995a, 1995b, 1996a, 1996b, 1997, 1998) - (Mayo, 2001) - (Bontis, 1998) - (Roos and Roos, 1997) - (Konrad Group, 1988) - (Sanchez-Canizares et. Al., 2007) - (Ordonez, 2002) - (Ordonez, 2004) - Bruggen et al. (2009)
Knowledge dissemination mechanisms	(Huang et. al., 2007) - (Darroch, 2003) - (Flostrand, 2006) - (CMA, 1999) - Bruggen et al. (2009)
Quality of IS used for knowledge dissemination	(Huang et. al., 2007) - (Darroch, 2003) - (Flostrand, 2006) - (Ordonez, 2002) - (Ordonez, 2005) - Bruggen et al. (2009)
Investment in infrastructure	(Huang et. al., 2007) - (Bose, 2004) - (Mouritsen et. al., 2002) - (Flostrand, 2006) -(Bozbura, 2004) - (Ordonez, 2002) - (Johanson et. al., 2006) - (Marr et. al., 2004) - (Guthrie and Petty, 2000) - (Brennan, 2000) - (Liebowitz and Suen, 2000) - (Ordonez, 2004)
Utilization of Project management techniques	(Mouritsen et. al., 2002) - (Seleim and Ashour, 2004)
Quality of processes for activities	(Bose, 2004) - (Liebowitz and Suen, 2000) - (CMA, 1999) - (Ordonez, 2004)
Qualitative performance of knowledge intensive processes	(Bose, 2004) - (Guthrie, 2001) - (Ordonez, 2002) - (Seleim and Ashour, 2004) - (Marr et. al., 2004) - (Pedrini, 2007) - (Ordonez, 2005) - (Liebowitz and Suen, 2000) - (CMA, 1999) - (Roos et. al., 1998) - (Engstrom et. al., 2003)
Quality of changes and renovations in the business processes	(Bose, 2004) - (Guthrie, 2001) - (Liebowitz and Suen, 2000) - (Roos et. al., 1998) - (Sanchez-Canizares et. Al., 2007)
The extent to which processes support innovation	(Mouritsen et. al., 2002) - (Huang et. al., 2007) - (Bose, 2004) -(Bozbura, 2004) - (Ordonez, 2002) - (Liebowitz and Suen, 2000) - (CMA, 1999) - (Roos et. al., 1998) - (Ordonez, 2004) - (Engstrom et. al., 2003)
The organizational attitude toward standardization	(Bose, 2004) - (Seleim and Ashour, 2004)
Knowledge reuse capabilities	(Bose, 2004) - (Seleim and Ashour, 2004) - (Jacobsen and Hofman, 2005) - (Ordonez, 2004)
Quality of investments for new products	(Huang et. al., 2007) - (Mouritsen et. al., 2002) - (Flostrand, 2006) - (Guthrie, 2001) -(Bozbura, 2004) - (Johanson et. al., 2006) - (Litschka et. Al., 2006) - (Liebowitz and Suen, 2000) - (Sanchez-Canizares et. Al., 2007) - (Engstrom et. al., 2003)
Quality of investments in exploring new markets	(Mouritsen et. al., 2002) - (Guthrie, 2001) - (Guthrie and Petty, 2000) - (Liebowitz and Suen, 2000) - (Roos et. al., 1998) - (Bontis, 2004)
Investments in analysis and design	(Huang et. al., 2007) - (Bose, 2004) - (Flostrand, 2006) - (Guthrie, 2001) - (Litschka et. Al., 2006) - (Guthrie and Petty, 2000) - (Liebowitz and Suen, 2000) - (CMA, 1999)
Training courses for supporting existing products	(Huang et. al., 2007) - (Bose, 2004) - (Liebowitz and Suen, 2000) - (Roos et. al., 1998)
Quality of investments for basic researches	(Bose, 2004) - (Liebowitz and Suen, 2000)
Quality of investments in new IS systems	(Huang et. al., 2007) - (Bose, 2004) - (Flostrand, 2006) - (Guthrie, 2001) - (Johanson et. al., 2006) - (Brennan, 2000) - (Guthrie and Petty, 2000) - (Liebowitz and Suen, 2000) - (Jacobsen and Hofman, 2005) - (Sanchez-Canizares et. Al., 2007) - (Bontis, 2004)
Creating equal chances for R&D	(Bose, 2004) - (Liebowitz and Suen, 2000)

continued on following page

Table 6. continued

index	Supporting articles
Quality of trainings in IT and new sciences	(Huang et. al., 2007) - (Bose, 2004) - (Liebowitz and Suen, 2000)
Resources available for R&D	(Bose, 2004) - (Liebowitz and Suen, 2000) - (Bontis, 2004)
Quality of investments in internal processes	(Bose, 2004) - (Guthrie, 2001) -(Bozbura, 2004) - (Johanson et. al., 2006) - (Pedrini, 2007) - (Liebowitz and Suen, 2000)

Table 7. Relational capital indicators

Measure	Supporting articles
Customer loyalty	(Johanson et. al., 2006) - (Ordonez, 2002) (Ordonez, 2003) - (Huang et. al., 2007)
Quality of customer acquisition and retention	(Liebowitz and Suen, 2000) - (Chen et al., 2004) -(Pedrini, 2007) - (CMA, 1999) - (Ordonez, 2002) - (Ordonez, 2003) - (Huang et. al., 2007)
Customer satisfaction	(Liebowitz and Suen, 2000)- (Chen et al., 2004) - (CMA, 1999) - (Konrad Group, 1988) - (Roos and Roos, 1997) - (Bontis, 1998) - (Mayo, 2001) - (Skandia, 1994, 1995a, 1995b, 1996a, 1996b, 1997, 1998) - (Cohen and Kaimenakis, 2007) - (Ordonez, 2002) - (Ordonez, 2003) - (Huang et. al., 2007)
Market share	(Liebowitz and Suen, 2000) - (Chen et al., 2004) - (Ordonez, 2002) - (Huang et. al., 2007)
Market demand for product or service	(Johanson et. al., 2006) - (Ordonez, 2002) - (Huang et. al., 2007)
Growth potential in market volume	(Chen et al., 2004) - (Huang et. al., 2007)
Opportunities for business alliances	(Johanson et. al., 2006) - (Huang et. al., 2007)
Quality of distribution channels	(Chen et al., 2004) - (Johanson et. al., 2006) - (Huang et. al., 2007)
Quality of customer relationships	(Pedrini, 2007) - (Chen et al., 2004) - (Ordonez, 2002) - (Huang et. al., 2007)
Quality of responses to Customer demands and complaints	(Liebowitz and Suen, 2000) - (Chen et al., 2004) - (Seleim et al., 2004) - (Huang et. al., 2007)
Quality of customers	(Konrad Group, 1988) - (Roos and Roos, 1997) - (Bontis, 1998) - (Mayo, 2001) - (Skandia, 1994, 1995a, 1995b, 1996a, 1996b, 1997, 1998) - (Cohen and Kaimenakis, 2007) - (Liebowitz and Suen, 2000)
Brand and Trademarks	(Pedrini, 2007) - (Johanson et. al., 2006) - (Ordonez, 2005) - (Ordonez, 2003) - (Chen et al., 2004)

the literature, an attempt has been made to review the most important indicators regarding the R&D in the context of IC. Many approved indicators have been stated and utilized by different authors in various fields which clearly states the pervasive usage of this component as an important factor. In R&D, human capital plays a significant role in triggering innovation and performance. Satisfied and highly educated technicians tend to improve organizational capital (e.g. process, culture, and brand value) and then engage in high R&D performances (Ellickson, 2002). In this way, innovation capital is considered to be a very important sub-component of organizational capital.

A total integration of indicators makes the interested researchers and practitioners capable of assessing the performance of intellectual capital in a comprehensive way and thus, preparing encompassing reports for the managerial levels. Considering the most accepted indicators also helps the IT departments select those knowledge management systems that best cover their intellectual weaknesses and therefore, assists in gradually gaining a knowledge-oriented competitive advantage over the competitors. Some of the indicators might seem to belong to more than one component as they are related to two or more of them at the same time, however, as the emphasis of each indicator is mostly on one of the aspects of intellectual capital, each one of them has been fallen under one of the IC components (Tables 5,6 and 7). Other scholars might prefer to categorize them under one of the other components. The innovation capital indicators are provided in grey in Table 6.

Since the measurement of indicators should take place periodically, then the results might be different and the IC management performance might improve or deteriorate. This feature of the intellectual capital shows that the nature of this concept is static since it's status changes from time to time. During the periods between IC assessments, it would be the knowledge management that addresses the knowledge being created or captured by intellectual capabilities of the organization. As the intellectual capital measurement outputs might illustrate changes in the performance of value creation processes, it can be shown that this change is due to the static nature of knowledge management, since the change in IC performance occurs according to the quality of managing knowledge during the mentioned period. The next section discusses the concepts relating to IC measurement.

IC MEASUREMENT

In a knowledge-based economy, the knowledge is included as a large part of a product's value as well as a company's wealth. Traditional accounting methods, which are based on tangible assets and historical, transaction-based information, are inadequate for valuing the intellectual capital, which is the largest and the most valuable asset for many enterprises (Chen et al., 2004). Also, present accounting standards do not require the recognition of IC in the financial statements and only a relatively low level of firms disclose IC in their annual reports (FASB, 2001). As a result, there is a growing level of information asymmetry between companies and users of financial statements (Bruggen et al., 2009). Measurement often refers to the relation between a phenomenon (such as intellectual capital management) and its expression, so that measurement captures the value(s) or inherent dimensions of the phenomenon. Measurement helps us establish a relation between phenomenon and our perception of it (Mattessich, 2003; Solomons, 1991). Measurement of intellectual capital is interesting because it is an input that starts action rather than a conclusion that stops action (Mouritsen, 2004).

In fact, it is the shortage of traditional accounting measures that have led to the gap between companies' book and market values: because the intellectual capital of organizations is not readily documented (Allee, 1999; Stewart, 1997). Market value is an appraisal based on an estimate of what a buyer would pay a seller for any piece of property. Book value is shareholders' equity shown in a firm's financial statement and reflects the value of corporate reported assets less liabilities. Recently, financial statements do not present relative market value, and the gap between firms' market and book value is increasing, with market value exceeding book value by up to 15 times (Cheng et al., 2008). The reduction of information asymmetry between a company and external users of information on these values, is one of the major reasons for voluntary IC disclosure (Bruggen et al., 2009). According to Andriessen (2004b), information asymmetry may result in the misallocation of capital, which eventually leads to social costs such as unemployment or reduced productivity.

The limits of monitoring systems based on financial criteria are found in the ever-increasing discrepancy in the price-to-book value observed in recent years, an indication of the incapacity to fully measure the value of a company's intangible resources (Lev, 1997, 2001; Andriessen, 2001, 2002; Pedrini 2007 ; Pike et al., 2002). Further, expenditure on the human component of intellectual capital is rarely included in corporate balance sheets, and if it is included some time then, it is not treated as an investment rather as a cost (Bassi, 1997; Guthrie et al., 1999; Carson et al. 2004).

There is no universally accepted IC measurement method among the 34 methods identified in the relevant literature (Andriessen, 2004b; Bontis, 2001; Sveiby, 2005). IC measurement is helpful in verifying the company's ability to achieve its strategic objective, laying out its R&D, providing background information for project readjustment, and confirming the emphases of a company's education and training program (Chen et al., 2004). As a result for strategic business and marketing management, the IC measurement can be seen as more useful for internal measurement than as an external communicative vehicle to shareholders or investors (Bontis, 2001).In a word, IC measurement is significant to IC management, i.e. effective management relies on effective measurement (Chen et al., 2004). While the financial value of intangible resources cannot be measured with ordinary accounting systems, a non-financial index expressing the value of intellectual capital can be drawn up (Pedrini, 2007). Some of the most important methods of this kind include:

- Balanced Scorecard (Kaplan and Norton, 1992, 1996);
- Dow's chemical value platform (Petrash, 1996);
- IC-Index (Roos and Roos, 1997);
- Skandia Navigator (Edvinsson and Malone, 1997);
- IC audit model (IFAC, 1998) and
- The Value Platform (Brennan and Connell, 2000);
- Danish guidelines (DATI, 1997, 2000);
- Intangible asset monitor (Sveiby, 1997, 2001);
- Value chain scoreboard (Lev, 2001).

Although, typically they refer to human, organizational and relational capitals without expounding upon them in detail (Carson et al., 2004). Kaplan and Norton's 'Balanced framework focuses not only on the capabilities and productivity of employees but also on employee satisfaction and retention. Edvinsson and Malone's Skandia Navigator (Edvinsson, 1997), measures an organization's human capital in terms of percentage of graduates; annual turnover of staff; average years of service with the company; and a leadership index measuring the proportion of leaders within an organization (Pedrini, 2007).

Some of the mentioned tools have gained more popularity comparing to others such as, Sveiby's (2001) Intangible Asset Monitor, which measures staff education levels, training and education costs, professional turnover, proportion of professionals to support staff, average age, seniority, and relative pay position. Guthrie and Petty's Modified Intangible Assets Monitor (Petty and Guthrie, 2000) which aims to measure employee know-how, education, vocational qualifications, work-related knowledge, work-related competencies, entrepreneurial spirit, innovations, proactive and reactive abilities, and changeability.

In addition, the Value Platform (Brennan and Connell, 2000) was developed in a collaborative effort of Edvinsson (Skandia), Onge (The Mutual Group) and Petrash (Dow Chemical) (Carson et al., 2004). Brennan and Connell define intellectual capital as the sum of human capital, organizational capital and relational capital and the model depicts the interrelationships between these three components and their management. This model does not specify what human and organizational capitals are, but points to the importance of considering dynamic relationships rather than attributes of each of the components of intellectual capital in isolation from each other (Carson et al., 2004).

Sveiby's intangible assets monitor and Kaplan and Norton's balanced scorecard models are excellent tools, but each represents a first attempt at solving the problem of visually representing (in numbers or other means) the IC of a specific organization (Petty and Guthrie, 2000). One of the most prominent classification of components is the Skandia Navigator, an accepted framework for representing the IC component classification. This navigator, as shown in Figure 2, divides the total value of the organization into two categories: Financial Capital and Intellectual Capital, and then breaks the IC down to the lower levels.

Skandia's model is particularly impressive in recognizing the role of relational capital in creating a company's value, as shown in figure 1 (Chen et al., 2004). It also provides a broad coverage of organizational process factors that has not been attempted before (Bontis, 2001; Chen et al., 2004; Lovingsson et al., 2000).

Being an important tool, Skandia Navigator has been relied extensively on by other companies to value their R&D and patent process. But like other simplifications of the reality, the Skandia Navigator also needs improvements and readjustments (Chen et al., 2004). For instance, it recognizes customer capital as a division of organizational capital while customer (relational) capital (as described in detail earlier) is mostly considered to be an independent and important component of IC in the recent literature.

Here, it is beneficial to have a brief description of IC measurement schools. According to Kannan and G.Aulbur (2004), there are four schools of measurement that help us in determining what items managers are going to measure, which strategic or fundamental business processes are to be included, and how organizations are going to report on IC components:

Figure 1. Skandia Navigator

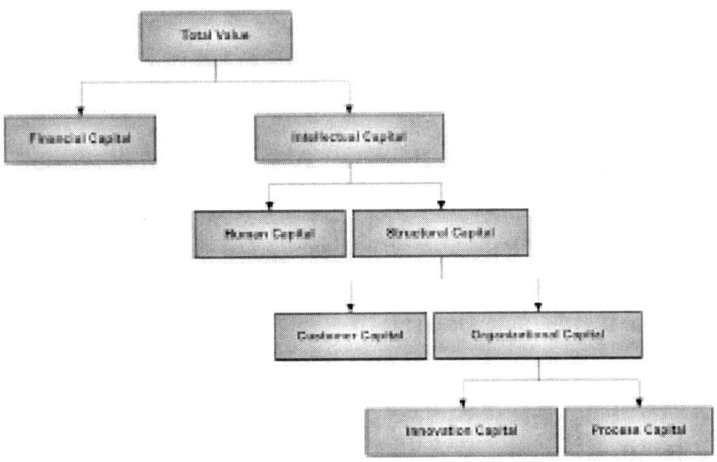

Figure 2. Value earned through IC

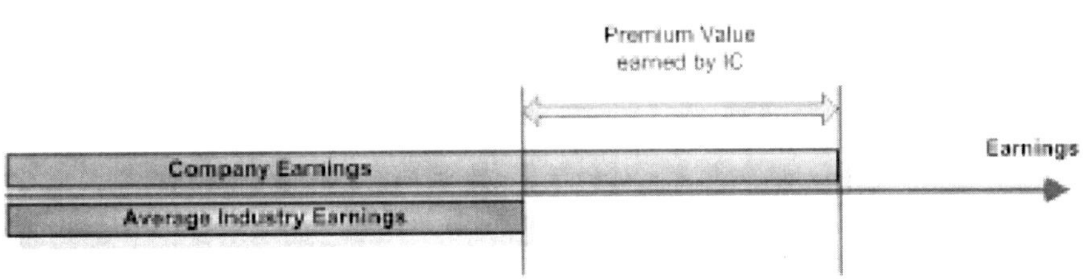

1. **Perceptual measures:** these measures concentrate on employees perceptions and their need for an effective knowledge management. These measures are mostly related to the indicators provided for human capital in table 5 and relational capital indicators stated in table 7.
2. **Financial measures:** these frequently appeared in financial statements as financial assets and liabilities of an organization. Since these measures are exposed in financial statements and mostly belong to the accounting side, they are not included in the non-financial indicators of intellectual capital provided in the chapter.
3. **Process measures:** these measures focus on the business processes and the knowledge workers who are working within the structure of these processes. They are related to the indicators provided for organizational capital in table 6.
4. **Other measures:** measurement techniques that cannot be classified as purely financial, perceptual, or process-based fall under this category.

For measuring the value of IC (especially the monetary value), which is of great importance for knowledge intensive organizations, there are few methods including the Calculated Intangible Value (CIV) (Stewart, 1997), the Intangibles Scoreboard developed by Lev (see, e.g. Hurwitz et al., 2002) and the Weightless Wealth Tool Kit (Andriessen, 2004a; Kujansivu and Lonnqvist, 2007). The CIV method is the only one that can be easily applied by an expert who is external to the organization and does not necessarily require the internal information (Kujansivu and Lonnqvist, 2007).

The CIV method is based on the assumption that a company's premium earnings, i.e. the earnings greater than those of an average return within the same business, result from the company's IC. The data for this method is acquired from the financial statements of companies except for the data for the average return on tangible assets in that business (Kujansivu and Lonnqvist, 2007). This concept can be depicted on a continuum (as shown in Figure 2).

The extra value created in comparison to other companies in the same business, is considered the value earned through the Intellectual capital. If a company has an earning less than the average, then it has encountered some deficiencies in its business processes that may lead to value loss. A method like CIV, however, calculates the monetary value of IC through a set of variables and fixed numbers. One of the limitations to such a method is the lack of enough support-base for the data driven from the financial statements. Are the presented items in the financial statements adequate for the calculations? This is an important question which needs more elaboration.

The CIV method is mainly used for investigating the value of IC in the firms. The value by itself, however, can not directly show the strength of IC management. To overcome this problem, some techniques have been proposed that measure the efficiency of intellectual capital. One of these techniques is Value Added Intellectual Coefficient (VAIC) developed by Ante Pulic. There has been an attempt by Sveiby (2005) to categorise various IC methods into four approaches, but this categorization did not include the one that is known as the Austrian approach or VAIC methodology. VAIC appears to be falling outside the four approaches. Proponents have been presented in parentheses following each approach (Chan, 2009):

- Market capitalization approach (Market-to-book value by Stewart (1997a));
- Direct IC measurement approach (Technology broker by Brooking (1997), Citation-weighted patents by Bontis (2001));
- Scorecard approach (Balanced score card by Kaplan and Norton (1992), Intangible asset monitor by Sveiby (2001), Skandia Navigatore by Edvinsson (1997));

- Economic value-added approach (Economic value added (EVA) by Stewart (1997a), Calculated intangible value by Stewart (1997); and
- VAIC methodology (Pulic, 2004).

VAIC is an analytical tool for measuring the performance of, e.g. a company (Pulic, 2000; Van der Zahn et al. 2004). While, CIV measures the value of IC, VAIC can measure the efficiency of IC within a company, i.e. it describes how a company's IC adds value to the company (Kujansivu and Lonnqvist, 2007). VAIC measures and monitors the total value creation efficiency in the company. One subordinate concept of VAIC, the Intellectual Capital Efficiency (ICE) describes the efficiency of IC within a company. The method is built on the premise that value creation is derived from two primary resource bases: physical capital resources and intellectual capital resources. Actually, VAIC indicates the total efficiency of value creation from all resources employed and ICE reflects the efficiency of value created by the IC employed (Kujansivu and Lonnqvist, 2007). The important factor here is that the data needed for the calculation can be found in the financial statement. The justification for adopting the VAIC methodology has been summarized by many scholars (Chan, 2009, Chen et al., 2005; Firer and Williams, 2003; Goh, 2005; Mavridis, 2004; Tseng and Goo, 2005; Williams, 2000).

INTELLECTUAL CAPITAL STATEMENT AND REPORTING

The source of companies' economic value no longer depends on the production of material goods but on the creation and manipulation of intellectual capital (Guthrie et al., 2004). Management of intellectual capital requires enough information about the current business situation as well as adequate data on the current and desired strategic state of the organization, especially in terms of intangibles. This is exactly what IC reporting is about. For this purpose, it will be useful to locate the exact position of IC in the network of inter-related knowledge-based concepts, as it appears in Figure 3. This figure makes the way clear in determining the position of the factors relating

Figure 3. Conceptual roots of intellectual capital

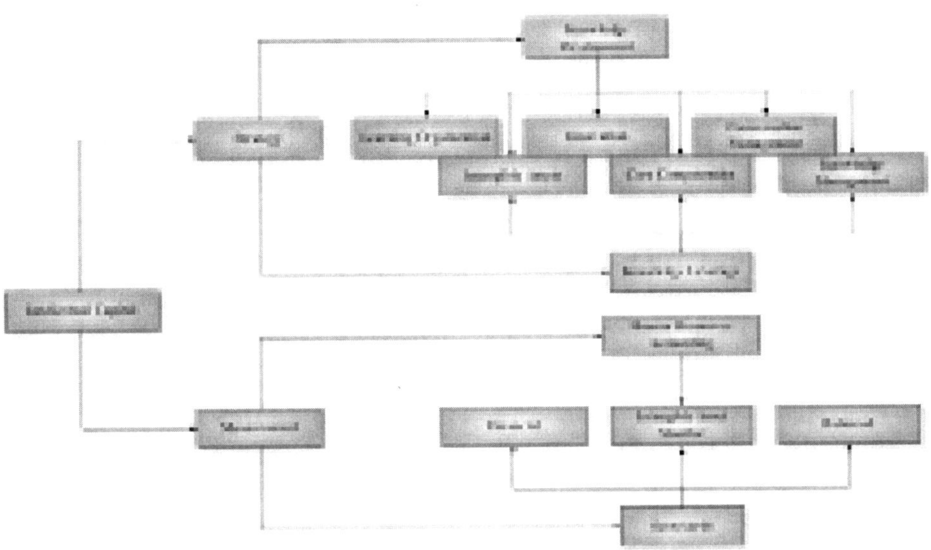

to intellectual capital, knowledge leverage and management and the IC measurement methods (Jashapara, 2003; Petty and Guthrie, 2000).

For measuring the concept of IC, there has been no workable recognition of intangible assets in the financial statements of organizations, albeit many attempts have been made. Financial statements fail to measure and show the "most significant building blocks of business", hence, these statements fail to communicate the state of the business in terms of IC development to the management and investors (Kannan and G.Aulbur, 2004). Recently, the companies have developed monitoring systems for intellectual capital, in response to the inadequacies of non-financial reporting mechanisms. The most important initial output, often referred to as the Intellectual Capital Report or Statement, communicates the coherence between corporate strategy and management practices (Dati, 1997, 2000; Edvinsson and Malone, 1997; Guilding and Pike, 1990; Haanes and Lowendhal, 1997; Hall, 1992; Hammerer, 1996; Hendriksen and van Breda, 1992; Kaplan and Norton, 1992, 2004; Mortensen et al., 1997; Pedrini, 2007; Petrash, 1996; Roos and Roos, 1997; Sullivan, 2000; Sveiby, 1997; Teece, 2000).

For the first step of IC reporting, IC statement can be applied as an instrument to encapsulate data about the intellectual potential of the organization. This instrument is devised and developed to better help the organizations understand and represent their non-financial performance objectives, efforts, challenges, overall indicators, and sometimes their detailed quantitative non-financial performance. In other words, the intellectual capital statement shows the firm's efforts to monitor, qualify and orchestrate its knowledge resources (Mouritsen, 2004). This statement can initially help the organizations to develop a comprehensive view about the final goal of intangible assets recognition and reporting and come to an overall consensus among the stakeholders and simultaneously, assist the top managers and supervisors to have a roadmap about how to proceed on reporting and managing intellectual capital (Mouritsen and Larsen, 2005):

Intellectual Capital Statement

1. **Knowledge narrative:** A narrative about the firm's ambition to create (use) value for its users and the required types of knowledge resources to accomplish:
    - What product or service does the company provide?
    - How does it make a difference for the user?
    - What knowledge resources are necessary to be able to supply the product or service?
    - How does the constellation of knowledge resources produce the service/product?
2. **Management challenges:** The durable challenges posed by the role of knowledge resources in the firm's business model:
    - How are the knowledge resources related?
    - Which existing knowledge resources should be strengthened?
    - What new knowledge resources are needed?
3. **Efforts:** The initiatives to compose, develop and procure knowledge resources:
    - What initiatives, actual and potential, can be identified?
    - What initiatives should be given priority?
4. **Indicators:** The mechanisms of monitoring the portfolio, development and effects of knowledge resources:
    - Effects – how do activities work?
    - Activities – what does the firm do to upgrade knowledge resources?
    - Resource mix – what is the composition of knowledge resources?

An intellectual capital statement is as how intellectual assets are created, developed and applied by the intellectual capabilities of the firm (Bontis, 2002; Edvinsson and Malone, 1997; Sveiby, 1997) It summarizes the firm's efforts to develop and use knowledge-based resources. The intellectual capital statement also creates distance to the ongoing affairs of the firm, and consequently facilitates evaluation (Mouritsen, 2004). To compare the IC statement and the financial statement, J. Mouritsen and H.T. Larsen (Mouritsen and Larsen, 2005) have proposed some questions that have been shown in Table 8.

The differences between the two approaches, however, are significant: the IC measurement is oriented towards the future while the financial accounting is supposed to look backwards. The IC measurement captures soft facts (qualities), while the financial accounting measures hard facts (quantities). Further, the IC measurement focuses on the value creation, while the financial accounting reflects the outcome of the past transactions and realizes cash flows (Chen et al., 2004). Another major difference is that we are accustomed to financial statements in spite of all of its deficiencies (Mouritsen, 2004). These questions seem to be excellent in providing an insight to the differences between the two mechanisms of reporting. But they seem to be inadequate as they do not elaborate on each of the statements and also, there is no distinction for the IC components. As an initial step toward the implementation of mechanisms for the IC measurement and management, managers should come to an agreement about the set of measures and variables to be taken into account according to the requirements and needs of their specific business or industry. The indicators presented in the integrative view section can hold a strategic position in determining the desired measures and moving toward the initial step of IC and KM management. Table 9 presents an equilibrium for an IC statement. For a balanced equilibrium, there should be an agreement at the managerial level on the list of measures in each of the columns.

Here, there is a continuum for the right side of the equilibrium. As the amount of acquired capitals (Assets) grows, the remaining capitals (on the right side) begin to decline. This can be increased by acquiring new capitals (through recruitment, purchases, ventures, etc.) which in turn, imposes some expenditures and liabilities to the firm. At the same time, it increases the assets possessed by the firm and this process continues in an ongoing manner. Each column can be filled with a set of IC indicators corresponding to the column concept and this can done through

Table 8. Financial and IC statement corresponding questions

Financial statement	*Intellectual capital statement*
What are the company's assets and liabilities?	Resources: How are the company's knowledge resources comprised?
What are the company's investments?	Activities: What does the company do to develop and strengthen its knowledge resources?
What is the company's profitability?	Effects: What are the effects of the company's knowledge resources?

Table 9. IC statement framework

Assets	=	*Capitals*	+	*Expenditures and Liabilities on Capitals or Assets*
Acquired R&D, human, organizational and relational capitals		R&D, human, organizational and relational capitals that are available to be acquired.		Liabilities on R&D, human, organizational and relational capitals

an investigation in the indicators listed in tables 5-8.After reaching an agreement upon each set of indicators, the equilibrium should be in balance until a new indicator is added. It should be noticed that useful accounts for each side should be specified and assigned under the supervision of knowledge management experts and professional accountants. Therefore, as a managerial and strategic technology, intellectual capital statements can:

1. Capture the on-going affairs of the business and transport them to a locality where they can be debated and assessed independently of the day-to-day concerns of operations.
2. Establish a distanced perspective on the actions that go on all the time to use and qualify the knowledge.
3. Induce evaluative and normative reflections by assessing knowledge management activities.
4. Allow decision making because there is time to contemplate on the future of knowledge resources. And then such decisions are implemented and the results turn up in a subsequent intellectual capital statement (Mouritsen, 2004).

The financial statement is not a direct representation of the firm's value. The equity of assets against liabilities and financial capitals is not a measure of the value of the firm to an investor. Likewise, the intellectual capital statement does not show the value of the firm to an investor (Mouritsen, 2004) unless we consider both of the statements as two complementing invaluable documents for representing the financial and non-financial performance of the firm. Besides, it is also worthwhile to consider that rare experts have been trained perfectly in understanding intellectual capital statements.

FUTURE RESEARCH DIRECTIONS

As argued in the chapter, a rather comprehensive consensus can be obtained through a deep investigation of various credible articles. The innovation capital has gained a very high popularity due to it's potential of assisting in acquiring a competitive advantage over time. Proven by the broad literature (but with no clear distinction), the following model can be proposed which displays the relationships among all of the IC components (see Figure 4):

Such a model can help the firms in following a roadmap toward the final goal of IC management. Although, there are many articles dealing with IC definitions and components, few of them properly provide a roadmap for initiating steps toward a practical IC measurement mechanism in order to effectively and efficiently manage knowledge and provide a better quality of knowledge management. The attempt of current chapter has been on deeply investigating various IC materials in order to provide a foundation for the future surveys. Considering the indicators provided, the practitioners and researchers can elaborate on each component and find the appropriate sub-indicators applicable for each market segment. It seems that each business might require it's exclusive and independent set of sub-indicators regarding the presented generic

Figure 4. IC components model

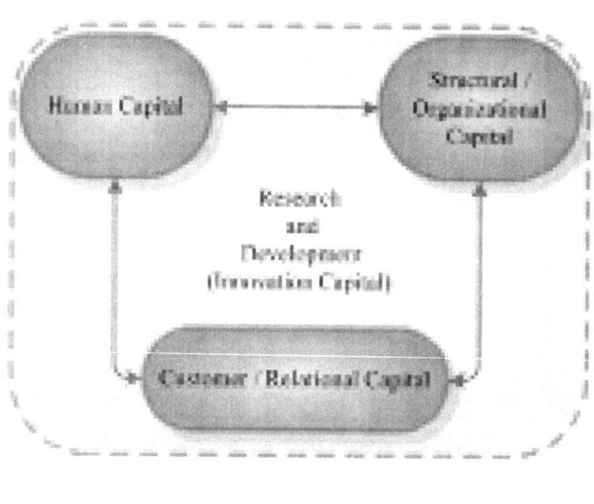

components and their relative indicators. In this regard, a stepwise roadmap has been provided for the managers and supervisors that are searching for managing their intellectual assets. This roadmap can help in exploiting the IC potentials toward a better quality of knowledge and intellectual assets management:

1. Assign a team for IC measurement.
2. Through a series of sessions, decide on the IC indicators approved to be useful for your business or industry. The list of indicators provided in this chapter can be a good starting point (most of these indicators have been tested and verified in different countries and for different businesses).
3. After reaching an agreement upon the list of indicators, formulate the specific IC statement of the firm.
4. After a comprehensive assessment of indicators, set a more clear boundary for each indicator and obtain KM systems that are capable of covering the weakly measured or supported indicators.
5. Revise the list of indicators frequently so as to keep the list up-to-date.
6. Repeat the steps 4 and 5 until you come to a practical model that is supported by the literature and other empirical studies.
7. Try to improve the model and integrate it with the strategic goals of the firm.
8. Repeat the steps 2 to 7 in order to keep the model and the indicators up to date.

We are at the beginning of the way toward mature IC models that comprise almost every part of the business. Since year 2000, a great deal of efforts have been made in order to conceptualize and make the knowledge collected through the years of studies on intellectual capital, more practical and applicable. Considering this fact and during the course of literature reviews, some important points emerged which are presented as suggestions for later conceptual and empirical studies:

1. Intellectual capital statements are important means of communicating the benefits of non-financial capitals. Working on the structure of these statements can help in converting them into a formal tool for assessing the intellectual status of any firm along with other financial measures.
2. Any attempt on finding the specific measures of different industries can help in determining the intellectual potential of businesses with more precision.
3. Alongside the measures, working on maturity models assists in providing a stepwise roadmap for the firms that are at the initial stages.
4. Providing software oriented mechanisms for the evaluation of IC status of the firms is of great importance and more attempts are required.
5. As there are state-of-the-art enhancements in the quality of information systems (specially for those systems that directly provide information and knowledge to the executive body of the firm), the need for a knowledge-oriented language also emerge. Developing the fundamental concepts of such a language alongside the analytical aspects of strategic systems development, assists in getting a more comprehensive and detailed understanding on how to extract the knowledge and to manage the intellectual potential of the firm.

CONCLUSION

This chapter focused on comprehensively reviewing the broad literature on intellectual capital. For covering this purpose, a great deal of effort has been made to pervasively review the importance, definitions, components and indicators applying to the concept of intellectual capital. The conceptual and empirical reviews followed a systematic process so as to bring the concepts together. Through the extensive reviews, a set of most important concepts and topics is extracted and it can help in directing the future studies toward a consensus. Also, the close relation among intellectual capital and knowledge management and the way companies report on intellectual capital and intellectual assets have been discussed. Conceptual surveys of literature revealed that there is a relative consensus on the components and measurement indicators of IC. Therefore, a comprehensive set of mostly accepted and applied indicators is provided in order to illuminate the course of future surveys for the researchers and practitioners in the field. Finally, a step-by-step guideline for intellectual capital management and a set of useful suggestions have been provided. Considering these suggestions makes the way clear for the scholars and managers who are in charge of converting the concepts of intellectual capital into more applicable terms. Additionally, it would be beneficial to work on methods that are applicable for measuring and implementing IC improvement programs and methods for assessing the level of intellectual capital strategic maturity in the knowledge-based organizations.

ADDITIONAL READING

There has been many empirical works intended for understanding the impacts of intellectual capital on the performance of organizations. Here, some of the important articles in the field of intellectual capital are provided. They represent useful attempts in starting or proceeding with the process of intellectual capital measurement and management:

- Human Capital and Europe Future (Groth, 1994)
- Intellectual capital- Australian annual reporting practices (Guthrie and Petty, 2000)
- Reporting intellectual capital in annual reports- evidence from Ireland (Brennan, 2001)
- Evaluating Intellectual Capital in Hotel Industry (Engstrom et. Al., 2003)
- Intellectual capital- management attitudes in service industries (Lim and Dallimore, 2004)
- Measurement and application of intellectual capital in Turkey (Buzbera, 2004)
- National IC index in Arab nations (Bontis, 2004)
- Intellectual capital performance of commercial banks in Malaysia (Goh, 2005)
- Intellectual capital and performance indicators- Taiwanese healthcare sector (Peng et. Al., 2007)
- Intangible measurement guidelines: a comparative study in Europe (Palacios et al., 2007)
- Intellectual capital and corporate performance in Indian pharmaceutical industry (Kamath, 2008)
- Intellectual capital statements in the Danish utility sector: materialization and enactment (Bukh and Jensen, 2008)
- Intellectual capital practices: a four-region comparative study (Tovstiga and Tulugurova, 2009)

REFERENCES

Ahn, J.-H., & Chang, S.-G. (2004). Assessing the Contribution of Knowledge to Business Performance: The KP3 Methodology. *Decision Support Systems*, *36*, 403–416. doi:10.1016/S0167-9236(03)00029-0

Alavi, M., & Leidner, D. E. (2001). Review: knowledge management and knowledge management systems: conceptual foundations and research issues. *Management Information Systems Quarterly*, *25*(1), 107–136. doi:10.2307/3250961

Allee, V. (1998). Creating value in the knowledge economy. *HR Monthly*, (April): 12–17.

Allee, V. (1999). The art and practice of being a revolutionary. *Journal of Knowledge Management*, *3*(2), 121–132. doi:10.1108/13673279910275576

Andriessen, D. (2001). Weightless wealth: four modifications to standard IC theory. *Journal of Intellectual Capital*, *2*(3), 204–214. doi:10.1108/14691930110399941

Andriessen, D. (2002). *The financial value of intangibles: searching for the Holy Grail*. Paper presented at the 5th World Congress on the Management of Intellectual Capital, McMaster University, Hamilton, 16-18 December.

Andriessen, D. (2004a). *Making Sense of Intellectual Capital: Designing a Method for the Valuation of Intangibles*. Oxford, UK: Elsevier Butterworth-Heinemann.

Andriessen, D. (2004b). IC valuation and measurement: classifying the state-of-the-art. *Journal of Intellectual Capital*, *5*(3), 230–242. doi:10.1108/14691930410533669

Andriessen, D. (2006). On the metaphorical nature of intellectual capital: a textual analysis. *Journal of Intellectual Capital*, *7*(1), 93–110. doi:10.1108/14691930610639796

Apostolou, D., & Mentzas, G. (1999a). Managing Corporate Knowledge: A Comparative Analysis of Experiences in Consulting Firms: Part 1. *Knowledge and Process Management*, *6*(3), 129–138. doi:10.1002/(SICI)1099-1441(199909)6:3<129::AID-KPM64>3.0.CO;2-3

Apostolou, D., & Mentzas, G. (1999b). Managing Corporate Knowledge: A Comparative Analysis of Experiences in Consulting firms: Part 2. *Knowledge and Process Management*, *6*(3). doi:10.1002/(SICI)1099-1441(199909)6:3<129::AID-KPM64>3.0.CO;2-3

Argyris, C. (1992). *On Organizational Learning*. Cambridge, MA: Blackwell.

Argyris, C., & Scho¨n, D. A. (1996). *Organizational learning II: Theory, Method and Practice*. Reading, MA: Addison Wesley.

Barney, J. B. (1991). Firm resources and sustained competitive advantage. *Journal of Management*, *17*(1), 99–120. doi:10.1177/014920639101700108

Barney, J. B. (1991). Firm resources and sustained competitive advantage. *Journal of Management*, *17*, 99–120. doi:10.1177/014920639101700108

Barney, J. B. (2001). Is the resource-based view a useful prospective for strategic management research? *Yes*. *Academy of Management Review*, *26*(1), 41–56. doi:10.2307/259393

Bassi, L. J. (1997). Harnessing the power of intellectual capital. *Training & Development*, *51*(12), 25.

Berends, H., Boersma, F. K., & Weggeman, M. P. (2001). *The structuration of organizational learning, Working Paper 01.12*. Eindhoven, The Netherlands: Eindhoven Centre for Innovation Studies.

Bontis, N. (1998). Intellectual capital: an exploratory study that develops measures and models. *Management Decision*, *36*(2), 63–76. doi:10.1108/00251749810204142

Bontis, N. (1999). Managing organizational knowledge by diagnosing intellectual capital: framing and advancing the state of the field. *International Journal of Technology Management*, *18* (5/6/7/8), 433-62.

Bontis, N. (2001). Assessing knowledge assets: a review of the models used to measure intellectual capital. *International Journal of Management Reviews*, *3*(1), 41–60. doi:10.1111/1468-2370.00053

Bontis, N. (2004). National Intellectual Capital Index: a united nations initiatives for the Arab region. *Journal of Intellectual Capital*, *5*(1), 13–39. doi:10.1108/14691930410512905

Bontis, N., Crossan, M., & Hulland, J. (2002). Managing and organizational learning system by aligning stocks and flows. *Journal of Management Studies*, *39*(4), 437–469. doi:10.1111/1467-6486.t01-1-00299

Bontis, N., & Fitz-enz, J. (2002). Intellectual capital ROI: a causal map of human capital antecedents and consequents. *Journal of Intellectual Capital*, *3*(3), 223–247. doi:10.1108/14691930210435589

Bower, J. L., & Christensen, C. M. (1995). Disruptive technologies: catching the wave. *Harvard Business Review*, *17*(1), 43–53.

Bozbura, F. T. (2004). Measurement and application of intellectual capital in Turkey. *The Learning Organization*, *11*(4/5), 357–367. doi:10.1108/09696470410538251

Brennan, N. (2001). Reporting intellectual capital in annual reports- evidence from Ireland. *Accounting, Auditing & Accountability Journal*, *14*(4), 423–436. doi:10.1108/09513570110403443

Brennan, N., & Connell, B. (2000). Intellectual capital: current issues and policy implications. *Journal of Intellectual Capital*, *1*(3), 206–240. doi:10.1108/14691930010350792

Brooking, A. (1996). *Intellectual Capital*. London: International Thompson Business Press.

Brooking, A. (1997). The management of intellectual capital. *Long Range Planning*, *30*(3), 364–365. doi:10.1016/S0024-6301(97)80911-9

Bruggen, A., Vergauwen, P., & Dao, M. (2009). Determinants of intellectual capital disclosure: evidence from Australia. *Management Decision*, *47*(2), 233–245. doi:10.1108/00251740910938894

Bukh, P. N. (2003). The relevance of intellectual capital disclosure: a paradox. *Accounting, Auditing & Accountability Journal*, *16*(1), 49–56. doi:10.1108/09513570310464273

Bukh, P. N., & Jensen, I. K. (2008). Intellectual capital statements in the Danish utility sector: materialization and enactment. *Journal of Human Resource Costing & Accounting*, *12*(3), 148–178. doi:10.1108/14013380810919831

Bukh, P. N., Larsen, H. T., & Mouritsen, J. (1999). *Developing intellectual capital statements: Lessons from 23 Danish Firms*. Paper for Workshop on Accounting for Intangibles and the Virtual Organization, Brussels, February 12-13.

Buzbera, F. T. (2004). Measurement and application of intellectual capital in Turkey. *The Learning Organization*, *11*(4), 357–367. doi:10.1108/09696470410538251

Canadian Management Accountants (CMA). (1999, April 16). *Focus Group Draft: Measuring Knowledge Assets*. Mississauga, Canada: Author.

Carson, E., Ranzijn, R., Winefield, A., & Marsden, H. (2004). Intellectual capital: Mapping employee and work group attributes. *Journal of Intellectual Capital*, *5*(3), 443–455. doi:10.1108/14691930410550390

Cascio, W. (1998). Valuing and costing intellectual capital. *HR Monthly*, (February): 12–16.

Castro, G. M., & Lopez, P. (2008). Intellectual capital in high-tech firms - The case of Spain. *Journal of Intellectual Capital*, *9*(1), 25–36. doi:10.1108/14691930810845786

Celemi. (1998). *Growing a knowledge company*. www.celemi.com Site includes Celemi intangible assets monitor, http://www.celemi.se/sbc/sbc2.html

Chan, K. H. (2009). Impact of intellectual capital on organizational performance, An empirical study of companies in the Hang Seng Index (Part 1). *The Learning Organization, 16*(1), 4–21. doi:10.1108/09696470910927641

Chatzkel, J. (2002). *Intellectual Capital*. Mankato, MN: Capstone.

Chen, J., Zhu, Z., & Xie, H. Y. (2004). Measuring Intellectual Capital: a new model and empirical study. *Journal of Intellectual Capital, 5*(1), 195–207. doi:10.1108/14691930410513003

Chen, M.-C., Cheng, S.-J., & Hwang, Y. (2005). An empirical investigation of the relationship between intellectual capital and firms' market value and financial performance. *Journal of Intellectual Capital, 6*(2), 159–176. doi:10.1108/14691930510592771

Cheng, M. Y., Lin, J. Y., Hsiao, T. Y., & Lin, T. W. (2008). Censoring model for evaluating intellectual capital value drivers. *Journal of Intellectual Capital, 9*(4), 639–654. doi:10.1108/14691930810913195

Cole, R. E. (1998). Introduction. *California Management Review, 45*(3), 15–21.

Cook, S. D. N., & Brown, J. S. (1999). Bridging epistemologies: the generative dance between organizational knowledge and organizational knowing. *Organization Science, 10*(4), 381–400. doi:10.1287/orsc.10.4.381

Cuganesan, S. (2006). Reporting organizational performance in managing human resources. *Journal of Human Resource Costing & Accounting, 10*(3), 164–188. doi:10.1108/14013380610718629

Darroch, J. (2003). Developing a measure of knowledge management behaviors and practices. *Journal of Knowledge Management, 7*(5), 41–54. doi:10.1108/13673270310505377

DATI (Danish Agency for Trade and Industry). (1997). *Intellectual Capital Accounts: Reporting and Managing Intellectual Capital*. Copenhagen: Danish Agency for Trade and Industry.

DATI (Danish Agency for Trade and Industry). (1998). *Intellectual Capital Accounts: New Tool for Companies, (English version)*. Copenhagen: DTI Council.

DATI (Danish Agency for Trade and Industry). (1999). *Developing Intellectual Capital Accounts. Experiences from 19 Companies*. Copenhagen: Ministry of Business and Industry.

DATI (Danish Agency for Trade and Industry). (2000). *A Guideline for Intellectual Capital Statements*. Copenhagen: Danish Agency for Trade and Industry.

Dell, O., C., & Grayson, C. J. (1998). If Only We Knew What We Know: Identification and Transfer of Internal Best Practices. *California Management Review, 40*(3), 154–174.

Edvinsson, L. (1997). Developing intellectual capital at Skandia. *Long Range Planning, 30*(3), 266–373.

Edvinsson, L., & Malone, M. S. (1997). *Intellectual Capital. Realizing your Companies True Value by Finding its Hidden Brainpower*. New York: Harper Business Publisher.

Edvinsson, L., & Malone, M. S. (1997). *Intellectual Capital*. London: Piatkus.

Ellickson, M. C. (2002). Determinants of job satisfaction of municipal government employees. *Public Personnel Management, 31*(3), 343–358.

Engstrom, T. E. J., Westnes, P., & Westness, S. F. (2003). Evaluating intellectual capital in the hotel industry. *Journal of Intellectual Capital, 4*(3), 287–303. doi:10.1108/14691930310487761

FASB. (2001). *Improving business reporting: insights into enhancing voluntary disclosures*. Norwalk, CT: Business Reporting Research Project.

Ferrier, F. (2001). *Managing, measuring and maximizing intangible assets: an eclectic view of some of the latest international developments*. Paper presented at the Intellectual Property: Managing, measuring and maximizing intangible assets, Sydney, 15 November.

Firer, S., & Williams, S. M. (2003). Intellectual capital and traditional measures of corporate performance. *Journal of Intellectual Capital*, *4*(3), 348–360. doi:10.1108/14691930310487806

Fischer, M. M. (2001). Innovation, knowledge creation and systems of innovation. *The Annals of Regional Science*, *35*, 199–216. doi:10.1007/s001680000034

Flostrand, P. (2006). The sell side – observations on intellectual capital indicators. *Journal of Intellectual Capital*, *7*(4), 457–473. doi:10.1108/14691930610709112

Goh, P. C. (2005). Intellectual capital performance of commercial banks in Malaysia. *Journal of Intellectual Capital*, 385–396. doi:10.1108/14691930510611120

Grant, R. M. (1991). A resource-based theory of competitive advantage: implications for strategy formulation. *California Management Journal*, *33*(3), 114–135.

Grant, R. M. (1996a). Prospering in dynamically-competitive environments: organizational capability as knowledge integration. *Organization Science*, *7*(4), 375–388. doi:10.1287/orsc.7.4.375

Grant, R. M. (1996b). Toward a knowledge-based theory in the firm. *Strategic Management Journal*, *17*, 109–122.

Grasenick, K., & Low, J. (2004). Shaken, not stirred defining and connecting indicators for the measurement and valuation of intangibles. *Journal of Intellectual Capital*, *5*(2), 268–281. doi:10.1108/14691930410533696

Groth, J. C. (1994). Human Capital and Europe Future. *European Business Review*, *94*(5), 20–27. doi:10.1108/09555349410068858

Guilding, C., & Pike, R. (1990). Intangible marketing assets: a managerial accounting perspective. *Accounting and Business Review*, *21*(18), 41–49.

Guthrie, J. (2001). The management, measurement and reporting of intellectual capital. *Journal of Intellectual Capital*, *2*(1), 27–41. doi:10.1108/14691930110380473

Guthrie, J., & Petty, R. (2000). Intellectual Capital: Australian annual reporting practices. *Journal of Intellectual Capital*, *1*(3), 241–251. doi:10.1108/14691930010350800

Guthrie, J., Petty, R., Ferrier, F., & Wells, R. (1999). *There is no accounting for intellectual capital in Australia: a review of annual reporting practices and the internal measurement of Intangibles*. Amsterdam: OECD.

Guthrie, J., Petty, R., & Yongvanich, K. (2004). Using content analysis as a research method to inquire into intellectual capital reporting. *Journal of Intellectual Capital*, *5*(2), 282–293. doi:10.1108/14691930410533704

Haanes, K., & Lowendhal, B. (1997). The Unit of Activity: Towards an Alternative to the Theories of the Firm. In *Strategy, Structure and Style*. London: John Wiley and Sons.

Hall, R. (1992). The strategic analysis of intangible resources. *Strategic Management Journal*, *13*, 135–144. doi:10.1002/smj.4250130205

Hammerer, G. (1996). *Intangible investments in Austria.* Paper presented at workshop, OECD New Indicators for the Knowledge-based Economy, OECD, Paris.

Hendriksen, E. S., & van Breda, M. F. (1992). *Accounting Theory* (5th ed.). Burr Ridge, IL: Irwin.

Huang, C. C., Luther, R., & Tayles, M. (2007). An evidence-based taxonomy of intellectual capital. *Journal of Intellectual Capital, 8*(3), 386–403. doi:10.1108/14691930710774830

Huang, K. T. (1998). Capitalizing on intellectual assets. *IBM Systems Journal, 37*(4), 570–583. doi:10.1147/sj.374.0570

Hurwitz, J., Lines, S., Montgomery, B., & Schmidt, J. (2002). The linkage between management practices, intangibles performance and stock returns. *Journal of Intellectual Capital, 3*(1), 51–61. doi:10.1108/14691930210412845

Hussi, T., & Ahonen, G. (2002). Managing intangible assets – a question of integration and delicate balance. *Journal of Intellectual Capital, 3*(3), 277–286. doi:10.1108/14691930210435615

IFAC. (1998). *The Measurement and Management of Intellectual Capital: An Introduction.* Financial and Management Accounting Committee IFAC 1998 Study 7.

Itami, H., & Roehl, T. W. (1987). *Mobilizing Invisible Assets.* Boston: Harvard University Press.

Jacobsen, K., & Hofman-Bang, P. (2005). The IC Rating model by Intellectual Capital Sweden. *Journal of Intellectual Capital, 6*(4), 570–587. doi:10.1108/14691930510628834

Jashapara, A. (2003). *Knowledge Management – An integrated approach (pp. 269).* New York: Prentice Hall.

Johanson, U. Martensson, M. & Skoog, M. (1999a). *Measuring and managing intangibles: eleven Swedish exploratory case studies.* Paper presented at the International Symposium Measuring Reporting Intellectual Capital: Experiences, Issues, and Prospects, OECD, Amsterdam, June.

Johanson, U., Eklov, G., Holmgren, M. & MaÊrtensson, M. (1999b). *Human resource costing and accounting versus the Balanced Scorecard: a literature survey of experience with the concepts.* Paper presented at the International Symposium Measuring Reporting Intellectual Capital: Experiences, Issues, and Prospects, OECD, Amsterdam, June.

Johanson, U., Koga, C., Skoog, M., & Henningsson, J. (2006). The Japanese Government's intellectual capital reporting guideline. *Journal of Intellectual Capital, 7*(4), 474–491. doi:10.1108/14691930610709121

Jørgensen, K. M., & Boje, D. M. (2006). Time, place and mind of intellectual capital: on Jørgensen's 'Conceptualising intellectual capital as language game and power.'. *Journal of Intellectual Capital, 7*(1), 78–88.

Kamath, G. B. (2008). Intellectual capital and corporate performance in Indian pharmaceutical industry. *Journal of Intellectual Capital, 9*(4), 684–704. doi:10.1108/14691930810913221

Kannan, G., & Aulbur, G., W. (2004). Intellectual Capital: Measurement effectiveness. *Journal of Intellectual Capital, 5*(3), 389–403. doi:10.1108/14691930410550363

Kaplan, R. S., & Norton, D. P. (1992). The balanced scorecard. Measures that drive performance. *Harvard Business Review, 70*(1), 71–80.

Kaplan, R. S., & Norton, D. P. (1996). *The Balanced Scorecard.* Cambridge, MA: Harvard Business School Press.

Kaplan, R. S., & Norton, D. P. (2004). Measuring the strategic readiness of intangibles assets. *Harvard Business Review, 82*(2), 52–63.

Kim, D. Y., & Kumar, V. (2009). A framework for prioritization of intellectual capital indicators in R&D. *Journal of Intellectual Capital, 10*(2), 277–293. doi:10.1108/14691930910952669

Konrad Group (The). (1988). *The invisible balance sheet*. Retrieved from www.sveiby.com/articles/IntangAss/DenOsynliga.pdf

Kujansivu, P., & Lonnqvist, A. (2007). Investigating the value and efficiency of intellectual capital. *Journal of Intellectual Capital, 8*(2), 272–287. doi:10.1108/14691930710742844

Lai, L. F. (2007). A knowledge engineering approach to knowledge management. *Journal of Information Sciences*. doi:10.1016/j.ins.2007.02.028 (available online at www.sciencedirect.com)

Leonard-Barton, D. (1992). Core capabilities and core rigidities: a paradox in managing new product development. *Strategic Management Journal, 13*, S111–S125. doi:10.1002/smj.4250131009

Lev, B. (1997). *The boundaries of financial reporting and how to extend them. Industrial Competitiveness in the Knowledge-Based Economy*. Stockholm: OECD.

Lev, B. (2001). *Intangibles: Management, Measurement, and Reporting*. Washington, DC: Brookings Institute Press.

Liebowitz, J., & Suen, C. Y. (2000). Developing Knowledge management metrics for measuring intellectual capital. *Journal of Intellectual Capital, 1*(1), 54–67. doi:10.1108/14691930010324160

Liebowitz, J., & Wright, K. (1999). Does measuring knowledge make "cents"? *Expert Systems with Applications, 17*(5), 99–103. doi:10.1016/S0957-4174(99)00027-5

Lim, L. L. K., & Dallimore, P. (2004). Intellectual capital: management attitudes in service industries. *Journal of Intellectual Capital, 5*(1), 181–194. doi:10.1108/14691930410512996

Litschka, M., Markom, A., & Schunder, S. (2006). Measuring and analyzing intellectual assets: an integrative approach. *Journal of Intellectual Capital, 7*(2), 160–171. doi:10.1108/14691930610661836

Lynn, B. E. (2000). Intellectual capital: unearthing hidden value by managing intellectual assets. *Ivey Business Journal, 64*(3), 48–52.

KPMG Management Consulting. Case Study: Building a Platform for Corporate Knowledge, *(1998a)*.

Marr, B. (2005). *Perspectives on Intellectual Capital*. Oxford, UK: Elsevier Butterworth-Heinemann.

Marr, B., Schiuma, G., & Neely, A. (2004). Intellectual capital – defining key performance indicators for organizational knowledge assets. *Business Process Management Journal, 10*(5), 551–569. doi:10.1108/14637150410559225

Marsili, O. (2000). *Technological regimes and sources of entrepreneurship*. Working Paper 00.10, Eindhoven Centre for Innovation Studies, Eindhoven.

Mattessich, R. (2003). Accounting representation and the onion model of reality: a comparison with Baudrillard's orders of simulacra and his hyperreality. *Accounting, Organizations and Society, 28*(5), 443. doi:10.1016/S0361-3682(02)00024-7

Mavridis, D. (2004). The intellectual capital performance of the Japanese banking sector. *Journal of Intellectual Capital, 5*(1), 92–115. doi:10.1108/14691930410512941

Mayo, A. (2001). *The Human Value of the Enterprise: Valuing People as Assets: Monitoring, Measuring, Managing*. London: Nicholas Brealey Publishing.

Meyer, A. D. (1991). What is strategy's distinctive competence? *Journal of Management, 17*, 821–833. doi:10.1177/014920639101700413

Miller, M., DuPont, B. D., Fera, V., Jeffrey, R., Mahon, B., Payer, B. M., & Starr, A. (1999). *Measuring and reporting intellectual capital from a diverse Canadian industry perspective: experiences, issues and prospects.* Paper presented at the International Symposium – Measuring and Reporting Intellectual Capital: Experience, Issues, and Prospects, 9-11 June, Amsterdam.

Mortensen, J., Eustace, C., & Lannoo, K. (1997). *Intangibles in the European economy: a review essay.* Paper presented at the Workshop on Intangibles in the European Economy, CEPS, Brussels.

Mouritsen, J. (2004). Measuring and intervening: how do we theorize intellectual capital management? *Journal of Intellectual Capital, 5*(2), 257–267. doi:10.1108/14691930410533687

Mouritsen, J., Bukh, P. D., Larsen, H. T., & Johansen, M. R. (2002). Developing and managing knowledge through intellectual capital statements. *Journal of Intellectual Capital, 3*(1), 10–29. doi:10.1108/14691930210412818

Mouritsen, J., & Larsen, H. T. (2005). The 2nd wave of knowledge management: the management control of knowledge resources through intellectual capital information. *Journal of Management Accounting Research, 16*, 371–394. doi:10.1016/j.mar.2005.06.006

Mouritsen, J., Larsen, H. T., & Bukh, P. N. (2005). Dealing with the knowledge economy: intellectual capital versus balanced scorecard. *Journal of Intellectual Capital, 6*(1), 8–27. doi:10.1108/14691930510574636

Mouritsen, J., Larsen, H. T., & Bukh, P. N. D. (2001). *Intellectual capital and the capable firm: narrating, visualizing and numbering for managing knowledge* (*Vol. 26*). Accounting, Organizations and Society.

Nonaka, I., Reinmoller, P., & Toyama, R. (2001). Integrated information technology systems for knowledge creation. In Dierkes, M., Berthoin-Antal, A., Child, J., & Nonaka, I. (Eds.), *Handbook of Organizational Learning and Knowledge*. Oxford, UK: Oxford University Press.

Nonaka, I., & Takeuchi, I. (1995). *The Knowledge-creating Company. How Japanese Companies Create the Dynamics of Innovation*. Oxford, UK: Oxford University Press.

Nordhaug, O. (1998). Competence specificities in organizations. *International Studies of Management and Organization, 28*(1), 8–27.

O'Donnell, D., & O'Regan, P. (2000). The structural dimensions of intellectual capital: emerging challenges for management and accounting. *Southern African Business Review, 4*(2), 14–20.

OECD. (2006). *Creating Value from Intellectual Assets*. Paris: Author.

Ordonez de Pablos, P. (2002). Evidence of intellectual capital measurement from Asia, Europe and the middle-East. *Journal of Intellectual Capital, 3*(3), 287–302. doi:10.1108/14691930210435624

Ordonez de Pablos, P. (2003). Intellectual capital reporting in Spain: a comparative view. *Journal of Intellectual Capital, 4*(1), 61–81. doi:10.1108/14691930310455397

Ordonez de Pablos, P. (2004). Measuring and reporting structural capital - Lessons from European learning firms. *Journal of Intellectual Capital, 5*(4), 629–647. doi:10.1108/14691930410567059

Ordonez de Pablos, P. (2005). Intellectual capital reports in India: lessons from a case study. *Journal of Intellectual Capital, 6*(1), 141–149. doi:10.1108/14691930510574717

Organization for Economic Co-operation and Development (OECD). (1999). Guidelines and instructions for OECD Symposium. In *International Symposium Measuring Reporting Intellectual Capital: Experiences, Issues, and Prospects*, June, Amsterdam, OECD, Paris.

Palacios, T. M. B., & Galvan, R. S. (2007). Intangible measurement guidelines: a comparative study in Europe. *Journal of Intellectual Capital*, *8*(2), 192–204. doi:10.1108/14691930710742790

Pearse, N. J. (2009). The role of experiences in creating and developing intellectual capital. *Management Research News*, *32*(4), 371–382. doi:10.1108/01409170910944317

Pedrini, M. (2007). Human capital convergence in intellectual capital and sustainability reports. *Journal of Intellectual Capital*, *8*(2), 346–354. doi:10.1108/14691930710742880

Peng, T. A., Pike, S., & Roos, G. (2005). Intellectual capital and performance indicators- Taiwanese healthcare sector. *Journal of Intellectual Capital*, *8*(3), 538–556. doi:10.1108/14691930710774902

Peppard, J., & Rylander, A. (2001). Using an intellectual capital perspective to design and implement a growth strategy: the case of Apion. *European Management Journal*, *19*(5), 510–525. doi:10.1016/S0263-2373(01)00065-2

Peteraf, M. (1993). The cornerstones of competitive advantage: a resource-based view. *Strategic Management Journal*, *14*, 179–191. doi:10.1002/smj.4250140303

Petrash, G. (1996). Dow's journey to a knowledge value management culture. *European Management Journal*, *14*(4), 365–373. doi:10.1016/0263-2373(96)00023-0

Petty, R., & Guthrie, J. (2000). Intellectual Capital: literature review – measurement, reporting and management. *Journal of Intellectual Capital*, *1*(2), 155–169. doi:10.1108/14691930010348731

Petty, R., & Guthrie, J. (2000b). Knowledge management and the measurement of intangibles. *Management Today*, March.

Pike, S., Fernstrom, L., & Roos, G. (2005a). Intellectual capital management approach in ICS Ltd. *Journal of Intellectual Capital*, *6*(4), 489–509. doi:10.1108/14691930510628780

Pike, S., Roos, G., & Rylander, A. (2002). Intellectual capital management and disclosure. In Bontis, N., & Choo, C. W. (Eds.), *The Strategic Management of Intellectual Capital and Organizational Knowledge (pp. 657-73)*. New York: Oxford University Press.

Priem, R. L., & Butler, J. E. (2001a). Is the resource-based 'view' a useful prospective for strategic management research? *Academy of Management Review*, *26*(1), 22–40. doi:10.2307/259392

Priem, R. L., & Butler, J. E. (2001b). Tautology in the resource-based view and the implications of externality determined resource value: further comments. *Academy of Management Review*, *26*(1), 57–66. doi:10.2307/259394

Prusak, L. (1998). *Working Knowledge: How Organizations Manage What They Know*. Cambridge, MA: Harvard Business School Press.

Pulic, A. (2000). *MVA and VAICTM Analysis of Randomly Selected Companies from FTSE 250*. Graz, London: Austrian Intellectual Capital Research Center.

Pulic, A. (2004). Intellectual capital: does it create or destroy value? *Measuring Business Excellence*, *8*(1), 62–68. doi:10.1108/13683040410524757

Rao, L., & Osei-Bryson, K. M. (2007). Towards defining dimensions of knowledge systems quality. *Expert Systems with Applications*, *33*, 368–378. doi:10.1016/j.eswa.2006.05.003

Rastogi, P. (2000b). Sustaining enterprise competitiveness – is human capital the answer? *Human Systems Management*, *19*(3), 193–203.

Reed, K. K., Lubatkin, M., & Srinivasan, N. (2006). Proposing and Testing an Intellectual Capital-Based View of the Firm. *Journal of Management Studies*, *43*(4), 867–893. doi:10.1111/j.1467-6486.2006.00614.x

Riegler, C., & Hollerschmid, C. (2006). Voluntary disclosure on project intangibles from R&D in business reporting. *Journal of Intellectual Capital*, *7*(4), 511–528. doi:10.1108/14691930610709149

Robinson, H. S., Anumba, C. J., Carrillo, P. M., & Al-Ghassani, A. M. (2006). STEPS: a knowledge management maturity roadmap for corporate sustainability. *Business Process Management Journal*, *12*(6), 793–808. doi:10.1108/14637150610710936

Romer, P. (1994). The origins of endogenous growth. *The Journal of Economic Perspectives*, *8*, 3–22.

Romer, P. (1998). Bank of America roundtable on the soft revolution. *Journal of Applied Corporate Finance*, *11*(2), 9–14.

Romijn, H., & Albu, M. (2001). *Explaining innovativeness in small high-technology firms in the United Kingdom*. Working Paper 01.01, Eindhoven Centre for Innovation Studies, Eindhoven University of Technology, Eindhoven.

Roos, J., & Roos, G. (1997). *Intellectual Capital: Navigating the New Business Landscape*. London: Macmillan.

Roos, J., Roos, G., & Edvinsson, L. (1997). *Intellectual Capital – Navigating the New Business Landscape*. Basingstoke, UK: Macmillan.

Saint-Onge, H. (1996). Tacit knowledge: the key to the strategic alignment of intellectual capital. *Strategy and Leadership*, *24*(2), 10–15. doi:10.1108/eb054547

Sanchez-Canizares, S. M., Munoz, M. A. A., & Lopez-Guzman, T. (2007). Organizational culture and intellectual capital: a new model. *Journal of Intellectual Capital*, *8*(3), 409–430. doi:10.1108/14691930710774849

Schiuma, G., Lerro, A., & Carlucci, D. (2008). The Knoware Tree and the Regional Intellectual Capital Index: An assessment within Italy. *Journal of Intellectual Capital*, *9*(2), 283–300. doi:10.1108/14691930810870346

Seleim, A., Ashour, A., & Bontis, N. (2004). Intellectual capital in Egyptian software firms. *The Learning Organization*, *11*(4/5), 332–346. doi:10.1108/09696470410538233

Senge, P. M. (1990). *The Fifth Discipline: the Art and Practice of the Learning Organization*. New York: Doubleday.

Shapiro, C., & Varian, H. R. (1999). *Information Rules: A Strategic Guide to the Network Economy*. Boston: Harvard Business School Press.

Skandia. (1994). *Visualizing intellectual capital in Skandia*. Supplement to Skandia's 1994 Annual Report, Skandia, Stockholm.

Skandia. (1995a). *Renewal and development: intellectual capital*. Supplement to Skandia's 1995 Interim Annual Report, Skandia, Stockholm.

Skandia. (1995b). *Value-creating processes: intellectual capital*. Supplement to Skandia's 1995 Annual Report, Skandia, Stockholm.

Skandia. (1996a). *Power of innovation: intellectual capital*. Supplement to Skandia's 1996 Interim Annual Report, Skandia, Stockholm.

Skandia. (1996b). *Customer value*. Supplement to Skandia's 1996 Annual Report, Skandia, Stockholm.

Skandia. (1997). *Intelligent enterprising*. Supplement to Skandia's 1997 Annual Report, Skandia, Stockholm.

Skandia. (1998). *Human capital in transformation.* Supplement to Skandia's 1998 Annual Report, Skandia, Stockholm.

Society of Management Accountants of Canada (SMAC). (1998). *The management of intellectual capital: the issues and the practice, (Paper No. 16).* Hamilton, Canada: The Society of Management Accountants of Canada.

Solomons, D. (1991). Accounting and social change: a neutralist view. *Accounting, Organizations and Society, 16*(3), 287–295. doi:10.1016/0361-3682(91)90005-Y

Spender, J. C. (1996a). Making Knowledge the Basis of a Dynamic Theory of the Firm. *Strategic Management Journal, 17*, 45–62.

Spender, J. C. (1996b). Organizational Knowledge, Learning, and Memory: Three Concepts in Search of a Theory. *Journal of Organizational Change Management, 9*, 63–78. doi:10.1108/09534819610156813

Stewart, T. A. (1997). *Intellectual Capital: The New Wealth of Organizations.* New York: Doubleday.

Subramaniam, M., & Youndt, M. A. (2005). The influence of intellectual capital on the types of innovative capabilities. *Academy of Management Journal, 48*(3), 450–463. doi:10.2307/20159670

Sullivan, P. H. (2000). *Value-driven Intellectual Capital. How to Convert Intangible Corporate Assets into Market Value.* London: John Wiley and Sons.

Sveiby, K. E. (1988). *Den nya AÊrsredovisningen [The New Annual Report]* (in Swedish). Stockholm, Pub. details, see www.sveiby.com.au for English translation.

Sveiby, K. E. (1997). *The New Organizational Wealth: Managing and Measuring Knowledge Based Assets.* San Francisco, CA: Berrett Koehler.

Sveiby, K. E. (1998). Intellectual capital: thinking ahead. Australian CPA, June (pp. 18-22).

Sveiby, K.E., (2001). A knowledge based theory of the firm to guide strategy formulation. *Journal of Intellectual Capital*, (2/4).

Sveiby, K. E. (2005). *Method of measuring intangible assets* (pp. 1-8). Retrieved January 2, 2006 from www.sveiby.com/articles/IntangibleMethods.htm

Swart, J. (2006). Intellectual Capital: disentangling an enigmatic concept. *Journal of Intellectual Capital, 7*(2), 136–159. doi:10.1108/14691930610661827

Teece, D. J. (2000). *Managing Intellectual Capital: Organizational, Strategic, and Policy Dimensions.* Oxford, UK: Oxford University Press.

The Swedish Coalition of Service Industries (SCSI). (1995). *Valuation of Service Companies.* Stockholm: Author.

Tovstiga, G., & Tulugurova, E. (2007). Intellectual capital practices and performance in Russian enterprises. *Journal of Intellectual Capital, 8*(4), 695–707. doi:10.1108/14691930710830846

Tovstiga, G., & Tulugurova, E. (2009). Intellectual capital practices: a four-region comparative study. *Journal of Intellectual Capital, 10*(1), 70–80. doi:10.1108/14691930910922905

Tseng, C., & Goo, Y. J. (2005). Intellectual capital and corporate value in an emerging economy: empirical study of Taiwanese manufacturers. *R & D Management, 35*(2), 187–201. doi:10.1111/j.1467-9310.2005.00382.x

Van Aken, J. E. (2001). *Mode 2 knowledge production in the field of management.* Working Paper 01.13, Eindhoven Centre for Innovation Studies, Eindhoven University of Technology, Eindhoven.

Van Buren, M. E. (1999). A yardstick for knowledge management. *Training & Development*, *53*(5).

Van Buren, M. E. (1999). *Making knowledge counts: knowledge management systems and the human element*. Unpublished manuscript.

Van de Ven, A. H. (1986). Central problems in the management of innovation. *Management Science*, *32*, 590–607. doi:10.1287/mnsc.32.5.590

Van der Zahn, M. J.-L. W., Tower, G., & Neilson, J. (2004). *Intellectual Capital and the Efficiency of Value Added: Trends in the Singapore Capital Market 2000-2002*. Burleigh, UK: Poseidon Books.

Wasko, M., & Faraj, S. (2005). Why Should I Share? Examining Social Capital and Knowledge Contribution in Electronic Networks of Practice. *Management Information Systems Quarterly*, *29*(1), 35–37.

Webster, J., & Watson, R. T. (2002). Analyzing the past to prepare for the future: Writing a literature review. *Management Information Systems Quarterly*, *26*(2), xiii–xxiii.

Wernerfelt, B. (1984). A resource-based view of the firm. *Strategic Management Journal*, *5*, 171–180. doi:10.1002/smj.4250050207

Wernerfelt, B. (1995). The resource-based view of the firm: ten years after. *Strategic Management Journal*, *5*(2), 171–174. doi:10.1002/smj.4250050207

Wiig, K. M. (2004). *People-focused knowledge management: how effective decision making leads to corporate success* (pp. 213–237). New York: Elsevier Inc.

Williams, S. M. (2000). *Is a company's intellectual capital performance and intellectual capital disclosure practices related?: Evidence from publicly listed companies from the FTSE 100* (pp. 1-41). Retrieved March 29, 2006 from www.vaic-on.net/start.htm

Wright, P. M., Dunford, B. B., & Snell, S. A. (2001). Human resources and the resource based view of the firm. *Journal of Management*, *27*, 701–721. doi:10.1177/014920630102700607

Youndt, M. A., Subramaniam, M., & Snell, S. A. (2004). Intellectual capital profiles: an examination of investments and returns. *Journal of Management Studies*, *41*(2), 335–361. doi:10.1111/j.1467-6486.2004.00435.x

KEY TERMS AND DEFINITIONS

Capital: In the information era, Capital is defined as the refined organizational knowledge that is utilized for creating value for stakeholders. This knowledge is being used in organizational processes and in relationships with outside world so as to improve the value creation process and eliminate or transform the value reducing activities. In economics, capital or capital goods or real capital refers to factors of production used to create goods or services that are not themselves significantly consumed (though they may depreciate) in the production process. Capital goods may be acquired with money or financial capital. In finance and accounting, capital generally refers to financial wealth, especially that used to start or maintain a business. Also, assets available for use in the production of further assets wealth in the form of money or property owned by a person or business and human resources of economic value are considered as capital.

Human Capital (HC): includes human resources within the organization (i.e. staff resources from a strategic point of view) and sometimes may include resources external to the organization, namely customers and suppliers. According to another definition, human capital is the capabilities of individuals who are the source of innovation and renewal within companies (Stewart, 1997).

Intellectual Capital (IC): refers to the non-financial value creation capabilities of a firm, which represents the difference between market and book values and can be further described by it's components (namely, Human, Organizational and Relational Capitals).

Knowledge: Knowledge is defined by the Oxford English Dictionary as (i) expertise, and skills acquired by a person through experience or education; the theoretical or practical understanding of a subject, (ii) what is known in a particular field or in total; facts and information or (iii) awareness or familiarity gained by experience of a fact or situation.

Knowledge Management: Knowledge Management (KM) comprises a range of practices used in an organization to identify, create, represent, distribute and enable adoption of insights and experiences. Such insights and experiences comprise knowledge, either embodied in individuals or embedded in organizational processes or practice.

Organizational Capital (OC): or "Structural Capital" (SC) is conventionally used to refer to the processes and procedures that are in some ways recorded and hence accessible to the organization as how things are done in order to maintain effectiveness. It is formed by the intellectual input of the employees, but it "belongs" to the firm.

Relational Capital (RC): or "Customer Capital" (CC) is one of the most important components for classifying and measuring the status of intellectual capital in a firm. Relational Capital refers to the value of an organization's external relationships with other organizations, suppliers, people, and even competitors with which, it has communications and goes through partnerships, alliances, purchase and sell contracts, etc.

Chapter 2
Globalization of Instruction:
Developing Intellectual Capital

Janet Holland
Emporia State University, USA

Joe Holland
Advatech Pacific, USA

ABSTRACT

The aim of the study was to find ways to embrace the current realities of a global society and use it to build the intellectual capital of students from elementary school through college then into the work environment. This took a concerted effort examining the literature in regards to globalization, the intellectual capital needed for globalized work environments, and the implications for globalized teaching and training for adapting to current economies. This chapter is designed to help you begin your own examination of these issues with an eye toward educational solutions at all levels. A pilot study was conducted of international college students at a small midwestern university, to serve as a model for conducting your own needs analysis. The analysis should help you to begin thinking differently about the educational needs for globalized classrooms and work environments while building the needed intellectual capital to survive and thrive under changing conditions.

"The self is not something ready-made, but something in continuous formation through choice of action." - John Dewey

INTRODUCTION

With the tremendous growth of culturally diverse populations and the ability to communicate with almost anyone in the world, anytime, about any topic, educators have the opportunity to provide the training and instruction needed to thrive in a global environment. Intellectual capital in this context is the building of knowledge to sustain and grow organizations, institutions, and governments. Efforts directed towards the cultivation of human capital- increasing individuals' knowledge and skill sets, organizational capital- keeping necessary skill sets and levels of knowledge within the organization, and social capital- relationship building

DOI: 10.4018/978-1-61520-875-3.ch002

and information sharing; will prepare us for the increasingly global knowledge-based environment. When intellectual capital is derived from so many unique perspectives, its growth will be accelerated and far richer in content.

In the current global knowledge-based economy, companies are finding their greatest asset to be the intellectual capital maintained by their knowledge-based workers. This points to the importance of preparing individuals in the educational environment for successful transitions into global work environments. This drives the need for educators and trainers to enable the acquisition of authentic knowledge and skills with expanded global perspectives. The merging and mixing of diverse backgrounds and experiences in education and the work environment offer opportunities to expand the knowledge base beyond current borders and in totally new directions. However, one big challenge is finding ways to bring educational systems and business training into alignment with the current realities of working and learning together effectively.

Defined

Globalization, in the current context, is identified as the process of transnational dissemination of ideas, languages, economies and cultures throughout the international community. With the increase in world travelers, immigration, and communication technologies the world is becoming more connected and interdependent. Globalization has accelerated the exchange of goods, capital, services, education, information, technology, and labor. Globalization, at its best, is represented as sharing unique ideas and perspectives to extend knowledge based assets for collective mutual gains. Historically, comparisons of countries with respect to globalization gains have been measured in terms of economic, social, and political indices. The problem with these measurements is, alone they can miss the mark on some of the more recent developments regarding the wider perspective of global intellectual capital.

Intellectual capital provides a structure for the different types of knowledge resource assets. Traditionally, it has been associated with knowledge for industry and profit, but perhaps with the current economy it is time to propose the term be expanded to encompass the educational community. The reason being, educators and trainers are charged with the responsibility of developing knowledge assets to prepare learners for employment knowledge and skills. Though intellectual capital can be limited to individual knowledge and skills, it can also include organizational processes and social relationships.

The literature defines intellectual capital as demonstrating how knowledge is viewed as a valuable and powerful component to successful businesses ventures. One leading author states, "Intellectual capital is the sum of everything everybody in a company knows that gives it a competitive edge" (Steward, 1997, p. xx). Another author states, "Their intangible assets are much more valuable than their tangible assets" (Sveiby, 1997, p. 19). Another leading author states it is more than just "the management and measurement of physical and financial assets", managers need to adjust to "the cultivation and leverage of knowledge as the most significant acts of value creation" (Edvinsson & Malone, 1997, p. 4). Human capital is defined as "The combined knowledge, skill, innovativeness, and ability of the company's individual employees to meet the task at hand" (Edvinsson & Malone, 1997, p. 11). Human capital includes knowledge, skill, and experiences, "[but] it must be more than simply the sum of these measures; rather, it also must capture the dynamics of an intelligent organization in a changing competitive environment" (Edvinsson & Malone, 1997, p. 34). Based on the past literature there is no doubt about the importance of educational development, hiring talent, and continual training of employees to foster the needed intellectual capital for thriving in current global economies.

Intellectual capital will be examined here as "three distinct categories: human, social, and organizational" knowledge assets (Youndt & Snell, 2004). Restating- Human capital refers to individual persons' knowledge and skills. Social capital is the relationship building and information sharing. Organizational capital points to the skill sets and levels of knowledge to be maintained within the organization. The intellectual capital definition, combined with the three sub-category definitions, aligns nicely with knowledge-based assets to target instructional efforts towards expanding global perspectives. Therefore, intellectual capital includes knowledge assets needed to "convert that knowledge into something that has value" (Lynn, 2000, p. 2).

Human Capital Issues

Human capital in the workplace encompasses the knowledge assets the individual brings with them from previous instruction and experiences. Companies are looking for employees with abilities to work smart, innovate new ideas, increase performance, quality, service, production, product uses, and reduce costs. Companies are working hard to attract and retain bright high-level knowledgeable talent by offering high salaries and stock benefits. Classically, in a formal education setting, human intellectual capital involves the type of knowledge needed to build a good foundation. Learners need to be able to examine problems, then determine underlying causes and effects, relationships, identify goals, innovate, test results, revise, and examine implications for proposed actions as these are skills commonly needed in the work environment. The educational foundation must provide the individual with the ability to think creatively and research solutions to problems. To capitalize on the knowledge base workers bring, the company needs to be "converting knowledge into value" by transforming ideas into a usable format (Cohen & Backer, 1999, p. 2). The technology tools used for the development of human capital ranges from content specific to broad base analysis, and presentation tools for transforming knowledge into innovative ideas for the development of products and services of value in global markets. Globalization efforts directed toward the development of human intellectual capital includes learning more about different cultures to broaden everyone's knowledge base and perspective. These skills are valuable everywhere, especially in the classroom, community, and work environment.

Social Capital Issues

Social capital in the workplace often includes collaborative efforts to identify and solve problems for leveraging knowledge gains. By removing competitive power positions, inequality, and status, the playing field is leveled. Leveling empowers open communications and sharing rather than protecting ones turf for job security. When employees have autonomy and authority, they are more open to influencing decision making within 'their' organization. Rewarding team success can promote needed "risk taking" and innovation to launch new ideas (Gnuschke, 2001, p. 7). Formal education supports the development of social intellectual capital through the use of teamwork and collaborative efforts to prepare students for future jobs. Having students working in teams helps to promote research, sharing, and problem solving. Education then enhances "knowledge sharing, and building and assimilation of mechanisms for the flow of knowledge" (Malhotra, 2000, p. 9). The technology tools used to support social capital focuses on sharing, collaborating, and networking so prevalent in innovative new technologies. Assessment tools should then be used to analyze needs for alignment with the service, policy, product, goals, and desired outcomes. Globalization efforts directed toward social intellectual capital include working collaboratively with international or transnational partners to broaden perspectives through interactions whether synchronous or asynchronous online conferencing by conduct-

Figure 1. Model of intellectual capital

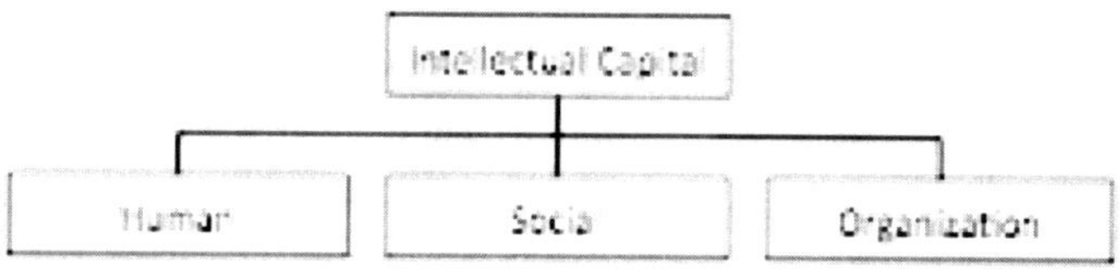

ing interviews, expert exchanges, focus groups, brainstorming, observations, and open dialogue for suggestions and feedback. Therefore, with the tremendous increase in diverse environments there is a need for learning how to successfully interact with individuals from across the globe.

Organizational Capital Issues

Organizational capital in the workplace can be maintained to a certain degree electronically, so when individuals retire or leave, corporate gains are maintained. Knowledge assets are found in the form of structures, processes, patents, copyright, trademarks, trade secrets, brands, licenses, company reports and documents; many of which are maintained or tracked within the corporate information technology system. Sometimes incentives or rewards are given to promote replicating important valuable company expertise to institutionalize it as company knowledge assets. In the work setting, the development of organizational intellectual capital tends to be developed internally within the company through training and experience, and is often targeted towards productivity and company profit. Technology tools used in the work environment can then help to develop organizational capital infrastructure management and codification focused on user-friendly, easy to access document database storage, intranet internal web access, and learning management systems to maintain and facilitate internal company knowledge and training. Globalization used for the development of organizational intellectual capital can be used to gain knowledge of different cultures, values, symbols, and images used for successful codification of documents for effective communications in global markets. In the formal education setting, learners use various technological tools to develop document structures, graphic organizers, processes, reports within databases, and Internet sites, while also working within learning management systems to gain practice in storing, sharing, and teaching others to prepare for eventual workplace knowledge and skills.

Globalization

We are currently experiencing more and more diversity in our classrooms and work environments. However, an understanding of this diversity, the forces driving it and its impact on instruction, and training are severely lacking in any level of education.

This is pointing the way to a "paradigm shift" in education on par with the 19th century goal of universal literacy and public education for all in the United States. Note, in the 19th century, public education was restricted to specific demographics with the masses only expected to have reading skills at the 6th grade level. Most students were then expected to acquire only one skill they would use the rest of their working life. Only a small fraction of this subset of the total population attempted college and fewer still needed to be innovators in their chosen profession. This education revolution occurred because it was considered "the right thing to do". Today, education policy is considered a "National Security" issue significant enough to be debated by the CEO's of Raytheon, and other industry leaders in technical trade journals like *Aviation Week and Space Technology* (Swanson,

2009, p. 58). The education these CEO's believe the United States needs are for larger percentages of the population to be trained to be creative and innovative lifelong learners while also being willing to work two to five specialties during the course of their working life. The bulk of these specialties have not even been conceptualized for the classic grade level tracks of middle school, high school, or college.

So, what needs to be taught, to whom, when, how, and for what reason is undergoing a total change from the classic model. Later in this chapter, we will examine the impact of globalization and its accompanying diversity, current limitations, benefits, and progress towards a quality education from the elementary level through college. Obstacles will be considered and policy suggestions offered with an eye towards future improvements.

BACKGROUND

Impact of Globalization

Over time we are finding ourselves merging and colliding with different nations over a vast array of global issues. Our increasingly globally interdependent paths are crossing at an accelerating rate due to cross border business, capital, goods, labor, services, social, cultural, technical, and educational interests. Our global interactions are increasingly, influencing local, national, and international cultures, societies, and governments. We find "education has been a national priority in many countries [and] is largely understood in terms of national economic survival in a fiercely competitive world" (Pang, 2006, p. 5). Our consumer/production-based economy is increasingly being overshadowed by our knowledge-based economy. Through our world's increasingly educationally enhanced human capital many countries are able to benefit from the higher level of knowledge needed to drive both local and world markets. On the down side, "globalization has led to greater economic and social inequality; and that educational access, whilst expanded, has also become more unequal in quality" (Pang, 2006, p. 12). Based on Ryan Wells research, his "study suggests trends from 1980-2000 toward more economic freedom, more within-country inequality, and higher levels of education" (Wells, 2006, p. 385). It is interesting to note, wealthy areas tend to "reap a disproportionate benefit from education and ultimately higher incomes" (Wells, 2006, p. 386). Individual countries' financial situations and cultures tend to influence ongoing education reform depending on their needs to be competitive, seek a financial advantage, or increase knowledge equity. According to David Bloom (2002), Harvard economist, each additional year of basic education provided to an individual corresponds to over a 10% rise in that individual's earning power. Again, for most of us, it is easy to see the financial advantages and opportunities for growth once we have been provided an extended education.

The demands to enhance workplace knowledge and skill requirements continue to keep raising the bar higher and higher. New technologies are being released and integrated at a very quick pace. All of this requires instructional training to meet current needs and future advancements. New initiatives are needed to mobilize educational resources to meet the huge demand for educational opportunities for all of our citizens.

Educational Limitations and Benefits

To date, global learning has failed in a number of areas, specifically, 1) engaging youth in learning, 2) properly educating and integrating immigrant youth, 3) in the variable and disparate quality of the education provided and 4) under-enrollment, the limiting of the number of students with access to a good education (Suarez-Orozco & Carolyn Sattin, 2007, p. 2-3). It seems strange to find this to be the case when education can be the strongest shield against poverty and the best way to increase those global knowledge and skill sets needed to create or maintain jobs and increase economic success.

So, how do we begin to address the failings with the goal of improving our educational systems and classrooms while meeting current demands for globalization? What knowledge and skills are critical for the realities of globalized learning? How do we begin to foster flexible lifelong learning to increase individuals' abilities to adapt to market and job shifts, such as those we are currently experiencing?

Current work skill sets often require learners to be curious, identify needs, address problems, conduct research from multiple perspectives, think critically about the issues, brainstorm solutions, think creatively, debate, work together collaboratively, use effective digital communications, evaluate options, synthesize findings, continue reflections, revisions, and presenting solutions. Teaching intercultural skills alongside basic content knowledge acquisition can assist learners with understanding others having different values, cultures, religions, and ethnicities from their own. Instruction requires a balanced curriculum with consideration to both the cognitive and affective domains. Expanding one's knowledge of different cultures, languages, traditions, history, and values is a good starting place and should begin at a young age and continue on.

Through understanding, acceptance, collaboration, and mutual engagement, subsequent benefits can be realized. These skill sets drive the engine moving students toward an important level of cultural sophistication needed for complex global markets. By making use of diverse talents, cultures, perspectives, and history we can expand our knowledge of global issues. Using integrated curriculums, team teaching, media rich instructional technologies, forming partnerships, and by fostering innovation, we can create knowledge and skills to prepare learners to work in future markets. The goal will be to continue to engage, educate, and integrate high-level knowledge for collective gains for jobs, cultures, and business market innovations.

As part of a quality education, why not engage all students in solving authentic problems to begin building interconnected, responsible, productive citizens to survive and thrive in our increasingly complex global economy? It is of utmost importance to provide an equitable quality education for all students in our classrooms as we all benefit from expanded perspectives.

Success in global markets demand successful interactions with a diverse range of individuals and cultures. It begins with intercultural knowledge, skills, and respect for our combined contributions and strengths. As educators we need to become international stewards sharing insights and preparing learners for the future.

Educational Progress

"Illiteracy in the developing world has fallen from 75% of the people a century ago to less than 25% today" (Bloom & Cohen, 2001, p. 1). We still have a large gap in the amount of education between developing and developed nations. In developed nations nearly all youth receive an education and a third to half receive postsecondary higher education. "Most nations have or are moving toward standardized curricula and assessments" in disciplines such as in science, math, technology, history, language, reading, and writing serving as "another indication of globalization's momentum" (Suarez-Orozco & Quin-Hillard, 2004, p. 238). The growing use of assessments such as the Third International Mathematics and Science Survey (TIMSS) and the growing use of International Baccalaureate programs point toward the increasing globalization of education.

Higher education has become more obtainable and can fit diverse learner needs for knowledge, degrees, eventual job skills, and specializations. "The escalation of economic globalization has driven a demand for instrumental education - that which can be clearly tied to the goals of production, productivity, and employment" (Stromquist, 2002, p. 105). Modern college educations can extend employment options. New trends in higher education studies include courses directed towards serving the community such as environmental and multicultural studies.

Benefits of Teaching Tolerance

The United States is in the "midst of the largest wave of immigration in its history, with over a million new immigrants per year" and in "Canada, Switzerland, and Australia the rates of immigration are nearly double the U.S. rate" (Suarez-Orozco, 2007, p. 9). As our classrooms become more diverse it begs the question, are we adequately preparing our administrators, teachers, and students for an increasingly integrated world? In an era of accelerated globalization, how can all of these different cultures coexist peacefully in an increasingly borderless world? (Bryan & Vavrus, 2005).

Online course offerings are also rising at a tremendous pace. The increase in global telecommunications now "connects people and facilitates transnational understanding" in ways not previously possible (Bryan & Vavrus, 2005, p. 184). Transnational discourse brings to the forefront differences in cultural identities, values, and life styles choices. So, how do we approach the teaching of tolerance? It may take a concerted effort of looking at "how educational curricula, textbooks and the social organization of schools" are designed (Bryan & Vavrus, 2005, p. 189). We need to stop and examine how we can increase equality, tolerance, inclusion, and acceptance in an era of globalization. It will take creative thinking and innovation to infuse needed changes into our instructional programs.

Benefits of Teaching Innovation

The educational knowledge and skills needed for "world sourcing of jobs – leveraging diverse global assets in an elegant array of physical, fiscal, and human capital" requires setting up the needed infrastructure (Milliron, 2007, p. 33). We need to think in new ways to leverage technology, talent, and tolerance. It means rethinking current structures in our schools, policies, and financial support to spark the needed innovations. Milliron refers to this shift as "Globalization 3.0" driven by "education, technology, and an ever-shrinking world of deep interconnections" (Milliron, 2007, p. 32). Also, he cites key skills needed as including "critical thinking, creativity, and courage" (Milliron, 2007, p. 34). Globalization requires taking a very hard look at current practices to determine effective approaches for teaching and learning to prepare students for the future.

Benefits of Teaching with Technology

Communications and information technology is inextricably linked to education and economic growth. "The engines of growth – capital, labor and technology – are increasingly globalized" (Good, 1996, p. 23). In fact, "leading economists estimate that technology has accounted for at least one-half of the economic growth in advanced industrial nations over the last 50 years" (Good, 1996, p. 23). The connections between technology, growth and jobs is reflected in the adoption of "high-technology products as a matter of national pride, and many have set their sights on joining the ranks of the world's technological leaders" (Good, 1996, p. 24). The most efficient way to introduce technology into a culture, society, or company is through exposure and interactions in practical applications through learning, and training.

In addition to high technology products, "technology is also flowing globally through people – widely acknowledged as one of the most effective technology transfer mechanisms" (Good, 1996, p. 24). This flow of dialogue and information sharing serves as the social capital driving the engine for innovation. Technology expansion is currently taking place through education, research and development, cooperation, collaboration, and partnerships developed. Exchanges of information are happening not only internally within a country, but also through international travel and digital communications such as through learning management systems.

In the classroom, technology integration "increases communication and collaboration, improves the quality of lesson plans and classroom instruction, and contributes to positive perceptions regarding the usefulness of the computer as a teaching tool" (Farenga & Joyce, p. 2). When the technology is matched well to the instructional objectives, learners benefit. There is a whole host of web-based resources to help instructors increase international exchanges within the classroom to deepen learners' knowledge and experiences with individuals from other cultures through global interactions whether text-based or live video exchanges. One of the current tools being used is Skype talks, chats, and video conferencing at http://www.skype.com. One interesting website posted by Passaic Valley High School has a schedule of their "Around the World" videoconference schedule at http://www.pvhs.k12.nj.us/atw/atw.asp?FeatureID=13 serving as a good model on how to integrate global studies into the curriculum. Web resources for setting up eTravel exchanges can be found on several different sites such as iEarn an international education and resource network at http://iearn.org/. GlobalSchoolNet is a site for linking students from around the world at http://globalschoolnet.org/. ePals is a global community of connected classrooms at http://www.epals.com/. Global Gateway is another site for setting up international partnerships. For expanding geographic studies and finding additional international projects, a good resource is the Association of American Geographers at http://www.aag.org/. Three interesting YouTube videos on globalization are "Shift Happens" with statistics on the impact of globalization, "Globalization What Does It Really Mean" from a students' perspective, and the "Global Gateway" project from the teachers' perspective (YouTube, 2009).

The other side of innovative globalization technology is the trend towards increased sharing, collaborating, and networking tools. The accelerating transnational or international communications opens up a whole new world of ideas and exchanges for expanding global perspectives. We have a new generation of digital natives growing up using technology to communicate as an integral part of their lives. The expectations for rich interactive exchanges are only increasing over time. In addition, we are seeing an increasing use of writing tools, image sharing sites, social bookmarking, social networking, digital write sharing boards, discussion blogs, collaborative wikis, audio podcasts, video vodcasts, video screen-casting, video storytelling, video creation and sharing, video conferencing, game-based environments, virtual reality simulations, and virtual worlds. The new generation is reflecting their sense of playfulness in sharing and collaborating as evidenced by the tools they use.

Today's learners are mobile and connected to new technologies at a level and complexity never before experienced. "The Internet is rapidly globalizing business, education and culture. It is transforming the economics of communication, so that geographic proximity no longer becomes a prime requirement for interaction" (Nabil, Awerbuch, Slonim, Wegner, & Yesha, 1997, p. 115). The benefit of the Internet is a "truly International information infrastructure" (Nabil, Awerbuch, Slonim, Wegner, & Yesha, 1997, p. 117). Learners can now interact and share multimedia resources easily with anyone across the globe. The new digital literacy includes international knowledge and skills. New digital technologies offer "tools for the production of information and aids for knowledge creation, as well as vehicles for community building and emotional support" (Apple, Kenway, & Singh, 2005, p. 21). In addition to shared content knowledge, equally important is the development of the social networks forged through current technologies. Digital knowledge and communication skills today require the ability to think critically and creatively, identify issues, conduct research, solve problems, and collaborate with others. The merger of technology with education can be used to bridge "curriculum innovation, knowledge production, and cultural formation" (Apple, Kenway, & Singh, 2005, p. 188).

With the increased time spent by individuals online, it carries the potential for increasing both learning and working productivity raising the education and technology bar even further. As a result, it will be important to work toward sustainable "efficient technology management, innovation and technological progress" (Huq, 2004, p. 159).

Technology Obstacles

We tend to assume everyone has current technologies with open Internet connections to the world, but in poor, remote, or restricted areas of the world this may not be the case. So, how can we reach out to remote areas? Small portable laptops such as the one developed by Nicholas Negroponte at the Massachusetts Institute of Technology are one possible solution (1997). Satellite connections can reach out to far away remote areas without an extensive technology infrastructure being put in place. For some cultures the completely open Internet is currently not acceptable to local values.

Some optional solutions to these local concerns include the use of filtering to block offensive sites. In even more extreme cases, secured educational systems can be used to provide the needed secure digital learning management systems. It is important for "nations and regions that do not have sophisticated ICT infrastructures and a wide spread of usage are likely to be disadvantaged economically because they do not have access to the vast pool of knowledge and information or enjoy the speed of communication and knowledge transfer that more fortunate nations have. Those who live on the wrong side of the digital divide will find themselves left further and further behind" (Lauder, Brown, Dillabough, & Halsey, 2006, p. 32). We can only hope the benefits of open discussions on all subjects will eventually break down some of the artificially imposed barriers to information sharing.

As access increases, one additional issue surfacing has been the lack of appropriate instruction and training using current digital media and communications technologies. The knowledge and skills, the instructors and trainers themselves possess, impact the quality of instruction the learner receives. By integrating instructional design and technology courses into pre-service and graduate level programs, it is a great way to leverage effective acquisition of technology based knowledge and skills. Another issue, connected to access and training, are the costs and time associated with the implementation of the hardware, software, instructors, and additional resources needed for effective implementation.

Policy Improvements

We are in the process of acknowledging the realities of an increasingly interdependent world. It is becoming evident through recent economic events and shared educational concerns. By choosing to address globalization we can work together to promote positive policies and effective educational programs. "We would be well-served by the development and cultivation of globally-focused educational programs at an early age" (Boyer, Brown, Butler, Niv-Solomon, Urlacher, Hudson, Johnson, & Lima, 2007, p. 174). Rather than waiting until college, interventions should begin much earlier and continue on. We need to focus on "innovative globally-focused educational programs and the development of training programs" (Boyer, Brown, Butler, Niv-Solomon, Urlacher, Hudson, Johnson, & Lima, 2007, p. 175). By ensuring appropriate curricula objectives and transnational educational policies are in place we can better prepare instructors for globalized teaching and learning. By establishing globalization priorities, educational funds can be targeted to facilitate needed changes. One additional source of funding, beyond the traditional campus experience at the college level, is through the extension of mediated distance course offerings. In addition to serving as another source of revenue, online distance learning facilitates diverse cultural interactions for knowledge building.

By analyzing our educational successes and failures on a regular basis we remain flexible to changing conditions to meet current education and policy needs. By transcending regional, local, and national interests alone and collaborating on effective global policies we develop the world community for mutual educational gains. This is not to say it will be an easy task when dealing with diverse cultural interests, as "relations of power – the social not natural kind – must be factored back in" (Marginson & Sawir, 2005, p. 307).

DESIGN AND IMPLEMENTATION

After informally interviewing many international graduate students at a small Midwestern University, it was impressive to learn they had traveled extensively and had command of up to five different languages. It was easy to tell they had been working to develop their global intellectual knowledge assets. When asked what they thought their instructors could do to help them to learn more, they supplied a varied and informative list of changes they would like to see in their courses. The list provided the foundation for creating a needs analysis survey to administer to international undergraduate and graduate students attending a small Midwestern University to better understand their learning needs. Understanding the instructional design needs of diverse learners will provide valuable insight into globalized course design for both learning and training to develop the intellectual capital. The data collected was a starting point with additional research still needed to continue the needs assessment at all levels of education and for different types of work environments for designing effective instruction for success in knowledge-based global markets.

The pilot survey included all 515 international students from a small Midwestern University. The students were e-mailed a request to participate in the online survey. The participants included 68% undergraduate and 32% graduate students. The ages ranged from 17 to 47 years old. The participants included 46% male and 54% female students. The countries of origin included Asia (76%), Middle East (14%), Africa (3%), Europe (3%), Latin America (2%), Canada (1%), and Caribbean (1%).

The pilot survey consisted of 11 items using a four-point Likert scale items rated as *Strongly Disagree* (1), *Disagree* (2), *Agree* (3), and *Strongly Agree* (4). The survey concluded with an open-ended question to gather additional information on students' perceptions about online learning needs. The open-ended question asked students for suggestions to help instructors design and develop improved courses for international learners. Table 1 presents the students' responses to the 11 survey items. Table 2 presents all of the students' responses to the open-ended question item on the survey.

DISCUSSION

Quantitative Assessment

The preliminary quantitative data indicated, international students preferred their learning environment to be well organized, have time zones included, and multimedia examples provided for additional content clarity. Most instructors are timely in providing learner feedback, but improvements can be made. The majority of international

Table 1. Summary of student quantitative survey results (N=38)

Question Items	Strongly Disagree	Disagree	Agree	Strongly Agree
1. Courses are well organized.	2.6%	2.6%	73.7%	21.1%
2. Time zone needs to be included for due dates.	5.3%	13.2%	47.4%	34.2%
3. I prefer group collaboration online.	7.9%	36.8%	52.6%	2.6%
4. Online Instructor feedback is clear.	2.6%	23.7%	50.0%	23.7%
5. Online Instructor feedback is timely.	2.6%	10.5%	63.2%	23.7%
6. Content specific glossaries are helpful.	2.6%	5.3%	68.4%	23.7%
7. I prefer instructors communicate by e-mail	7.9%	18.4%	47.4%	26.3%
8. Instructors need to participate in discussion forums.	2.6%	7.9%	52.6%	36.8%
9. Prefer student only communications in discussion forums.	13.2%	55.3%	26.3%	5.3%
10. I would like to have live virtual meetings of the class.	7.9%	15.8%	39.5%	36.8%
11. I prefer multimedia examples on topics.	0%	13.2%	50.0%	36.8%

students preferred instructor communications to take place through e-mail and within the course management program. Content specific glossaries would be helpful for most second language learners whether instructor or student generated. International students were open to using live virtual meetings to bring in more direct interactions for improving language skills.

International learners were split on their preference for/against using online collaboration. Possible reasons for the varied responses may be due to differing collaboration structures, grouping, communication, extended time involved online, equitable division of responsibilities, acceptance, respect, and credit for efforts.

Many of the international students felt instructor feedback was not clear and this area needs to be monitored for improvement. International students preferred instructors be engaged with learners in the discussion forum rather than relying only on student interactions, to be sure the content provided was accurate.

Qualitative Assessment

Student qualitative feedback provided additional insights to consider in designing instruction. International students preferred having access to dictionaries to overcome potential language barriers and to assure success in the learning tasks. It takes more time for second language learners to read the course materials and this should be considered in the course design.

Instructors should not assume all students have the same level of prior knowledge and experience. It is important to design instruction with this in mind. One possible solution might be to place international students together for group work so they are more comfortable and can work at a slower pace. The group work process should allow everyone to contribute equally providing a valuable and expanded perspective.

Learners find it helpful to have international examples included so they can relate to the content and make comparisons for improved learning and retention of new material. Questions and multimedia postings in the learning environment need to be in a very clear concrete format so learners know how to interact with the materials to meet desired expectations.

Live interactions would be valuable to assist in improving communication skills by offering an opportunity for practice. Honest, open, clear, detailed feedback helps learners to learn, grow,

Table 2. Summary of student qualitative survey results (N=38)

Student Excerpts
Do you have any suggestions to help instructors design and develop improved courses for international learners? 1. Dictionaries of all kinds are necessary. Sometimes, it's hard to understand concepts just because of one word that I don't know. Also, for international students, instructors should be careful to talk about other cultures or their cultural difference. 2. Student-centered traditional classroom is more advantageous for international students. In such classrooms teacher plays an important role on how the course should or even can be taught, depending on the students enrolled. 3. It is important to have the desired questions in a clear format. 4. Some of the courses really have disadvantages when offered online/web. As international student who came here as fulltime student, I don't expect to have online courses and pay extra money for that. I expected it to be all traditional courses because we need real interaction with the teacher...courses online not helping international students to improve their communication skills. 5. Providing a good environment to improve communication skill / interaction. 6. International student needs honest, concrete, and helping feedback. 7. Cost effective - give options for courses to be offered online or offline. 8. I hope instructors communicate more to help us choose classes each semester, and I wish we could talk with instructors about our future careers. 9. Treat us as same as American students. That is what we want. 10. Understand that the international students don't have the American education background like making projects or researching. 11. I think if the instructor gave more multimedia examples. 12. Because of the language, the instructors might be slower. 13. When making a group assignment, it should put the international students in one group, because the international student can not read as soon as the other American students and it will make it hard to do the job and hard to discuss. 14. To consider cultural differences. 15. The instructor can bring some topics related to international settings, such as e-Learning in Asia. 16. Have a detailed syllabus. 17. More practice. 18. To keep more communications during the beginning of the course. The international students usually get confused at the beginning and then they move very quickly and perfectly. 19. Give them more time and instructions to comprehend the topics. 20. More communication is really necessary sometimes. 21. Globalization as a topic in class and how the course material would relate to different parts of the world. 22. Professors willing to offer extra time to answer questions like explaining a word, sentence, phrase for American specific phenomena mentioned in the classes, that would be helpful. Not necessary all the time, but at least a certain amount of time. 23. Give clear guidelines. 24. Awareness of cultural differences 25. Consider the feelings of international students. 26. More face-to-face instructions when something is not clear. 27. Grades should not consider as much about grammar. After all, it is a second language. I can not use it as good as local people. Especially when the course is not a language course. 28. Group work is one thing of concern because I hesitate to express opinion. At the same time I understand we still need to deal with it as a student and that improves our skills for discussion in America. 29. Pay attention to learning difficulties 30. Don't assume we know everything about laws and policies in the US. 31. Give more examples to explain, and lower requirements for grammar mistakes.

and improve the quality of the work produced. International students simply want to be treated with equity.

Overall, international students thought what was needed was improved communication and clearer guidelines between instructors and students. Instructors should take additional time to answer questions or respond more slowly when answering questions to help international students grasp unique phrases or concepts unfamiliar to them. International students thought it would also be helpful to relate topics to international settings and discuss cultural differences in perceptions of a topic. International students thought it might be useful to have them work together when there was group work but still wanted to be able to interact with other students and the instructor.

FUTURE RESEARCH

We are just beginning to consider how "accelerating transnational dynamics are affecting education" (Suarez-Orozco & Quin-Hillard, 2004, p. 1). To prepare for global perspectives will involve formulating cultural knowledge, developing flexibility and tolerance, improving communication skills, using current technologies, forming partnerships, social networking, and international collaborations at an unprecedented pace. "These global transformations, we believe, will require youth to develop new skills that are far ahead of what most educational systems can now deliver" (Suarez-Orozco & Quin-Hillard, 2004, p. 2).

It is time to examine our curricula, standards, administration policies, and teacher preparation courses to integrate quality programs beginning at the elementary level and continuing on through college. Global perspectives in education will require instructors and trainers to embrace integrated international topics throughout the curriculum rather than taught in isolation. Policy must be in place to support international efforts.

A "new kind of citizenship education" will be needed to address global issues such as the environment and human rights, transnational in scope, and mutually beneficial (O'Meara, Mehlinger, & Newman, 2001, p. 215). Historically, providing incentives has served as one of the most powerful and lasting motivators for sustained improvements in the "pursuit of legitimated goals" for the enactment of progress seeking or justice-caring (Anderson-Levitt, 2003, p. 252). Globalization offers the opportunity to contribute to the well-being of the individual, collectively to the nation, and globally to enhance international stability and growth.

CONCLUSION

With the shifting global landscape and increasing need for intellectual capital, we are reminded of the importance of placing thought into the development of our knowledge assets. It is time to put policies recognizing global issues in place to meet the growing diversity found in our classrooms and work environments. We need to better understand each other to learn and work together effectively. We need to integrate quality global studies beginning at the elementary level and continue on through the college level then into the work environment. It will be important to infuse appropriate technologies and global perspectives in education and training to support the development of the intellectual capital. At the college level, in addition to preparing future teachers, we also need to prepare our instructional designers and corporate trainers to be able to provide quality knowledge assets in an era of globalization. "The competition for top talent remains strong" and will require "exciting work and challenge, career growth, learning, and development" along with salary and other benefits to retain and realize long-term intellectual capital benefits (Risher & Stopper, 2002, p. 1-2). With this in mind, it points to the need for attracting employees with the appropriate knowledge, skills, and talent needed to be successful in global knowledge-based economies.

REFERENCES

Anderson-Levitt, K. M. (2003). *Local meanings, global schooling: Anthropology and world culture theory*. New York: Palgrave MacMillan. doi:10.1057/9781403980359

Apple, M. W., Kenway, J., & Singh, M. (2005). *Globalizing education: policies, pedagogies, & politics*. New York: Peter Lang Publishing, Inc.

Bloom, D. E., & Cohen, J. E. (2001). *The unfinished revolution: Universal basic and secondary education*. Paper presented at the American Academy of Arts and Sciences, Cambridge, MA, July.

Bloom, D. E., & Cohen, J. E. (2002). Education for all: An unfinished revolution. *Daedalus*, (summer), 84-95.

Boyer, M. A., Brown, S. W., Butler, M. J., Niv-Solomon, A., Urlacher, B., & Hudson, N. F. (2007, July). Experimenting with global governance: understanding the potential for generational change. *Globalisation, Societies and Education*, *5*(2), 153–180. doi:10.1080/14767720701425727

Bryan, A., & Vavrus, F. (2005, July). The promise and peril of education: the teaching of in/tolerance in an era of globalization. *Globalisation, Societies and Education*, *3*(2), 183–202. doi:10.1080/14767720500167033

Cohen, S. L. & Backer, N. K. (1999). Making and mining intellectual capital: method or madness? *Training and Development, September* (5), 46.

Edvinsson, L., & Malone, M. S. (1997). *Intellectual capital: Realizing your company's true value by finding its hidden brainpower*. New York: Harper Collins.

Farenga, S. J., & Joyce, B. A. (2001). Hardware versus brainware: Where are technology dollars being invested? *Journal of Technology and Teacher Education*, (Autumn): 313.

Gnuschke, J. (2001). Obtaining market value from intellectual capital. *Business Perspectives*, *6*(Summer-Fall), 4.

Good, M. L., (1996). The globalization of technology. *Physics Today, August, Part 1, 49*(8), 23-27.

Huq, M. (2004). Building technological capability in the context of globalization: opportunities and challenges facing developing countries. *International Journal of Technology Management and Sustainable Development*, *3*(3). doi:10.1386/ijtm.3.3.155/0

Lauder, H., Brown, P., Dillabough, J., & Halsey, A. H. (2006). *Education, globalization, and social change*. New York: Oxford University Press.

Lynn, B. E. (2000). Intellectual capital. *Ivey Business Journal*, (January).

Malhotra, Y. (2000). Knowledge assets in the global economy: Assessment of national intellectual capital. *Journal of Global Information Management*, (July-September): 5.

Marginson, S., & Sawir, E. (2005, November). Interrogating global flows in higher education. *Globalisation, Societies and Education*, *3*(3), 281–309. doi:10.1080/14767720500166878

Milliron, M. D. (2007). Transcendence and globalization: our education and workforce development challenge. *New Directions for Community Colleges, Summer* (138).

Nabil, A., Awerbuch, B., Slonim, J., Wegner, P., & Yesha, Y. (1997). Computer database: globalizing business, education, culture through the Internet. *Communications of the ACM*, *40*(2).

O'Meara, P., & Mehlinger, H. D. D.H & Newman R. M., (2001). Changing perspectives on international education. Bloomington, IN: Indiana University Press.

Pang, N. S. (2006). *Globalization: Educational research, change and reform. SHA TIN, N.T.* Hong Kong: The Chinese University Press.

Publishers, Inc.

Risher, H., & Stopper, W. G. (2002). Attracting and retaining intellectual capital. Human Resource Planning, (March), 9.

Stewart, T. A. (1997). *Intellectual capital: The new wealth of organizations*. New York: Doubleday.

Stromquist, N. P. (2002). *Education in a globalized world: The connectivity of economic power, technology, and knowledge*. Lanham, MD: Rowman & Littlefield.

Suarez-Orozco, M. (2007). *Learning in the global era: international perspectives on globalization and education*. Los Angeles, CA: University of California Press.

Suarez-Orozco, M., & Sattin, C. (2007). *Learning in the global era: International perspectives on globalization and education*. Los Angeles, CA: University of California Press.

Suarez-Orozco, M. M., & Quin-Hillard, D. B. (2004). *Globalization culture and education in the new millennium*. London: University of California Press, Ltd.

Sveiby, K. E. (1997). *The new organizational wealth: Managing and measuring knowledge-based assets*. San Francisco, CA: Berrett-Koehler.

Swanson, W. H. (2009, Jan. 5). A plea to upgrade education. [Letter to the editor]. *Aviation Week & Space Technology*, 58.

Wells, R. (2006). Education's effect on income inequality: an economic globalization perspective. *Globalisation, Societies and Education*, *4*(3), 371–391. doi:10.1080/14767720600955428

Youndt, M. A., & Snell, S. A. (2004). Human resourse configurations, intellectual capital, and organizational performance. *Journal of Managerial Issues, Fall* (26), 337.

ADDITIONAL READING

Aslund, A. & Dabrowski (2008). The challenges of globalization: Imbalances and growth. Washington, DC: Peter G. Peterson Institute for International Economics.

Augustine, N. R. (2007). Is America falling off the flat earth? *National Academies Press*, 1-93. Retrieved March 2, 2009, from http://www.nap.edu/catalog.php?record_id=12021

Block, D., & Cameron, D. (2002). *Globalization and language teaching*. New York, NY: Routledge.

Cheng, Y. C. (2005). New paradigm for re-engineering education: Globalization, localization, and individualization. Dordrecht, The Netherlands: Springer.

Dubie, B. (2009, Feb. 2). Invest in America's Future. *Aviation Week & Space Technology*, 42.

Harrison, S. S., & Sullivan, P. H. (2006). *Einstein in the boardroom: Moving beyond intellectual capital to I-Stuff*. Hoboken, NJ: John Wiley & Sons, Inc.

Kirkbride, P., & Ward, K. (2001). *Globalization, the internal dynamic*. New York, NY: John Wiley & Sons, Ltd.

Kirkland-Tucker, T. F. (2009). *Visions in global education: The globalization of curriculum and pedagogy in teacher education and schools: Perspectives from Canada, Russia, and the United States*. New York, NY: Peter Lang Publishing, Inc.

Moore, L., & Craig, L. (2008). *Intellectual capital in enterprise success: Strategy revisited*. Hoboken, NJ: John Wiley & Sons, Ltd.

Nermien, A. A. (2003). *Comprehensive intellectual capital management: Step-by-step*. Hoboken, NJ: John Wiley & Sons, Inc.

Roo, G., Pike, S., & Fernstrom, L. (2005). *Managing intellectual capital in practice*. Burlington, MA: Butterworth-Heinemann Imprint of Elsevier Ltd.

Rustan, P. L. (2009, Jan. 12). Refocus Darpa beyond defense. *Aviation Week & Space Technology*, 62.

Spring, J. (2001). *Globalization and educational rights: An intercivilizational analysis*. Mahwah, NJ: Lawrence Erlbaum Associates, Inc.

Spring, J. (2006). *Pedagogies of globalization: The rise of the educational security state*. Mahwah, NJ: Lawrence Erlbaum Associates, Inc.

Spring, J. (2009). *Globalization of education: An introduction*. New York, NY: Routledge.

Stewart, T. (1999). *Intellectual capital: The new wealth of organizations*. New York, NY: Bantam Doubleday Dell Publishing Group Inc.

Sullivan, P. H. (2000). *Value driven intellectual capital: How to convert intangible corporate assets into market value*. New York, NY: John Wiley & Sons, Inc.

Chapter 3
Measuring Science & Technology in Panama:
Towards a National Intellectual Capital Framework

Yuan-Chieh Chang
National Tsing Hua University, Taiwan, R.O.C.

Carlos Kan
National Tsing Hua University, Taiwan, R.O.C.

ABSTRACT

This chapter examines the notion of national intellectual capital, which could bring new insights to the existing national science and technology policy thinking. This notion leads that the chapter proposes a framework to measure national intellectual capital, and the investigation based on the framework was applied in Panama. The results revealed that the Panamanian intellectual capital faces the decreasing supply of human capital in S&T field, the service-dominant market capital, weaker and less indigenous innovation capital, and a steady increase of process capital. Some intellectual capital policy implications are drawn for Panama and other developing countries.

INTRODUCTION

As economies continue to become more knowledge intensive, intellectual capital has become the competitive edge for people, corporations, regions, and nations. Higher levels of intellectual capital have been associated with higher standards of living, improved health, and continued increases in national competitiveness. As a result, nations and firms have to effectively manage their intellectual capital in order to outperform other nations and firms (Bornemann et al., 1999; Johanson et al., 1999). However, a systemic framework to measure and to report national intellectual capital is understudied. This highlights a need to effectively measure intellectual capital at the macro level.

Previous research on the intellectual capital community has illuminated the intellectual capital perspective as a useful framework to explain abnormal growth and development in cities, regions and nations (Bontis, 2004; Bounfour and Edvisson, 2005). The interest of intellectual capital in weighting the intangible content at macro spaces

DOI: 10.4018/978-1-61520-875-3.ch003

is growing. In addition, it is notable that not only the products have become more knowledge intensive, but so have services and business processes (Andriessen, 2004). The chapter seeks to develop a preliminary framework for measuring and reporting national intellectual capital, especially for developing countries such as Panama. It provides an insightful framework to plan and develop intellectual capital in nations.

The chapter proceeds as follows. The chapter reviews the definition, theoretical foundations, types of intellectual capital and its relation to knowledge, innovation and strategy. Further, existing indexes for measuring national intellectual capital are examined and a research framework is proposed. The chapter measures the national human capital, national innovation capital, national market capital and national process capital in Panama. The results are discussed. We reach its conclusions. Some policy implications are made to Panama and other developing countries.

BACKGROUND

Theories of Intellectual Capital

Intangible assets include not only R&D (research and development) but also human capital, innovation in products and in organizations, trademarks and patents, information technology, networking and software. This chapter will use the terms intangible assets and intellectual capital interchangeably. Edvinsson and Malone (1997) clearly identify the four dimensions of intellectual capital in terms of human capital, customer capital, innovation capital and process capital.

Resource-based view (RBV) is the current dominant school of strategic management applied by firms to explain abnormal profits resulting from its core competence and distinctive resources. Resource-based views highlight intellectual capital as heterogeneous and firm-specific. Therefore, developing universal intellectual capital indicators for various firms could be infeasible. In other words, each firm has to develop their own firm-, nation-specific intellectual capital measurements. Although the importance of tangible resources, such as financial and physical resources, is undeniable, RBV tends to place more stress on the importance of intangibles include human resources (Barney, 1991), organizational resources (Barney, 1991; Wenerfelt, 1989; Grant, 1991), brand, reputation and market position (Wenerfelt, 1989), and technological and innovative capabilities (Wenerfelt, 1989). It is worth noting that the resource-based perspective pinpoints the most important intangible resources which firms can distinguish themselves from others.

The national system of innovation (NSI) study highlights the importance of indigenous intellectual capital but also the influence of external partnership forces that contribute to the structure of in-house intellectual capital (Freeman, 1987; Lundvall, 1992). For that reason, the interactive processes between firms, universities and R&D institutions are one of main mechanisms to create intellectual capital of organizations, regions and nations. The NSI contributes to relational capital embedded in collective learning among actors. Moreover, inter-firm and inter-organizational knowledge links are crucial. Despite OECD has developed systemic guidelines to measure national science, technology and innovation through Frascati Manual (OECD, 1981) and Oslo Manual (OECD, 1992: OECD, 1997). These STI indicators can be found in the series publication of STI Outlook. Currently, STI Outlook (OECD, 2008b) mainly provides S&T information related to national R&D spending, patents, scientific publications, and human resource in S&T in OECD member countries. Regarding the Information and Communication technologies, OECD ICT Outlook will report national ICT hardware, systems and service indictors in the member countries (OECD, 2008a).

Knowledge management highlights the importance of intangibles. Firms effectively creates organizational capability to systemically identify, reserve, share, use and develop knowledge that will be a key to sustain their competitive advantage (Nonaka and Takeuchi 1995; Leonard-Barton, 1995). We have observed that intellectual capital is still in the infancy stage; nevertheless, intellectual capital framework presents a promising path to integrate the resource-based view, innovation studies and knowledge management as a whole. However, the intellectual capital approach has a natural weakness. That is, it tends toward static and forms of knowledge stocks rather than knowledge flows. These three theories form the major building blocks of intellectual capital. Their relations with intellectual capital can be summarized: (1) Resource-based theory as strategy developer and enabler; (2) Systems of innovation as innovation facilitator; (3) Knowledge management as provider of knowledge metrics. Pertinently speaking, the future of intellectual capital studies is to link firm strategy, innovation and knowledge structure (Figure 1), as they complement each other and contribute to intellectual capital.

Conceptual Models for National Intellectual Capital

Intellectual capital is used as a generic term to denote all intellectual resources (e.g., knowledge and information databases), assets (e.g., processes), and properties (e.g., patents and trademarks) that an organization owns, controls, or has access to (Al-Ali, 2003). An emerging stream of studies has documented the need for improved reporting of the intangibles of stakeholders outside and inside organizations. Human resource accounting (Flamholtz, 1985; Johanson, 1999), balanced scorecard (Kaplan & Norton, 1992), intellectual capital statement (Edvinsson and Malone, 1997), and the intangible asset monitor (Sveiby, 1997) have been launched to measure and report intangibles. However, these IC reporting approaches tend to focus on firms and organizations rather than macro levels such as regions and nations

According to Bontis (2004) the intellectual capital of a nation includes covert values of individuals, enterprises, institutions, communities and regions that are the existing and potential sources for the creation of national wealth. The overall structure drawn from academics and practitioners has appointed three pillars of capital: human capital (Edvinsson and Malone, 1997; Petty and Guthrie, 2000; Mouritsen et al., 2002; Ordóñez de Pablos, 2003), relational capital (Ordóñez de Pablos, 2003), and organizational capital (Edvinsson and Malone, 1997; Petty and Guthrie, 2000; Mouritsen et al., 2002; Ordóñez de Pablos, 2003).

Emphasis on the importance of human capital as the generator and facilitator of value (Edvinsson and Malone, 1997), and as a source of innovation and renewal (Bontis, 2004) suggests the necessity of appointing this type of capital into our conceptual model. The concept of structural capital encompasses market capital and organizational capital. The former quite simply represents the relations for the externalization of the nation,

Figure 1. Three theoretical building blocks of intellectual capital

and so, the World Bank (1999; 2004) refers to this kind of externalization in exports and FDI (foreign direct investment) capacity to extract the maximum output of nations.

Last, but not the least important is the organizational capital. The structure of this capital is twofold including innovation capital and process capital. Innovation capital is synchronized with other capitals, and so, its multiplicative is aided by the process capital. As a result, it would be too naive to state that intellectual capital components do not interact with each other. Studies on IC reporting in Malaysia (Bontis, 2000) and Spain (Ordóñez de Pablos, 2003) shows that some components of intellectual capital do interact with each other.

In a study by Bontis (2004), he developed four dimensions of national intellectual capital indexes: national human capital, national market capital, national process capital, and national renewal/innovation capital. Each national intellectual capital dimension consists of many metrics (Table 1). National human capital is mainly measured by education and qualified teachers. National process capital is measured by the ICT national wide adopters such as telephones, computers, mobile phones and internets. National market capital is measured by national exports, patents granted by the USPTO (US Patent Office) and international conferences. Finally, national renewal capital is mainly measured by national R&D expenditure, personnel and university funding. Bontis (2004) deployed seven metrics of national human capital, eight metrics of national process capital, three metrics of national market capital, and seven metrics of national renewal capital.

Thus, we modified the national intellectual capital model proposed by Bontis (2004). The model consists of four dimensions of intellectual capital: human capital, process capital, market capital and innovation capital. However, due to the limitations in data availability and the specificities of Panama context, we elaborated the metrics of intellectual capital slightly differently from Bontis (2004). The chapter mainly focused on human capital on tertiary education rather than literacy rate and students in take. We argue that it is more appropriate to move the metrics of "public education funding" and "S&T personnel" from the renewal capital to the human capital because these two metrics are more directly related to human capital. The chapter treated "future human capital" as human capital rather than innovation capital in Bontis (2004). Regarding to national process capital, the chapter mainly concentrated on the telephone, mobile phone and internet technologies. The rest of process capital such as television, radio and newspapers were excluded. For the market capital, we mainly explored the GDP by sector, trade by sector and international trade relationship. The number of patents issued by the US PTO was excluded in market capital. Alternatively, we consider patents as one of key innovation capital. The details of national intellectual capital research framework are listed as follows:

1. The metrics of national human capital include: (1) future human capital, (2) national education expenditure, (3) number of first-level university degrees, (4) number of master graduate degrees, and (5) types of R&D personnel.
2. The metrics of national market capital include: (1) gross domestic product by sectors, (2) trade by sectors, and (3) international trade partnership.
3. The metrics of national innovation capital include: (1) expenditure on science and technology activities, and (2) number of patent applications and patent grants by Panama patent office, and
4. The metrics of national process capital: the metrics of process capital include: (1) telephone services, (2) mobile phone services, and (3) internet services.

Table 1. National intellectual capital: Dimensions and metrics

Dimensions	Metrics
National human capital	Literacy rate
	Number of tertiary schools per capita
	% of primary school teachers with required qualification
	No. of tertiary students per capita
	Cumulative tertiary graduates per capita
	% of male grade 1 net intake
	% of female grade 1 net intake
National process capital	Telephone mainlines per capita
	Personal computers per capita
	Internet hosts per capita
	Internet users per capita
	Mobile phone per capita
	Radio receivers per capita
	Television sets per capita
	Newspaper circulation per capita
National market capital	High-tech exports as a percent of GDP
	No. of patents grated by the USPTO per capita
	No. of meetings hosted per capita
National renewal/innovation capital	Book imports as a percent of GDP
	Periodic imports as a percent of GDP
	Total R&D expenditure as a percent of GDP
	No. of ministry employees in R&D per capita
	No. of university employees in R&D per capita
	Tertiary expenditure as percent of public education funding

Source: adapted from Bontis (2004)

MEASURING NATIONAL INTELLECTUAL CAPITAL IN PANAMA

A Sketch of Panama

Panama is a 78,200 square km country situated on a natural land bridge between the North American and South American continents. The isthmus is in the middle of America, between Costa Rica at the west and Colombia in the east. The north touches the Caribbean Sea, part of the Atlantic Ocean, and at the south it is enriched by the Pacific Ocean. The Isthmus, as a geographical bridge of the Americas, is one of the most important epicenter of terrestrial, tropical sciences and navies of the world and one of the most important centers of economic activity in Latin America.

The capital is Panama City, which is located in the south entrance of the Panama Canal, surrounded by another main city known as Colon City, located in the north entrance of the Panama Canal. Panama has little over 3 million people in population, with Panama City having around 1.5 million of them. That population is well distributed in gender, having 50.5% male and 49.5% female.

The life expectancy for a Panamanian is 74.9 years (Comptroller General of the Republic, 2005; World Bank, 1999, 2004).

Panama has reached high economic growth in Central America over decades. The unit of currency used in Panama is the Balboa, which is pegged at parity to the US dollar. There is no Panamanian paper currency and the US dollar is the de facto official currency for all but minor transactions. As a result, the Government cannot print money, and inflation is low, estimated at 2% in 2004 (CIA, 2006).

Based on WEF (2009) report, a list of 7 nations in Latin America and the Caribbean situated the efficiency-driven stage (GDP per capita: US$ 3,000~9,000) were selected. Seven nations include Argentina, Brazil, Colombia, Costa Rica, Ecuador, Panama and Peru. Panama was ranked 59th of Global Competitiveness Index in 2009 (WEF, 2009), behind Costa Rica (Rank 55), Brazil (Rank 56) and ahead rest of the countries, Colombia (Rank 69), Peru (Rank 78), Argentina (Rank 85), and Ecuador (Rank 105). In term of basic requirements, Panama relatively had good macroeconomic stability (Rank 46), weak health and primary education (Rank 64) and infrastructure (Rank 65) comparing to its overall ranking, 59. In terms of efficiency enhance, Panama relatively had super financial market (Rank 22) and market efficiency (Rank 54), acceptable technological readiness (Rank 59), weak higher education and training (Rank 80), market size (Rank 81) and labour market efficiency (Rank 89). Finally, Panama had good business sophistication (Rank 45) and weak innovation (Rank 66). Overall speaking, the strength and weakness of Panama intellectual capital can be summarized as follows: (1) strong national market capital in terms of macroeconomic stability, financial market and business sophistication environment; (2) the fair national process capital in term of technological readiness and infrastructure; (3) weaker national innovation capital; (4) and the weakest national human capital (e.g., weak higher education and training, low labor market efficiency).

The GDP per capita in Panama was US$ 6,900 in 2004. Panama has experienced a stable economy growth 7.6% in 2004 and 6% in 2005 (World Bank, 2004; CIA, 2006). The economic growth was attributed for favorable great environments and buoyant domestic demand. The labor market continued to improve by virtue of the considerable buoyancy of the economy. However, the urban unemployment has increased relatively compared to the country side measures. The employment rate has increased due to the high demand for labor in the service-based sectors, port and tourist activities, where is experiencing a positive balance. All of these brought down a national unemployment rate to 9.6% in 2004*(Comptroller General of the Republic, 2005)*. We further measure and report national intellectual capital in Panama in the following sections.

National Human Capital

The educational system of a nation is considered one of the fundamental source and more effective mechanisms to generate favorable conditions for the development of national human capital. In the same instance, the USAID refers to human capital as an important cornerstone complementary to labor, boosting productivity and the potential for income generation.

Future human capital represents a nation's future intellectual wealth (Bontis, 2004). Although Bontis (2004) has referred to this as renewal capital, the chapter categorizes future human capital as part of human capital because it provides the necessary impact for future generations of individuals. According to the Comptroller General of the Republic (2005), in the future human capital of Panama, the population of people between the ages of 0 to 14 represents approximately 31 percent (Table 2). The human capital figures indicate an even and healthy Panamanian population distribution in terms of future human capital and gender.

In response to develop national human capital, the central government has structured a budget

Table 2. Future human capital in Panama (2004)

Age	Male	%	Female	%	Population	%
0-4	174,072	5	166,835	5	340,907	11
5-9	166,386	5	159,501	5	325,887	10
10-14	156,219	5	149,942	5	306,161	10
Other	1,104,202	35	1,095,203	35	2,199,405	69
Sub-total	1,600,879	50	1,571,481	50	3,172,360	100

Source: Adapted from Comptroller General of the Republic (2005)

for education allocated to the Ministry of Education, Universidad de Panamá, Universidad Tecnológica de Panama, Universidad Autónoma de Chiriqui, and Universidad Especializada de Las Américas; which are Panama's four state universities. The government expenditure devoted to education has increased consistently from US$ 549 billion in 2001 to US$ 604 billion in 2005 (Table 3). The annual growth rare of government education expenditure reached the highest at 4.4% in 2005. These increases indicated the higher government's emphasis on quality improvement in higher education.

On the other hand, there are over twenty five private universities operating on Panamanian soil (Franco, 2003). On this ever increasing list, we can highlight two universities from the USA with branches in the Republic of Panama, Florida State University and University of Louisville. These universities have grown during recent years due to the high surplus demand of students who work part-time during the day to pay for their tuition in order to attend night classes. It is evident that the main income resource of the private universities is the tuition and credit fees imposed on the student community.

Based on the 1990 UNESCO World Education Report, Panama was one of the less developed regions of Latin America and the Caribbean in terms of literacy. Thus, in 2005, Panama was ranked 56 out of 177 countries on the United Nation's Human Development Index (HDI) with only Costa Rica as a Central America country ranking higher. Although the education levels of the Panamanian knowledge labor force are decreasing, according to the estimates of the number of persons from private and public universities by fields between the years 2000 and 2003, Tables 4&5 provide information of individuals holding first level university degree and master graduate degrees.

Table 4 shows that the total number of individuals with first level university degrees seems to have fluctuated during the period of 2000-2003, and to have decreased to the some extent. In 2003,

Table 3. National education expenditure (US$ million)*

Years	2001	2002	2003	2004	2005[(P)]
Net Total[(1)]	548,584	555,537	562,363	579,179	604,408
Operations	444,675	466,696	477,949	500,037	533,327
Investment	49,558	36,751	24,217	31,622	60,391
Annual growth rate	--	1.3%	1.2%	3.0%	4.4%

[(P)] Predicted figures
[(1)] It includes current expenditures and value-added tax (VAT)
* It only considers the public education budget.
Source: Adapted from Comptroller General of the Republic (2005)

Table 4. First level university degrees from public and private universities by fields

Year	2000		2001		2002		2003	
Fields	No.	%	No.	%	No.	%	No.	%
Social Sciences	9,036	65	10,513	81	8,603	66	9,770	74
Sciences	1,366	9	1,107	9	1,793	14	1,782	13
Natural Sciences	588	4	469	4	618	5	285	2
Medical Sciences	612	4	568	4	1,013	8	849	6
Agricultural Sciences	166	1	70	1	162	1	648	5
Engineering and Technology	2,090	15	1,138	9	1,403	11	1,274	10
Humanities	1,360	10	220	2	1,145	9	343	3
Total	13,852	100	12,978	100	12,944	100	13,169	100

Source: Adapted from RICYT (2005)

Table 5. Master graduate degrees by field

Year	2000		2001		2002		2003	
Field	No.	%	No.	%	No.	%	No.	%
Social Sciences	614	78	548	82	476	56	555	78
Sciences	71	9	39	6	304	36	74	10
Natural and Exact Sciences	57	7	21	3	24	3	20	3
Medical Sciences	10	1	18	3	280	33	38	5
Agricultural Sciences	4	1	0	0	0	0	16	2
Engineering and Technology	32	4	22	3	50	6	61	9
Humanities	71	9	60	9	24	3	21	3
Total	788	100	669	100	854	100	711	100

Source: Adapted from RICYT (2005)

most of the degrees were given in the social sciences field with 9,770 (74%) degrees, following engineering and technology field with 1,274 (10%) degrees, medical science with 849 (6%) degrees, agricultural sciences with 648 (5%) degrees, humanities with 343 (3%) degrees, and natural sciences with 285 (2%) degrees. The first degree graduates in science, engineering and technology were only 23% of the total first degrees in 2003. It is special for first degree graduates that most of them took social sciences as their first degrees, rather than sciences, engineering & technology in Panama.

The number of master graduate degrees in the period of 2000-2003 by field has shown in Table 5. The total number of master graduates by year was fluctuated and reached the highest with 854 graduates in 2002 and the lowest with 669 graduates in 2001. It follows the similar pattern as the number of first degrees by field (Table 4), which master degrees were dominant by social sciences (56~82%), following by natural sciences (6~36%), engineering and technology (3~9%) and humanities (3~9%). All figures in Tables 4&5 confirm that the human capital of higher education is dominant by social sciences, following Sciences, engineering and technology and the least, humanities in Panama.

S&T personnel have increased in the period of 2000-2003, from 1,659 persons in 2000 to 2,621

Table 6. Science & technology personnel

S&T personnel (Head count)	1999		2000		2001		2002		2003	
	No.	%	No.	%	No.	%	No.	%	No.	%
Researchers	495	30	446	27	841	41	416	17	432	16
PhD Scholars/Post doctoral	0	0	0	0	0	0	0	0	0	0
Technicians	717	43	754	45	689	34	1,272	52	1,363	52
S&T services personnel	447	27	476	28	509	25	780	32	826	32
Subtotal	1,659	100	1,676	100	2,039	100	2,468	100	2,621	100

Source: Adapted from RICYT (2005)

persons in 2003 (Table 6). However, there were no doctoral graduates undertaken as S&T research. One of major reasons is because few people completed their doctoral degrees in Panama. Technicians are the major S&T personnel indicating the major development-oriented S&T activities. The number of researcher has dramatically increased since 2002. It is because many private universities expanded to recruit new faculty and researchers.

National Market Capital

The market capital of a nation is defined by Bontis (2004) as the intellectual capital embedded in national capabilities to develop, export and build trade relationships. Therefore, we will explore the GDP by sector, trade by sector and international trade relationship in national market capital. The nation's market capital demonstrates its capabilities and success to provide an attractive and competitive framework to meet the needs of its international partners, customers and market. Panama economy rests primarily on a well-developed services sector that accounts for three-fourths of the GDP (Table 7). Services include operating the Panama Canal, banking (this sector has over seventy different international banks), the Colon Free Zone (CFZ), insurance, container ports, flagship registry, and tourism. A slump in the CFZ and agricultural exports, the global slowdown, and the withdrawal of US military forces held back economic growth from 2000 to 2003; growth picked up in 2004 and 2005 led by export-oriented services and a construction boom (e.g., Panama Canal expansion) stimulated by tax incentives. The government has implemented tax reforms, as well as social security reforms, and the development of tourism and regional trade agreements.

These trade agreements enroll a list of commercial partners who have proven, in our view, to enrich the social and commercial networks of the nation. From the list of nations that signed the FTA (free trade agreement) include: Taiwan, El Salvador, and Singapore. Recently, the central government also has listed other future FTA countries, such as: USA, Chile, and the Central America block. These countries are currently carrying out evaluations, conversations and further deliberations prior to co-signing.

Sectors providing services to the rest of the world have benefited from the buoyancy of the United States and Asian economies, and the steady growth of Latin America. Domestic demand has expanded thanks to household consumption (boosted by the increase in employment and credit) and abundant private investment in sectors such as ports and tourism. FTA talks with Singapore were concluded successfully in lately 2005, while those with the United States were stalled, although a final round of negotiations was supposed to be scheduled for the beginning of 2006. However, it was slowed down due to the Panama Canal expansion project. Negotiations with Chile were reopened, and an agreement was signed with Mexico on the reciprocal promotion and protection of investment.

Table 7. Share of GDP by sector (in current prices)

Indicator	2001	2002	2003	2004(P)
Primary activities				
Agriculture(1)	7.3	7.2	7.4	7.1
Mining	0.7	0.7	1	1
Total	7.9	8	8.4	8.1
Secondary activities				
Manufacturing	8.8	7.9	7.5	7.2
Construction	3.9	3.5	4.6	4.8
Total	12.6	11.4	12.1	12
Services				
Public Utilities	2.8	2.9	2.8	3
Commerce, Restaurants, and Hotels	8.9	9	8.9	8.9
Transportation and Communications	8.8	8.9	8.8	9.3
Colon Free Zone	7.4	6.5	6	6.8
Panama Canal	4.8	5	5.5	5.9
Financial Intermediation	9.8	9.6	8.9	8.4
Real Estate	17.5	18.2	18.4	17.8
Public Administration	10.9	11.2	10.5	10.5
Other Services	6.1	6.1	6.1	5.9
Total	77.1	77.3	76.1	76.3
Plus Import Taxes(2)	5.8	6.3	6.5	6.4
Less Imputed Banking Services	-3.4	-3	-3.1	-2.8
Gross Domestic Product	100	100	100	100

P) Preliminary figures
(1) It includes fishery
(2) It includes value-added tax (VAT)
Note: The corrected series based on the change the base year of national accounts to 1996, by the Comptroller General'
Source: Adapted from Ministry of Economy and Finance (2005)

The services occupied more than three-quarters of GDP in Panama, following secondary sectors (manufacturing and construction) 12% and the primary sectors 8.1% in 2004 (Table 7). It is obvious that the Panamanian economy is service-based. Among services, the most buoyant sectors were still real estate 17.8%, public services 10.5%, and transport, storage and communications 9.3%. The manufacturing sector is shrinking in term of share of GDP. However, construction sector increased its importance of GDP in the period of 2001-2004 due to Panama Canal expansion. Agriculture sector remained stable and occupied 7~8% of GDP in recent years. Thanks to the positive results of port activities, railways, the Panama Canal and mobile phones. The Comptroller General of the Republic and CEPAL (2005a) report that the growth of agriculture and fishing sectors performed well (7.1%), based mainly on the strength of increases in fishing, livestock and export crops such as coffee and pineapples. Commerce also turned in a fine performance growth (8.9%), both at the strictly domestic level and in the CFZ.

In summary, the national sector and trade are key dimensions to measure national market capital. The statistics above indicates that Panama

Table 8. Trade by sector (US$ Million)

Trade	2003	2004	2005[(P)]
Total export	3306.668	3417.324	4261.45
Export of Goods	804.992	890.295	963.764
Petroleum derived products[(1)]	5.37	4.832	7.063
Bananas	112.787	108.221	96.515
Shrimps	56.457	53.819	57.18
Shrimp larvae's exports	4.495	6.439	7.164
Frozen yellow fin tuna, fresh and frozen fish filet	262.633	280.699	264.656
Coffee	12.158	10.922	13.511
Clothing	11.034	8.795	10.279
Cattle Meat	9.555	13.856	12.417
Others	330.503	402.712	494.979
Export of Services	1276.384	1419.534	1668.522
Canal Commission Tolls (millions of US$)	690.268	766.897	886.805
Entrance of passengers (millions)	1.43	1.617	1.865
Travelers expenditures (millions of US$)	584.686	651.02	779.852
Colon Free Zone	420.3	217.2	665.4
Total Imports[(2)]	3,122.275	3,592.209	4,155.293
Consume goods	1,294.015	1,547.226	1,860.389
Not durable	545.539	596.127	636.922
Semi durable	270.85	303.539	339.217
Domestics utensil	117.633	131.177	147.632
Fuels and lubricants	359.994	516.384	736.618
Intermediate goods	994.442	1,119.183	1,215.052
Raw materials for agriculture	74.27	88.53	98.416
Raw materials for the industry	671.317	735.895	796.142
Construction materials	216.01	254.873	275.713
Others intermediate goods	32.845	39.885	44.781
Capital goods	833.818	925.8	1,079.851
For agriculture	24.019	25.151	25.035
For industry, construction and electricity	254.38	277.752	340.275
Transportation equipment and telecommunication	467.991	525.142	585.862
Others capital goods	87.428	97.754	128.68
Balance of trade	184.393	(174.885)	106.157

[(P)] Preliminary figures; [(1)] It does not include sales of oil to ships and aircrafts; [(2)] Import figures for 2004 are for the months of January to November.

Source: Adapted from the Ministry of Economy and Finance (2005)

had trade surplus over in the period of 2003-2005 (Table 8). Export of services was more than export of goods. Export of services was mainly derived from Panama Canal service occupying about four-fifths of total export. It confirmed a strong service-based market capital in Panama.

However, the export of Colon Free Zone plays a decisive role to balance the trade deficit. Panama suffered a trade deficit in 2004 because the net trade surplus declined to almost half of figure in 2003. An export recovery is made by the CFZ whose net trade value reached more than three times growth of 2004 in 2005.

National Innovation Capital

The concept of national innovation capital refers to the value of what is achieved through knowledge sharing and the investment in knowledge and R&D; these are the key metrics that shape the development of the national capabilities. These investigating parameters include: government expenditure on S&T activities and patents reviewed and granted by Panama patent office.

Table 9 shows that between the years 2001 and 2002. A decrease is observed of approximately US$ 15 million in the total investment in activities of science and technology. However, the figures were underestimated due to the following factors: (1) Budgetary differences among it reported by some institutions for the 2001 survey and the 2002 survey, due mainly to the culmination of projects and budgetary cuts at the governmental level; (2) Two institutions in the higher education sector and three in the government sector did not provide information for the years 2002 and 2003. These institutions had reported an expenditure of approximately seven million US$ the previous year. (3) Shutting down of entities, mainly in the higher education sector (due to closing of operations at national level) and not governmental agencies.

According to the Science and Technology Activities Report (SENACYT, 2004), expenses in Science and Technology Activities are the joint conglomerate of investment carried out by the government, higher education and non-governmental organizations in activities narrowly related to the production, promotion, diffusion and application of scientific know-how and technicians in all the fields of science and technology. Nevertheless, expenditures were estimated because the S&T statistics of all private sectors was not available. Panamanian government spent about US$ 100 million on S&T annually in the period of 2000-2003. However, the lion share of budget was spent on the technical and scientific service, following by R&D activities, and education and training. In term of national government S&T intensity was about 0.3~.0.4% of GDP, about half of the Taiwanese, 0.8% of GDP. Overall, Taiwan spent 2.33% of GDP on national R&D expenditure in 2004 (NSC, 2005). The Taiwanese government S&T expenditure occupied 36% of national R&D expenditure in 2004. There is a room for improvement in government spending in S&T in Panama.

WIPO (2004) divides intellectual property into four major mechanisms: patent, copyright, trade secret and trademark. Especially, patents are regarded the best proxy for innovation because patents are widely available and easily retrieved.

Table 9. Government expenditure on science & technology (US$ Million) (1)

Indicator	**2000**	**2001**	**2002**	**2003**
Government S&T Expenditure (A)	101.58	116.12	103.97	95.13
Research & Development	44.63	45.11	44.47	43.97
Scientific Education and Technical Training	34.3	41.13	41.7	17.8
Technical and Scientific Services	95.13	43.97	34.13	17.03
Gross Domestic Product (B)	N.A.*	11,807.5	12,272.4	12,933.2
National Government R&D Intensity (A/B) (%)	N.A*	0.382	0.3624	0.34

(1) This figure includes the STRI expenditures, *: Not available
Source: Adapted from SENACYT (2004)

The further application of patents is high (e.g., licensing) so that patents provided a huge impact on innovation, productivity and intellectual capital. Table 10 shows patent applications and patent grants by Panama Patent Office (PPO) in the period of 1999-2003. PPO received about 180-266 patent applications and granted about 64-274 patents annually. Although domestic Panamanian applicants and patent assignees have gradually increased over years, however, more than 80% of patent applications and more than 95% of patent grants were done by non-Panamanians, e.g.,, MNEs. Some of the MNEs in Panamanian soil worth mention include: Cable & Wireless (C&W) from the UK; Panama Ports under Hutchinson Group from Mainland China; Telefónica from Spain; Union Fenosa from Spain; HSBC Holdings PLC from the UK; DELL from the USA, etc. This indicated that foreign firms/organizations more emphasize on intellectual property protection than domestic firms, in turn, protect their innovations in Panamanian market. Moreover, this demonstrated low indigenous innovation undertaken by Panamanian firms and organizations.

National Process Capital

Process capital is defined by Bontis (2004) as the non-human storehouses of knowledge in a nation which are embedded in its technological, information and communication systems as represented by its hardware, software, databases, laboratories and organizational structures. International experience shows that technological progress in ICT today enables us to process, store, retrieve and communicate information in whatever form it may take, unconstrained by distance, time, volume and increasingly, by cost (UNDP, 2005). At the same time, WSIS demonstrated that many developing countries are preparing policies and strategies for the deployment and adoption of ICT in their societies as they acknowledge that continuous developments in ICT are opening up a world of new opportunities for harnessing knowledge for development.

This trend brings urgent threats, especially for developing countries. Countries with inadequate computers, Internet access and telecommunications are at risk of falling even farther competitors in the world market (Bontis, 2004). With national policy intent on ICT development, Panama created the ERSP (Public Utility Regulated Entity) in 1996 which has provided legal regulation faculties with defined purposes for the telecommunication sector over the years. Previously state owned, the ERSP body dictated guidelines for the privatization of the telecommunication market in 1997, and it was not until 2003 when the market was de-regulated.

Table 10. Patent applications and patent grants by Panama intellectual property office

Indicator	1999		2000		2001		2002		2003	
	No.	%	No.	%	No.	%	No.	%	No.	%
Patent Applications										
Residents	10	5.5	25	11.7	24	9.1	26	9.8	33	12.4
Non Residents	172	94.5	189	88.3	241	80.1	240	80.2	240	87.6
Total	182	100	214	100	265	100	266	100	266	100
Granted Patents										
Residents	1	1.6	4	23.5	7	4.4	6	2.2	9	5.2
Non Residents	63	98.4	13	76.5	152	95.6	268	97.8	163	94.5
Total	64	100	17	100	159	100	274	100	172	100

Source: Adapted from RICYT (2005)

Table 11 presents the 2000-2005 development of national process capital in terms of telephone line services, mobile phone services, and internet services. Overall, the Panamanian process capital has experienced a substantial growth through the years. The telephone landline services in working and in large cities was saturated in 2000 and stable in the period of 2001-2005. Declaring the new era of mobile phone services, 15.8 annual mobile phone subscribers per 100 people exceeded 12.6 telephone line subscribers per 100 people, 12.6 in 2001. Increasing Internet connectivity is an important step that Panama must undergo to access the myriad store of knowledge available through this technology. Internet subscribers have been double in 2005 comparing to those in 2000. Moreover, collaboration both within the country and across borders, like with Panama's trade partners, can be facilitated through shareware technologies that leverage the Internet. In return, the sharing and access of Internet has given an engine for development. Moreover, the drop of bandwidth prices demonstrates the high penetration of process capital development in Panama.

The Panamanian telecommunications infrastructure has greatly improved as geographical location and the abundant supply of submarine fibber optic communications circuits on both coasts makes possible interconnection to North and Latin America, Asia, Europe, Africa and the world. The numerous routes have provided competitive advantage over the years.

DISCUSSIONS

Panama is a developing country with as a middle income country (GDP US$ 6,200 in 2004), an efficiency -driven economy (WEF, 2009), having literacy rate of 92%, and a population about 3 million. In term of national human capital, Panama government had a very keen commitment on education with a stable 2.5% growth in the period of 2001-2005. The share of future human capital (ages of 0-14) was 31 percent and presented a healthy proportion in total population. The gender of the population has been balanced development. However, the human capital growth

Table 11. Panama's process capital

Indicators	2000	2001	2002	2003	2004	2005
Telephone Landline Service:						
Telephone mainlines working	429,135	381,912	386,904	381,569	410,185	440,129
Telephone mainlines index (per 100 people)	15.1	12.6	13.6	12.2	11.9	13.9
Telephone mainlines in largest city	285,000	347,000	221,575	216,728	214,393	230,044
Mobile Service:						
Mobile phone subscribers	410,401	475,141	525,845	692,406	1,259,948	1,351,924
Mobile phone annual index (per 100 people)	14.5	15.8	17.2	27.12	39.7	41.9
Population covered by mobile telephony (%)	0.744	0.785	0.843	0.868	0.868	0.889
National territory covered (%)	17	18	19	20	23	26
Internet Service:						
Internet subscribers	42,982	48,570	57,985	69,234	78,619	82,471
Internet subscribers annual index (per 100 people)	1.5	1.6	1.9	2.2	2.5	2.6

Source: Adapted from ITU and ERSP (2005)

in the higher education degrees has been stagnant and strongly biased towards social sciences field.

According to our understanding, such behavior is quite related closely to the fact that education, from a macro level perspective, is not taken to consideration within domestic priorities. This problem is part of a group of fallacies undertaken by the central government for several years. For instance, the preliminary release of the National Strategic Plan for the Development of Science, Technology and Innovation (SENACYT, 2002) took more than two governmental periods. The length of time required to release the preliminary plans does not mean the plan is till in an infant stage or ready to be fully released. In fact, the authors were lucky to interview Professor Humberto Alvarez from Universidad Tecnológica de Panama, who was part of the commission in charge of the aforementioned document, and was initially appointed firstly to share his view of the proceedings of the document in mention. The overall message is that the plan lacks also from effective communication and also suffers from the diffusion within the Panamanian society. For these reasons, the plan has not yet taken off.

Times goes by and despite the buoyancy indicators of the UNDP and World Bank, it is time to put into practice a strategic national plan for education development. Currently, Panama's human capital also lacks of a goal. Moreover if we make comparisons with the opinions of the CRP assembly, we find the resulting opinions are not different but do add one interesting fact, the human capital supply does not meet the demands of the market. Thus, initiatives should go first priority from the public sector which, in our understanding, lacks financial resources, which is reflected in the low levels of knowledgeable individuals that estimates revealed. Consequently, national initiative should focus on specific areas comprised of market and the domestic priorities to meet the demands of the present and the future.

Despite the HDI ranking places Panama as a country with high levels of human development country in Latin America and the Caribbean. However, primary education (rank 64) still has a room to improve compare to developed countries (WEF, 2009). Especially, Panama has experienced decreasing levels of achievements over the years in the number of degrees awarded (WEF, 2009). The proposed national human capital indexes are complementary to UNDP human development index which consists of life expectancy, educational attainment and adjusted real income (UNDP, 2005). In short, government programs to promote people to pursue S&T careers are needed, especially for postgraduate degrees.

Panama's geographic location has always played an important role over the centuries in the global, regional and national commerce; the enormous amounts of goods going through in the isthmus are reflected in the Canal's unique competence. This competence works jointly with the dollarized economy and the service sector. Such openness to the world markets makes it easier to acquire international goods and for ideas to grow faster. Panama has not escaped from that reality, as international experience has shown that both trade in goods, services and FDI have provided important movements of people as the principal channels for acquiring imported knowledge (CEPAL, 2006).

According to the World Bank (1999), national exports expose firms to global benchmarks of quality and design. In other words, they allow firms to realize the scale of economies by expanding production beyond what is possible in the domestic market. These trends show how nations induce efficiency to compete in the global marketplace through an export orientation. Our results are consistent with WFE (2009), that Panama has created super national market capital in terms of sector, trade and international connectivity. Moreover, through their commercial partners and FTAs (free trade agreements), nations can also achieve such good development. Panama has welcomed to enforce its position and ensure its role as a service provider. Enabling a positive atmosphere of collaboration for the development within a country and across borders creates important social networks and market institutions. Apparently, Panama has been attached to market capital since the discovery of the isthmus in 1501 by Rodrigo Galván de Bastidas (History of Panama, 2002). Therefore, the Panama Canal resembles a geographical position that is attached to a great nation asset, The Panama Canal.

For the national market capital, Panama has a strong service-based, declining manufacturing and active agriculture economy. Due to location advantage, the Panama Canal service and Colon Free Zone are the main exports of services in the Panamanian economy. Panama has built many FTAs not only with their neighbor countries in North America, Central America, and some Latin America but also with distant Asian countries. Three future challenges ahead to enhance the Panamanian national market capital. One is maintaining a good balance among three economic development blocks: services, manufacturing and agriculture. The declining manufacturing sectors highlights that high-tech manufacturing sectors such as biotechnology, ICTs and advanced materials sectors should be promoted to adjust high dependence on services in the Panamanian market capital. Secondly, a slowdown of canal service and decline of FDI in Colon Free Zone signalized how to sustain their canal service and investment services (FDIs) becomes imperative for the future market capital. Some service innovations (e.g., Panama Canal expansion) and tax reforms are the key to sustain national market capital in canal services and free trade zone. Finally, the enlargement of FTAs with emerging economies in Latin America, such as Brazil is a good starting to extend their international market capital.

Panama shows a weak national government R&D intensity accounting only for 0.34 percent of the GDP in 2003 comparing 0.9 of GDP in Taiwan. In collecting national R&D expenditure in Panama, the national business R&D expenditure was not available. It is urged that Panamanian government should annually annual expenditure on R&D by following the indicators of *OECD Science, Technology and Innovation Outlook*. By doing so, it is allow to monitor the changes of the Panamanian S&T systems and to benchmark other countries.

The pattern of patent application and patents grants dominated by MNEs signify a great threat of the Panamanian innovation capital development. The resulted also confirmed by WEF (2009) report that Panama lagged in innovation (rank 80). We understand that Panama's economy is strong service-based and service innovations are not used to patents to protect their innovation. However, national innovation development which is heavily reliant on MNEs, can weak profitability and indigenous innovative capabilities development of local firms. We urge that Panama government can create R&D programs to support local firm's R&D activities and cultivate their innovative capabilities as soon as they can. Government also can create a great environment to facilitate innovation capital by claiming a strong intellectual property right regime. Without doing so, the innovation dependence on MNEs can not be alleviated.

It was a counter-global trend that the Panamanian innovation capital had a decrease of government S&T intensity in the developing countries in the period of 2001-2003. This could possibly erode the national innovation capital in a long run. How government demonstrated their commitment on S&T activities will be a decisive role to sustain the Panamanian long-term competitive advantage and economic growth.

To compete in the new global knowledge-based economy, developing countries must see the development and effective use of their information infrastructure as a key objective for national process capital intent. In Panama, development of an information society and the spread of the ICTs (information and communication technologies) have penetrated increasingly into businesses and households, specifically in Panama City and Colon City, as both cities are at the end points of the Panama Canal. The international experience shows that critical factors to foster greater economies include broad access to the Internet through low-cost, high-quality connections, high computer literacy and comfort in using the Internet, and useful content and innovative Internet applications to create value (Mustafa, 2003). On the other hand, the Panamanian estimates reveal that in general, ICTs not completely penetrated the nation which has in return left gaps. This phenomenon may be due to abnormal income distribution within Panamanian society.

Panama has well expanded mobile phone services and internet services. Maintaining a good internet diffusion policy to facilitate internet technology adoption in households, business and government is crucial to embrace the era of information society. Moreover, the duty of the regulators provided an enduring regulatory strategy for cooperative behavior among the actors. However, the actors seem to be sailing in different directions from what the regulators have stipulated. Thus the regulators should take a permissive approach to create an innovative and free competition environment for process capital.

CONCLUSION

This chapter proposed a refined national intellectual capital framework to diagnosing the national intellectual capital in Panama. Four dimensions of national IC: national human capital, national market capital, national innovation capital, and national process capital were investigated. Traditionally, the existing S&T indicators can be spread anywhere. There are three major contributions in the chapter. One is to propose a new national intellectual capital framework that is a refined the model based on Bontis (2004). Another is a holistic approach. This framework is allowed to synthesize the existing national S&T activities in a holistic, intellectual view and provide many potential S&T policy insights. The other is to identify the problems and difficulties in adopting national intellectual capital framework, especially in a developing country context.

Through the investigation of Panamanian intellectual capital, the chapter identified the characteristics of Panamanian intellectual capital. The chapter is complementary to WEF (2009) global competitiveness report, which allows examine national resources, technologies and knowledge from intangible perspective. We find that Panama has a service-dominant market capital specializing in canal services, financial, and business services. Panama has steadily increased process capital or ICT readiness. This increasing ICT penetration will enable higher data, information and knowledge transaction efficiency. However, the weaknesses of Panama's IC are decreasing supply of human capital in S&T fields, weak S&T commitment and lack of indigenous innovation capital.

The chapter suggests that Panama can create a sound national IC policy, in terms of human capital, market capital, process capital and innovation capital, respectively. It is crucial to increase labour market efficiency by developing market-oriented labor market. Panama can subsidize some S&T-oriented postgraduate programs in order to adjust its shortage supply of national human capital in S&T field. Panama has created a strong national market capital, especially in services and trade. However, a declining manufacturing sector signifies the importance of developing a balancing service-manufacturing-agriculture sector portfolio as a strategic imperative. Panama's innovation was dominant by MNEs, it is crucial for government to provide patient finance and support for domestic R&D and innovative activities. Otherwise, indigenous innovation will be never realized. Panama has gradually developed national process capital in communication and computer infrastructure. However, it is a must to maintain a good diffusion-oriented policy to continuously upgrade national process capital. For those nations who want to develop a national IC strategy, it is impossible that there are no statistics of national

IC or S&T available, specially, in some developing countries. We hope that this chapter motives policy makers to take measuring, monitoring and reporting national innovation capital more serious than before.

ACKNOWLEDGMENT

The earlier draft of the chapter was presented in the *European Conference of Intellectual Capital*, IN-Holland University of Applied Sciences, Haarlem, the Netherlands, 28-29, April, 2009. The authors would like to thank two anonymous reviewers for their insightful comments on this chapter. The financial support from National Science Council, Taiwan (Project No.: NSC95-2416-H-007-002) and National Tsing Hua University, Taiwan (Project No.: 98N-2925E1) is gratefully acknowledged.

REFERENCES

Al-Ali, N. (2003). *Comprehensive intellectual capital management: Step-by-step*. Mahwah, NJ: John Wiley & Sons.

Andriessen, D. (2004). *Making sense of intellectual capital: Designing a method for the valuation of intangibles*. Boston: Butterworth-Heinemann.

Barney, J. (1991). Firms, resources, and sustained competitive advantage. *Journal of Management*, *17*(1), 99–120. doi:10.1177/014920639101700108

Bontis, N. (2000). Intellectual capital and business performance in Malaysian industries. *Journal of Intellectual Capital*, *1*(1), 85–100. doi:10.1108/14691930010324188

Bontis, N. (2004). National intellectual capital index: a Uunited Nnations initiative for the Arab region. *Journal of Intellectual Capital*, *5*(1), 13–39. doi:10.1108/14691930410512905

Bornemann, M., Knapp, A., Schneider, U., & Sixl, K. I. (1999, June). *Holistic Measurement of Intellectual Capital*. Paper presented at the International Symposium Measuring and Reporting Intellectual Capital: Experiences, Issues, and Prospects, Amsterdam, Netherlands.

Bounfour, A., & Edvinsson, L. (2005). *Intellectual capital for communities: Nations, regions and cites*. Burlington, MA: Butterworth Heinemenn.

CEPAL. (2005a). *Preliminary overview of the economies of Latin America and the Caribbean 2005*. Santiago, Chile: CEPAL.

CEPAL. (2005b). *La inversión extranjera en América Latina y el Caribe*. Santiago, Chile: CEPAL.

CEPAL. (2006). *América Latina y el Caribe: proyecciones 2006-2007*. Santiago, Chile: CEPAL.

CIA. (2006). *The world factbook*. Washington, DC: Central Intelligence Agency.

Comptroller General of the Republic. (2005). *Panama in numbers: Years 2000-2004*. Panama: General Comptroller's of the Republic.

Dahlman, C., Routti, J., & Ylä-Anttila, P. (2005). *Finland as a knowledge economy: Elements of success and lessons learned*. Washington, DC: World Bank.

Edvinsson, L., & Malone, M. (1997). *Intellectual capital: Realizing your company's true value by finding its hidden brainpower*. New York: Harper Business.

Flamholtz, E. (1985). *Human research accounting*. Los Angeles: Jossey-Bass.

Franco, H. (2003). *Private universities*. Panama: UNESCO.

Freeman, C. (1987). *Technology policy and economic performance: lessons from Japan*. London: Pinter.

General Comptroller's of the Republic. (2005). *Panama in numbers: Years 2000-2004*. Panama: General Comptroller's of the Republic.

Grant, R. M. (1991). The resource-based theory of competitive advantage: implications for strategy formulation. *California Management Review*, (Spring): 114–135.

Johanson, U. (1999). Why the concept of human resource costing and accounting do not work. *Personnel Review*, *28*(1-2), 91–107. doi:10.1108/00483489910249018

Johanson, U., Martensson, M., & Skoog, M. (1999, June). *Measuring and Managing Intangibles: Eleven Swedish Qualitative Exploratory Case Studies*. Paper presented at the International Symposium on Measuring and Reporting Intellectual, Amsterdam, Netherlands.

Kaplan, R., & Norton, D. (1992). The balanced scorecard: measures that drive performance. *Harvard Business Review*, (January-February): 71–79.

Kaufmann, L., & Schneider, Y. (2004). Intangibles: A synthesis of current research. *Journal of Intellectual Capital*, *5*(3), 366–388. doi:10.1108/14691930410550354

Leonard-Barton, D. (1995). *Wellsprings of knowledge: Building and sustaining the sources of innovation*. Boston: Harvard Business School Press.

Lundvall, B.-Å. (Ed.). (1992). *National systems of innovation: Towards a theory of innovation and interactive learning*. London: Pinter.

Ministry of Economy and Finance. (2005). *Economic statistics*. Panama: Ministry of Economy and Finance.

Mouritsen, J., Larsen, H., & Bukh, P. (2002). Understanding intellectual capital statements: designing and communicating knowledge management strategies. In N. Bontis (Ed.), *World Congress on intellectual capital readings*, (pp. 179-202). Boston: Butterworth-Heinemann.

Mustafa, M. (2003). Internet Access. *World Bank Public Policy for the Private Sector*, 256.

Nonaka, I., & Takeuchi, H. (1995). *The knowledge-creating company: How Japanese companies create the dynamics of innovation*. New York: Oxford University Press.

OECD. (1981). *Proposed standard, practice for surveys of research and experimental development- Frascati Manual*. Paris: OECD.

OECD. (1992). *OECD proposed guidelines for collecting and interpreting technological innovation data- Oslo manual*. Paris: OECD.

OECD. (1997). *OECD proposed guidelines for collecting and interpreting technological innovation data- Oslo manual*. Paris: OECD.

OECD. (2008a). *OECD ICT (Information and Communication Technology) outlook*. Paris: OECD.

OECD. (2008b). *OECD STI (Science, Technology and Innovation) outlook*. Paris: OECD.

Ordóñez de Pablos, P. (2003). Intellectual capital reporting in Spain: a comparative view. *Journal of Intellectual Capital*, *4*(1), 61–81. doi:10.1108/14691930310455397

Petty, R., & Guthrie, J. (2000). Intellectual capital literature review: measurement, reporting and management. *Journal of Intellectual Capital*, *1*(2), 155–176. doi:10.1108/14691930010348731

RICYT. (2005). The source for this adaptation is on: http://www.ricyt.org/interior/interior.asp?Nivel1=1&Nivel2=1&Idioma

SENACYT. (2002). *National Strategic Plan for the Development of Science, Technology and Innovation. Unpublished preliminary proposal*. Panama: SENACYT.

SENACYT. (2004). *Science and Technology Activities Report. Unpublished preliminary report*. Panama: SENACYT.

Sirilli, G., & Evangelista, R. (1998). Technological innovation in services and manufacturing: results from Italian surveys. *Research Policy*, *27*, 881–899. doi:10.1016/S0048-7333(98)00084-5

Sveiby, K.-E. (1997). *The new organizational wealth: Managing and measuring knowledge-based assets*. San Francisco: Berrett-Koehler.

UNDP United Nations Development Program. (2005). *World development report 2005: International cooperation at a crossroads, aid, trade, and security in an unequal world. United Nations Development Program*. New York: Hoechstetter Printing Co.

WEF. (2009). *The global competitiveness report 2009-2010*. Geneva: World Economic Forum.

Wenerfelt, B. (1989). From critical resources to corporate strategy. *Journal of General Management*, *14*(3), 4–12.

WIPO. (2004). *WIPO intellectual property handbook: Policy, law and use*. Geneva: World Intellectual Property Organization.

World Bank. (1999). *Knowledge for development*. New York: Oxford University Press.

World Bank. (2004). *A better investment climate for everyone. World Development Report*. Washington, DC: World Bank and Oxford University Press.

APPENDIX

Table 12. Acronyms

Acronym	Explanation
CRP	CRP stands for "Consejo de Rectores de Panama" [University President Council of Panama, UPCP]
CEPAL	Comisión Económica para América Latina y el Caribe [Economic Commission for Latin America and the Caribbean, ECLAC]
CFZ	Colon Free Zone
ERSP	Ente Regulador de los Servicios Públicos [Public Utilities Regulating Entity]
FDI	Foreign Direct Investment
FTA	Free trade agreements
FTE	Full Time Equivalent
GDP	Gross Domestic Production
HD	Head Count
HDI	Human Capital Index
IC	Intellectual Capital
ICT	Information and Communication Technology
NSI	National System of Innovation
PPO	Panama Patent Office
R&D	Research and Development
RBV	Resource-based view
RICYT	Red de Indicadores de Ciencia y Tecnología [Iberoamerican Network of Science and Technology]
SENACYT	Secretaría Nacional de Ciencia, Tecnología e Innovación [National Science, Technology and Innovation Secretariat]
STI	Science, Technology and Innovation
STRI	Smithsonian Tropical Research Institute
UNDP	United Nations Development Program
UNESCO	United Nations Educational, Scientific & Cultural Organization
USAID	United States Agency for International Development
VAT	Value-added Tax
WIPO	World Intellectual Property Organization
WSIS	World Summit on the Information Society

Chapter 4
A Study on the Relations Between Organizational Identity and Intellectual Capital:
Empirical Evidence in New Technology Based Firms at Madrid Science Park

Eduardo Bueno Campos
Universidad Autónoma de Madrid, Spain

Mónica Longo Somoza
Universidad Complutense de Madrid, Spain

M. Paz Salmador
Universidad Autónoma de Madrid, Spain

ABSTRACT

This chapter studies the relations between the emergent concept of Organizational Identity and the concept of Intellectual Capital. Specifically, the chapter focuses on the idea that when new technology-based firms develop the social process of construction of their Organizational Identity, at the same time they develop the social interaction needed to define variables of their Organizational Capital. This proposition is grounded in a theoretical review of the concept of Organizational Identity and Organizational Capital in the Intellectus Model, and it is empirically tested in five case studies of new technology-based firms created at Madrid Science Park. Finally, the limitations and suggestions for future research as well as conclusions of the study are presented.

INTRODUCTION

Nowadays knowledge has become a key asset for firms to manage in order to gain a sustainable competitive advantage. The organization's environment changes quickly and knowledge is a critical factor to take advantage of the opportunities of these changes. Authors such as Conner and Prahalad (1996), Grant (1996), Kogut and Zander (1996) and Spender (1996), among others, have developed the Knowledge-based Theory of the firm remarking the relevance of knowledge as a

DOI: 10.4018/978-1-61520-875-3.ch004

resource and the view of the firm as an entity that stores knowledge. In this current knowledge-based economy, new technology-based firms (NTBFs) have a relevant role. In this chapter we analyze some of the dynamics developed by these firms. We focus on the process developed by the NTBFs created at Madrid Science Park to construct their Organizational Identity and how, at the same time, they construct their Intellectual Capital, and more specifically their Organizational Capital. In the area of the Organizational Identity, this concept has been studied as a key factor in the process of strategic change (Glynn, 2000; Nag, Corley & Gioia, 2007; Scott & Lane, 2000). However, the relation between the process of construction of the identity of a recent founded firm and its Intellectual Capital has not been explored enough yet.

When people commit themselves with organizations and contribute with their knowledge, organizations acquire this knowledge which can become technology if it is developed and transmitted. Therefore, individual knowledge can be transformed into social or collective knowledge and shared by the members of an organization when transferred through oral or written language (Argyris and Schön, 1978; Bueno, 2005; Cook and Brown, 1999; De Geus, 1997; Quinn, 1992; Spender, 1996; Von Krogh and Roos, 1995). Following the autopoiesis theory (Maturana and Valera, 1984), two prerequisites are required to create and develop organizational knowledge: People need to interact to allow communication; and a self-description of the organization must exist. Albert and Whetten (1985) proposed the first definition of the term "Organizational Identity", proposing that it is a self-reflective question (Who are we as an organization?) that captures central, enduring and distinctive features of the organization. From an interpretive paradigm, Organizational Identity is constructed by members of an organization through the mentioned interaction in order to agree a set of meanings about "who they are as organization", that is, they develop a social process of self-description which constructs their Organizational Identity (Gioia, 1998). Hence, the development of organizational knowledge and the construction of Organizational Identity share two core elements: interaction to allow communication among members; and a self-description of the organization. Besides, when members express the central characteristics of their organization, they are creating filters to select what knowledge is and what it is not, and they are also creating guidelines to coordinate opinions, practices, decisions, acts, and strategies (Brown and Humphreys, 2006; Bürgi, Roos y Oliver 2002; Bürgi and Oliver 2005; Dutton and Dukerich, 1991; Gioia and Thomas, 1996; Nag, Corley and Gioia, 2007; Reger et al., 1994).

In this chapter, we propose that during the social process of construction of Organizational Identity the NTBFs also construct their Organizational Capital. First, in the section "Background" we set up the theoretical propositions about Organizational Capital and Organizational Identity. We review the concept of Intellectual Capital, some models to measure and manage Intellectual Capital, and its variables from the point of view of the Intellectus Model. We also review the concept of Organizational Identity and its paradigms and research approaches as point of reference to choose the proper one for our empirical research. Next, in the section "Relations between Organizational Capital and Organizational Identity", we empirically test our proposition and explain our findings in five NTBFs created at Madrid Science Park. We take an interpretive paradigm and a shared meaning research approach and use a case study methodology. Later, we discuss the limitations of our research and future research directions. Finally, we provide a discussion of the chapter in the section "Conclusion".

The five new technology-based firms created at Madrid Science Park are small and micro firms in a process of growth. We choose these firms to test our proposals empirically because they have been recently founded and asked for technical assistance in order to understand "who they are" as well as to develop successful ways of work in their critical first years. These firms are based on the exploitation of an invention or technological innovation, employ a high proportion of qualified employees, and use the Intellectus Model to measure and manage their intangible assets. In consequence, they were of great interest for our empirical analysis. We follow the definitions of NTBF proposed by Butchart (1987)Butchart, R., 1987., A new UK definition of high technology industries. Econ. Rev., pp. 82–88and Shearman and Burrel (1988) and the definition of small and micro firms adopted by the European Commission in 2003. These definitions are stated in the section "Research approach and methods".

BACKGROUND

Intellectual Capital: Organizational Capital in the Intellectus Model

In the last decade of the 20th century, a great interest in knowledge management in the Strategic Management area emerged as a way of levering knowledge as a strategic resource for the firm (Bueno, Morcillo & Salmador, 2006; Teece, 2000). This perspective is developed by the Knowledge-based Theory of the firm, as it was remarked in the 'Introduction' section (Conner & Prahalad 1996, Grant, 1996; Kogut & Zander 1996; Spender, 1996). Nowadays traditional tangible assets continue being important to produce goods and services. Nevertheless, knowledge has become a key asset to manage in order to gain a sustainable competitive advantage (Boulton, Libert, & Samek, 2000; Lev, 2001; Low, 2000) and wealth creation (Edvinsson & Malone, 1997; Stewart, 1997). Firm's environment changes quickly and, in this context, knowledge turns into a key resource to take advantage of the opportunities that changes may bring. Following McDermott (1999), the human act of knowing belongs to communities. People learn by participating in communities where knowledge circulates in many ways. It circulates through articles or written procedures, and also through unwritten artifacts such as stories, specialized language, and common wisdom about cause-effect relationships. People observe and discuss for example informal work routines and doing so they exchange their experience, make sense of the information and share and use their knowledge. Levering and managing knowledge involves getting people together in order they share insights they do not know they have. So, strategies to manage knowledge must identify the community that cares about a topic, enhance their ability to think together, stay in touch with each other, and connect with other communities. Through this social process of interaction and communication members of the community creates knowledge through a conversion process where tacit and explicit knowledge (Polanyi, 1969) expands. Nonaka (1994) and Nonaka &Takeuchi (1995) called this process SECI which is composed of four modes of knowledge creation: socialization (from tacit knowledge to tacit knowledge); externalisation (from tacit to explicit knowledge); combination (from explicit knowledge to explicit knowledge); internalisation (from explicit to tacit knowledge).

In Economy the study of knowledge as a source of wealth creation and the ability of immaterial assets to increase the productivity of traditional resources of production was introduced in the 19th century by Marshall (1890) (Bueno, 2002a) and developed during the 20th century by thinkers such as Knight (1921), Hayek (1945), Drucker (1965) and Machlup (1980). Nowadays, the critical role of knowledge has been contrasted by the influence of intangible activities carried out by the firms that generate intangible assets based on knowledge. These assets have been recognized by the market and had brought the concept of Intellectual Capital (Bueno, 2005, Bueno, Morcillo & Salmador, 2006).

The concern for measuring and managing intangible assets appears in the end of the 20th century and the first holistic Intellectual Capital model to do it in a systematic way was the Skandia Navigator (Edvinsson & Sullivan, 1996; Edvinsson & Malone, 1997). Some Other basic models to measure and manage knowledge were Technology Broker (Brooking, 1996), Western Ontario Universy Model (Bontis, 1996), Canadian Imperial Bank Model (Saint Onge, 1996), Intangible Assets Monitor (Sveiby, 1997), Modelo Nova (Camisón, Palacios y Devece, 1998), Stewart Model (Stewart, 1997) Balanced Scorecard (Kaplan & Norton, 1992) and Intellectus Model (CIC, 2002).

Definitions of Intellectual Capital (IC) were proposed in 1990s by authors such as Bueno (1998), Collins et al. (1997), Edvinsson & Malone (1997), Lev (1998), Lev & Zarowin (1998), Roos et al. (1998), Stewart (1997) and Sveiby (1997). IC is generally defined as the intellectual material that can be put to use to create wealth. It includes organization's processes, technologies, patents, employees' skills and information about customers, suppliers and stakeholders (Stewart, 1997). The IC categories differ slightly among researchers and Intellectual Capital Models (Kaufmann & Schneider, 2004), however internationally it is accepted three basic dimensions: Human, Relational and Structural Capital (CIC, 2002). Human Capital is concerned with the accumulated value or wealth generated by the values, knowledge and abilities of people (Human Intelligence) and it represents the stock of knowledge within an organization rather than in the minds of individual employees (Bontis, Crossan, & Hulland, 2002). Structural Capital expresses the accumulated value or wealth generated by the value of the existing knowledge which is property of the organization that generates its knowledge base. This knowledge is the combination of shared values, culture, routines, protocols, procedures, systems, technological developments and intellectual property of an organization which make up the collective know how and which remain in the entity whether people leave (Organizational Intelligence). Relational Capital expresses the accumulated value or wealth generated by the value of the knowledge which comes to the organization through the relationships and actions shared with external or social agents (Social and Competitive Intelligence) and it refers to customers, social capital, and stakeholders (Bukh, 2002; Bueno, Salmador & Rodriguez, 2004; Johanson, Martensson & Skoog, 2001; Ordoñez de Pablos, 2003; Sanchez, Chaminade & Olea, 2000; Stewart, 1997).

Within Relational Capital it is necessary to distinguish between Business Capital and Social Capital. The former is directly related to the agents linked to the business process, and the latter is connected with the remaining agents (Bueno, 2002b; Cohen & Prusak, 2001; Coleman, 1988; McElroy, 2001; Nahapiet & Goshal, 1998). Also, within Structural Capital it is necessary to distinguish between Organizational Capital which is a combination of intangibles that structure and develop the organizational activity of the organization, and Technological Capital which is a combination of intangibles directly linked to the development of activities and functions of the technical system of the organization's operations which are responsible for obtaining products, developing efficient production processes and advancing the knowledge base necessary for future innovations in products and processes. These two structures of Relational Capital and Structural Capital were incorporated to the Intellectus Model for the measurement and management of intangible assets (CIC, 2002).

Focusing our attention on the Organizational Capital, relevant to test the research problem of this chapter, the Intellectus Model proposes the following elements or homogenous groups of intangible assets to be measured and managed: Culture, Structure, Organizational Learning and

Table 1. The main paradigms on "organizational identity" research and denominations

DENOMINATION (Gioia, 1998)					
Functionalist		Interpretive		Postmodern	
ALTERNATIVE DENOMINATIONS & REFERENCES TO BE CONSULTED					
Realist	Cornelissen (2002b); Gioia, Schulz & Corley, (2002a)	Social constructionist	Corley et al. (2006); Bürgi, Roos & Oliver (2002); Bürgi & Oliver (2005); Golden-Biddle & Rao (1997)	Postmodern	Corley et al. (2006); Brown, Humphreys & Gurney (2005); Bürgi, Roos & Oliver (2002)
Normative/Positivist	Porter (2001)	Constructionist	Cornelissen (2002b); Gioia, Schulz & Corley, (2002a)		
		Qualitative	Porter (2001)		

Source: Own elaboration

Processes. These groups or elements are (CIC, 2002):

- **Culture:** A combination of values, norms and ways of behaviour which are assumed and shared by the majority of the members of the organization and condition the organization behaviour and corporate results.
- Structure: Modes and processes of formal organization.
- **Organizational learning:** The capacity of the organization to acquire new competencies and knowledge with the aim of responding to answer to the dynamics of organizational development and change.
- **Processes:** Combination of activities which make up the organisational operations directed to internal and external customers and suppliers.

In the Intellectus Model the four aforementioned groups are made of the following variables or intangible assets (CIC, 2002):

- **Culture:** Cultural homogeneity, evolution of cultural values, social climate and business philosophy.
- **Structure:** Organisational design and organisational development.
- **Organizational learning:** Learning environments, organisational guidelines, capture and transmission of knowledge and creation and development of knowledge.
- **Processes:** Directed towards the internal customers, directed towards the external customers and directed towards the suppliers.

Organizational Identity: Paradigms and Research Approaches

The original definition of "Organizational Identity" states that it represents a self-reflective question (Who are we as an organization?) which captures central, enduring and distinctive features of the organization (Albert & Whetten, 1985). Since this first conceptualization, different research approaches and definitions of this concept have been presented but there is not consensus about the paradigms[1] to conceptualize it and its main features (Albert, Ashforth & Dutton, 2000; Bouchikhi et al., 1998; Corley, et al., 2006, Cornelissen, 2006; Gioia, 1998). Gioia (1998) proposes three main lenses that emerge from differing paradigmatic assumptions about the nature of organizations when theorists and researchers analyse Organizational Identity: functionalist, interpretive and postmodern. When reviewing the literature about

Organizational Identity others names to these three paradigms can be found. Table 1 shows the main different proposals:

As it can be observed in Table 1, choosing Gioia's proposal as the main one, functionalist paradigm is also named realist or normative/ positivist (Cornelissen, 2002b; Gioia, Schultz & Corley, 2002a; Porter, 2001) and interpretive paradigm is named social constructionist, constructionist[2] or qualitative paradigm (Bürgi, Roos & Oliver, 2002; Bürgi & Oliver, 2005; Corley et al., 2006; Cornelissen, 2002b; Gioia, Schulz & Corley, 2002a; Golden-Biddle & Rao, 1997; Porter, 2001). In this chapter, we have chosen Gioia's proposal to name the paradigms after having reviewed Berger & Luckman (1968), Cooper & Burrel (1988), Gioia (1998), Gioia & Pitre (1990), Knorr-Cetina, (1981), Rabinow & Sullivan, (1979), Ritzer (2003), Schultz & Hatch (1996), Schwandt (1998, 2000) and Thiétart (2001).

From an empirical perspective and in relation with the concept of Organizational Identity, first empirical researches were carried out by Dutton & Dukerich and by Gioia & Chittipeddi in 1991 (Gioia, Schultz & Corley, 2002a). Analysing empirical research on Organizational Identity since that date, several research approaches can be identified that are framed in these three paradigms. It is important to disclose these research approaches in order to develop an empirical investigation about Organizational Identity because they can offer a point of reference to orientate our research about the construction and emergence of Organizational Identity in new technology-based firms at the Science Park in Madrid. Following Corley et al (2006), we propose to set a reference point that may guide future empirical work on Organizational Identity and to make explicit the research approach chosen. So, our objective in this section of the chapter is to choose the right Organizational Identity approach to develop our empirical work. To do it we are going to describe briefly the proposal of the main current research approaches in the field of Organizational Identity stated by Bueno, Salmador & Longo (2008).

In the mentioned proposal, each research approach is described through its particular definition of Organizational Identity, main research objectives, the data to analyse and the research methods. The proposal is based on the review of twenty-three main empirical investigations about the Organizational Identity of different kind of organizations (universities, port organizations, orchestras, non profit organizations, big public accounting firms, high-technology R&D organization, etc.) from 1991 to 2007. As mentioned above, although the main conceptual statement of Organizational Identity was made in 1985 by Albert & Whetten it was not until 1991 when the first empirical investigations took place. This is the reason why the review of papers to make this proposal starts in 1991.

According Bouchikhi et al. (1998) and Gioia (1998) and in order to frame each approach in one of the three paradigms, the proposal begins summarizing these paradigms by explaining the definition of Organizational Identity, core features of this concept, research problem and data to analyse. Later, following Bueno, Salmador & Longo (2008), we review the main empirical investigations in the field of Organizational Identity to frame them in a paradigm, analysing elements which usually underlie either theoretical explanations or research methodology: definition and core features of Organizational Identity, objective of the research and methodology used (data and instruments). These elements lead to frame each paper in a paradigm or even some of them in a multiparadigmatic approach. Five research approaches emerge from this analysis. Next, Table 2 and 3 summarizes the main characteristics of the three paradigms and the five research approaches.

Table 2 shows the main characteristics of each paradigm. Functionalist paradigm is realist and objectivist and it is based on positivism. Organizational Identity is something that an organization has, a product of the members' beliefs about "who

Table 2. Main paradigms on Organizational Identity (OI) research

	PARADIGMS		
	FUNCTIONALIST	INTERPRETIVE	POSTMODERN
	OI DEFINITION		
OI DEFINITION	Beliefs about "who we are" that have been institutionalized and taken-for-granted. Identity is a product of these beliefs that manifest themselves in objective characteristics	The continuously renegotiated set of meanings about "who we are" shared by members of organization. Identity is a socially-constructed process	Momentary and fragmented statements about who we take ourselves to be that may be contradictory. Identity is an invention of the power groups in order to get their aims
ORGANIZATION FEATURES WHERE OI RESIDES			
CENTRAL	Core values and beliefs of the organization	Meanings that members agree are central	Loyalties and claims typical of social interactions are impermanent so there cannot be either central essence or enduring characteristics in an organization. Identity is continuously deconstructed and reconstructed
ENDURING	What is hard to change but not immutable	What will remain while the social context affirms the projected identity. Identity is in flux, there is a continuous process of construction and reconstruction	
DISTINCTIVE	¿Which are the sources of distinctiveness, where they exist and how can they be manage?	They derive from a comparison process that members of the organization make about what is different and what is similar to other organizations	There are fragmented, multiple identities and distinctiveness of one identity is defined with respect to the others
	CENTRAL PROBLEM TO STUDY		
CENTRAL PROBLEM TO STUDY	To analyse how Organizational Identity shapes cognitions of organizational members and how these cognitions in turn shapes members individual and organizational actions	To discover and disclose the meanings about "who are we?" that are negotiated among organizational members because these meanings can facilitate or limit organization management	To problematize the notion of "Organizational Identity" which mean to accept multiple identities and analyse claims of identities and existing power relations
	DATA TO ANALYSE		
DATA TO ANALYSE	Objective characteristics of organizations as for example demography data, patterns of decisions or statements of managers.	Symbols (specific language of the organization, mannerisms, artefacts, etc), cognitive schema or worldview, interpretive schemes that members use to make sense of those symbols, interpretive schemes that researchers use to make sense of what they see	Language and discourse focusing on ruptures and absences in the discourse

Source: Own elaboration fromBouchikhi et al. (1998)andGioia (1998)

they are as organization" that have been institutionalized and taken-for-granted and manifest themselves in objective characteristics to be analysed by an impartial researcher. Interpretive paradigm is subjectivist and rooted in the hermeneutic tradition. The Organizational Identity is socially constructed by the members of an organization through a continuous process of negotiation of meanings about "who they are as organization". These meanings shared by members must be studied analysing symbols such as organization language, cognitive schemes and also interpretive schemes that members use to make sense of those symbols. Researchers that investigate Organizational Identity from this point of view made an insightful explanation because the views of the organization's members are fundamental in understanding their meanings as a system. So these researchers approach organization and its members as close as possible in order

Table 3. Main paradigms, research approaches on Organizational Identity (OI), research objectives and empirical analysis

MAIN PARADIGMS	**Funcionalist**		**Interpretive**		**Postmodern**
RESEARCH APPROACH-ES	**Shared beliefs**	**Institutional theory**	**Shared meaning**	**Hybrid identity**	**Multiple identities**
OI DEFINITION	Characteristics of the organization (values, practices, etc) that are rooted in tacit shared beliefs about what is central, enduring and distinctive	Characteristics of the organization that members believe to be central, enduring and distinctive and give it a status as an actor in its social environment. This identity is constructed through a social process	Meanings shared by members about OI, constructed through a claim-making process that depends on the context and becomes a collective "cognitive frame" to sensemaking	Meanings shared about OI, constructed through a claim-making process that depends on the context and becomes a collective "cognitive frame". Social groups can develop distinct "Organizational Identity" dimensions resulting in a hybrid identity that can be controversial	Identities are unstable social constructions constituted through acts of language. Multiple OI emerge because different groups of power define the organization in particular ways according to their interests to impose their identity version. Other groups develop processes of resistance
RESEARCH OBJECTIVES	To uncover shared beliefs and features where identity of the organization resides and to manage changes in those features and beliefs to get changes in identity and avoid it be an internal barrier for planned organizational change	To analyse an organization's symbolic attributes that represent its identity comparing the resemblance of them to those of other organizations within its institutional field	To describe how OI is a key element in the processes of strategic change and its influence in actions taken by the members and in their sensemaking when the external context changes	To study how in an organization with a hybrid identity, conflicts among the dimensions of the OI and the legitimacy of one over the others filters core capabilities and strategic resources	To describe how the dominant group of the organization makes use of power/knowledge strategies (discursive techniques) to transmit employees its "Organizational Identity" version

Source: Own elaboration

to understand and describe adequately Organizational Identity as a socially-constructed process. Finally, postmodernism paradigm is subjectivist in extreme. As the interpretive paradigm, postmodernism proposes that identity is a socially-constructed process however from its point of view it is also an "illusion", an invention of the power groups of the organization in order to get their aims. The Organizational Identity is different for each power group, momentary, fragmented, made of contradictory statements about who they take themselves to be as an organization and used to set hegemony. It is continuously deconstructed and reconstructed. Postmodernism researchers think that e Organizational Identity is a social construction and that the source of social phenomenon is language, so the proper data to analyse are organization language and discourse and the absences and ruptures in them.

Tables 3 and 4 outline the relation between the general paradigms accepted in the study of Organizational Identity, disclosed in Table 2, and Bueno, Salmador & Longo (2008) proposal of research approaches. The last row in Table 4 discloses the papers or references reviewed on which the approaches presented are based. Each paper was classified in one or two paradigms after having analysing in them: definition and core features of

Table 4. Main paradigms, research approaches on Organizational Identity (OI), research objectives and empirical analysis

MAIN PARADIGMS	Funcionalist		Interpretive		Postmodern
RESEARCH APPROACHES	Shared beliefs	Institutional theory	Shared meaning	Hybrid identity	Multiple identities
DATA TO ANALYSE	Organization characteristics that represent cultural and economic values, practices and theories	Members perceptions, actions and strategies, and symbols	Interpretive schemes, symbols and rhetoric techniques	Interpretive schemes, symbols and rhetoric techniques	Rhetoric techniques
METHODS	Cognitive maps, repertory grid technique and self-discrepancy scales	Analytical narrative approach, graphical and statistical models	Case study, ethnographic study, and statistic analysis	Case study and ethnographic study	Case study and ethnographic study
REFERENCES REVIEWED	Cornelissen (2002a); Cornelissen (2002b); Fiol & Huff (1992); Reger et al. (1994)	Glynn & Abzug (2002); Rao, Monin & Durand, (2003); Whetten & Mackey (2002); Zuckerman et al. (2003)	Dutton & Dukerich (1991); Fiol (2002); Gioia & Chittipeddi (1991); Gioia, Schultz & Corley (2002a); Gioia, Schultz & Corley (2002b); Gioia & Thomas (1996); Nag, Corley & Gioia (2007); Scott & Lane (2000)	Bürgi & Oliver (2005); Bürgi, Roos & Oliver (2002); Glynn (2000); Golden-Biddle & Rao (1997)	Brown & Humphreys (2006); Brown, Humphreys & Gurney (2005); Covaleski et al. (1998)

Source: Own elaboration

Organizational Identity, objective of the research, data to analyse and methods. The denominations given to the approaches outline the key element of each one and were chosen to represent the definition of the Organizational Identity. Three are the research approaches framed in Gioia (1998) paradigms: shared beliefs approach is framed in functionalist paradigm, shared meaning approach in interpretive and multiple identities approach in postmodern. Two approaches are multiparadigmatic as they share features of two paradigms: Institutional theory approach share elements of the functionalist and interpretive paradigms, hybrid identity share elements of interpretive and postmodern.

RELATIONS BETWEEN ORGANIZATIONAL CAPITAL AND ORGANIZATIONAL IDENTITY

Research Problem

After having described how the Intellectus Model reflects Organizational Capital and having reviewing the main research approaches in the area of Organizational Identity, our next step is to specify the proposal of this chapter about the relations between both concepts.

As we present in the "Introduction" section, two are the main links between Organizational Identity and organizational knowledge:

1. The processes of construction of Organizational Identity and the development of organizational knowledge have two

elements in common: interaction to allow communication among members, and a self-description of the organization.

2. When members express the central characteristics of their organization that define their Organizational Identity, they create filters to select what knowledge is and what it is not.

Following these links, we propose that when the members of the NTBFs created at Madrid Science Park construct their Organizational Identity through a social process of claims and counterclaims, at the same time they are constructing the firm strategy and Organizational Capital. Also, when they create and develop organizational knowledge through interaction, they are constructing their Organizational identity. This relation between Organizational Capital and Organizational Identity has been understudied until this moment, however it is interesting to explore in order to help NTBFs to understand "who they are", define their strategy and set the bases of success in their initial years.

Research Approach and Methods

To empirically test the aforementioned research problem, we take an interpretive paradigm and shared meaning research approach. Following Whittington (2001), "the objectives and practices of strategy depend on the particular social system in which strategy-making takes place" (p. 4), so strategy and Organizational Capital have a social construction nature (Bueno, Morcillo & Salmador, 2006), and so does Organizational Identity.

We use a case study methodology particularly suitable for answering "how" and "why" questions (Yin, 1984) and that also enables to use "controlled opportunism" to respond flexibly to new discoveries made while collecting new data (Eisenhardt, 1989). Making this choice, we ensure that data collection and analysis meet tests of construct validity, reliability, and internal and external validity by carefully considering Yin's (1984) tactics. Construct validity is enhanced by using the multiple sources of evidence (interviews, observations and secondary data sources) and by establishing a chain of evidence as we concluded the interviews. Reliability was promoted by: (a) Using a case-study protocol in which all firms and all informants were subjects to the same entry and exit procedures and interview questions; (b) by creating similarly organized case data bases for each firm we visited. External validity was assured by the multiple-case research design itself, whereby all cases were NTBFs of Madrid Science Park. Finally, we addressed internal validity by the pattern-matching data-analysis method described in "Data Analysis Procedure" section.

The case study methodology provided a real-time study of this chapter research problem in the natural field setting by investigating five new technology-based firms created at Madrid Science Park. These five firms were of great interest for our empirical work because: (1) NTBFs play a relevant role in the current knowledge-based economy (Kauser, Oakey & During, 2005, 2004 & 2002); (2) These new five firms asked for assistance in order to set the best strategies, structure an procedures to warrant their success in their first years so they offered their collaboration in our research; (3) They employed a high proportion of qualified employees so when analyzing the elements of the Organizational Capital and the Organizational Identity it was easy to make them understand this last emergent concept and its possible relation with IC what made our work as researchers easier and fruitful; (4) These firms belong to different industries, what allowed us to treat this element as a *ceteris paribus* variable and to focus our attention on elements they share as NTBFs; (5) They use the Intellectus Model in order to set their strategy and manage their Intellectual Capital.

The term "new technology-based firms" (NTBFs) was coined by the Arthur D. Little Group (Little, 1977) stating that a NTBF was "an independently owned business established for not more than 25 years and based on the exploitation of an invention or technological innovation which implies substantial technological risks". Also, Butchart (1987) and Shearman & Burrell (1988) define this kind of firms. They focus on sectors which have higher than average expenditures on R&D as a proportion of sales or which employ proportionately more qualified scientists and engineers than other sectors. This last definition has been widely used however these authors call these firms "high tech SMEs" and distinguish them from NTBFs which are both newly established and independent. There are other interpretations about the term NTBFs as for example: Smilor et al. (1990) identifies this kind of firms with the Research-based spin-offs which commercialize an invention as resulted of an academic investigation; Bollinger et al. (1993) defines these firms as independent business establish by four or five people that produce their own technology and use processes, goods and services where the technology is new or innovative.

We follow Little (1977), Butchart (1987) and Shearman & Burrell (1988)'s definitions to develop the empirical research on NTBFs established at Madrid Science Park because these definitions approaches closer to the features of these firms than the other definitions. These firms have been established by a group of entrepreneurs, based on exploitation of an invention or technological innovation and employ a high proportion of qualified employees.

We take the European Commission definition of micro and small firms adopted in 2003 in the recommendation C (2003) 1422. A small firm is defined as "an enterprise which employs fewer than 50 persons and whose annual turnover and/or annual balance sheet total does not exceed EUR 10 million". A micro firm is defined as "an enterprise which employs less than 10 people and whose annual turnover and/or annual balance sheet total does not exceed EUR 2 million".

The sample of five NTBFs was not random however it reflects a representative selection of this kind of firms established at the Science Madrid Park that, as we mentioned in the "Introduction", are new research small and micro firms (see Table 5). Besides, as it was mentioned above, these five NTBFs have been developed the Intellectus Model in order to set their strategy and manage their Intellectual Capital so they are of great interest for our research. To determine our number of cases, we used theoretical saturation – that is, we stopped adding cases when our incremental learning diminished (Glaser and Strauss, 1967; Sutton and Callahan, 1987). Our decision was also guided by the range of minimum and maximum cases recommended (Eisenhardt, 1989). The comparison of case studies within the same industrial context enabled "analytic generalization" through the replication of results, either literally (when similar responses emerged) or theoretically (when contrary results emerged for predictable reasons) (Yin, 1984), thus ensuring that the evidence in one well-described setting was not wholly idiosyncratic (Miles & Huberman, 1984).

Although space prevents our providing "thick descriptions" of each case (McClintock, Brannon & Maynard-Moody, 1979), Table 5 makes a brief description of the firms studied at the time of our analysis. To maintain confidentiality the names of the firms are fictional names. Also, Table 5 presents the technical record of the cases studies showing the period and average durations of the interviews, the fictional selected names to each NTBF, the legal entity, their activity sector, date of establishment, the number of employees, informants, jobs and qualifications. As it is shown in this table, the firms that have taken part in the case studies were established between 2000 and 2007 as Limited Companies and belong to activity sectors based on the exploitation of an invention or technological innovation. These sectors are: Information, Technology & Communications, Biotechnology & Agroalimentation and Environment & Renewable Energies. They employ qualified people with a PhD, Master or Bachelor

Table 5. Study cases technical file

INTERVIEWS	DATE:					JUNE-AUGUST, 2008
	AVERAGE DURATION:					60 MINUTES
NTBF					INFORMANTS	
NAMES	LEGAL ENTITY	ACTIVITY SECTOR	ESTABLISH.	N. OF EMPLOYEES	JOBS	QUALIFICATIONS
NTBF A	Limited Company	Information, Technology & Communications	2005	19	President and Founder Shareholder	Bachelor Degree
					Sales & Marketing Department	Bachelor Degree
NTBF B	Limited Company	Biotechnology & Agroalimentation	2005	8	General Manager	Master Marketing & Management
					Sales & Marketing Department	Bachelor Degree
					Business Development Area	MBA
NTBF C	Limited Company	Environment & Renewable Energies	2004	5	Founder Shareholder and collaborator in the Development Area	Bachelor Degree
					Research & Development Manager	PhD
NTBF D	Limited Company	Information, Technology & Communications	2007	7	Promoter and in charge of the Organization and Consulting Services Area	PhD
					Promoter and in charge of the Technological Implementation Area	Bachelor Degree
NTBF E	Limited Company	Environment & Renewable Energies	2000	4	Founder Shareholder and Technical Manager	Bachelor Degree
					Taxonomic Identifications and Reports Area	Bachelor Degree

Source: Own elaboration

Degree and following the European Commission definition and they are micro and small firms as they have from 4 to 19 employees.

The data-collection process took place in the period June-August, 2008. We used several data-collection methods (Eisenhardt, 1989). We collected data through interviews, observations, and secondary sources. The underlying rationale is "triangulation", which it is possible by using multiple data sources providing stronger substantiation of constructs and propositions (Webb, Campbell, Schwartz & Serchrest, 1996).

INTERVIEWS

It was developed a case-study protocol in order to pursue reliability in the findings. A pilot study was carried out too to refine our data-collection plan with respect to both the content of the data and the procedures followed. The primary source of initial data collection came from semi-structured interviews with eleven informants which lasted sixty minutes on average per case. To obtain various points of view and to avoid slants these interviews were conducted with the General Manager and/or founder shareholder or promoter and one or two employees. Comparisons are made among people holding same positions in different firms, and people holding different positions in the same firm. They took the form of focused interviews that remained open-ended and assumed a conversational manner. We began the interviews by asking the respondents to take the role of spokesperson for the organization to focus on organizational level issues, following we explained the concept of Organizational Identity and that the aim of the interview was to study how this identity was being constructed and the relation of this construction with the construction of the Organizational Capital. In the first stage of the interviews we asked the respondents to describe his or her job in the firm and also open questions about the history of the firm, activity sector, structure, core characteristics, strengths or customers. In the second stage of the interview we focused on areas such as the feeling of being a community, ways of share knowledge, climate between members, business philosophy, share values, breaks during a day of work, the employees' features, the communications ways between them, departments or formal functions.

All the interviews were recorded and transcribed immediately afterward (Eisenhardt, 1989). In the transcriptions, we included all data, regardless of its apparent importance in the interview. We then checked facts and ended the transcription notes with our lingering impressions, trying to supplement the transcribed interviews and trying to sharpen them by asking ourselves questions (e.g.What did I learn? How does this interview compare to prior interviews?). We completed the interview notes and impressions within a day of the interview (Yin, 1984).

OBSERVATIONS AND SECONDARY SOURCES

We used secondary sources to collect background information about the cases. Such sources included annual reports, internal documents provided by the interviewees, agendas for meetings, minutes of past meetings, internal newsletters and intranets, industry reports, annual reports of Madrid Science Park, websites, and articles in magazines and newspapers about the situation and evolution of the industry in general and of the five NTBFs in particular. Also, along the visits to the five NTBFs, we kept a record of our impressions and observations we made when we participated in activities such as coffee breaks and lunches. Whenever possible, we attended meetings as passive note-takers. These observations provided real-time data. We used the secondary sources and data to supplement the data obtained from the interviews.

Data Analysis Procedure

To analyse the colleted data we have set the general analytic strategy called "relying on theoretical propositions" (Yin, 1984). To follow this strategy first we have described the theoretical propositions about the concepts of Organizational Capital and Organizational Identity in section "Background" and second, these theoretical propositions will be the guide to analyze the empirical evidence (see "Findings" section) to answer the research questions stated in the "Research problem" section. Also, we have followed the explanation-building data-analysis method, which is a special type of pattern-matching method. We have chosen this method to analyze data because it is a relevant procedure for explanatory case studies where casual links are in narrative form (Yin, 1984).

To sum up, the final explanation of the research problem of this multiple-case research is the result of: (1) The theoretical propositions initially established about Organizational Capital and Organizational Identity; (2) An iterative process of comparisons between these propositions and the findings; (3) A continuous revision of the propositions.

Findings

Considering the perspective of the interpretive paradigm and the shared meaning research approach, Organizational Identity is constructed through a social process of claims and counterclaims used by the members of the firm to agree what is core in the organization. This social process means continuous interactions between members and involves the use of symbols, interpretive schemes and language. So, through conversations, as a way of social interaction, members define who they are as an organization and agree core elements of the firm (Bürgi & Oliver 2005; Bürgi, Roos & Oliver, 2002, Dutton & Dukerich, 1991; Fiol, 2002; Gioia & Chittipeddi, 1991; Gioia, Schultz & Corley, 2002a; Gioia, Schultz & Corley, 2002b; Gioia & Thomas, 1996; Nag, Corley & Gioia, 2007; Scott & Lane, 2000; Stacey, Griffin and Shaw, 2000; White, 2008). These core elements are culture, structure, organizational learning and processes that are also elements of the Organizational Capital.

Following, we outline examples of descriptions made by some informants of the five case studies about the continuous formal and informal interaction and communication processes they use to exchange their point of views as members of the firms. These processes are conversations and a continuous sensemaking process of claims and counterclaims. Informants' firms and jobs are stated:

- NTBF A - Sales & Marketing Department: "*There is something we call "coffee time" that consists of everybody having breakfast at the same time everyday. We have a kitchen that allows us to do this and sometimes we speak about our jobs and tasks, other times about the sales or about the weather, or about..., then this is more familiar [...] Members of Sales & Marketing department we meet every Monday and also the General Manager with us and The Technical Department with us*"
- NTBF B - Sales & Marketing Department: "*We interact constantly. From a small comment we can make a meeting, from an e-mail I receive we can put the chairs together and discuss what to answer. With the laboratory I interact constantly [...] We speak, speak, speak, speak, speak very much. We discuss, discuss very much [...] Also, The General Manager, Sale & Marketing, Business Development and the Laboratory Manager meet every Monday*"
- NTBF C - R&D Manager: "*We have at least one coordination meeting every week to establish the week plan [...] During*

Table 6 NTBFs' Organizational identity (OI) >

NTBF	A	B	C	D	E
INTERWIEWEE	President & Founder Shareholder	General Manager	R&D Manager	Promoter and in charge of the Organization and Consulting Services Area	Founder Shareholder and Technical Manager
CLAIMS ABOUT OI	"We define our firm as an innovation model. We tell our costumers: we are your technological partner and we solve your technological problems"	"We are a biotechnological firm. I found a laboratory. I removed that concept and now we are a firm with different formal areas"	"We are not manufacturers, we are technologic developers. We define our NTBF as a R&D firm"	"We are a research team, we are innovative and very oriented to our clients"	"Our identity would be laboratory of taxonomic determinations. We are a micro-firm"

Source: Own elaboration

the coffee time and lunch we never speak about our job because the data protection policy the firm follows [...] about work we speak in the work place and during breaks we speak about things of breaks"

- NTBF D - Promoter and in charge of the Organization and Consulting Services Area: "*We have a web with a ShareSpace where we create our knowledge base [...] We share coffee maybe too much but I think this is part of the work. Thinking people is not working when they are having a coffee or smoking is wrong. It is during those breaks when people get deeply in touch and make report and they tell private affairs too*"
- NTBF E - Founder Shareholder and Technical Manager: "*Weekly we have a technical meeting [...] We share our knowledge everyday because things arise and we say 'Look, I know how to do it!', it is like this. Also our activity has got a peculiarity, we go to the countryside a lot and in the countryside we spend weeks with the same person, sharing breakfast, lunch, hotel, conversations and it emerges a special relationship of camaraderie*"

The above statements point out the importance not only of the formal communication channels such as meetings every week but also of the informal ones. The members remark the importance of the informal interaction in their companies. They develop continuous informal processes of claims to transfer knowledge. For example in the NTBF A, a member remarks the continuity of this process saying: *"There is something we call "coffee time" that consists of everybody having breakfast at the same time everyday [...] sometimes we speak about our jobs and tasks, other times about the sales or about the weather"*. The same happens in NTBF B where a member of the Sales & Marketing Department says: *"From a small comment we can make a meeting, from an e-mail I receive we can put the chairs together and discuss what to answer"*. In the NTBF D the interviewee remarks the importance of informal meetings to transfer knowledge: *"Thinking people is not working when they are having a coffee or smoking is wrong. It is during those breaks when people get deeply in touch and make report"*. The interviewee in NTBF E remarks how usual the process is: *"We share our knowledge everyday because things arise and we say "Look, I know how to do it!"*. NTBF C follows a data protection policy that controls when to discuss about work and who with, so informal communication processes are not promoted. They get under control these processes, trying to transfer knowledge only through formal channels.

The members of the NTBFs are constructing their Organizational Identity through these continuous formal and informal communication processes. Table 6 summarizes claims of the Organizational Identity that some informants of the firms made during the interviews. These claims are self-reflective questions that answer who they are as an organization, question that characterize the term "Organizational Identity" from the point of view of Albert and Whetten (1985). These comments are summarized in Table 6:

All the statements in Table 6 express a feeling of community or a feeling of the organization as a whole. All the statements made by the members start "*we are/we define/our identity*" so they have arrived to an agreement of what they are as an organization. Based on the information presented in Table 6 we can extract the following organizational identities of each firm:

- NTBF A: *"We define NTBF A as an innovation model"*
- NTBF B: *"We are a firm of security alimentary services"*
- NTBF C: *"We define our NTBF as a small firm of R&D"*
- NTBF D: *"We are an research team"*
- NTBF E: *"Our identity would be laboratory of taxonomic determinations"*

To obtain several points of view and avoid slants we conducted interviews with the General Manager and/or founder shareholder or promoter and one or two employees of the NTBFs selected (see section "Interviews"). So, we can check the comments shown above with the ones made by the other informants:

- NTBF A - Sales & Marketing Department: *"A is a business innovation model"*
- NTBF B - Sales & Marketing Department: *"We are a firm that tries to solve security in alimentation"*
- NTBF B - Business Development Area: *"B is a firm of the health area but from the point of view of the alimentation"*
- NTBF C - Founder Shareholder and collaborator in the Development Area: *"C is the fruits of a passion for the research"*
- NTBF D - Promoter and in charge of the Technological Implementation Area: *"People who belong to this research team that is D have acquired our philosophy"*
- NTBF E - Taxonomic Identifications and Reports Area: *"E is a team specialized in taxonomic determinations"*

We can see the coincidence between the statements of Table 6 and the last ones and so the agreement between the informants about the Organizational Identity of their firms. In order to test this agreement, we checked these statements about the Organizational Identity with the background information from secondary sources such as the record of the impressions and observations we made when we participated in some firms' activities (coffee breaks and lunches) and also, we checked the websites of the five cases where they state "who we are", some internal and confidential documents about their strategy such as Intellectus Model, the advertisements inserted in the annual report of Madrid Science Park and interviews in newspapers and magazines recollected by NTBF A and B in their press kit. The rationale is "triangulation" used to check the data obtained from the interviews:

- In July, 2008, in their web sites firm A and B stated "who we are" saying:
    - NTBF A – *"We are software manufacturer through an innovative model"*
    - NTBF E – *"We are a team of professionals in the environment area"*
- NTBF A and B make a press kit with interviews in magazines and newspapers where they save the following statements:

  - NTBF A - the Commercial Manager stated in July, 2008 in a newspaper: *"Our firm proposes a business model totally new in Spain"*
  - NTBF B – This firm stated in its press kit in July, 2008: *"B is a bio-technological firm specialized in alimentary health and security"*
- NTBC & D do not stated "who they are" in their web site and nor had they a press kit, however they inserted an advertisement in the annual report, 2008 of Madrid Science park where they said:
  - NTBF C - *"C is a firm of R&D"*
  - NTBF D - *"D is a firm of research"*

To sum up, the statements obtained from the interviews about the Organizational identity and the statements obtained from the secondary sources show the members' agreement about the Organizational identity and the processes of construction of it. However, analyzing again these statements we can see that through these social interaction and communication processes members of the firms create and develop organizational knowledge too, in this case, elements of the Organizational Capital. Following the Intellectus Model, we observe that they define variables of the culture such as the business philosophy and the social climate; variables of the structure such as the organizational design; variables of the organizational learning such as learning environments and capture and transmission of knowledge; and variables of the processes towards internal customers and external customers. Next, we analyse these variables in the five NTBF of the case studies:

- The claims about Organizational Identity offer the vision of the business the organization carries out, variable that in the Intellectus Model is called business philosophy and it is part of the culture of the organization.
- During the social process of construction of Organizational Identity members also define the working atmosphere and employees' willingness to participate actively, which is the variable of the culture called in the Intellectus Model social climate. The members' statements at the beginning of this section are not only examples of processes of construction of Organizational Identity, they also describe the social climate of each firm too: NTBF A develops a familiar atmosphere, NTBF B an atmosphere of constant interaction, NTBF C a very formal one, NTBF D an informal one and NTBF E a relationship of camaraderie.
- About the structure, the five NTBF define ways of configuring the organizational structure and the formal relations between its elements, that is, they configure the organizational design. This design makes the process of claims to agree the Organizational Identity easier. Thus, reviewing the statements presented at the beginning of this section, the five NTBF have defined formal areas and formal meetings between these areas that, in some way, can be related to the structure of the organization. For example, NTBF A has at least three formal areas (Sales & Marketing, the General Manager and Technical Department) that meet every Monday; NTBF B has four formal areas (the General Manager, Sale & Marketing, Business Development and the Laboratory Manager) that also meet every Monday; in NTBF C there is at least one coordination meeting every week; NTBF D has a web with a sharespace; and NTBF D has a technical meeting per week too.
- These five firms have defined formal and informal learning environments that are the organizational contexts in which the dynamics of change are produced and

developed and that give rise to the acquisition of competencies and knowledge. Also, these environments are the ones where claims about the Organizational Identity arise. Thus, NTBF A, B, D and E promote informal meetings to exchange knowledge: NTBF A and D during the coffee time and the few members of NTBF B and E working as an informal team exchanging ideas constantly. NTBF C prefers the formal channels because of their data protection policy.

- The principal aim of the in-depth interviews was to understand the claims processes of construction of the Organizational Identity so we asked about the internal communication between members of the organizations. By doing so, we obtained information about the capture and transmission of knowledge that is a variable of the Intellectus Model. This variable refers to the processes and procedures through which the organization perceives and communicates knowledge. In NTBF A the President & Founder Shareholder stated *"There is not still any computerized system to knowledge exchange. We manage it by the traditional system, that is, by paper documents or Internet chats"*. The NTBF B General Manager said *"the individual knowledge is a question of projects, that is, if X tells us how to do something this is gathered in files of her department which can be consulted by all members of the firm"*. In NTBF C the R&D Manager stated *"We make separate meetings in each area of activity. People of different areas are not in contact because of the data protection policy"*. The Promoter and in charge of the Organization and Consulting Services Area of the NTBF D stated *"We have a web with a ShareSpace where we create our knowledge base"*. Finally, in NTBF E the Founder Shareholder and Technical Manager added *"We have a technical library and a budget to buy technical books. We have organized all the references"*.
- In the Intellectus Model processes directed towards the internal customers are defined as the combination of phases of the organizational operation whose results are directed to members of the organization. Analyzing again the statements about the internal communication it was the General Manager of the NTBF B who expresses a clear procedure for welcoming new employees in order they get the firm philosophy: *"When a new employee arrives to the laboratory, the Laboratory Manager gives him a welcoming training. In the other areas we agree a learning plan too"*.
- Finally, other variable of the element "processes" of the Intellectus Model it is the processes directed towards the external customers. These processes are the combination of phases of the organizational operation whose final result is directed to a consumer situated in the external market. When asking in the interviews about the firm business model and activity, NTBF A and B explain procedures to get new clients: The President & Founder Shareholder in NTBF A said *"to contact with new clients we make mailings, meetings in collaboration with professional associations and we use personal contacts"*; the NTBF B General Manager said *"to get new clients we make a direct action and also we are in contact with specializing associations"*.

LIMITATIONS AND FUTURE RESEARCH DIRECTIONS

As every empirical research, our study is not free of limitations. These limitations could serve as guidelines for future research in the field of Organizational Identity and its relations with the Intellectual Capital as follows:

- **Generalizations:** We have tested the research problem in five new technology-based firms created at the Madrid Science Park, so our findings are based on an "analytic generalization" (Yin, 1984) from the cases studied. Nevertheless, the conceptual perspective proposed in this research should be viewed with some caution, because of the methodological limitations of the present study. Because the purpose was to build, rather than test, theory, we drew on detailed descriptions of a small number of organizations. The extent to which the local explanation proposed by our research develops into a more general theory depends on how well it, or its descendants, explains the "phenomena" in other settings.
- **Findings transferability:** Therefore, a question for future work is whether these findings can be applied in other settings or they are idiosyncratic to the firms that we studied. It could be interesting to study if the grounded theory about Organizational Capital and Organizational Identity can be applicable in NTBFs of other sciences parks different from Madrid Science Park or even in other kind of new organizations different from NTBFs
- **Knowledge - management strategies, knowledge creation process and firms success:** We have focused this research in describing the relationship between the process of construction of Organizational Identity and the creation and development of Organizational Capital in five NTBFs. However, we have not analysed in depth the relation of these concepts with other matters of potential interest, such as the success of the firms or the relation between Organizational Identity and the knowledge management strategies implemented by these firms. As future research, it would also be interesting to study the relation between these firms' process of knowledge creation and the construction of their Organizational Identity.

CONCLUSION

We have conducted a multiple-case study to analyse the relation between the process of construction of Organizational Identity in five new technology-based firms created at Madrid Science Park, and the creation and developing of their Organizational Capital. To provide the explanation of the research problem described in the section "Findings", we have described the processes of claims and counterclaims developed in these five firms, processes they use to construct their Organizational Identity and their Organizational Capital as well. After analysing theoretically Organizational Capital and the Intellectus Model, as a model for the measurement and management of Intellectual Capital, and the main paradigms and research approaches in the field of the Organizational Identity, we concluded that the most adequate paradigm to develop our research was the interpretive one and that the most appropriate research approach was the share meaning one. We selected a case study methodology and we used as primary data collection instrument semi-structured interviews, and as secondary data collection instruments observation and secondary resources.

The interviews objective was to understand the meanings shared by members of the five firms about what was central in their organizations to answer the question "who they were as an organization". Doing this we found:

1. The formal and informal channels of communication members used to interact and construct their Organizational Identity and Organizational Capital.

2. The answer given by the firms to the self-reflective question "Who are we as an organization?", that is, the Organizational identity of each NTBF.
3. The processes of claims and counterclaims they used to construct their Organizational Identity and simultaneously their Organizational Capital.
4. The variables of the Organizational Capital created and developed at the same time that the Organizational Identity.

These findings allow us to conclude that during the processes of construction of Organizational Identity the five new technology-based firms of the case study also construct their Organizational Capital and vice versa. Following Intellectus Model, variables of Organizational Capital we have identified in these firms are:

- **Culture:** business philosophy and social climate.
- **Structure:** Organizational design.
- **Organizational learning:** Learning environments and capture and transmission of knowledge.
- **Processes:** Processes towards internal customers and external customers.

Therefore, in the five NTBFs analysed there is a relation between the process of construction of the Organizational Identity and the social interaction to create and develop the Intellectual Capital. They are small and micro firms with a high proportion of qualified employees and researchers that think the best way to develop and exploit an invention and technological innovation is working in group, exchanging and sharing knowledge between all members through conversations. There is only one exception, NTBF C that follows a strict data protection policy. To facilitate this process, they promote informal relations and design formal channels of communication.

In the field of Organizational Identity it has been studied how this concept is a key element in the processes of strategic change and in situations of external context changes. However, past studies have not explored the relations between Organizational Identity and Organizational Capital in new organizations. This chapter analyzes these relations by making a theoretical proposal and testing it empirically in the context of five NTBFs created at Madrid Science Park. Our investigation opens opportunities for future research in directions that we mention in the section "Limitations and future research". We think these directions could be object of deeper future analyses.

REFERENCES

Albert, S., Ashforth, B. E., & Dutton, J. (2000). Organizational identity and Identification: Charting new waters and building new bridges. *Academy of Management Review*, *25*(1), 13–17.

Albert, S., & Whetten, D. (1985): Organizational Identity. In L. L. Cummings, B. M. y Staw, (Eds.), Research in Organizational Behavior, (pp. 263-295). Greenwich, CT: JAI Press.

Argyris, C., & Schön, D. (1978). *Organizational Learning: A Theory of Action Perspective*. Reading, MA: Addison-Wesley.

Berger, P. L., & Luckman, T. (1968). *La construcción social de la realidad*. Buenos Aires, Argentina: Amorrortu Editores.

Bollinger, L. (1983). A Review of Literature and Hypotheses on New Technology-Based Firms. *Research Policy*, *12*(1), 1–14. doi:10.1016/0048-7333(83)90023-9

Bontis, N. (1996). Intellectual Capital: an exploratory study that develops measures and models. *Working paper* 96-11, University of Western Ontario. Published in *Management Decision (1998)*, 36 (2), 67-76.

Bontis, N., Crossan, M., & Hulland, J. (2002). Managing an organizational learning system by aligning stocks and flows. *Journal of Management Studies*, *39*, 437–469. doi:10.1111/1467-6486.t01-1-00299

Bouchikhi, H., et al. (1998). The Identity of Organizations. In Whetten, D. A. & Godfrey, P. C. (Eds), Identity in Organizations: Building Theory Trough Conversations (pp. 33-80). Thousand Oaks, CA: Sage Publication.

Boulton, R., Libert, B., & Samek, S. (2000). A business model for the new economy. *The Journal of Business Strategy*, *21*, 29–35. doi:10.1108/eb040102

Brookin, A. (1996). *Intellectual Capital. Core asset for the third millennium enterprise*. London: International Thomson Business Press.

Brown, A. D., & Humphreys, M. (2006). Organizational Identity and Place: A Discursive Exploration of Hegemony and Resistance. *Journal of Management Studies*, *43*(2), 231–257. doi:10.1111/j.1467-6486.2006.00589.x

Brown, A. D., Humphreys, M., & Gurney, P. M. (2005). Narrative, identity and change: a case study of Laskarina Holidays. *Journal of Organizational Change Management*, *18*(4), 312–326. doi:10.1108/09534810510607029

Bueno, E. (1998). El capital intangible como clave estratégica en la competencia actual. Boletín de estudios económicos, 3(August), 207-229.

Bueno, E. (2002a). Enfoques principales y tendencias en dirección del conocimiento (knowledge management). In Hernández, R. (Ed.), *Gestión del conocimiento: Desarrollos teóricos y aplicaciones* (pp. 21–54). Cáceres, Spain: Ediciones La Coria-Fundación Xavier Salas.

Bueno, E. (2002b). El capital social en el nuevo enfoque del capital intelectual de las organizaciones. *Revista de Psicología del Trabajo y de las Organizaciones*, *2/3*, 177–345.

Bueno, E. (2005). Fundamentos epistemológicos de dirección del conocimiento organizativo: desarrollo, medición y gestión de intangibles. *Economía Industrial*, *357*, 13–26.

Bueno, E., Morcillo, P., & Salmador, M. P. (2006). *Dirección estratégica: Nuevas perspectivas teóricas*. Madrid: Pirámide.

Bueno, E., Salmador, M. P., & Longo, M. (2008, September). *Marco conceptual y enfoques principales del concepto identidad organizativa: Una aproximación empírica*. Paper presented at the XVIII meeting of Asociación Científica de Economía y Dirección de Empresas, León, Spain.

Bueno, E., Salmador, M. P., & Rodriguez, O. (2004). The role of social capital in today's economy: empirical evidence and proposal of a new model of intellectual capital. *Journal of Intellectual Capital*, *5*(4), 556–574. doi:10.1108/14691930410567013

Bukh, P. N. (2002). The relevance of intellectual capital disclosure: a paradox? *Accounting, Auditing & Accountability Journal*, *16*(1), 49–56. doi:10.1108/09513570310464273

Bürgi, P., & Oliver, D. (2005). Organizational Identity as a Strategy Practice. *Imagination Lab Foundation*. Retrieved July 26, 2008, from http://imagilab.org/publications_wp.html#61

Bürgi, P., Roos, J., & Oliver, D. (2002). Organizational Identity and Strategy. *Imagination Lab Foundation*. Retrieved July 26, 2008, from http://imagilab.org/publications_wp.html#14

Butchart, R. (1987). A new UK definition of high technology industries. *Economy Trends*, *400*(February), 82–88.

Camisón, C., Palacios, D., & Devece, C. (2000). *Un nuevo modelo para la medición del capital intelectual: el modelo Nova.* Paper presented at X meeting of Asociación Científica de Economía y Dirección de Empresas, Oviedo, Spain.

CIC. (2002). *Guías y Directrices de Utilización del Modelo Intellectus (Documento Intellectus-abril).* Madrid: Universidad Autónoma de Madrid, Society Research Centre.

Cohen, D., & Prusak, L. (2001). *In good company: how Social Capital make organizations work.* Boston: Harvard University School Press.

Coleman, J. (1988). Social Capital in the creation of Human Capital. *American Journal of Sociology, 94*, 95–120. doi:10.1086/228943

Collins, D. (1997). Changes in the value relevance of earnings and book values over the past forty years. *Journal of Accounting and Economics, 24*(1), 39–67. doi:10.1016/S0165-4101(97)00015-3

Conner, K. R., & Prahalad, C. K. (1996). A Resource-based Theory of the Firm: Knowledge Versus Opportunism. *Organization Science, 7*(5), 477–501. doi:10.1287/orsc.7.5.477

Cook, S. D. N., & Brown, J. S. (1999). Bridging epistemologies: The Generative Dance Between Organizational Knowledge and Organizational Knowing. *Organization Science, 10*(4), 381–400. doi:10.1287/orsc.10.4.381

Cooper, A. C. (1971). *The Founding of Technology-Based Firms.* Milwaukee, WI: Centre for Venture Management.

Cooper, R., & Burrel, G. (1988). Modernism, Postmodernism and Organizational Analysis: An Introduction. *Organization Studies, 9*(1), 91–112. doi:10.1177/017084068800900112

Corley, K. G. (2006). Guiding Organizational Identity through Aged Adolescence. *Journal of Management Inquiry, 15*(2), 85–99. doi:10.1177/1056492605285930

Cornelissen, J. P. (2002a). On the "Organizational Identity" Metaphor. *British Journal of Management, 13*(3), 259–288. doi:10.1111/1467-8551.00242

Cornelissen, J. P. (2002b). The Merit and Mischief of Metaphor: A Reply to Gioia, Schultz and Corley. *British Journal of Management, 13*(3), 277–279. doi:10.1111/1467-8551.00244

Cornelissen, J. P. (2006). Metaphor and the Dynamics of Knowlege in Organization Theory: A Case Study of the Organizational Identity Metaphor. *Journal of Management Studies, 43*(4), 683–709. doi:10.1111/j.1467-6486.2006.00607.x

Covaleski, M. A. (1998). The Calculated and the Avowed: Techniques of Discipline and Struggles Over Identity in Big Six Public Accounting Firms. *Administrative Science Quarterly, 43*, 293–327. doi:10.2307/2393854

De Geus, A. (1997). *The Living Company: Habits for survival in a Turbulent Business Enviroment.* Boston: Harvard Business School Press.

Drucker, P. (1965). *The Future of Industrial Man.* London: New American Library.

Dutton, J., & Dukerich, J. M. (1991). Keeping an Eye on the Mirror: Image and Identity in organizational Adaptation. *Academy of Management Journal, 34*, 517–554. doi:10.2307/256405

Edvinsson, L., & Malone, M. S. (1997). *Intellectual Capital.* New York: HarperBusiness.

Edvinsson, L., & Sullivan, P. (1996). Developing a model for managing intellectual capital. *European Management Journal, 14*(4), 356–364. doi:10.1016/0263-2373(96)00022-9

Eisenhardt, K. M. (1989). Building theories from case study research. *Academy of Management Review, 14*, 532–550. doi:10.2307/258557

Fiol, C. M. (2002). Capitalizing on paradox: The role of language in transforming organizational identities. *Organization Science, 13*(6), 653–666. doi:10.1287/orsc.13.6.653.502

Fiol, C. M., & Huff, A. S. (1992). Maps for Managers: Where are we? Where do we go from here. *Journal of Management Studies, 29*(3), 267–285.

Gioia, D. A. (1998). From Individual to Organizational Identity. In Whetten, D. A., & Godfrey, P. C. (Eds), Identity in Organizations: Building Theory Trough Conversations (pp. 17-31). Thousand Oaks, CA: Sage Publication.

Gioia, D. A., & Chittipeddi, K. (1991). Sensemaking and sensegiving in strategic change initiation. *Strategic Management Journal, 12*(6), 443–458. doi:10.1002/smj.4250120604

Gioia, D. A., & Pitre, E. (1990). Multiparadigm perspectives on theory bulding. *Academy of Management Review, 15*(4), 584–602. doi:10.2307/258683

Gioia, D. A., Schultz, M., & Corley, K. G. (2002a). On Celebrating the Organizational Identity Metaphor: A Rejoinder to Cornelissen. *British Journal of Management, 13*(3), 269–275. doi:10.1111/1467-8551.00243

Gioia, D. A., Schultz, M., & Corley, K. G. (2002b). Metaphorical Shadow Boxing: A Response to Cornelissen's Reply to our Rejoinder. *British Journal of Management, 13*(3), 281–281. doi:10.1111/1467-8551.00245

Gioia, D. A., & Thomas, J. (1996). Identity, image and issue interpretation: Sensemaking during strategic change in academia. *Administrative Science Quarterly, 41*, 370–403. doi:10.2307/2393936

Glaser, B. G., & Strauss, A. L. (1967). *The Discovery of Grounded Theory: Strategies for Qualitative Research.* New York: Aldine Publishing Company.

Glynn, M. A. (2000). When Cymbals Become Symbols: Conflict Over Organizational Identity Within a Symphony Orchestra. *Organization Science, 11*(3), 285–298. doi:10.1287/orsc.11.3.285.12496

Glynn, M. A., & Abzug, R. A. (2002). Institutionalizing identity: Symbolic isomorphism and organizational names. *Academy of Management Journal, 45*(1), 267–280. doi:10.2307/3069296

Golden-Biddle, K., & Rao, H. (1997). Breaches in boardroom: Organizational Identity and Conflicts of Commitment in a Nonprofit Organization. *Organization Science, 8*(6), 593–609. doi:10.1287/orsc.8.6.593

Grant, R. M. (1996). Toward a Knowledge-Based Theory of the Firm. *Strategic Management Journal, 17*(Winter Special Issue), 109-122.

Hayek, F. A. (1945). The use of knowledge in society. *The American Economic Review, 35*(4), 519–530.

Johanson, U., Martensson, M., & Skoog, M. (2001). Measuring to understand intangible performance drivers. *European Accounting Review, 10*(3), 407–437. doi:10.1080/09638180120081562

Kaplan, R. S., & Norton, D. P. (1992). The Balanced Scorecard-measures that drive performance. *Harvard Business Review*, (January-February): 71–79.

Kaufmann, L., & Schneider, Y. (2004). Intangibles: a synthesis of current research. *Journal of Intellectual Capital, 5*(3), 366–388. doi:10.1108/14691930410550354

Kauser, S., Oakey, R., & During, W. (2002). *New Technology-Based Firms in the New Millenium* (*Vol. II*). London: Pergamon/Elsevier Science Ltd.

Kauser, S., Oakey, R., & During, W. (2004). *New Technology-Based Firms in the New Millenium* (*Vol. III*). London: Pergamon/Elsevier Science Ltd.

Kauser, S., Oakey, R., & During, W. (2005). *New Technology-Based Firms in the New Millenium* (*Vol. IV*). London: Pergamon/Elsevier Science Ltd.

Knight, F. H. (1921). *Risk, Uncertainty and Profit*. New York: Hart, Schaffner & Mark.

Knorr-Cetina, K. (1981). *The Manufacture of Knowledge: An Essay on the Constructivist and Contextual Nature of the Science*. New York: Pergamon Press.

Kogut, B., & Zander, U. (1992). Knowledge of the Firm, Combinative Capabilities and the Replication of Technology. *Organization Science*, *3*(3), 383–397. doi:10.1287/orsc.3.3.383

Lev, B. (1998). *Intangibles and gains from insider trading*. New York: New York University.

Lev, B. (2001). *Intangibles: Management, Measurement, and Reporting*. Washington, DC: Brookings Institution Press.

Lew, B., & Zarowin, P. (1998). The boundaries of financial reporting and how to extend them. *Working Paper*. New York: New York University

Lincoln, Y. S. (Ed.). (1985). *Organizational Theory and inquiry: The paradigm revolution*. Beverly Hills, CA: Sage.

Little, A. D. (1997). *New Technology-Based Firms in the United Kingdom and the Federal Republic of Germany*. London: Wilton House.

Low, J. (2000). The value creation index. *Journal of Intellectual Capital*, *1*, 252–262. doi:10.1108/14691930010377919

Machlup, F. (1980). *Knowledge: Its Creation, Distribution and Economic Significance*. Princeton, NJ: Princeton University Press.

Maturana, H., & Valera, F. (1984). *El árbol del conocimiento*. Santiago de Chile, Chile: Universitaria.

McClintock, C., Brannon, D., & Maynard-Moody, S. (1979). Applying the logic of sample surveys to qualitative case studies: The case cluster method. *Administrative Science Quarterly*, *24*(4), 612–629. doi:10.2307/2392367

McDermott, R. (1999). Why Information Technology Inspired But Cannot Deliver Knowledge Management. *California Management Review*, *41*(4), 103–117.

McElroy, M. V. (2001). Social innovation capital draft. *Macroinnovation Associates*, July, 1-14.

Miles, M. B., & Huberman, A. M. (1984). *Analyzing qualitative data: A source book for new methods*. Beverly Hill, CA: Sage.

Nag, J., Corley, K. G., & Gioia, D. A. (2007). The intersection of Organizational Identity, Knowledge, and Practice: Attempting strategic change via knowledge grafting. *Academy of Management Journal*, *50*(4), 821–847. doi:10.2307/20159892

Nahapiet, J., & Goshal, S. (1998). Social Capital, Intellectual Capital and the organisational advantage. *Academy of Management Review*, *23*, 242–266. doi:10.2307/259373

Nonaka, I. (1994). A Dynamic Theory of Organizational Knowledge Creation. *Organization Science*, *5*(1), 14–37. doi:10.1287/orsc.5.1.14

Nonaka, I., & Takeuchi, H. (1995). *The Knowledge-Creating Company*. New York, NY: Oxford University Press.

Ordoñez de Pablos, P. (2003). Intellectual capital reporting in Spain: a comparative review. *Journal of Intellectual Capital*, *4*(1), 61–81. doi:10.1108/14691930310455397

Polanyi, M. (1969). *Knowing and being*. Chicago: University of Chicago Press.

Porter, T. B. (2001, August). *Theorizing Organizational Identity*. Paper presented at the meeting of Academy of Management Meeting, Washington, D. C.

Quinn, J. B. (1992). *Intelligent Enterprise: A Knowledge and Service Based Paradigm for Industry*. New York: The Free Press.

Rabinow, P., & Sullivan, W. M. (1979). *Interpretive Social Science: A reader*. Berkeley, CA: University of California Press.

Rao, H., Monin, P., & Durand, R. (2003). Institutional change in Toque Ville: Nouvelle cuisine as an identity movement in French gastronomy. *American Journal of Sociology, 108*(4), 795–843. doi:10.1086/367917

Reger, (1994). Reframing the organization: Why implementing total quality is easier said than done. *Academy of Management Review, 19*(3), 565–584. doi:10.2307/258939

Ritzer, G. (2003). *Teoría sociológica moderna: Quinta edición*. Madrid: McGrawHill.

Roos, J. (1998). *Intellectual capital: Navigating in the New Business Landscape*. New York: University Press.

Saint Onge, H. (1996). Tacit knowledge: the key to the strategic alignment of Intellectual Capital. *Strategy and Leadership, 24*(2), 10–14. doi:10.1108/eb054547

Sanchez, P., Chaminade, C., & Olea, M. (2000). Management of intangibles: An attempt to build a theory. *Journal of Intellectual Capital, 1*(4), 312–327. doi:10.1108/14691930010359225

Schultz, M., & Hatch, M. J. (1996). Living with multiple paradigms: The case of paradigm interplay in organizational culture studies. *Academy of Management Review, 21*(2), 529–527. doi:10.2307/258671

Schwandt, T. A. (1998). Constructivist, Interpretivist Approaches to Human Inquiry. In Denzin, N. K., & Lincoln, I. S. (Eds.), *Handbook of Qualitative Research* (pp. 221–259). Thousand Oaks, CA: Sage Publication.

Schwandt, T. A. (2000). Three epistemological stances for qualitative inquiry. In Denzin, N. K., & Lincoln, I. S. (Eds.), *Handbook of Qualitative Research* (2nd ed., pp. 189–213). Thousand Oaks, CA: Sage Publication.

Scott, S. G., & Lane, V. R. (2000). A stakeholder approach to organizational identity. *Academy of Management Review, 25*(1), 43–62. doi:10.2307/259262

Shearman, C., & Burrell, G. (1988). New technology-based firms and the emergence of new industries: some employment implications. *New Technology, Work and Employment, 3*(2), 87–99. doi:10.1111/j.1468-005X.1988.tb00092.x

Smilor, R. W. (1990). University spin-out companies.-Technology Starts ups from university of Texas at Austin. *Journal of Business Venturing, 5*(1), 63–76. doi:10.1016/0883-9026(90)90027-Q

Spender, J. C. (1996). Making Knowledge the Basic of a Dynamic Theory of the Firm. *Strategic Management Journal*, (17): 45–62.

Stacey, R. D., Griffin, D., & Shaw, P. (2000). *Complexity and management: Fad or radical challenge to systems thinking?* London: Rutledge.

Stewart, T. A. (1997). *Intellectual Capital: The New Wealth of Organizations*. New York: Doubleday.

Sutton, R., & Callahan, A. (1987). The Stigma of Bankruptcy: Spoiled Organizational Image and Its Management. *Academy of Management Journal, 30*, 405–436. doi:10.2307/256007

Sveiby, K. (1997). *The New Organizational Wealth*. New York: Berret-Koehler.

Teece, D. J. (2000). Strategies for managing knowledge assets: the role of firm structure and industrial context. *Long Range Planning, 33*, 35–54. doi:10.1016/S0024-6301(99)00117-X

Thiétart, R. A. (2001). *Doing Management Research. A Comprehensive Guide*. London: SAGE Publications.

Von Krogh, G., & Ross, J. (1995). *Organizational epistemology*. New York: MacMillan and St Martin's Press.

Webb, E., Campbell, D. T., Schwartz, R. D., & Serchrest, L. (1996). *Unobtrusive measures: Nonreactive research in the social sciences*. Chicago: Rand McNally.

Whetten, D. A., & Mackey, A. (2002). A social actor conception of organizational identity and its implications for the study of organizational reputation. *Business & Society, 41*(4), 393–414. doi:10.1177/0007650302238775

White, H. C. (2008). *Identity and Control: How Social Formations Emerge*. Princeton, NJ: Princeton University Press.

Whittington, R. (2001). *What Is Strategy and Does It Matter?*London: Thomson Learning Business Press.

Yin, R. K. (1984). *Case Study Research: Design and Methods*. San Francisco: Sage Publications.

Zuckerman, E. W. (2003). Robust identity or nonentities? Typecasting in the feature-firm market. *American Journal of Sociology, 108*(5), 1018–1074. doi:10.1086/377518

ADITIONAL READING

Balmer, J. M., & Greyser, S. (2002). Managing the multiple identities of the organization. *California Management Review, 44*(3), 72–86.

Bueno, E. (1999). *La gestión del conocimiento en la nueva economía, Gestión del Conocimiento y capital intelectual*. Madrid: Instituto Universitario Euroforum Escorial.

Bueno, E., Longo, M., & Salmador, M. P.(forcoming). CoPs and Organizacional Identity. Five case studies of NTBFs. In Bueno, E., & Rivera, O. (Eds.), *Handbook of Research on Communities of Practice for Organizational Management and Networking: Methodologies for Competitive Advantage*. Hershey, PA: IGI Global.

Bueno, E., Morcillo, P., & Salmador, M. P. (2006). *Dirección estratégica: Nuevas Perspectivas Teóricas*. Madrid: Pirámide.

CIC. (2003). *Model for the measurement and management Intellectual Capital: Intellectus Model (Intellectus Document-June)*. Madrid: Universidad Autónoma de Madrid, Society Research Center.

Cornelissen, J. P., Haslam, S. A., & Balmer, J. M. T. (2007). Social Identity, Organizational Identity and Corporate identity: Towards an Integrated Understanding of Processes, Patternings and products. *British Journal of Management, 18*(s1), S1–S16. doi:10.1111/j.1467-8551.2007.00522.x

Fiol, C. M. (1991). Managing culture as a competitive resource: an identity-bases view of sustainable competitive advantage. *Journal of Management, 17*(1), 191–211. doi:10.1177/014920639101700112

Fiol, C. M., Hatch, M. J., & Golden-Biddle, K. (1998). Organizational Culture and Identity: What's the Difference Anyway? In Whetten, D. A. & Godfrey, P. C. (Eds.), Identity in Organizations: Building Theory Trough Conversations (pp.56-59). Thousand Oaks: Sage Publication.

Foucault, M. (1980). *Power/Knowledge: Selected Interviews and Other Writings 1972–1977* (Gordon, C., Ed.). Brighton, Sussex: Harvester.

Gorey, R.M., & Dobat, D.R. (1996). Managing on the knowledge era. *The systems thinker,* 7(8), 1-5.

IADE-CIC (2003). Modelo Intellectus: medición y gestión del capital intelectual.

Kaplan, R. S., & Norton, D. P. (1996). The Balanced Scorecard. Harvard, Massachussets: Harvard Business School Press.

Kaplan, R. S., & Norton, D. P. (2001). *The Strategy-Focused Organization: How balance scorecard companies thrive in the new business environment.* Chicago: University of Chicago Press.

Kuhn, T. S. (1970). *The structure of scientific revolutions*. Chicago: University ofChicago Press.

Marshall, A. (1890). *Principles of economics*. London, U.K.: MacMillan and Co.

Meritum Project. (2002). *Guidelines for managing and reporting on intangibles*. Madrid: Fundación Airtel-Vodafone.

Polanyi, M. (1969). *Knowing and being*. Chicago: University of Chicago Press.

Porter, M. (1985). *Competitive advantage*. New York, NY: The Free Press.

Pratt, M. G. (1998). To be or not to be: Central questions in organizational identification. In Whetten, D. A. y Godfrey, P. C. (Eds.), Identity in Organizations: Building Theory Trough Conversations (pp. 171-208). Thousand Oaks: Sage Publication.

Roos, J., & Victor, B. (1999). Towards a new model of strategy-making as serious play. *European Management Journal, 17*(4), 348–355. doi:10.1016/S0263-2373(99)00015-8

Scott, S. G., & Lane, V. R. (2000). A stakeholder approach to organizational identity. *Academy of Management Review, 25*(1), 43–62. doi:10.2307/259262

Spender, J. C. (1996). Making Knowledge: The basic of a dynamic theory of the firm. *Strategic Management Journal, 17*, 45–62.

Stewart, T. A. (1994). Intellectual Capital: Your company's most valuable asset. *Fortune Magazine, 10*, 68–73.

Tsoukas, H. (1991). The missing link: a transformational view of metaphors in organizational science. *Academy of Management Review, 16*(3), 566–585. doi:10.2307/258918

KEY TERMS AND DEFINITIONS

Intellectual Capital: IC is generally defined as the intellectual material that can be put to use to create wealth which includes organization's processes, technologies, patents, employees' skills and information about customers, suppliers and stakeholders (Stewart, 1997).

Micro Firm: The European Commission in its recommendation C 1422/2003 defined it as an enterprise which employs less than 10 people and whose annual turnover and/or annual balance sheet total does not exceed EUR 2 million.

New Technology-Based Firms (NTBFs): An independently owned business based on the exploitation of an invention or technological innovation which implies substantial technological risks (Little, 1977).

Organizational Capital: A combination of intangibles that structure and develop the organizational activity of the organization. It is part of the Structural Capital (CIC, 2002).

Organizational Identity: A self-reflective question (Who are we as an organization?) which captures central, enduring and distinctive features of the organization (Albert & Whetten, 1985).

Paradigm: A general perspective or way of thinking that reflects fundamental believes and assumptions about the nature of organization (Gioia & Pitre, 1990).

Small Firm: The European Commission in its recommendation C 1422/2003 defined it as an enterprise which employs fewer than 50 persons and whose annual turnover and/or annual balance sheet total does not exceed EUR 10 million.

Structural Capital: The combination of knowledge and intangible assets derived from the processes of action which are property of the organization and which remain in people leave (CIC, 2002).

ENDNOTES

1 Gioia & Pitre (1990: 585), based on Kuhn (1970) and Lincoln (1985), assess that "a paradigm is a general perspective or way of thinking that reflects fundamental believes and assumptions about the nature of organization".

2 Schwandt (1998: 221) asserts that "*constructivist, constructivism, interpretivist, and interpretivism* are terms that routinely appear in the lexicon of social science methodologists and philosophers. Yet, their particular meanings are shaped by the intent of their users". According to Schwandt (1998, 2000) both perspectives share the goal of understanding the complex world of lived experience from the point of view of those who live it, but constructivism and interpretivism are unique in the manner in which each answer questions as "what is the purpose and aim of human inquiry?, or how can we know about the world of human action?

Section 2
Technological Innovation

Chapter 5
Problem Finding and Solving:
A Knowledge-Based View of Managing Innovation

Bruce A. Heiman
San Francisco State University, USA

Pia Hurmelinna-Laukkanen
University of Oulu & Lappeenranta University of Technology, Finland

ABSTRACT

In this chapter the authors focus on a problem-oriented view of managing innovation. In particular, they ask what it means to manage innovation well, so that innovation outcomes are enhanced. They start by briefly reflecting on the rise in value of intangibles as an asset class (notably knowledge), and by defining terms. They then briefly review selected literature on strategic management as an example of the typical manner in which innovation has been previously approached. Looking at selected macro- and micro-organizational theories the authors reflect on several shortcomings of extant literature. They then introduce the knowledge-based view and identify a particular sub-field, the problem finding/problem solving (PF/PS) perspective, as one promising way to address these issues. They identify several challenges in the present formulation of the perspective, and discuss further research opportunities. Though the PF/PS perspective presently lacks compelling empirical evidence, it shows promise for developing into a useful logic for managing and enhancing innovation.

INTRODUCTION

This chapter briefly reviews several conversations regarding managing innovation. We seek to identify trends and promising directions for both scholars and managers. In particular, we argue for a connection between the role of knowledge in innovation, creation and management, and the emerging problem finding/problem solving literature—a recent development with strong potential for shedding light innovation management.

DOI: 10.4018/978-1-61520-875-3.ch005

The transition from an industrial to an information economy has had notable effects on the ways in which companies operate and seek to establish stronger competitive positions. While the industrial economy was to a large extent driven by economies of scale, the rise of the information/services economy has made it natural to put more emphasis on knowledge. Rapid changes, uncertainty over technological trends, emerging new markets, cultural and technical convergence, and increasing vagueness of industrial boundaries force companies to manage both external and internal knowledge (e.g., Rothwell, 1994, Tidd et al., 2005, Chesbrough et al. 2006). Deregulation and sharpened competition along with the rise of complex networks and more transparent, low-cost, high bandwidth channels of communication (e.g., internet, GSM), require companies to continuously meet changing customer needs. This underscores the challenge of generating, obtaining, deploying and using knowledge for innovation. The nature of what comprises a valuable asset has fundamentally changed, and is increasingly determined by intangible rather than tangible assets (Sveiby, 1997, Johannessen et al. 1999). Knowledge assets may well soon overtake traditional resources such as labor, natural resources and capital stock as a primary source of long-term increasing returns.

In this context, knowledge-related processes, and decisions made by managers regarding these processes are central in determining innovation outcomes, and ultimately firm success. The constantly and rapidly changing business environment raises the minimum acceptable innovation performance level for most, if not all firms; technology-driven firms are particularly strongly affected. This type of highly selective environment creates a need for continuous search for new business opportunities and subsequent revenues. Put differently, the need exists in present day organizations to better understand finding and choosing appropriate problems to solve and economically exploit. This particular aspect of value creation, though, has not received all the attention that it deserves. One notable exception is Teece (2007, p. 13) who discusses actions like "sensing" and "seizing" opportunities—notions evocative of problem finding. Teece, while identifying the microfoundations of performance that suggest the importance of problem finding, stops short of suggesting a logic of these microfoundations.

BACKGROUND

Creating and maintaining competitive advantage based on knowledge assets means that knowledge needs to be efficiently and frequently generated, accessed, transferred, stored, used, modified, or otherwise deployed in the service of innovation. Consequently, activities that link innovativeness to knowledge are receiving increasing attention both in industry and academia. The focus of efforts to date, however, seems a bit skewed. Emphasis for management of innovation has frequently been on the side of managing value capture from innovation (essentially squeezing every last drop of profit out of an existing product or service) rather than value creation (see, e.g., Teece, 2006). Value creation is often pre-assumed in a substantial fraction of the existing literature (Nickerson, Silverman and Zenger, 2007). The importance of capturing value is easy to comprehend, and with many related issues currently under-studied, this aspect of managing innovation and technology naturally attracts a lot of research attention. Among prominent approaches that focus on firm performance and say something about innovation outcomes, the literature on strategic organization (macro-organizational behavior theory) is focused primarily on value capture activities, and has historically neglected discussion of value creation activities (see March, 1991, on over-focus on exploitation-related activities at the expense of exploration). Similar challenges exist for the micro-organizational behavior literature. Despite the evolutionary "flavor" offered by modern organizational theorists like Nelson and Winter (1982), an assumption persists that management of innovation is a matter of developing routines which are iteratively refined and optimized over time. Recently, for example Chesbrough's (2003) empirical work explaining entry mode choices in the disk drive industry focuses fundamentally on exploitive innovation (disk drive refinement). Yet, as noted by Levinthal and March (1993), exploration is also vital, hinting that more research efforts should be targeted in the area of unpacking value creation. Value creation is a necessary precursor to any value capturing activities (see Amit and Zott, 2001). The capability to create value and valuable knowledge assets can hardly be taken as given, and thus a deeper understanding of the premises underlying the value creation aspect of innovation is needed.

We assert that the problem solving/problem finding perspective (PF/PS) offers a superior lens for choosing and changing innovation-related processes by focusing on achieving better innovation outcomes. The PF/PS view serves the needs of innovators (and their managers) by offering a context-sensitive framework for understanding the factors affecting innovation outcomes (problem finding outcomes, problem solving outcomes, and ultimately value capture outcomes, e.g., profits). These outcomes are achieved by using a logic of problem finding and solving, and identifying where key managerial decision points occur. Innovation managers match processes well to biases and obstacles in the problem environment to enhance innovation outcomes. Through facilitating understanding of problem finding and problem solving processes, and the issues they raise for firms engaging in innovation, innovation outcomes (our dependent variable of interest) are improved.

Key Questions and Scope of the Chapter

Broadly, we see the capability to create value and subsequent valuable (exploitable) knowledge assets as evidence of "good" management of innovation. In this chapter, we rather narrowly choose to ask what it means to manage innovation well given a specific organizational and problem context. We prefer this question to the very challenging and not very context-sensitive question of "what are the best way(s) to manage innovation?"

Since the notion of management of innovation casts a wide net (e.g., identifying, measuring, controlling, enhancing and developing innovation inputs, processes and outputs), we start by offering simple, tractable definitions of some central terms to add clarity and tractability.*

1. 'Processes' are policies adopted or actions taken in a firm or a team that result in a change of state (not necessarily positive or negative) for a project.
2. 'Managing' refers to making or participating in decisions aimed at achieving a change of state in a project or organization—managers choose innovation-related processes and make decisions that affect innovation processes. Management of innovation processes affects innovation outcomes.
3. 'Innovation' is problem-finding and problem solving, or creativity in the service of commercial ends (most typically profits). Innovation comprises, for example, commercially applied fluency in idea generation, openness to new ideas (e.g., respect for the ideas of others), and the ability to generate ideas, test ideas, and render good decisions regarding many ideas at the individual and team levels.

*Other important terms are defined in the Key Terms & Definitions section at the end of the chapter.

Using the above terms, we approach our research question by presenting (and critiquing) a logic of management of innovation, one that uses knowledge-related issues as a unifying thread to predict innovation outcomes. The Problem Finding/Problem Solving Perspective asserts that in addition to other existing resources, the state of one's (or one's team's) knowledge base fundamentally affects the ability to innovate. Von Hippel (1994) and Von Hippel and Tyre (1995) show that knowledge transfer for the sake of problem solving typically occurs under stressful conditions (e.g., something is broken), and not by design. We consider awareness of and deliberate management of innovation to be more useful than a more passive or reactive mode of putting out fires. Actively managing sticky, difficult-to-handle knowledge in order to (especially) find and solve good problems is relevant in the pursuit of an organization capable of superior innovation.

Below, we discuss problem finding and solving in relation to a logic of managerial choice, one form of which is discriminating alignment (see Williamson, 1985). Discriminating alignment is the matching of organizing choices to people, events, and situations that might inhibit innovation (For the seminal thinking on inhibitors and enablers of innovation, see Mintzberg, 1979). The complexity of a problem, personal and group biases, and the work environment in which problem finding and solving occurs all affect innovation outcomes. Managerial attention in the form of decision making (choice) manifests via how managers allocate, educate, direct, wield authority, or promote consensus as they try to drive innovation (Bowen and Ostroff, 2004).

Nickerson and Zenger (2004) offer a logic of managerial choice that effectively recognizes idiosyncratic people and situations (the problem environment). Their approach has merit because others ignore the conditions under which particular high-powered (but costly and time-consuming) forms of organizing might be deployed. For example, fostering a work environment that supports innovation or the proper management of people for innovation, one may ask what the limits to the useful spreading and promoting of design thinking skills are? Should the accounting department be taught to brainstorm? Should the firm's project teams all be multidisciplinary in character? (see, e.g., Anderson et al. 2004 on potential effects of conflicts). Marginal analysis seems lacking in addressing these questions; common sense (economic rationality) suggests that we do not generally need to invest much time and effort in brainstorming about, for example, what office supplies to order. Similarly, there is widespread agreement that developing new biotechnology-based drugs requires commitment of substantial innovation resources. Much of the academic literature, however, lacks a convincing logic appropriate for handling the gray areas in between these two polar examples. The problem finding/problem solving perspective is a promising vehicle for filling this gap.

The coverage of a new topic is always problematic. Specifically, we are subject to the tradeoff between, on the one hand, wanting to explain the latest and greatest framework for thought and action, and, on the other hand, offering a broad survey of the existing orthodox thinking on innovation, fragmented though it may be for a relatively new field. In this chapter, we have unambiguously chosen the former option: We argue for a particular point of view, the PF/PS perspective, and in so doing, we restrict our review of literature to related foundational work that helps build our arguments. In line with this, we resist the idea of "more is better" regarding training for and deploying innovation-related skills. Rather, we seek to articulate a simple logic that identifies fundamental drivers of how and when, for example, brainstorming, research, problem-finding, prototyping, iteration or design review (critique) are best matched to particular people or environmental situations. We do not directly address such general questions as "how can one best manage people's innovation activities in all situations?" or "how can I create a work environment that generally promotes innovation?" These are the wrong questions. The context specificity of innovation problems suggests that a micro-analytic, context-sensitive logic would be more useful for managers of technological innovation.

In line with this, we critique and thereby advance incrementally, Nickerson, Silverman and Zenger's (2007) alluring idea of a logic that can be applied to determine situation- and environment-specific responses to issues that occur in the innovation process. Though this book is concerned with both intellectual capital and technological innovation, our focus is primarily on innovation, specifically innovation outcomes given the existence of bias in the early stages of value creation (i.e., problem finding). Innovation outcomes (as a function of problem finding/solving efficacy) are the main dependent variables of interest in this chapter. In addition to innovation, intellectual capital also plays a role, but as one of (several discussed) causal variables. We use the issues around protection of intellectual capital as part of an extended example of a factor that Heiman, Nickerson and Zenger (2008) neglect in proposing a research agenda for the study of problem finding. We suggest that other as yet unspecified factors affecting innovation outcomes exist. We offer and example of addressing this shortcoming by asserting that neglected intellectual property protection issues affect problem finding outcomes and hence innovation outcomes.

We also pursue an important goal of paradigm-based research—cumulatively building on previous work (Kuhn, 1970). Our approach is also useful from a more practical managerial perspective: in its 2005 ITC Survey of Senior Management, Boston Consulting Group (2005) noted that worldwide, less than 50% of executives surveyed were satisfied with the present return on investment in innovation. This motivates us to focus on a novel way of thinking regarding management of innovation—we see no reason to re-articulate, in great detail, accepted approaches that have left managers unsatisfied. The next part of the chapter discusses an admittedly narrow view of some major literatures, but nonetheless identifies some gaps in their usefulness for understanding value creation. Rather than dwelling in detail on previous perspectives and prescriptions, we choose to argue for a very promising, but emerging, view of managing and enhancing innovation. In this manner, we seek to offer what we perceive as the greatest utility to the reader seeking direction.

ISSUES CONTROVERSIES, PROBLEMS

Recently, macro-organization theorists have dominated thinking about the determinants of high performance in organizations. This also applies to performance in the area of innovation. Prior research has often approached innovativeness and innovation management, for example, from the perspectives of Transaction Cost Economics, Theories of Organizational Power, and Industrial Organization Economics. In the search for best practices, quite a few approaches identify factors improving innovativeness, or seek to show the effects of innovations on a variety of outcomes such as firm profitability. Notably, substantial efforts have been made under the umbrella of management research in the areas of strategic management and micro-organizational behavior. Below, we briefly discuss some efforts and outcomes of these research threads.

Strategic Management

The main approaches in strategic management "...include structure-conduct-performance based on the five forces approach of Michael Porter, the resource-based view, and the capabilities perspective" (Teece 2004, p. 11). Discussion of these issues has been carried on by Porter (1980), Rumelt (1984, 1987), Wernerfelt (1984), Nelson (1991), Williamson (1991, 1999), and Winter (2006) among others. They focus on understanding how competitive advantage can be achieved and sustained at the firm- or transaction-levels, and why and how firms end up performing differently even in the same industry and in similar environments. Though not always accorded the importance it deserves, innovation has been part of the dialogue in these discussions.

Knowledge and Value Capture in Strategic Management Approaches

Defensive value-capture oriented issues are considered under the Resource Based View (RBV), according to which the deployment and protection of unique knowledge competencies is the basis for sustainable competitive advantage (Penrose, 1959, Wernerfelt, 1984, Dierickx and Cool, 1989, Barney, 1991, Mahoney and Pandian, 1992, Peteraf, 1993). Further, an extension of Nelson and Winter's (1982) Evolutionary Economics combined with the resource perspective, the Dynamic Capabilities View, suggests that competencies need to change over time so as to preserve their value, and that the acquisition of new knowledge, as well as the ability to absorb and act on it, are important (Teece et al., 1997, Christensen, 1997, Eisenhardt and Martin, 2000, Cohen and Levinthal, 1990, and Zahra and George, 2002). Also, the Knowledge Based View of the firm (e.g., Kogut and Zander, 1992, Grant, 1996, Demsetz, 1988, Conner, 1991, Foss, 1996, Heiman and Nickerson, 2004) emphasizes the integration, replication, and recombination of knowledge and competencies. Common to these approaches is that they concentrate to a large extent on existing resources as well as on their access, exchange and utilization for capturing value (see, e.g., Amit and Zott 2001 regarding the RBV). Although the above theories enhance understanding of innovation and knowledge in many ways, they still are evolving. They have limitations and imperfections, reflecting a tendency in the field of strategic management "to encourage fragmentation and to favor new ideas rather than the integration and consolidation of old ones" (Teece 2004, p. 12). Addressing these limitations has the potential to diminish disintegration of the field and advance practice.

Issues with the Strategic Management Approach

While the above approaches have shown some utility in relation to the management of innovations, they also make substantial assumptions regarding the pre-existence of knowledge, skills, and results from value creation activities, i.e., innovation. Extant literature focuses to a large extent on the determinants of how to best operate pre-existing economically rewarding activities, and not so much on questions such as which economic activities should be brought into existence by finding new problems to solve and from which rents may be subsequently realized.

The above theory camps are macro-organizational theories of the firm that typically predict outcomes at the level of the firm, industry, transaction, or network. This directs attention away from crucially important questions such as "what problems should I choose to solve?" Macro-organizational theories discuss value creation, but in many cases the discussion is restricted because of limitations imposed by a lack of agreement on the unit of analysis. Existing theories tend to concentrate on questions such as what actions companies should take as they strive for improved performance, and the operative logics tend to be profit maximization and/or cost efficiency-driven. Similar challenges are found at the micro-level as well.

Micro-Organizational Behavior

The micro side of organizational behavior (OB) theory stems primarily from the field of psychology, and also draws from other disciplines, such as sociology, anthropology, and political science. Micro-OB deals with individual, interpersonal, and group level issues within a firm (or the workplace, or a team). Deriving from as well as benefiting from various behavioral science disciplines, micro-organizational behavior nonetheless fails to fully address questions of innovation management.

Managerial Issues in Micro-OB

For managers, on the micro-organizational behavior level, the most influential existing managerial wisdom regarding managing innovation comes to a large extent from books and practitioner journals rather than peer-reviewed, academic work (consider, e.g., 1974's "Conceptual Blockbusting" by Adams, which is still in print—other examples of classic books include "Design Yourself" (Hanks et al., 1977) and "Universal Traveler" (Koberg and Bagnall, 1974)). Practitioner journals include MIT's Sloan Management Review, and California Management Review (e.g., Beckman and Barry, 2007). A downside to this literature—and simultaneously an opportunity—is its orientation towards the individual and neglect of group and context-related factors that affect management of innovation. For instance, Porter (1996, p. 264) notes that while the "B" in OB has been emphasized, the "O" has been largely ignored. Also, many of the recipes offered by this literature have not been rigorously empirically tested for their effects on organizational performance.

Micro-OB: Scholarly Work

Many studies on organizational behavior have mainly aimed at finding out how different problems can be solved (see, e.g., Cyert and March, 1963 on the behavioral theory of the firm, Porter, 1996, Dent, 2002, and Nickerson and Zenger, 2004). Today's organizational behavior research is based on a fundamentally Taylorist approach, which sees rationalizing the organization with precise instructions and management as important for firm performance (e.g., Taylor, 1911, Hersey and Blanchard, 1988, Conner and Prahalad, 1996). This direction represents a good start because it advances understanding of what managerial choices lead to high performance output. However, this approach emphasizes the critical role of manager as teacher or ultimate source of knowledge in all hierarchies, which is an oversimplification (Conner and Prahalad, 1996 are notable for articulating this view). Further, it only considers direct costs (e.g., of labor and materials) and neglects the costs of organizing for particular tasks. Also, it assumes that managerial choices are obvious because they are tractable to particular scientific management principles, for example, time-motion studies. However, direction from above or watching people work does not necessarily lead to obtaining superior innovative output. As discussed by Nickerson and Zenger (2004), a logic of discrete choice is preferable to a logic of "when in doubt, ask the manager, who has the answers." This may be especially true for innovation-related activities. While some existing literature brings up the idea that structuring in a particular way can support innovation (see, e.g., Tushman and Nadler, 1986, Teece, 1996), explanations are still needed regarding issues such as which design-related initiatives are comparatively superior for enhancing innovative performance and why. The question of when (and when not) to invest time and effort in creative resources and endeavors calls for further examination. Optimizing activities analytically (scientific management) is not enough in the complex realm of innovation.

Though it is beyond the scope of this chapter to discuss the full extent of the impact of increased network access and transparency (e.g., the internet) on innovation, we readily acknowledge that network effects as well as the rise of knowledge aggregators and seekers (for example, web- and email-based discussion groups, search engines, special interest sites, wikis, "social networking" media) have impacted the manner in which innovation is carried out. Enhanced network bandwidth and access affect ways of organizing for innovation. The focus of existing relevant work on innovation and the enhancements offered by networks is largely on location-based factors (e.g., Malhotra and Majchrzak, 2004 on far-flung teams, and Von Hippel, 1994 on locations and the stickiness of problem solving activities). This work increases our understanding of network effects on the feasibility of innovation, generally finding that network effects in many firms are facilitated by co-location. Network effects, in turn, facilitate innovation by lowering costs (for example, of searching for relevant previous work, or of maintaining contact with major players on a project). The questions of what sorts of problems and what sorts of problem finding processes are best chosen by managers to "benefit" from these processes are not, however, fully addressed by this literature.

Common Gaps in the Literatures

In both schools of thought discussed above, micro- and macro-organizational behavior, there is little convergence as to the best unit of analysis for understanding and managing innovation. Lack of convergence on the unit of analysis makes cumulative research progress—as well as combining micro- and macro-level frameworks and concepts—difficult. Should the unit of analysis be the industry? the organization? the team (group)? the project? the transaction? the task? the activity? The literature is ad hoc and inconclusive on this question. Consequently, existing theory provides little direction for examination of value creation and shows little evidence of cumulative paradigm-oriented research related to value creation. The strategic management literature focuses mostly on value capture side of innovation management at the expense of unpacking value creation and facilitation of innovations. The micro-organizational behavior literature, on the other hand, while more oriented towards determining optimal organizational configurations and measures for innovation, lacks marginal analysis and many studies underemphasize the importance of performance as an important outcome variable.

A DIFFERENT APPROACH: FOCUSING ON PROBLEMS

Considering the shortcomings and blind spots of existing theories, shifting the focus from capturing value to creating value means that the frameworks utilized for understanding and managing innovation need to be approached differently and perhaps revised. Instead of abandoning existing models and theories (which naturally would lead to further undesirable fragmentation), selected work can be built upon. Existing theories and approaches offer some useful tools that can be adopted. Some examples of these include the following: For (micro) Organizational Behavior, cognitive frames are a useful way to model snapshots of a cognitive state and the associated changes wrought by implementing processes (Witt, 2000). For (macro) Organizational Theory, the analytical techniques of network analysis have potential to shed light on the dispersion of knowledge within and between firms (Dyer and Nobeoka, 2000). Knowledge dispersion is an important aspect of the problem finding environment. Further, Transaction Cost Economics' core logic of economizing via managerial choice of organizing options underlies one highly promising research direction on which we subsequently build. Identifying and understanding the critical moments of managerial choice is central to the logic we promote in this chapter. Discriminating alignment of discrete managerial choices is a central tenet carried forward from Williamson's (1985) work.

A Promising Emergent Direction in Managing Innovation

Nickerson and Zenger (2004) have started the work of addressing the gaps in prior views by proposing a Knowledge Based Theory, the Problem Solving Perspective (PSP). Compared to much prior research, the PSP takes one relevant step back: It does not concentrate on how solving problems can lead to capturing value and improved performance, but rather takes a closer (but still somewhat cursory) look at how the problems come up in the first place. One of the most stimulating ideas in this work is that "value creating firms relentlessly uncover 'strategic problems' to solve, whether embedded with customers (or potential customers), with suppliers or within their own firms." (Nickerson et al. 2007, p. 212) The PSP primarily considers the challenges related to problem solving and presents remedies for these. Building on this work, and that of Nickerson, Silverman and Zenger (2007), in this chapter the focus turns from knowledge-based efficient problem solving in firms to knowledge-efficient problem finding for innovation (problem solving for commercial purposes), and to unpacking managerial/team biases as well as environmental impediments that can influence problem finding efficiency. This work is relevant for management of innovation because understanding the nature and impact of management's choices on knowledge in organizations facilitates effective and unbiased managerial choices regarding innovation activities.

Unpacking Value Creation

In the PSP framework, in line with prior research, value creation is based on generation of new knowledge (e.g., Nelson and Winter, 1982, Nonaka, 2002). However, there are differences in emphasis. While the problem solving perspective "shares the assumption of the Knowledge Based View that the key managerial objective is to create valuable new knowledge" (Nickerson et al. 2007, p. 213), it also departs from prior research that has emphasized supporting exchange, exploitation, and protection of knowledge through organizing as an important managerial task (see, e.g., Arrow, 1974; Oxley, 1997; Teece, 2006). Under a problem solving lens, the type(s) or substance of knowledge that need to be generated cannot in many cases be known beforehand (Arrow, 1974). Accordingly, problem finding is treated as a relevant question, and a central element of deliberate value creation. In Figure 1, value creation is shown split into problem finding and problem solving—two distinct activities. Figure 1 helps position this thread in the context of selected prior literature. Beckman and Barry (2007) perceptively point out that often, in the course of problem finding, many possible problems are uncovered, and there needs to be an intermediate stage of the value creation process between problem finding and problem solving: problem selection. We find this idea intriguing, but do not treat it in depth in this chapter.

The Problem as Unit of Analysis

In addition to putting value creation under intensive scrutiny, the Problem Solving Perspective aims at creating clarity by taking the problem as the unit of analysis. The focus is on the characteristics of problems and on the organizational structures that foster successful solution of problems (and, subsequently, capturing value). Managing innovation successfully depends on how well a firm is able to align different types of organizational forms with problem attributes to most efficiently solve problems Nickerson and Zenger (2004) have noted that the complexity (and resultant degree of decomposability) of problems is positively related to the comparative cost and effectiveness of intra-firm organization choices (i.e., authority- versus consensus-based management; see also Macher, 2006), which has an effect on the success of problem solving. (See Figure 2(a) below, showing the internal governance-choice aspect of the problem solving perspective as laid out in Nickerson and Zenger, 2004).

In a similar vein, for problem finding, conscious discriminating alignment (matching) of organizing choice to some problem-aspect is called for. In this case, the emphasis is not on matching governance choices to problems based on degree of decomposability/complexity (i.e., choosing among markets, authority-based hierarchies or consensus-based hierarchies per Nickerson and

Figure 1. Problem finding in the problem solving perspective.

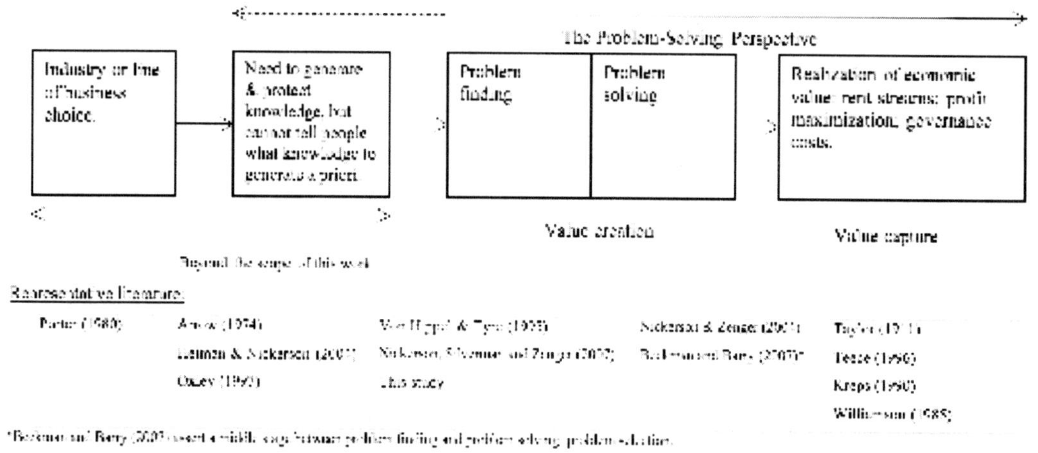

Zenger, 2004). Instead, problem finding emphasizes choosing appropriate processes for avoiding or overcoming different obstacles (biases) that may lead to sub-optimal problem-finding. Considering the "problem" of problem finding, Nickerson et al. (2007) state that processes (implementation of which result in measurable changes of state in projects and teams) need to be matched to individual biases, group biases and problem-environment-based impediments to finding good problems. Figure 2(b) shows problem finding as a special case of problem solving. Via a discriminating, economizing logic, problem finding matches the right process to individual and group biases and environmental aspects of problem finding.

Processes for Finding and Solving Problems

In the Problem Solving Perspective, different elements of problem finding and solving, as well as the characteristics of the problems themselves, play an important role. Different kinds of obstacles (and enhancing factors) exist in problem solving and problem finding, which means that various remedies (processes) are needed (see also Jensen and Harmsen, 2001). While problem solving concentrates on finding solutions to the problems at hand and seeks to deal with complexity issues, problem finding looks at processes that deal mainly with overcoming obstacles that make finding relevant problems ineffective.

Nickerson et al. (2007, p. 216) define processes as "individual, group, or organizational activities and efforts that aid in discovering problems that, if solved, create value." When considering processes that innovators use for finding and solving problems, one should keep in mind the various outcomes associated with each aspect of the total chain of causality. This includes intermediate innovation outcomes as opposed to profit. For example, the OSLO manual (OECD, 1997) suggests using share of turnover from new innovations as a measure. Outcomes of value capture, as noted, are the more conventional financial measures of performance, e.g., ROI, profitability, sales, ROA. Broadly, our outcome factor of interest is innovation outcomes, but these outcomes take several forms. Two outcomes (of problem finding and problem solving processes) are part of value creation. Another deals with value capture issues.

For clarity we briefly specify each outcome-type in the proposed causal chain. We envision a causality chain that starts with problem finding, which has as its outcome the quality of the

Figure 2. (a, b): Problem finding as a special case of problem-solving.

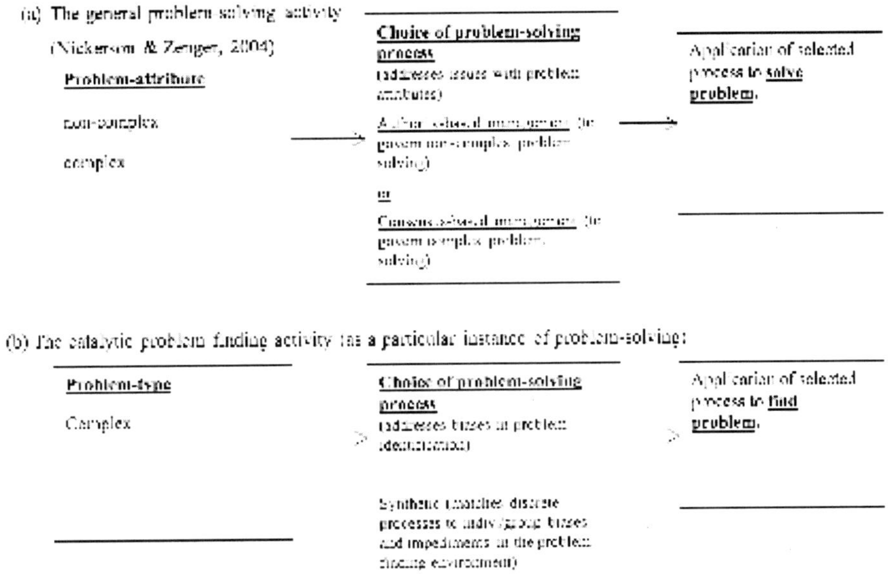

problem(s) found. Once an efficacious (or not) problem has been found, this becomes an input to the problem solving process, which has as its outcome the quality of problem solution(s). Once problems are solved, firms engage in earnest in value capture processes designed to exploit the solution(s) from problem solving. Though our main emphasis is on problem finding logic, the reader should understand the "larger picture:" we assert that solid problem finding outcomes (quality of problem found) lead to high performance problem solving (quality of solution), which leads to high performance value capture (profits). Two intermediate innovation outcomes (quality of problem(s) found and quality of solution(s) generated) are important. Since problem finding occurs at the start of the chain of reasoning, it seems especially crucial to offer a compelling logic of problem finding performance, one type of innovation outcome that has remained neglected until recently.

We see processes as simpler and less bound to high performance than Nickerson et al. (2007) do. To us, processes result in a change in the status of a problem or problem-solvers, for better or for worse. Processes can be modified, which means that if firms and teams can carry out needed process fixes without excessive costs, there is potential for improved performance. Processes can be seen as sequences of deductive or inductive steps through which desired outcomes are achieved (or not).

Processes that enable a firm to find relevant and valuable problems for innovation can be categorized as primarily either analytical or synthetic processes: taking deductive steps is a characteristic of analytical processes, whereas inductive, more exploratory approaches are used in relation to synthetic processes. While analytical processes more easily lead to finding (and later solving) incremental problems, synthetic processes may be comparatively more cost effective for generating more radical problems, and hence innovations (see Figure 2(b) above).

Impediments to Creativity and Innovation in Problem Finding

Processes chosen by managers should ideally deal with several elements that make problem finding less efficient. Such elements include individual and group biases, or attributes of the problem solving context, such as the extent of dispersion of essential knowledge across people (and organizations).

For Nickerson et al. (2007), the biases that can be overcome by appropriate application of *analytical processes* consist of anchoring, conflict of interest, and dominance. *Anchoring* refers to sticking to the way a process has always been done–letting path-dependence drive thinking. For instance, if a manufacturing process has always employed standard plastics casting techniques, and a new product is difficult to create using these techniques, an anchored group or individual would resist considering the use of specialized rotational molding techniques (regarding resistance, see also Fosfuri and Ronde 2007). *Conflict of interest* arises when a team member(s) pursues sub-goals that do not align with the goals of the project or whole team. For instance, a team member may feel that his or her prospects for promotion are improved if the team leader fails at an assigned project. Finally, *dominance* refers to a manager or influential team member asserting his or her authority in such a manner as to discourage or forbid fluid discourse regarding problems, inhibiting, for example, fluency in problem finding (finding lots of potential problems to solve). Dominance issues emerge, in particular, if a top manager sees him/herself as a controller and not an innovator. Per Kirton's Adoption-Innovation Inventory model (Chan, 1996, Kirton, 1980, 2003), under these conditions, innovation would be unlikely to flourish.

Analytical Processes for Overcoming Biases

We agree with Nickerson et al. (2007) that individual and group biases are likely to impact problem finding and that analytical (deductive) processes suffice to overcome biases in efforts to find problems involving less complex issues such as decreasing waste, reducing manufacturing costs, second-sourcing materials, or increasing product/service quality. Anchoring and dominance can be addressed by use of disciplined idea generation and filtering techniques, e.g., brainstorming. Genuine top management enthusiasm for a rules-based approach to deliberate innovation is also helpful. Dominance can be addressed by sharing or rotating team leadership (researchers at IBM, designers at Stanford University, IDEO and others are increasingly referring to this as radical collaboration). The rule, "rotate leadership appropriately, and help the current leader succeed" creates important group dynamics for performance (see also Davis and Eisenhardt, 2007). Conflict of interest is best addressed by aligning the incentives of all team members prior to project commencement. Conflicts of interest often involve private information and may be very difficult to detect without careful discussions and evaluations ex ante and during projects. Development of acceptance of others' views, which is a function of perception of others' competence and differences in individual cognitive frames, may also be relevant in overcoming these biases. For instance, managers should make sure that people are not included in teams that have poor perceptions of each other's competences. In such situations the likelihood of conflicts of is high.

The Problem of Finding Good Problems

Catalytic problems (a term recently informally originated by Jeff Dyer), are exemplified by problems faced by early-stage entrepreneurs. This formulation has roots in Christensen's (1997) examination of disruptive technologies and the need to find new problems to solve. Typical questions associated with problem finding in this realm include "what industry should my firm solve problems in?" "what problems do my potential customers need solved?" and "what problems will secure a growing stream of profits over time?" The highly complex problem of finding good catalytic problems is vulnerable to bias and requires particular *synthetic processes* for efficacy. While anchoring, conflict of interest, and dominance can be remedied with analytical processes, there exist other biases that may require different approaches. Christensen (1997) observes that an entrepreneur, armed with a disruptive technology, is already inherently biased against finding/recognizing/selecting other non-compatible approaches over his/her disruptive technology. For example, why start a new, for-profit educational organization to educate people regarding, say new recycling technology, when this information may be freely available to consumers via the World Wide Web?

Nickerson et al. (2007) identify five types of bias that hamper finding catalytic problems: perceptual, information distortion, groupthink, primacy and satisficing. *Perceptual bias* refers to peoples' distinct cognitive frames and their lack of ability to see beyond the frames they possess--rose-colored glasses. For example, the cognitive frame of an accountant might make it unlikely for that person to start a nanotechnology company. Second, *information distortion* refers to the inaccuracy of internal or external data available to a project team. For example, if a marketing manager possesses inaccurate or outdated knowledge of the direct and indirect costs of manufacturing a product, the pricing of that product may be problematic. Third, *groupthink* is a well-known phenomenon: groupthink-based decisions embody irrelevant compromises and use the results of sub-optimal interactions to find problems. For example, if brainstorming is used to find problems, people may collectively ignore the rule of "suspend judgment" and criticize new, offbeat ideas inappropriately. This deprives a team of potentially interesting fuel for exploration of problem finding. *Primacy* is similar to dominance in character but more informal: a team member engaging in primacy is overly insistent or vociferous regarding his or her ideas for good problems. This introduces bias into a team's problem finding, possibly colored by a desire to placate or fight the overbearing person (regardless of hierarchical authority—primacy is distinct from dominance). Finally, *satisficing*, identified by Simon (1945), means that people cease search activities, in this case the search for a good problem to solve, prior to finding the optimal problem to solve. For Simon, and later Williamson (1985), satisficing is an efficient, economizing behavior. Though we agree this is true for Williamson's transactional context, we note that satisficing is also a source of potential bias in finding good catalytic problems, and hence a potentially negative factor. For example, when managers consider results "good enough," the search for continuing profits through innovation becomes dampened. Related to this, new technology may be generated that lacks future markets (see Christensen, 1997): instead of taking another look at the technology's inherent potential future value, the search may be directed only to meet the needs of existing markets.

Matching Synthetic Process to Biases

Addressing the above-noted issues associated with catalytic problem finding entails the use of synthetic processes. Typically, these processes involve unstructured, non-systemic elements, such as ethnographic observation of leading edge or marginal users or critical reviews focused on user satisfaction. Thus, they are not directly related to marketing-driven priorities. Primacy and satisficing lie beyond the scope our present discussion, but we discuss in a general manner processes for remedying groupthink bias as well as information and perceptual distortion.

Dealing with perceptual and information distortion, as well as groupthink biases may be partly handled by developing and deploying multidisciplinary teams that also possess a high degree of fluency (Heiman and Burnett, 2007). Multidisciplinary teams bring a necessary diversity of perspectives to bear on many biases, especially perceptual and information distortion (Gupta et al. 1992, Christensen and Overdorf, 2000). Multidisciplinary teams also have elements in common with loosely coupled inter-firm innovation networks. The ways in which such inter-firm networks are governed (e.g., through facilitating knowledge mobility, stability of the network/team, and innovation appropriability) reflect wrestling with and hopefully overcoming biases within a firm with respect to a partner-firm (Dhanaraj and Parkhe 2006). Supporting fluency (the ability of individuals and groups to generate ideas and be open to the ideas of others) ensures that everyone understands and respects multiple perspectives, and this allows dealing with groupthink issues. Groupthink may also be dealt with via a so called skunk works model of innovation, where the resistance of manufacturing towards radical innovation is effectively avoided by isolating the R&D function for an adequately long time (see Fosfuri and Ronde 2007). We offer the reader a caveat here, because groupthink may also be fomented by a firm's formal or informal norms for decision making. Firm-level groupthink may also be difficult to recognize and correct (though this is one argument for bringing in outside contractors for finding new problems—they notice assumptions behind groupthink and will remark on it).

Problem Finding: Issues

Though we applaud Nickerson et al's (2007) ideas and framework, they admit that their logic is not complete; we feel compelled to point out several wrinkles in their approach. For example, why is primacy a bias associated solely with catalytic problems? People may act in an over-zealous manner in the finding (or prioritization) of non-catalytic problems as well, particularly if several problems of seemingly similar severity are seen to exist. Similarly, why are catalytic problem finding tasks not subject to dominance issues? Since senior managers often possess comparatively larger personal stakes in the outcomes of problem finding activities than junior managers, it seems possible, if not likely, that a senior manager would not hesitate to loudly voice his/her preferences or exert authority to ensure that his/her favorite catalytic problem is "found" and chosen. The possibility that senior managers may seek to torpedo a new direction is very real. Senior level managers can often kill an idea for a new problem to work on if they perceive it conflicts, for example, with personal or departmental goals. This reflects a conflict of interest bias. Dominance seems to overlap somewhat with the conflict of interest bias—dominant behavior may be a manifestation of an underlying conflict of interest. Biases, when specified, should be more clearly mutually exclusive.

We additionally observe that Nickerson et al's (2007) list of biases regarding finding catalytic problems seems incomplete. There are likely many other, as yet unspecified biases to which a group or individual may be subject. The notion of bias may have widespread, overlap with other threads of research. For example, the consideration of culture is one of the most prominent threads in micro-OB and international business (with roots in, e.g., Litwin and Stringer, 1986, Smircich, 1983, Barney, 1986, Hofstede, 1991), but what prevents us from claiming that firm-level culture simply reflects firm-level biases? It is tempting to turn up bias under every rock. Therefore, it is important to specify biases concretely.

Recognizing the need to usefully describe any new bias, we assert (and use as an example) the existence of a *"legalistic" bias*, pertaining to how intellectual property issues are perceived by individuals and teams. First, team members or individuals may be hesitant to search for problems in such areas where others have already worked owing to fear of running into legal problems with obtaining and maintaining secure rights to any resultant intellectual property (over-caution). A legalistic bias may also be a result of the desire to capture higher returns from value capture by doing things alone (also an instance of over-caution). Over-caution may result from employee agreements concerning rights to residual payments from inventions and ideas. Potential innovators may fear they will not be adequately rewarded for expressing their own (individual) best ideas for problems. They have agreed contractually that their firm "owns" the rights to their best ideas. This incentive structure partially explains why some entrepreneurially-oriented individuals (or teams) leave established firms to start new ones: their best efforts at problem finding will be more highly rewarded. Over-cautious non-disclosure agreements may create problems for efficient knowledge exchange and impede problem finding: people may hesitate to share useful knowledge of problem areas for fear of violating an agreement. This is a particularly acute problem for inter-firm knowledge sharing (see Heiman and Nickerson, 2004 for a discussion of the costs of knowledge diffused across firms). On the other hand, a team or manager may decide to use a trade secrets approach instead of seeking formal legal IP protection for cost reasons (an example of under-caution). This issue may be particularly acute for smaller firms with limited resources.

Like over-caution, under-caution in the legal realm may have a negative effect on the efficiency of knowledge sharing as choosing secrecy over more public forms of protection can be seen as a sign of distrust (see, e.g., Hannah, 2005). While over-cautious legal biases may result in over-cautious behavior towards, for example, potential collaboration partners, under-cautious biases may result in less than the minimum protection required to prevent leakages of intellectual property. *Synthetic processes* are needed to mitigate legalistic bias—it is not a simple matter of refinement or error/waste reduction (or other issues susceptible to deduction) to find good new problems. Building a complete understanding of the existing legal status of a prospective problem area through briefings is one process with potential. One synthetic process might be to arrange meetings with legal teams to discuss limits imposed by existing patents or other IP. Creating awareness of the need to limit invention to exclude certain areas can resolve many problems.

Legal bias can be very relevant in practice: Many people in modern organizations have legal content-exclusivity possibilities in mind when they begin developing "new" content. For example, in 1940, when choosing music to accompany the breakthrough animated film, Fantasia, Disney's team chose to use music of long-dead composers to avoid paying royalty fees. Contemporary Disney also recognizes the value of IP by vigorously protecting its own original characters and themes. Trademarks, royalties, copyrights and patents fall across both the value capture and the value creation realms. For value creation, the legalistic bias affects managerial decision making: with what problems do I avoid contracting/paying for usage rights? What problems should I attack so solutions are most likely to be protectable? For value capture, the legal environment of problem finding and solving often drives decisions. Realizing economic value from ownership of intellectual property via exploiting the rules of the legal system is paramount: how can I minimize the costs of obtaining substantial protection while profiting from my IP? What form of governance optimizes my returns on my IP? The degree of protection conferred by intellectual property rights, as well as the costs of obtaining/maintaining protection, are critical to outcomes. These are "classic" transaction cost/property rights questions; we see them as affecting problem finding outcomes.

In this section, we have briefly proposed and described some forms of legalistic bias, distinguishable roughly as under- and over-caution biases. We strongly suspect additional biases exist, and these should be articulated in at least as much detail as our above example. Nickerson et al's (2007) list of biases, comprised of anchoring, conflict of interest, dominance, perceptual, information distortion, groupthink, primacy and satisficing—in addition to being incomplete—is subject to refinement to eliminate overlap between biases. Also, the logic of how biases map to remedial processes (analytical vs. synthetic) is not yet perfectly clear. Identification of additional bias-types using empirical evidence represents a potentially major research thread for those interested in exploring this area further, and it should help managers a great deal: without a taxonomy of biases, the roles of individual biases are difficult to identify. Addressing biases and corresponding processes (remedies) enhances readiness to select proper processes and the chances of success (see Hurmelinna-Laukkanen and Puumalainen, 2007). For academics, developing new taxonomic frameworks for classifying biases also represents a substantial research opportunity for enhancing understanding of problem finding. Indeed, developing taxonomies of impediments that originate from the problem solving environment (i.e., disturbances not of a cognitive origin) offers yet another tantalizing chance to nurture this important and emerging research thread.

Imperfect Problem Finding

We suggest that it is plausible to treat problem finding as a special case of problem solving that involves matching remedial processes to biases. The "solutions" to finding good synthetic problems are all processes and result in changes of state after applied. A remedy for one or more biases either succeeds or fails. Figure 3(a) shows the prevalent view of problem finding (related to emphasis on value capture), and the actual situation (3(b)) as we see it, wherein the choice of process to overcome particular discrete bias types results in either successful (unbiased) problem finding or bias-contaminated problem finding. Framing problem finding in this manner holds promise for innovation management, particularly as testable hypotheses are feasible.

SOME TESTABLE HYPOTHESES

This section offers empirically tractable assertions regarding problem finding efficacy and summarizes the discussion presented above. First, we state that it is necessary for managers and team members to recognize that varying biases exist in people and organizations. Managers who pay attention to the existence of bias in problem finding will find valuable problems with greater efficacy by more often appropriately matching appropriate process type to the bias(es) present in the individual, team or problem finding environment. We suggest that awareness is the first step in successful problem finding:

Hypothesis 1: Awareness of individual/team and organizational biases is positively associated with high performance problem finding.

Awareness of biases may be at least partly instinctive, but it also can be deliberately constructed. Careful examination of prior successes and failures provides managers with one tool for making biases more visible. It also helps reveal the ways in which biases can and should be handled.

Indeed, in addition to the effects of understanding that biases exist and have an effect on problem finding outcomes, we expect direct effects on problem finding performance to be a result of the correct matching of process choice with bias(es). As the process choices made by managers and teams are increasingly well-aligned with biases inherent in individuals/teams/firms, we expect problem finding to improve. For high performance problem finding, processes that mitigate biases are chosen based on their comparative costs and competencies using an economizing orientation.

Hypothesis 2: Alignment of distinct process choices with organizational and individual biases is positively associated with high performance problem finding.

Aligning processes with biases requires managers to identify types of bias and to know the remedial properties of various processes. Random utilization of processes is not efficient. In order to get the matching right, a manager needs to know

Figure 3. (a,b): Problem finding: Conventional assumption (a) and the actual situation (b)

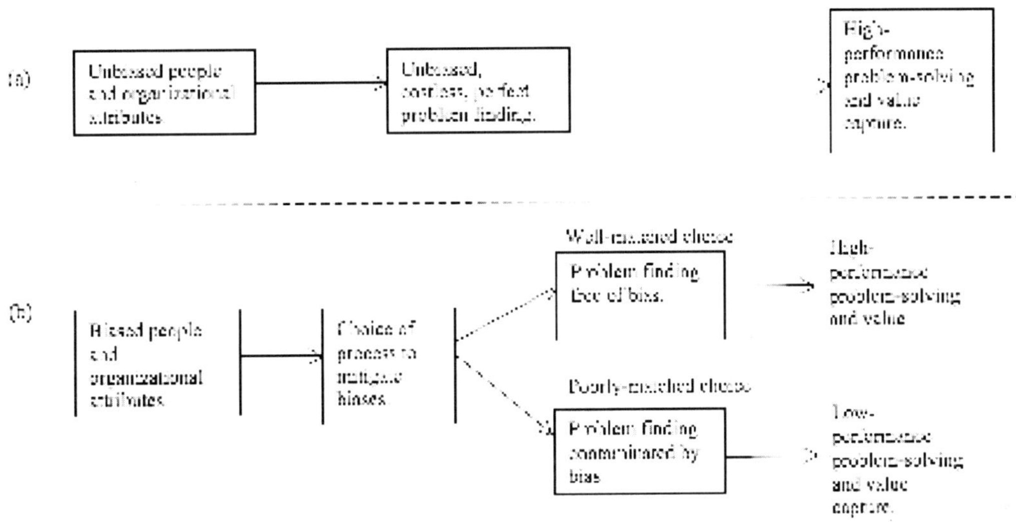

the composition of the team and individuals in it, be familiar with the problem finding environment, and keep up-to-date on the actions taken and their results (i.e., how well certain processes deal with biases in practice).

Team members' assigned and perceived roles in a firm are also important—job duties are often seen as limiting the scope of employees' efforts, including efforts at problem finding. Also, assigning duties according to the capabilities of individuals is important (see, e.g., Rizova, 2006). People and teams perform different designated roles in organizations—and teams—based on their views of what they are hired to do (not necessarily what they are best at). Often actors are required by their jobs' definitions (or induced by incentive systems) to focus their attention on incremental improvements over radical innovation (e.g., they are rewarded for as little disturbance in production and product quality as possible), or vice-versa. Successful and unbiased problem finding is related to assigned roles of people or groups undertaking activities. To achieve high performance innovation, individuals' and groups' roles need to be aligned with process choices: radical innovators deploy synthetic processes and incremental innovators use analytical processes.

Hypothesis 3: People in radical (incremental) innovator roles choose synthetic (analytic) processes to overcome biases in finding problems.

The three hypotheses above look at the impact of bias on problem finding performance (the outcome of interest). High performance problem finding, in turn, should result in problems that, once found, are tractable to a firm's particular (distinct) abilities to solve problems. In other words, the best companies are good at finding suitable problems for solving using the firm's existing capabilities and resources.

Finding problems that match the capabilities of the firm enables problem solving to proceed smoothly. The firm ideally has (access to) resources needed to solve the problem. If problems have emerged in problem finding, challenges are more likely to emerge in problem solving. In line with this, in Hypothesis 4, the interest shifts to problem solving from problem finding as outcome of interest—good problem finding should lead to good problem solving.

Hypothesis 4: High performance problem finding is positively associated with problem-solving performance.

An additional characteristic of a "good" problem is the organization's ability to solve it in a manner that is economically exploitable in relation to well-established, credible, feasible value capture mechanisms, for example, writing a long term contract so as to maximize licensing income from an innovation. High performance problem finding should not only lead to efficient problem solving, but should also ultimately result in high-performance (profitable) value capture activities. Finding problems which are highly amenable to value capture is critical. For example, Google has identified the mega-problem of organizing the world's information in a way that is economically rewarding (so far with great success). Finding important problems relatively quickly (especially if the operations environment of a firm is very turbulent and dynamic), and finding problems that the organization can economically exploit, gives a firm competitive advantage. Hypothesis 5 again shifts the outcome of interest, this time to value capture: We expect that good problem finding practices lead to value capture performance:

Hypothesis 5: High performance problem finding is positively associated with high value-capture performance (e.g., project profitability).

FUTURE RESEARCH DIRECTIONS

Specific opportunities for research include, as noted above, discovering new biases, refining the logic of how and why a particular bias maps to a particular remedial process, creating taxonomies of biases and environment-based impediments, and empirically testing our hypotheses. Most importantly, empirical evidence for the PF/PS perspective remains to be convincingly established—opportunities abound for confirmation, refutation, and refinement studies. For a more in-depth discussion regarding the nature of potential rewarding research directions, see Heiman, Nickerson and Zenger (2008) who highlight current research opportunities in substantial detail.

Here, we merely note that the study of the alignment of processes to biases is in its early days, and has been constructed with empirical tractability in mind. The logic employed is economizing in nature, and follows Williamson's (1985) transaction cost economics assumptions regarding human behavior: rationality is bounded, and people may misbehave to the detriment of outcomes. Managerial choice of process to match biases in project teams is a "Williamsonian" notion, but the applicable unit of analysis is the problem, rather than the transaction (Nickerson and Zenger, 2004). The transaction as the unit of analysis is deceptively alluring; the logic is attractive, but the focus of transaction cost thinking is almost exclusively on value capture. By refocusing on the problem as the unit of analysis, the unpacking of value creation as a relevant part of innovation management in a useful, actionable manner becomes possible.

CONCLUDING REMARKS

This chapter has sought to redirect the question of "how to" manage innovation towards understanding what it means to manage innovation well from a problem finding/problem solving perspective. This approach focuses on the importance of particular knowledge-based factors' (biases' and processes') influence on innovation outcomes. The chapter initially briefly discussed the rise of knowledge and its impact on the environment: continuous innovation is increasingly necessary to firms' survival. Following Nickerson, et al. (2007), by distinguishing between value capture and value creation, and subsequently unpacking value creation, we raised the question, "what comprises high performance innovation using the problem as unit of analysis?" Subsequent to defining terms, we touched on the literature of managing innovation from the strategic management and micro-organizational behavior perspectives. We then introduced the problem finding and problem solving approach as a response to the shortcomings of these two views. We recognized Nickerson and Zenger's (2004) contribution of a logic for problem solving, but argued for treating problem finding as a special case of problem solving (diverging from Nickerson et al. 2007). We offered critical reflections on the current conceptualization of problem finding and the specification of biases, and briefly argued for and stated testable hypotheses.

For enhancing and managing innovation, managers need to make decisions that allow many different ideas to be found. They need to ensure that teams consist of people with both different knowledge bases and different cognitive frames, yet avoid conflicts that inhibit good problem finding. Dominance and conflict of interest might be expected to be particularly detrimental to high performance problem finding. Further, managers need to select solvable problems that also yield profit when solved—problems must be tractable to available resources and skills in the team/firm. Broadly, managing innovation involves a logic of managerial choice: matching organizing approaches to impediments to performance; the core logic is economizing and highly Williamsonian: an economizing, discriminating alignment of discrete managerial process choices to mitigate bias issues and thereby foment superior innovation outcomes.

The emerging problem finding/problem solving perspective (PF/PS) represents one way to proceed in research aimed at deepening our understanding management of innovation. Its benefits include empirical tractability, a concrete unit of analysis (the problem), and a cumulative paradigmatic orientation. The relative lack of empirical research in this relatively new research thread should not be seen as a flaw, but rather as an opportunity for scholars interested in making meaningful progress towards advancing knowledge of managing innovation. Relevance for managers starts with developing an awareness of the existence of biases in teams and individuals as well as in the environment in which problem finding occurs. For example, deliberately addressing potential legal bias issues at the start of a project seems sensible. Also, consciously building multidisciplinary teams helps mitigate perceptual distortion bias in innovation: diverse points of view, when articulated, offer more objective (multiple) perspectives on problems than a very few, narrow points of view. More broadly, improved innovation outcomes are initially driven by overcoming biases in problem finding; this sets the stage for subsequent efficient problem solving and value capture activities. The PF/PS perspective holds promise for practitioners and academicians, but awaits extensive empirical confirmation.

REFERENCES

Adams, J. L. (1974). *Conceptual blockbusting: A guide to better ideas*. New York: W.H. Freeman & Co Ltd.

Amit, R., & Zott, C. (2001). Value creation in e-business. *Strategic Management Journal, 22*, 493–520. doi:10.1002/smj.187

Anderson, N., Dreu, C., & Nijstad, B. (2004). The routinization of innovation research: a constructively critical review of the state-of-the-art-science. *Journal of Organizational Behavior, 25*, 147–173. doi:10.1002/job.236

Arrow, K. J. (1974). *Organization and Information. The Limits of Organization*. New York: W.W. Norton & Co.

Barney, J. B. (1986). Organizational Culture: Can It Be a Source of Sustained Competitive Advantage. *Academy of Management Review, 11*(3), 656–665. doi:10.2307/258317

Barney, J. B. (1991). Firm resources and sustainable competitive advantage. *Journal of Management*, *17*, 99–120. doi:10.1177/014920639101700108

Beckman, S. L., & Barry, M. (2007). Innovation as a learning process: Embedding design thinking. *California Management Review*, *50*(1), 31.

Boston Consulting Group. (2005). *ITC Survey of Senior Executives*. Retrieved from http://www.signallake.com/innovation/BCG2005ITCSurveyReport.pdf

Bowen, D. E., & Ostroff, C. (2004). Understanding HRM-Firm performance linkages: The role of the "strength" of the HRM system. *Academy of Management Review*, *29*(2), 203–221. doi:10.2307/20159029

Chan, D. (1996). Cognitive misfit of problem-solving style at work: A facet of person-organization fit. *Organizational Behavior and Human Decision Processes*, *68*(3), 194–207. doi:10.1006/obhd.1996.0099

Chesbrough, H. Vanhaverbeke, W. & West, J. (2006) Open Innovation: Researching a New Paradigm. New York: Oxford University Press, USA.

Chesbrough, H. W. (2003). Environmental influences upon firm entry into new sub-markets Evidence from the worldwide hard disk drive industry. *Research Policy*, *32*(4), 659–678. doi:10.1016/S0048-7333(02)00033-1

Christensen, C. M and Overdorf, M. (2000). Meeting the challenge of disruptive innovation. *Harvard Business Review, 78*(2), 66-76.

Christensen, C. M. (1997). *The Innovator's Dilemma*. Cambridge, MA: Harvard Business School Press.

Conner, K. R. (1991). A historical comparison of resource-based theory and five schools of thought within industrial organization economics: Do we have a new theory of the firm? *Journal of Management*, *17*(1), 121–154. doi:10.1177/014920639101700109

Conner, K. R., & Prahalad, C. K. (1996). A resource-based theory of the firm: Knowledge versus opportunism. *Organization Science*, *7*(5), 477–501. doi:10.1287/orsc.7.5.477

Cyert, R. M., & March, J. G. (1963). *A Behavioral Theory of the Firm*. Englewood Cliffs, NJ: Prentice-Hall.

Davis, J. P., & Eisenhardt, K. (2007). *Rotating leadership and symbiotic organization: relationship processes in the context of collaborative innovation*. MIT Sloan School of Management Working Papers.

Demsetz, H. (1988). The theory of the firm revisited. *Journal of Law Economics and Organization*, *4*(1), 141–162.

Dent, E. B. (2002). The messy history of OB&D: How three strands came to be seen as one rope. *Journal of Management History*, *40*(3), 266–280.

Dhanaraj, C., & Parkhe, A. (2006). Orchestrating innovation networks. *Academy of Management Review*, *31*(3), 659–669. doi:10.2307/20159234

Dierickx, I., & Cool, K. (1989). Asset Stock Accumulation and Sustainability of Competitive Advantage [Comment; Reply]. *Management Science*, *35*(12), 1504–1514. doi:10.1287/mnsc.35.12.1504

Dyer, J. H., & Nobeoka, K. (2000). Creating and managing a high-performance knowledge-sharing network: the Toyota case. *Strategic Management Journal*, *21*(3), 345–367. doi:10.1002/(SICI)1097-0266(200003)21:3<345::AID-SMJ96>3.0.CO;2-N

Eisenhardt, K. M., & Martin, J. A. (2000). Dynamic capabilities: What are they? *Strategic Management Journal*, *21*(10-11), 1105–1121. doi:10.1002/1097-0266(200010/11)21:10/11<1105::AID-SMJ133>3.0.CO;2-E

Fosfuri, A., & Ronde, T. (2007). *Leveraging resistance to change and the Skunk Works model of innovation.* Corporate intellectual property (RIPE) Conference, 7-8 December 2007, Copenhagen, Denmark.

Foss, N. (1996). Knowledge-based approaches to the theory of the firm: Some critical comments. *Organization Science, 7*(5), 470–476. doi:10.1287/orsc.7.5.470

Grant, R. M. (1996). Toward a knowledge-based theory of the firm. *Strategic Management Journal, 17*, 109–122.

Gupta, A. K., Brockhoff, K., & Weisenfeld, U. (1992). Making trade-offs in new product development process: A German/US comparison. *Journal of Product Innovation Management, 9*(1), 11–18.

Hanks, K., Belliston, L., & Edwards, D. (1977). *Design Yourself. Los Altos, CA*. W.: Kauffman Inc.

Hannah, D. R. (2005). Should I keep a secret? The effects of trade secret protection procedures on employees' obligations to protect trade secrets. *Organization Science, 16*(1), 71–84. doi:10.1287/orsc.1040.0113

Heiman, B. A., & Burnett, W. R. (2007). *The Isolation of Design Thinking in Firms and Management Education: Definitions, Impediments, and Ways Forward. Society for Global Business and Economic Development*. Kyoto, Japan: SGBED.

Heiman, B. A., & Nickerson, J. A. (2004). Empirical evidence regarding the tension between knowledge sharing and knowledge expropriation in collaborations. *Managerial and Decision Economics, 25*(6-7), 401–420. doi:10.1002/mde.1198

Heiman, B. A., Nickerson, J. A., & Zenger, T. (2008). *Knowledge governance: Multidisciplinary perspectives* (Michailova, S., & Foss, N., Eds.). Oxford, UK: Oxford University Press.

Hersey, P., & Blanchard, K. (1988). *Management of Organizational Behavior*. Englewood Cliffs, NJ: Prentice Hall, Inc.

Hofstede, G. (1991). *Cultures and Organizations: Software in the Mind*. Maidenhead, UK: McGraw Hill.

Hurmelinna-Laukkanen, P., & Puumalainen, K. (2007). The nature and dynamics of appropriability – Strategies for appropriating returns on innovation. *R & D Management, 37*(2), 95–112. doi:10.1111/j.1467-9310.2007.00460.x

Jensen, B., & Harmsen, H. (2001). Implementation of success factors in new product development – the missing links? *European Journal of Innovation Management, 4*(1), 37–52. doi:10.1108/14601060110365565

Johannessen, J.-A., Olsen, B., & Olaisen, J. (1999). Aspects of innovation theory based on knowledge-management. *International Journal of Information Management, 19*, 121–139. doi:10.1016/S0268-4012(99)00004-3

Kirton, M. J. (1980). Adaptors and Innovators in Organizations. *Human Relations, 3*, 213–224. doi:10.1177/001872678003300401

Kirton, M. J. (2003). *Adaption-Innovation in the Context of Diversity and Change*. New York: Routledge.

Koberg, D., & Bagnall, J. (1974). *The Universal Traveler*. Los Altos, CA: William Kaufmann, Inc.

Kogut, B., & Zander, U. (1992). Knowledge of the firm, combinative capabilities and the replication of technology. *Organization Science, 3*(3), 383–397. doi:10.1287/orsc.3.3.383

Kreps, D. (1990). A Course in Microeconomic Theory. In *Modeling competitive situations*. Princeton, NJ: Princeton University Press.

Kuhn, T. (1970). *The Structure of Scientific Revolutions*. Chicago: Chicago University Press.

Levinthal, D. A., & March, J. G. (1993). The Myopia of Learning. *Strategic Management Journal, 14*, 95–112. doi:10.1002/smj.4250141009

Litwin, G. H., & Stringer, R. A. (1968). *Motivation and Organizational Climate*. Boston: Harvard University.

Macher, J. T. (2006). Technological Development and the Boundaries of the Firm: A Knowledge-Based Examination in Semiconductor Manufacturing. *Management Science, 52*(6), 826–843. doi:10.1287/mnsc.1060.0511

Mahoney, J. T., & Pandian, R. J. (1992). The Resource-based view within the conversation of strategic management. *Strategic Management Journal, 13*(5), 363–380. doi:10.1002/smj.4250130505

Malhotra, A., & Majchrzak, A. (2004). Enabling Knowledge Creation In Far-Flung Teams: Best Practices For IT Support And Knowledge Sharing. *Journal of Knowledge Management, 8*(4), 75–88. doi:10.1108/13673270410548496

March, J. (1991). Exploration and Exploitation in Organizational Learning. *Organization Science, 2*(1), 71–87. doi:10.1287/orsc.2.1.71

Mintzberg, H. (1979). *The Structure of Organizations*. Englewood Cliffs, NJ: Prentice-Hall.

Nelson, R. R. (1991). Why do firms differ, and how does it matter? *Strategic Management Journal, 12*(8), 61–74. doi:10.1002/smj.4250121006

Nelson, R. R., & Winter, S. G. (1982). *An Evolutionary Theory of Economic Change*. Cambridge, UK: Belknap Press.

Nickerson, J. A., Silverman, B. S., & Zenger, T. R. (2007). The 'problem' of creating and capturing value. *Strategic Organization, 5*(3), 211–225. doi:10.1177/1476127007079969

Nickerson, J. A., & Zenger, T. R. (2004). A knowledge-based theory of governance choice—A problem-solving approach. *Organization Science, 15*(6), 617–632. doi:10.1287/orsc.1040.0093

Nonaka, I. (2002). A Dynamic Theory of Organizational Knowledge Creation. In Choo, C. W., & Bondis, N. (Eds.), *The Strategic Management of Intellectual capital and Organizational Knowledge* (pp. 437–462). New York: Oxford University Press.

OECD. (1997). *Oslo Manual - proposed guidelines for collecting and interpreting technological innovation data*. Paris: OECD.

Oxley, J. (1997). Appropriability hazards and governance in strategic alliances: A transaction cost approach. *Journal of Law Economics and Organization, 13*(2), 387–409.

Penrose, E. T. (1959). *The Theory of the Growth of the Firm*. Oxford, UK: Basil Blackwell.

Peteraf, M. A. (1993). The cornerstones of competitive advantage – a Resource-based view. *Strategic Management Journal, 14*(3), 179–191. doi:10.1002/smj.4250140303

Porter, L. W. (1996). Forty years of organizational studies: Reflections from a micro perspective. *Administrative Science Quarterly, 41*, 262–269. doi:10.2307/2393718

Porter, M. E. (1980). *Competitive Strategy*. New York: The Free press.

Rizova, P. (2006). Are you networked for successful innovation? *MIT Sloan Management Review, 47*(3), 49–55.

Rothwell, R. (1994). Towards the fifth-generation innovation process. *International Marketing Review, 11*(1), 7–31. doi:10.1108/02651339410057491

Rumelt, R. P. (1984). Towards a Strategic Theory of the Firm. In Lamb, R. (Ed.), *Competitive Strategic Management* (pp. 556–570). Englewood Cliffs, NJ: Prentice-Hall.

Rumelt, R. P. (1987). Theory, Strategy, and Entrepreneurship. In Teece, D. (Ed.), *The Competitive Challenge: Strategies for Industrial Innovation and Renewal* (pp. 137–158). Cambridge, MA: Ballinger.

Schumpeter, J. A. (1942). *Capitalism, Socialism and Democracy*. London: G. Allen & Unwin Ltd.

Simon, H. (1945). *Administrative Behavior*. New York: The Free Press.

Smircich, L. (1983). Concepts of Culture and Organizational Analysis. *Administrative Science Quarterly*, *28*(3), 339–358. doi:10.2307/2392246

Sveiby, K. E. (1997). *The new organizational wealth: Managing and measuring knowledge-based assets*. San Francisco: Berrett-Koehler.

Taylor, F. W. (1911). *The Principles of Scientific Management*. New York: Harper & Brothers.

Teece, D. J. (1996). Firm Organization, Industrial Structure, and Technological Innovation. *Journal of Economic Behavior & Organization*, *31*(2), 193–224. doi:10.1016/S0167-2681(96)00895-5

Teece, D.J. (2004). *The emerging science of Strategic Management*. Prepared for a panel discussion at Lappeenranta University of Technology May 13, 2004.

Teece, D. J. (2006). Reflections on "Profiting from innovation.". *Research Policy*, *35*(8), 1131–1146. doi:10.1016/j.respol.2006.09.009

Teece, D. J. (2007). Explicating Dynamic Capabilities: The nature and microfoundations of (sustainable) enterprise performance. *Strategic Management Journal*, *8*(13), 1319–1350. doi:10.1002/smj.640

Teece, D. J., Pisano, G., & Shuen, A. (1997). Dynamic capabilities and strategic management. *Strategic Management Journal*, *18*(7), 509–533. doi:10.1002/(SICI)1097-0266(199708)18:7<509::AID-SMJ882>3.0.CO;2-Z

Tidd, J., Bessant, J., & Pavitt, K. (2005). *Managing Innovation: Integrating Technological, Market and Organizational Change*. Chichester, UK: John Wiley & Sons.

Tushman, M., & Nadler, D. (1986). Organizing for Innovation. *California Management Review*, *28*(3), 74–92.

Von Hippel, E. (1994). 'Sticky information' and the locus of problem solving: Implications for Innovation. *Management Science*, *40*(4), 429–438. doi:10.1287/mnsc.40.4.429

Von Hippel, E., & Tyre, M. (1995). How Learning by Doing is Done: Problem Identification in Novel Process Equipment. *Research Policy*, *24*(1), 1–12. doi:10.1016/0048-7333(93)00747-H

Wernerfelt, B. (1984). A Resource-Based View of the Firm. *Strategic Management Journal*, *12*, 75–94.

Williamson, O. E. (1985). *The Economic Institutions of Capitalism*. New York: Macmillan Inc.

Williamson, O. E. (1991). Strategizing, economizing, and economic organization. *Strategic Management Journal*, *12*(8), 75–94. doi:10.1002/smj.4250121007

Williamson, O. E. (1999). Strategy research: governance and competence perspectives. *Strategic Management Journal*, *20*(12), 1087–1108. doi:10.1002/(SICI)1097-0266(199912)20:12<1087::AID-SMJ71>3.0.CO;2-Z

Winter, S. G. (2006). The logic of appropriability: From Schumpeter to Arrow to Teece. *Research Policy*, *35*(8), 1100–1106. doi:10.1016/j.respol.2006.09.010

Witt, U. (2000). Changing cognitive frames - changing organizational forms: an entrepreneurial theory of organizational development. *Industrial and Corporate Change*, *9*(4), 733–755. doi:10.1093/icc/9.4.733

Zahra, S. A., & George, G. (2002). Absorptive capacity: A Review, reconceptualization, and extension. *Academy of Management Review*, *27*(2), 185–203. doi:10.2307/4134351

ADDITIONAL READING

Arrow, K. J. (1974). *Organization and Information. The Limits of Organization*. New York: W.W. Norton & Co.

Barney, J. B. (1991). Firm resources and sustainable competitive advantage. *Journal of Management*, *17*, 99–120. doi:10.1177/014920639101700108

Beckman, S. L., & Barry, M. (2007). Innovation as a learning process: Embedding design thinking. *California Management Review*, *50*(1), 31.

Christensen, C. M. (1997). *The Innovator's Dilemma*. Cambridge, MA: Harvard Business School Press.

Cohen, W. M., & Levinthal, D. A. (1990). Absorptive capacity: A new perspective on learning and innovation. *Administrative Science Quarterly*, *35*(1), 128–152. doi:10.2307/2393553

Cyert, R. M., & March, J. G. (1963). *A Behavioral Theory of the Firm*. Englewood Cliffs, NJ: Prentice-Hall.

Eckhardt, J. T., & Ciuchta, M. P. (2008). Selected variation: The population-level implications of multistage selection in entrepreneurship. *Strategic Entrepreneurship Journal*, *2*(3), 209–224. doi:10.1002/sej.48

Foss, K., & Foss, N. J. (2008). Understanding opportunity discovery and sustainable advantage: The role of transaction costs and property rights. *Strategic Entrepreneurship Journal*, *2*(3), 191–208. doi:10.1002/sej.49

Foss, N. (1996). Knowledge-based approaches to the theory of the firm: Some critical comments. *Organization Science*, *7*(5), 470–476. doi:10.1287/orsc.7.5.470

Heiman, B. A., & Burnett, W. R. (2007). *The Isolation of Design Thinking in Firms and Management Education: Definitions, Impediments, and Ways Forward. Society for Global Business and Economic Development*. Kyoto, Japan: SGBED.

Heiman, B. A., & Nickerson, J. A. (2004). Empirical evidence regarding the tension between knowledge sharing and knowledge expropriation in collaborations. *Managerial and Decision Economics*, *25*(6-7), 401–420. doi:10.1002/mde.1198

Heiman, B. A., Nickerson, J. A., & Zenger, T. (2008), Knowledge governance: Multidisciplinary perspectives. Chapter 2. Snejina Michailova and Nicolai Foss, Eds., Oxford: Oxford University Press.

Klein, P. G. (2008). Opportunity discovery, entrepreneurial action and economic organization. *Strategic Entrepreneurship Journal*, *2*(3), 175–190. doi:10.1002/sej.50

Kogut, B., & Zander, U. (1992). Knowledge of the firm, combinative capabilities and the replication of technology. *Organization Science*, *3*(3), 383–397. doi:10.1287/orsc.3.3.383

Macher, J. T. (2006). Technological Development and the Boundaries of the Firm: A Knowledge-Based Examination in Semiconductor Manufacturing. *Management Science*, *52*(6), 826–843. doi:10.1287/mnsc.1060.0511

Mahoney, J. T., & Pandian, R. J. (1992). The Resource-based view within the conversation of strategic management. *Strategic Management Journal*, *13*(5), 363–380. doi:10.1002/smj.4250130505

Malhotra, A., & Majchrzak, A. (2004). Enabling Knowledge Creation In Far-Flung Teams: Best Practices For IT Support And Knowledge Sharing. *Journal of Knowledge Management*, *8*(4), 75–88. doi:10.1108/13673270410548496

March, J. (1991). Exploration and Exploitation in Organizational Learning. *Organization Science*, *2*(1), 71–87. doi:10.1287/orsc.2.1.71

Michailova, S., & Foss, N. (Eds.). (2008). *Knowledge governance: Multidisciplinary perspectives*. Oxford: Oxford University Press.

Mitchell, R. K., Mitchell, J. R., & Smith, J. B. (2008). Selected variation: The population-level implications of multistage selection in entrepreneurship. *Strategic Entrepreneurship Journal*, *2*(3), 225–242. doi:10.1002/sej.51

Nelson, R. R., & Winter, S. G. (1982). *An Evolutionary Theory of Economic Change*. Cambridge: Belknap Press.

Nickerson, J. A., Silverman, B. S., & Zenger, T. R. (2007). The 'problem' of creating and capturing value. *Strategic Organization*, *5*(3), 211–225. doi:10.1177/1476127007079969

Nickerson, J. A., & Zenger, T. R. (2004). A knowledge-based theory of governance choice—A problem-solving approach. *Organization Science*, *15*(6), 617–632. doi:10.1287/orsc.1040.0093

Simon, H. (1945). *Administrative Behavior*. New York: The Free Press.

Teece, D. J. (2007). Explicating Dynamic Capabilities: The nature and microfoundations of (sustainable) enterprise performance. *Strategic Management Journal*, *8*(13), 1319–1350. doi:10.1002/smj.640

Teece, D. J., Pisano, G., & Shuen, A. (1997). Dynamic capabilities and strategic management. *Strategic Management Journal*, *18*(7), 509–533. doi:10.1002/(SICI)1097-0266(199708)18:7<509::AID-SMJ882>3.0.CO;2-Z

Von Hippel, E. (1994). 'Sticky information' and the locus of problem solving: Implications for Innovation. *Management Science*, *40*(4), 429–438. doi:10.1287/mnsc.40.4.429

Von Hippel, E., & Tyre, M. (1995). How Learning by Doing is Done: Problem Identification in Novel Process Equipment. *Research Policy*, *24*(1), 1–12. doi:10.1016/0048-7333(93)00747-H

Williamson, O. E. (1985). *The Economic Institutions of Capitalism*. New York: Macmillan Inc.

Williamson, O. E. (1999). Strategy research: governance and competence perspectives. *Strategic Management Journal*, *20*(12), 1087–1108. doi:10.1002/(SICI)1097-0266(199912)20:12<1087::AID-SMJ71>3.0.CO;2-Z

Zahra, S. A. (2008). The virtuous cycle of discovery and creation of entrepreneurial opportunities. *Strategic Entrepreneurship Journal*, *2*(3), 243–258. doi:10.1002/sej.47

KEY TERMS AND DEFINITIONS

Creativity: Fluency in idea generation and openness to new ideas, in particular, openness to the ideas of others.

Design: A set of tools or processes that are deployed in the service of creativity. Design often entails the ability to communicate new ideas to self, designers and/or the outside world.

Innovation: Creativity and design applied for the purposes of commercialization; problem finding and problem solving are key elements of innovation.

Process: An activity, policy or rule that, when implemented, results in a change of state in a team or project outcome(s).

Governance: Refers to the mode of organization of economic activities (e.g., a firm or a market, see Williamson, 1985). Has recently been extended by Nickerson and Zenger (2004) to also encompass variation in the way activities are organized within a firm.

Synthetic Process: A process that involves inductive, often unstructured methods to limit the impact of biases on problem finding efficacy.

Analytic Process: A process that involves deductive, highly structured, systemic methods to limit the impact of biases on problem finding efficacy.

Anchoring: Sticking to the way a process has always been done, historically, especially with respect to the medium and long-term history of a team or firm (contrast to primacy).

Conflict of Interest: Pursuing individual subgoals that do not align with the goals of others, the project or whole team.

Dominance: A manager or team member asserting his or her authority in a manner so as to discourage, impede, or forbid fluid discourse regarding problems.

Perceptual Distortion: Peoples' inability to see beyond their own personal cognitive frames.

Information Distortion: The inaccuracy of internal or external data available to a project team.

Groupthink: Groupthink-based decisions embody irrelevant compromises and use the results of sub-optimal interactions to find problems.

Primacy: Recent prior approaches to action influence current/future approaches (contrast to anchoring).

Satisficing: People/teams cease search activities prior to finding the optimal problem to solve.

Chapter 6
Innovation Dynamics of Materials Technology:
An Empirical Analysis and Recommendations

Masahiro Nakagawa
Japan Science and Technology Agency, Japan

Chihiro Watanabe
Tokyo Seitoku University, Japan

ABSTRACT

Technological capital drives technological innovation if it is appropriately used; in reality, however, many kinds of management constraints hamper its effective use. This chapter elucidates changes in the use of technological capital from the technology spillover perspective. Tracking patent applications and analyzing the relationships among them have enabled us to conduct microscopic analyses of technology spillovers. The authors classify technology spillover structures by organization and technological field, forming a matrix of two-by-two arrays and applying it to Japan's R&D in compound semiconductor materials. In the 1980s, technology management was affected by the characteristics of an industrial society and spillover was therefore limited to individual firms and technological fields. In the 1990s, technology spillover shrank, reflecting organizational inertia in technology management. In the early 2000s, technology spillover broadened across a wider range of industries and technological fields, reflecting changes in science and technology policy and technology management that was ripe for open innovation.

INTRODUCTION

In the first decade of the 21st century, with the ongoing rapid emergence of service innovation, technological innovation is becoming a matter of common interest to economists, business leaders, and even consumers. For example, society has eagerly embraced desirable services such as those offered by Google and YouTube. These new services are, needless to say, built on information technology (IT) infrastructure. As IT infrastructure is based on electronic device technology, service innovation can in turn deliver new functions for material technology, and especially technology related to nonferrous metal materials. On the other

DOI: 10.4018/978-1-61520-875-3.ch006

hand, Internet services have resulted in goals being set for cutting-edge materials innovation. In other words, materials innovation has supported service innovation, and service innovation has boosted materials innovation. This means that there has been a degree of interplay between service technology, information technology, and materials technology. Moreover, facing an unprecedented economic crisis, the economy is increasingly reliant on technological innovation. Therefore, now is an appropriate time to elucidate the dynamics of technological innovation.

The relationship between innovation and economic growth has been demonstrated from the intellectual capital perspective. Kennedy (1964) shows that innovation can not only act as a substitute for or save on capital and labor, but also that equilibrium between capital saving and labor saving innovations tends to be maintained in the long run. Griliches (1979) demonstrates that the stock of technological capital[1] can be calculated by aggregating R&D investment, taking into account depreciation by obsolescence. Moreover, Jaffe (1986) quantifies the effects of technological opportunity and technology spillovers on R&D productivity. These studies demonstrate that innovation plays a significant role in economic growth; technology capital stock is now recognized as a key success factor for innovation.

Although the above mentioned studies focus on cross-industry analysis, we seek to highlight the details of a specific firm in a specific industry and elucidate microscopic phenomena in which innovation is actually created. From the economics and innovation perspective, the nonferrous metal industry is one of the best cases for empirical analysis for three reasons. First, because it has supplied a variety of materials to the IT industry, it has been a major technology spillover source for that industry, and has subsequently acted as a spillover source for the service industry. Second, because the nonferrous metal industry has been highly R&D-intensive in comparison with other materials industries, it represents a good industry to analyze in this context; and third, the nonferrous metal industry has a longer history than either the IT industry or the service industry. These features suggest that this industry meets the general requirements of a proxy.

We conduct an empirical analysis on the nonferrous metal industry in Japan over the past quarter of a century, elucidating changes in the relationship between technological capital, technology spillover and technology management. Changes in firm profits can largely be attributed to changes in technology spillover management. Moreover, we confirm that closed and open spillover structures induce each other on a recurrent basis and result in new innovations. The conclusions outlined above provide the basis for some useful recommendations for technological innovation leaders.

Background

Since Solow (1957) confirmed that in addition to capital and labor, technology also contributes to industrial productivity, technological capital and innovation have been major concerns for economists and business leaders. Kennedy (1964) shows that innovation can not only act as a substitute for capital and labor, but also it tends to lead to a certain level of equilibrium between those two factors of production. Griliches (1979) proposes a method for calculating technological capital stocks. These studies confirm that the creation of technological capital is equally as important as capital and labor.

On the other hand, the utilization of technological capital is just as important as its creation. Jaffe (1986) examines circumstantial evidence of technology spillover by analyzing patents granted to 1,700 firms. Ornaghi (2006) suggests that knowledge spillover plays an important role in improving the quality of products. Griliches & Lichtenburg (1984), Bernstein & Nadiri (1988, 1989), Goto & Suzuki (1989) and Nakanishi (2002) demonstrate the positive effects of technology spillover on technological capital and the economy. Watanabe, Zhu, Griffy-Brown & Asgari (2001), Hur & Watanabe (2001), Watanabe, Takayama, Nagamatsu & Tagami (2002) and Watanabe & Ane (2003) underline the important roles of assimilation capacity in innovation. Nakagawa (2008), and Nakagawa, Watanabe & Griffy-Brown (2009) propose a new method for classifying technology spillovers based on both technological field and organization; however, they do little to examine such spillovers from the point of view of technological capital. Given that technological diversification can promote innovation (Garcia-Vega, 2006), diversification strategies are favorable to innovation. For example, diversification strategies have successfully created new businesses, supported by entrepreneurship and the intra-firm venture business system (Hirota, 1994; Hirota, 1995).

Some of the other factors that foster innovation are reported. A recent empirical analysis conducted in the Netherlands concludes that the survival probability of innovators is higher than that of non-innovators, independent of size and age class (Cefis & Marsili, 2006). Watanabe & Wakabayashi (1996) note the positive effects of the marginal productivity of technology on technological capital. An empirical analysis of the R&D-patents relationship concludes that R&D activities organized in-house are the main source of more significant innovations, whereas externally contracted R&D services are more productive in terms of incremental innovations (Beneito, 2006). Another analysis shows that investors expect market-stealing effects to dominate spillover effects when external inventors develop important innovations that are applicable to a focal firm's industry, other than under conditions of low appropriability or highly complementary assets (McGahan & Silverman, 2006). Manufacturers should look systematically at the new product opportunities that user innovations, user innovation communities and related emerging markets represent (Baldwin, Hienerth & von Hippel, 2006). Complementarities have been found in joint cooperation strategies between competitors and customers and those between customers and universities (Beldebos, Caree & Lokshin, 2006). In addition, it has been demonstrated that government support can induce learning in firms (Fukuda & Watanabe, 2008) and that government has a role to play in ensuring market competitiveness by enhancing technology spillover (Jin, Perote-Pena & Troege, 2004). However, an empirical study concludes that the majority of firms have yet to conduct closed innovation activities (Lichtenthaler, 2008).

On the other hand, the fall in the marginal productivity of technology that occurred in Japan in the 1990s can be attributed to the exhaustion of technology spillover pools within firms (Nakagawa, 2008; Nakagawa & Watanabe, 2007). In addition to the exhaustion of technology spillover pools within firms, organizational inertia is generally considered an impediment to a firm's sustainable growth (Matsumoto, Ouchi, Watanabe & Griffy-Brown, 2002; Watanabe & Tokumasu, 2003). Larsen & Lomi (2002) define organizational inertia as the tendency of formal organizations to resist internal change in response to external change. Thus, the existence of this type of inertia constrains a firm's ability to move towards emerging opportunities and thereby increase the potential for new ventures to exploit market opportunities (Dean & Mayer, 1996). Roller & Sinclair-Desgagne (1996) identify differences in initial capabilities and organizational inertia as sources of divergence in firms' capabilities. Greenwood & Jovanovic (1999) find that sources of organizational inertia can be attributed to three factors: awareness and skill, vintage capital, and vested interests. Hobjin & Jovanovic (2001) attribute the decline of the US stock market in the 1970s to organizational inertia, whereby firms could not adjust themselves to the information technology revolution..

While these studies provide useful results on the issue of innovation and technological capital, few of them elucidate the dynamics of innovation creation by observation inside a firm where R&D is actually conducted. Actually, in many firms, such figures as operating income to sales (OIS) and R&D expenditures of internal business sectors are not disclosed by internal sectors; therefore, technological capital in internal sectors is hidden in an opaque black box. However, a method for estimating technological capital in a firm's internal sector has been proposed (Nakagawa & Watanabe, 2007; Nakagawa, 2008). This method suggests the number of patent applications or publications can be calculated as a production function of technological capital stocks or R&D expenditure. Therefore, once the production function for the entire firm has been computed, you can apply this function to any of its internal sectors. Consequently, if the number of patent applications in the aggregate and internal sectors of a firm over a certain period can be counted, and R&D expenditure of the firm is disclosed, you can estimate R&D expenditures in internal sectors.

HYPOTHESES AND METHODOLOGY

Hypotheses about the Use of Technological Capital

A profit of a firm is largely affected by the characteristics of contemporary socio-economic systems. Nakagawa & Watanabe (2007) and Nakagawa (2008) examined the trend in operating income to sales (OIS) in six major firms of Japan's nonferrous metal industry from the year 1980 to 1999. These studies demonstrate that the trend in OIS shows a turning point in 1991, when an industrial society changed to an information technology society. We highlighted Sumitomo Electric Industries, Ltd. (SEI): the most representing firm amongst those six firms, and examined its OIS and technological capital trends extended to 2007. Figure 1 shows the trends in operating income to sales and technological capital stock in SEI from 1980 to 2007.

It is noteworthy that Figure 1 shows two turning points on the operating income to sales (OIS) in 1991 and 2001. Although technological capital stock shows an almost monotonic increase, OIS increased in the 1980s, it began to fall in 1991 and 2001 following the bursting of the bubble economy and the IT bubble economy, respectively.

Nakagawa (2008) and Nakagawa, Watanabe & Griffy-Brown (2009) calculate the technological capital stocks of SEI's internal business sectors from 1980 to 2004. Figure 2 illustrates the estimated technological capital stocks of SEI's internal business sectors (1980-2004). SEI's internal sectors can be divided into four groups: 1) the Wire and Cables Sector; 2) the Special Steel Wire Sector; 3) the Powder Alloy Products Sector; and 4) the New Business Sectors. While the three former sectors deal in relatively low-tech traditional products, such as power, communication wires and cables, stainless steel wires, and sintered alloy cutting tools, the New Business Sectors cover a broad range of products including optical fibers, information technology systems, and compound semiconductor materials.

Figure 2 demonstrates two turning points in the trend in technological capital stocks in new business sectors. The first is 1991 when new business sectors outpaced the ordinary wire and cable business sector, and the second is 2001 when it turned upward.

It is noteworthy that Figures 1 and 2 highlight 1991 and 2001 as the turning points in terms of business profit (operating income to sales) and technological capital stock. Thus, it is reasonable to divide the period analyzed into three: 1980 to 1990, 1991 to 2000, and 2001 to today.

In addition, from a socio-economic point of view, 1991 and 2001 are the turning points that saw transformations from an industrial society to an information technology society and from an information society to a post-information society; marking as they do the collapses of the bubble and IT bubble economies.

Figure 1. Trends in operating income to sales and technological capital stocks at Sumitomo Electric Industries, Ltd. (1980-2007) Billons of yen at 2000 fixed prices.

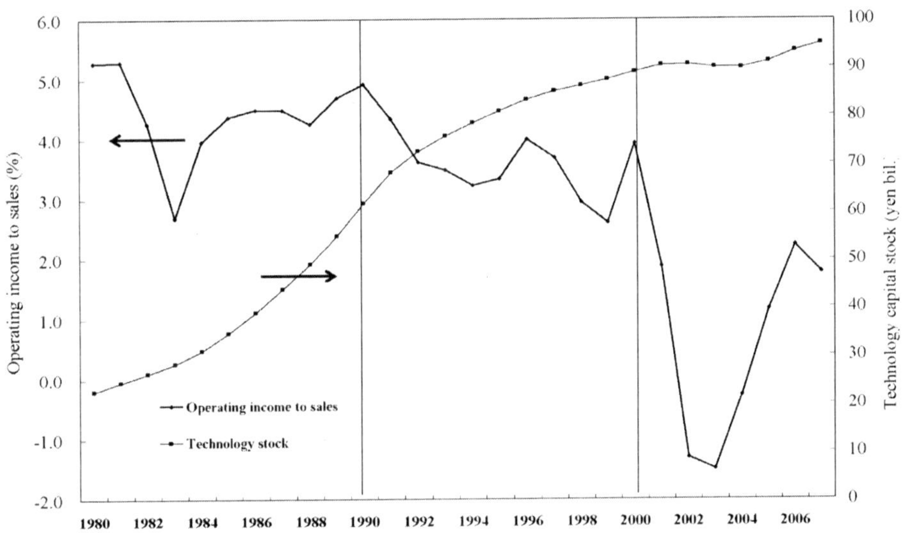

Figure 2. Trends in estimated technological capital stocks in internal business sectors at SEI (1980-2004).

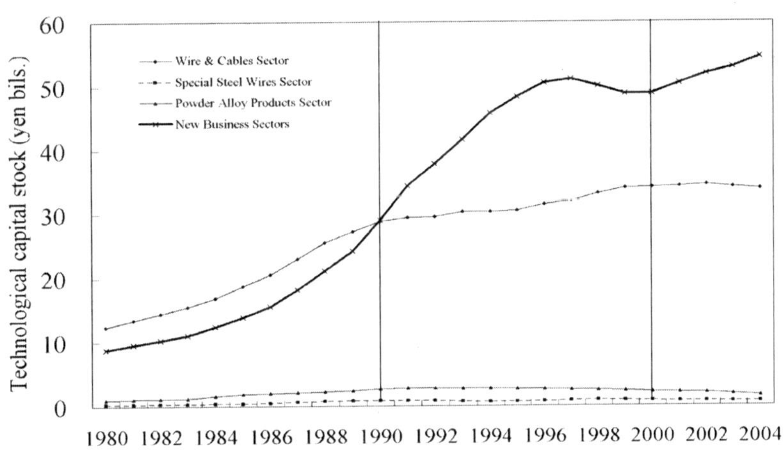

Taking into account these two turning points, we will go on to examine changes in the management of technological capital from the technology spillover point of view, we propose the three following hypotheses:

- In the industrial society of the 1980s, the appropriate use of technological capital was confined to a limited part of that society;
- Technological capital was not effectively used in the lost decade of the 1990s, in spite of managerial efforts; and
- In the early 2000s, technological capital was effectively used in combination with acquired technology, which helped spur an upturn in firm profits.

Selection of a Proxy Firm and Technology

The case we select for examination is a leading nonferrous metal firm in Japan (Sumitomo Electric Industries, Ltd. (SEI)[2]). SEI is chosen partly because it has a longer business history and a higher level of R&D investment than any of its competitors, and partly because its R&D on compound semiconductor materials is recognized as a successful case of its intra-firm venture business system (Hirota, 1994; Hirota, 1995); consequently its main product, gallium arsenide (GaAs) has maintained its position in the world's top market share over the last quarter century. Thus, SEI meets the general requirements of a proxy.

Our empirical analysis focuses on research and development work on compound semiconductor materials technology as such materials are the key materials in information network devices. Compound semiconductors, particularly gallium arsenide (GaAs), are the key materials as equipped in optical and wireless telecommunication devices, and in home-use AV products. In addition, they have been researched since before information technology society had started: their R&D has passed all three societies highlighted in this chapter: an industrial, an information technology, and a post-information society.

Classification of Technology Spillover

As the two-dimensional classification of technology spillover (Nakagawa, 2008; Nakagawa, Watanabe & Griffy-Brown, 2009) gives a clear-cut view of technology spillovers and socio-economic changes, let us now review an outline of it. Some technologies spill over between firms, whereas other technologies spill over within a particular firm. Some technologies spill over between technological fields, while other technologies spill over within a particular field. Based on these two spillover directions, *i.e.*, between or within technological fields and organizations, we can form a two-dimensional matrix of technology spillovers.

Figure 3 illustrates a schematic view of the two-dimensional matrix of technology spillover in relation to product development.

Each of the spillover types has distinctive characteristics.

1. **Intra-intra technology spillover** (intra-firm, intra-technology spillover): Technology spillover within a firm, within a technological field can commonly be observed because researchers often share the knowledge and experiences they gain through their daily R&D activities.
2. **Intra-inter technology spillover** (intra-firm, inter-technology spillover): Technology spillover within a firm, between technological fields can also be observed within a firm, particularly one whose business is well diversified.
3. **Inter-intra technology spillover** (inter-firm, intra-technology spillover): Technology spillover between firms, within a technological field is rarely observed, other than in the case of basic research. Firms that own technology in the same field are natural competitors.
4. **Inter-inter technology spillover** (inter-firm, inter-technology spillover): Technology spillover between firms, between technological fields can be observed in joint research with other firms. If technologies used in two firms complement each other, the two firms can cooperate in their R&D activities.

Investigation of Patent Application Data

A patent application is not necessarily an innovation per se. However, patent applications allow us to view information on new technology, time,

Figure 3. A Two-dimensional matrix of technology spillover structures.

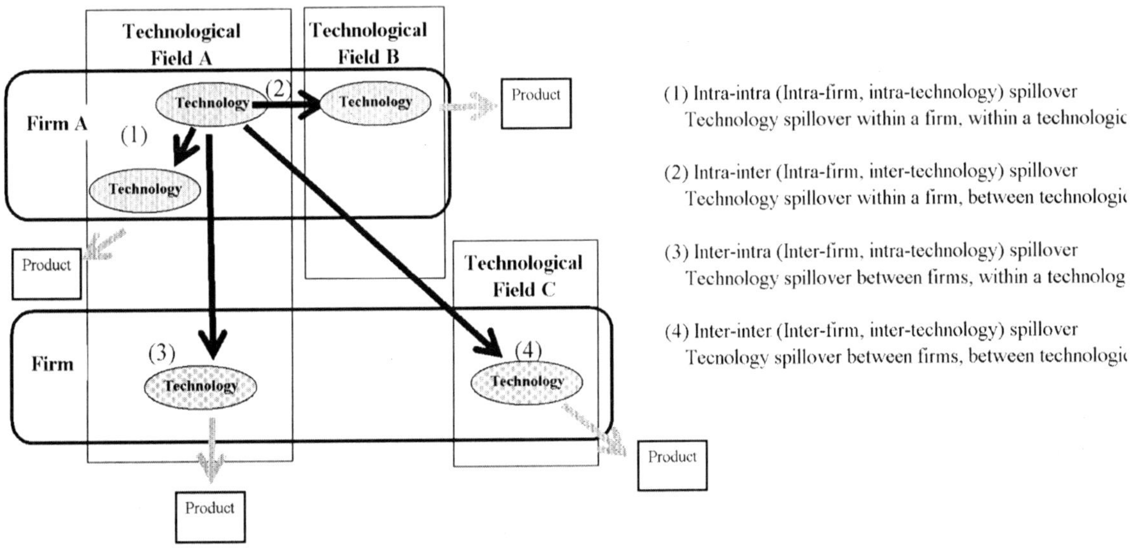

inventors, and their affiliations. Thus, by comparing such information across a range of patent applications, we can interpret progress in and the propagation of new technologies. In other words, we can track the footprints of innovation by relating one patent application to another. As Jaffe (1986) explains in his analysis of the patent-innovation relationship[3], one of the effects of technological opportunity and spillovers on R&D productivity is that we can relate patent applications to technological capital.

Nakagawa (2008) draws up a long list of 918 patent applications on compound semiconductor materials filed by SEI from January 1980 to June 2006[4]. The list includes patent application IDs and application dates, the primary materials which the claims relate with, the inventors, and the applicants. In addition, the Industrial Property Digital Library[5] was searched for more technological information.

Our analysis was conducted by tracing inventors and technological contents claimed in these patent applications. For instance, a researcher, Hirota, and his colleagues at Sumitomo Electric Industries, Ltd. filed a patent application (JP1998-4857[6]) in 1998 in which he sought a patent for a manufacturing technology for a material, ZnSe. In 2000, the same researcher (Hirota) and his colleagues filed another patent application (JP2000-157036) in which the claim related to another material, GaN. A comparison between technologies claimed in these two patent applications shows that Hirota spilled ZnSe technology over GaN technology in 2000.

CHANGES IN TECHNOLOGY SPILLOVER MANAGEMENT

Let us now move on to an analysis of changes in technology spillover management. Figure 4 summarizes changes in technological capital in new business, operating income to sales, and patent applications made in relation to compound semiconductor materials at Sumitomo Electric Industries, Ltd. (SEI).

Figure 4. Technological capital in new business,operating income to sales, and patent applications for compound semiconductor materials at SEI (1980 – 2006).

	1980 - 1990	1991 - 2000	2001 - 2006
Technology capital in new business	increase	increase/ decrease	increase
Operation income to sales	increase	decrease	decrease/ increase
Patent applications in compound semiconductor materials	515	211	192

Technology Spillover Management from 1980 to 1990

First, in the eleven years from 1980 to 1990, SEI filed 515 patent applications with the Japan Patent Office. Looking carefully at these applications, we note that none of them were made in conjunction with other industrial firms. Only a few joint applications were filed in tandem with basic research scientists from universities and Nippon Telephone and Telegram, a government-owned public corporation. Thus, it can be confirmed that intra-intra technology spillover was the major spillover structure in this period. In fact, at SEI during the 1980s, research into compound semiconductor materials was undertaken by a group in the Advanced Basic Research Laboratory, where more than ten researchers shared an office, a laboratory and experimental apparatus. They attended the same meetings. Thus, although they were researching different materials, their knowledge was freely exchanged in their daily work, leading to intra-intra technology spillover. Figure 5 illustrates major materials, the years of patent applications, patent application IDs and inventors listed in the patent applications for a kind of compound semiconductor materials manufacturing technology: VCZ technology.

Figure 5 illustrates the dynamics of the knowledge flows resulting from intra-intra technology spillover observed in the course of developing compound semiconductor materials. The VCZ technology for GaAs was developed in this group, and was soon propagated to InAs, GaP, InP and CdTe materials. In fact, looking carefully at Figure 5, we note that all researchers listed other than Namikawa researched two or more materials within the space of only four years. Thus, VCZ technology spilled over from a material to the other within this group at SEI. Consequently, SEI released high quality GaAs substrate in 1985 and InP substrate in 1989. The technologies used in these products can be connected to the patent applications, JP1984-117845 and JP1987-14087, respectively.

Looking back at technology spillovers in this period, the technological capital created at SEI was used purely on an internal basis by SEI itself. Specific technologies were simply used to develop new products in the same technological field. Thus, technological capital investment affected technological innovation and increased company profits. Thus, the first hypothesis that in the industrial society of the 1980s, the appropriate use of technological capital was confined to a limited part of that society, is confirmed.

Technology Spillover Management from 1991 to 2000

In the next decade from 1991 to 2000, 211 patent applications on compound semiconductor materials were filed with the Japan Patent Office. In comparison with the previous period, this represented a sharp fall in the number of applications. However, in common with the previous period, few joint applications were filed in conjunction with industrial firms - five cases in total. One was filed in conjunction with a semiconductor manufacturer; four were filed in tandem with R&D venture firms. Over the course of this decade, in addition to a large number of intra-intra spillovers,

Figure 5. Intra-intra technology spillover of VCZ technology at SEI.

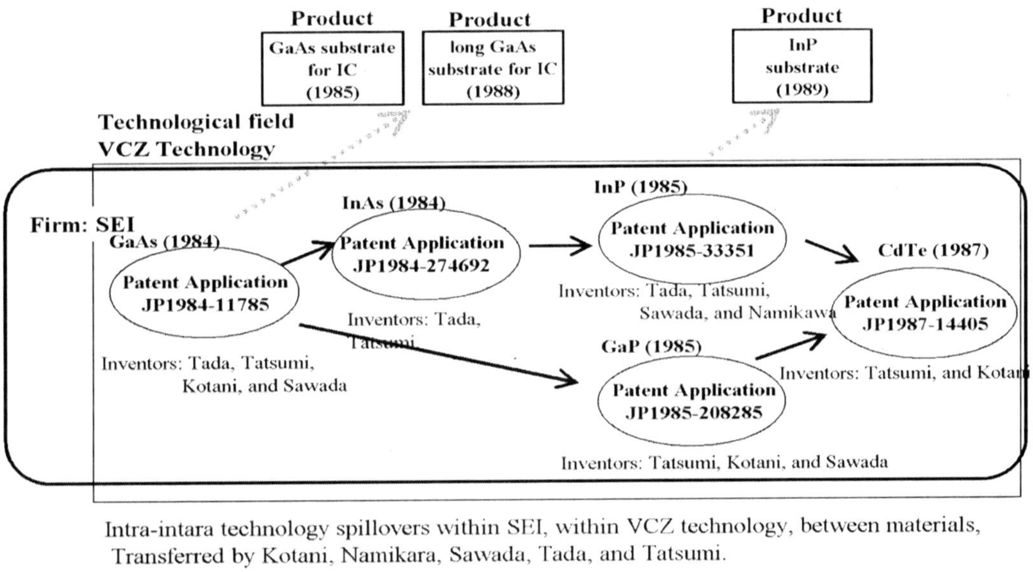

intra-inter technology spillovers can be observed in this decade. For example, Researcher Motoki applied a manufacturing technology for ceramics products (*e.g.* JP1985-159621) to GaN substrate (*e.g.* JP1998-171276), Researcher Shikata applied GaAs device technology (*e.g.* JP1991-76112) to a Diamond Surface Acoustic Wave (SAW) filter device (*e.g.* JP1991-245645), Researcher Yano filed patent applications on fiber-optics lasers (*e.g.* JP1991-76112) and GaN devices (*e.g.* JP2004-109296) as well, and Researcher Hirota utilized ZnSe technology (*e.g.* JP1998-4857) in GaN (*e.g.* JP2000-157036). Figure 6 illustrates the relationship between technology spillovers, patent applications and major products related to these intra-inter technology spillovers at SEI.

The development of the diamond SAW filter is an application of synthetic diamond technology. Since the successful synthesis of the world's largest (1.2-carat) diamond in 1982, various applications have been developed at SEI. A SAW filter device is one of these applications. The diamond SAW filter is a thin-film device fabricated on a bulky synthetic diamond substrate. Its development requires the concurrent use of diamond synthesis and device fabrication technologies. In this case, a GaAs device researcher transferred device technology to diamond synthesis technology. In other words, the development of the diamond SAW filter device can largely be attributed to the integration of diamond synthesis and device fabrication technologies. The diamond SAW filter device was finally released in 1999. Business related to this device has now been transferred to a subsidiary firm.

Ceramics and ZnSe technologies contributed to the development of GaN substrate at SEI. Although both ZnSe and GaN are compound semiconductors, GaN is a hard, ceramic-like material, whereas ZnSe is relatively soft and fragile. As a result, GaN researchers found it beneficial to collaborate with hard ceramics technology researchers. While ZnSe does not have physical characteristics similar to those of GaN, like GaN, it is a semiconductor which emits blue light. This meant that blue light emission technology could be transferred from ZnSe to GaN substrate. It is noteworthy that both SAW filters and GaN are important materials used for key information technology devices. Thus, it is not surprising that

Figure 6. Intra-inter technology spillovers at SEI.

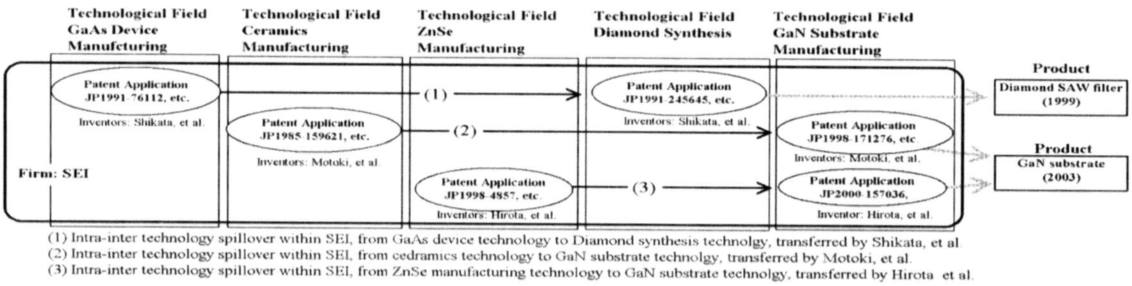

SEI developed these IT products using their own materials technologies. It is also noteworthy that there were long time lags from the beginning of the R&D process to the launch of a new product. For SAW filters, it took nine years from the time the JP1991-76112 (1991) application was made until the launch in 1999.

In summary, less technological capital was created at SEI in this period than in the previous period, and the technological capital that was created contributed little to company profits. Thus, the second hypothesis that technological capital was not effectively used in the lost decade of the 1990s, in spite of managerial efforts, is confirmed.

Technology Spillover Management from 2001 to 2006

SEI filed 192 patent applications on compound semiconductor materials from 2001 to June 2006. The most noteworthy difference from the previous two periods is that many joint applications were made with other industrial firms. In particular, SEI filed 12 patent applications on gallium nitride compound semiconductors in conjunction with Sony. In addition to intra-intra, intra-inter, and inter-intra technology spillovers, inter-inter spillover is a feature of technology management in this period. As for inter-inter technology spillover, patent applications filed jointly by SEI and Sony from 2001 to 2003 show that a laser fabrication technology developed by Sony and GaN substrate material technology developed by SEI spilled over into each other. That is, technologies of GaN material and GaN laser crossed over in joint research activities between SEI and Sony. As a result of these activities, SEI released GaN substrate for lasers in 2003 and Sony released a Blu-Ray game machine: PlayStation 3, which featured a GaN laser, in 2006.

Figure 7 illustrates technology spillovers observed in the development of GaN laser substrate.

Consequently, SEI and Sony managed inter-inter technology spillover, which resulted in the successful launch of Blu-Ray products. This change in technology management was triggered by the exhaustion of internal technology spillover sources, supported by science and technology policy, and fostered by technology transactions with external entities.

Thus, we can confirm that SEI and Sony used each other's technological capital for their own R&D; consequently, these efforts resulted in new products.

Moreover, in the period 2001 to 2006, during and after the joint patent activity with Sony, 52 researchers at SEI filed substrates patent applications, 57 researchers filed device patent applications, and 28 among those researchers filed both substrate and device patent applications (Nakagawa, 2008; Nakagawa & Watanabe, in press). Figure 8 presents examples of these technology spillovers.

Figure 8 shows that inter-inter technology spillover (1) triggers intra-intra (2), and intra-inter (3) technology spillovers. First of all, the

Figure 7. Technology spillovers between SEI and Sony, and between GaN Substrate and laser technologies.

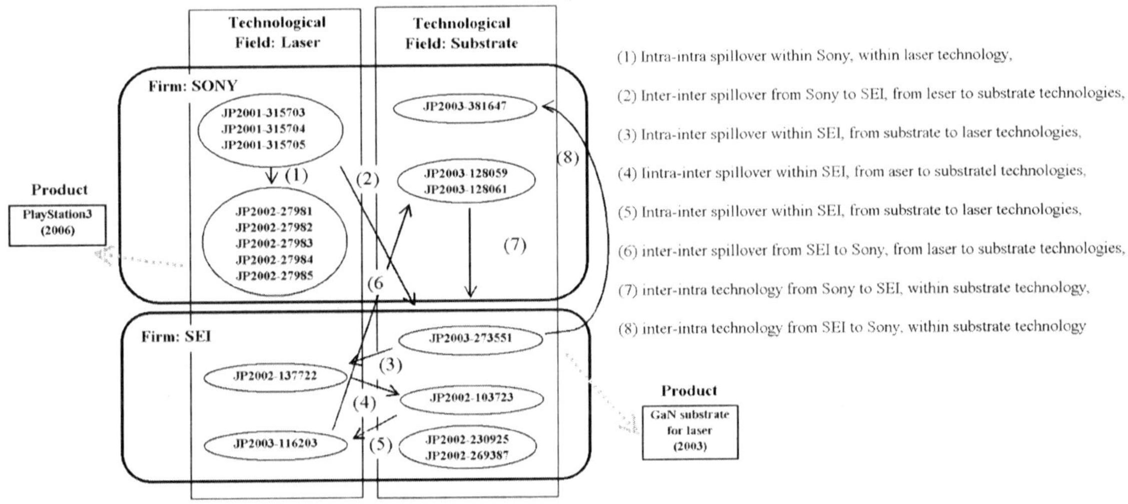

feature of intra-inter technology spillover here is that the same researcher conducted development work on both substrates and laser technologies in SEI. This activity suggests that Sony's laser technology can be transferred to SEI via substrate technology and consequently domesticated in SEI. Inter-intra spillover (Figure 8 (4)) can be broken down into inter-inter spillover (Figure 8 (1)) and intra-inter spillover (Figure 8 (3)). In other words, inter-intra spillover involves a merger of inter-inter and intra-inter spillovers, which involves domestication of acquired technology.

Thus, the third hypothesis that in the early 2000s, technological capital was effectively used in combination with acquired technology, helping to spur an upturn in firm profits, is confirmed.

Changes in Technological Capital, Technology Spillover and Society

In this subsection, we consider the results of our analysis in the context of paradigm changes in society. First, in the industrial society of the 1980s, Japanese firms were likely to protect their internally developed manufacturing technologies, as they acted as sources of competitive advantage. This is one reason that technology spillovers were confined within individual firms and technological fields. Industrial innovations were made as part of the daily R&D activities undertaken by an in-house R&D group by applying a process technology to other new products. As manufacturing technologies were sources of firms' competitive advantages, closed spillover was an appropriate strategy for manufacturing firms, though an excessive reliance on internal development, known as the "NIH syndrome," arose. It is quite natural that intra-intra technology spillover was the dominant structure in the 1980s. In short, the mainstream form of technology spillover management adopted in this period was intra-intra technology spillover, which was adjusted to meet the needs of an industrial society. In short, firms effectively used their own technological capital within the firm.

In 1990s in Japan, at the same time the Internet began to be used commercially, Japan became an information society. During this decade, Japan suffered a period of stagnation called the "lost decade." Operating income to sales and the marginal productivity of technology eventually fell abruptly. This period of stagnation was partly attributable to a decrease in the Total Factors of

Figure 8. Intra-intra technology spillovers, inter-inter technology spillovers, and domestication of technologies at SEI.

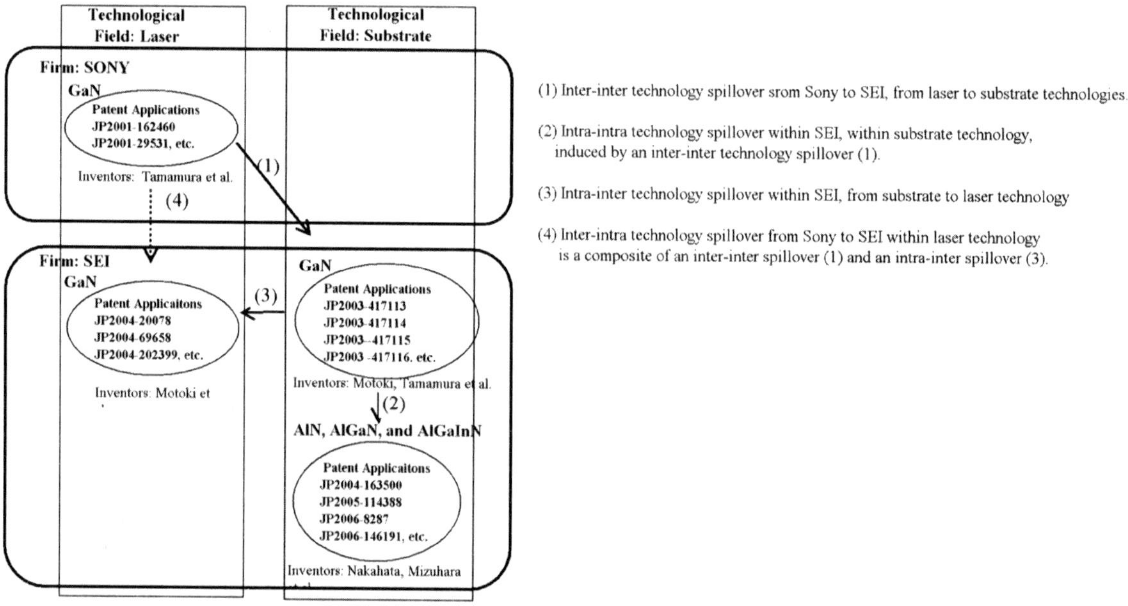

Productivity (TFP) growth rate in Japan's economy (Hayashi & Prescott, 2002). Firms in Japan should have adapted their technology management practices to the new information society, utilizing information technology to realize technological innovations. However, in practice, partly because of organizational inertia in which firms clung to old paradigms, and partly because of finance restrictions, firms' technology spillover sources were exhausted and technology spillover shrank (Nakagawa, 2008; Nakagawa & Watanabe, 2007; Nakagawa, Watanabe & Griffy-Brown, 2009). As we demonstrate in this chapter, intra-inter technology spillovers alone were not sufficient to maintain business profits.

On the other hand, it must be pointed out that there were excellent firms that remained active in technological innovation. For example, Canon successfully developed new types of printers, applying technology developed for old-fashioned models to high performance, brand new models (Matsumoto, Ouchi, Watanabe & Griffy-Brown, 2002). Canon also utilized a development project and the diffusion of the market for personal computers for its printer business (Watanabe & Ane, 2003). Canon's case shows that inter-inter technology spillover could occur in the 1990s. Thus, comparing SEI with Canon, it can be confirmed that the success of an innovation can be determined by whether or not the firm utilizes external resources. In short, the critical factor for success in this period was whether or not the firm could continuously utilize technological capital from external firms.

In the first decade of the new millennium, firms which have already exhausted internal sources of technology spillover have had to search for new sources outside their own firms. In addition, the decade-long period of stagnation induced changes in government science and technology policy. The Japanese government enacted the second and third science and technology basic plans in 2001 and 2006, which required universities and public research institutes to contribute to economic revitalization. Consequently, firms actively tried to establish joint research initiatives either between themselves or between firms and universities or public institutions. An agreement concluded in 2005 between Sumitomo Electric Industries, Ltd. (SEI) and the National Institute of Advanced Science and Technology (AIST) is a typical example of this trend. One study concludes that in 2002, 70% of Japanese firms participated in some form of collaboration with external organizations and 40% participated in university-industry cooperation programs (Motohashi, 2005). The joint patent applications made by Sony and SEI are examples of successful efforts induced by this policy move. These two firms used both their own and each other's technological capital at the same time.

Recurrent Inducement of Closed and Open Technology Spillovers

Let us now look back at Figure 8. The first technology spillover from GaN to other materials appeared in 2004 (JP2004-163500); just after the SEI-Sony joint activity came to an end. This indicates that termination of the joint activity stimulated intra-intra technology spillover.

More than 80% of the total patent applications on AlN, AlGaN and AlGaInN materials were filed by GaN device researchers, while 50% were filed by GaN substrate researchers. Thus, technologies on AlN, AlGaN, and AlGaInN materials can largely be attributed to GaN device technology (Nakagawa, 2008; Nakagawa & Watanabe, in press).

Recall that technology diffusion from GaN to other nitrides was an intra-intra technology spillover and that GaN device technology had been acquired as a result of inter-inter technology spillover. It can be concluded that intra-intra technology spillover followed inter-inter technology spillover. In addition, as inter-inter technology spillover follows intra-intra technology spillover, these two types of spillover follows each other.

As the most closed, intra-intra spillover and the most open, inter-inter spillover follows each other, it is suggested that close innovation and open innovation follows each other; they are complementary and form a recurrent cycle.

Fusions of imported and endogenous technologies and mutual inducement of intra-intra and inter-inter spillover suggest that such fusion establishes a new starting point for internal innovation, which is mainly driven by intra-intra technology spillover. It is likely that open and closed innovations have their own peculiar advantages, and that innovators should choose the best option according to the stage of development.

New Findings

On the basis of the foregoing sections, our noteworthy findings include the following:

- The strategy of intra-intra technology spillover was a key factor for success in the 1980s. This is because in the industrial society of that time, the ability to gain a competitive advantage relied on manufacturing technologies created as a result of internal development activities. Firms effectively used their own technological capital within the firm.
- Technology spillover shrank in the 1990s. Because of organizational inertia, in which firms clung to the industrial society of the past, technology management practices in the materials industry resembled those of a manufacturing society. On the other hand, some firms successfully developed technological innovations utilizing external technologies as complementary resources. The difference between these two examples indicates that overcoming organizational inertia is a critical factor for innovation. Technological capital leads to technological innovation so long as it is combined with external resources.
- The utilization of external resources boosted the revitalization of technological innovation in the early 2000s. Changes in

the mindsets of firm managements and government policy on science and technology stimulation helped to enhance open innovations. The sharing of technological capital by firms can lead to technological innovation.

- Thus, inter-inter and intra-intra technology spillovers can induce each other. When this occurs, continuous innovation will result.
- Furthermore, mutual inducement of intra-intra and inter-inter technology spillovers suggests that closed innovation and open innovation are complementary and recurrently occur.
- It also suggests that closed and open innovations reflect paradigm and had their peculiar advantage related with business environments.

The findings discussed in this chapter were obtained as a result of analyses of a firm in Japan's nonferrous metal industry. However, these findings are applicable to other firms in other industries. For example, the black box in technological capital stock can be opened in this way, regardless of the industry. The two-dimensional matrix of the technology spillover structure is also applicable to all other firms. Furthermore, given the technology management changes seen in all industries as a result of socio-economic changes, the management approach adopted must have changed from closed to open management in the last quarter century. Thus, to varying degrees, the findings presented in this chapter are applicable to other industries.

Recommendations for Technological Innovators

We can safely recommend the following actions for technological innovators who seek to use technological capital in an effective manner.

- Technology managers should carefully consider the effective use of third party technological capital, as well as of their own. The mutual and effective use of knowledge in conjunction with external organizations is one of the critical factors for success.
- For this purpose, technology managers should endeavor to establish cooperative relationships with external organizations, preferably forming a cooperative network of research organizations including not only corporate laboratories, but also universities and public research laboratories;
- Technology managers should examine their external and internal technologies to determine whether or not they can complement each other;
- Technology managers should carefully observe their own technologies and business environments, which include all aspects of institution. Then, technology managers should choose the best option in between closed and open innovations.

The foregoing recommendations should be applied not only to firms in the manufacturing industry, but also to those in a broad range of other sectors.

FUTURE RESEARCH DIRECTIONS

As this chapter has focused on the nonferrous metal industry, comparative analyses on the similarities and disparities across different materials and industries could be the focus of future studies. Such studies would generalize our conclusions more widely.

In addition, further studies should focus on knowledge spillover and the co-evolution of various kinds of technological innovation and the manifold dimensions of institutions. The co-evolution of innovations and institutions, not only at the R&D stage, but also at the marketing, manufacturing and consuming stages, should also be addressed. Furthermore, given that service innovation is becoming more and more important, knowledge spillover between manufacturing technologies and services could also be the focus of future studies.

As the significance of materials technology to service innovation is increasing in today's world, it is suggested that, in contrast to the information society model whereby service innovation leads materials innovation, materials innovation is likely to lead service innovation in the coming post-information society. Further studies on the relationship between service and materials innovation might also be conducted.

Moreover, it is certain that the present economic crisis will not only result in falling industrial R&D investment, but will also result in the emergence of a new paradigm after the downturn has ended. It is expected that further research in this area, including research on the impact of the current economic downturn, will be conducted.

CONCLUSION

This study presents the results of an empirical analysis on technology spillover structure that involved tracking the patent applications of a specific firm, thereby demonstrating changes in the use of technological capital that resulted from socio-economic changes.

In the 1980s, as manufacturing technology was a key factor for developing competitive advantage, technology spillover was limited to spillover within specific firms. A diversification strategy required a broad application of firm technology. Thus, intra-intra spillover was an appropriate strategy for the use of firms' technological capital.

In the 1990s, in order to extend its product line-up to IT devices, the materials industry adopted intra-inter spillover as an R&D strategy. However, organizational inertia and long time lags from the R&D stage to product launch acted as constraints on the effective use of technological capital.

In the early 2000s, with the help of government science and technology policy, firms changed their technology strategy to one of open innovation. The resulting inter-inter spillover strategy was effective, and business profits recovered.

Thus, the following hypotheses are confirmed:

- In the industrial society of the 1980s, the appropriate used of technological capital was confined to a limited part of that society;
- Technological capital was not effectively used in the lost decade of the 1990s, in spite of managerial efforts; and
- In the early 2000s, technological capital was effectively used in combination with acquired technology, which helped spur an upturn in firm profits.

To summarize our conclusions, industry profits can be attributed not only to the quantity of technological capital stock, but also to its use; technological capital stock should be distributed in a manner that takes account of business strategy and contemporary paradigms. The recurrent inducement of intra-intra spillover and inter-inter spillover is a promising model for the effective use of technological capital. Therefore, firms should strive to absorb and embrace acquired technologies to prepare themselves for the next cycle.

Today's world faces an unprecedented economic crisis, which has also led to an abrupt change in society. Many people including policy makers and business leaders do not appear certain of the best way forward. However, the new findings and recommendations discussed in this chapter are all derived from analyses undertaken on a firm representing the entire industry, across three paradigms: an industrial society, an information society, and a post-information society. Therefore, these recommendations are useful to technological innovators who are now struggling with unprecedented challenges. We hope that our study will help such people meet this challenge.

REFERENCES

Baldwin, C., Hienerth, C., & von Hippel, E. (2006). How user innovations become commercial products: A theoretical investigation and case study. *Research Policy, 35*, 1291–1313. doi:10.1016/j.respol.2006.04.012

Beldebos, R., Caree, M., & Lokshin, B. (2006). Complementary in R&D cooperation strategies. *Review of Industrial Organization, 28*, 402–426.

Beneito, P. (2006). The innovative performance of in-house and contracted R&D in terms of patents and utility models. *Research Policy, 35*, 502–517. doi:10.1016/j.respol.2006.01.007

Bernstein, J. I., & Nadiri, M. I. (1988). Interindustry R&D spillovers, rates of return, and production in high-tech industries. *The American Economic Review, 78*(2), 429–434.

Bernstein, J. I., & Nadiri, M. I. (1989). Research and development and intra-industry spillovers: An empirical application of dynamic duality. *The Review of Economic Studies, 56*, 245–269. doi:10.2307/2297460

Cefis, E., & Marsili, O. (2006). Survivor: The role of innovation in firm's survival. *Research Policy, 35*, 626–641. doi:10.1016/j.respol.2006.02.006

Dean, T. J., & Mayer, G. D. (1996). Industry environments and new venture formations in U.S. manufacturing: A conceptual and empirical analysis of demand determinations. *Journal of Business Venturing, 11*, 107–132. doi:10.1016/0883-9026(95)00109-3

Fukuda, K., & Watanabe, C. (2008, October). *Catalyst role of government R&D inducing hybrid management.* Paper presented at 1st International Conference on Technology Management, Tokyo, Japan.

Garcia-Vega, M. (2006). Does technological diversification promote innovation? An empirical analysis for European firms. *Research Policy, 35*(2), 30–246.

Goto, A., & Suzuki, K. (1989). R&D capital, rate of return on R&D investment and spillover of R&D in Japanese manufacturing industries. *The Review of Economics and Statistics, 71*(4), 555–564. doi:10.2307/1928096

Greenwood, J., & Jovanovic, B. (1999). The information-technology revolution and the stock market. *The American Economic Review, 89*(2), 116–122.

Griliches, Z. (1979). Issues in assessing the contribution of R&D to productivity growth. *The Bell Journal of Economics, 10*, 92–116. doi:10.2307/3003321

Griliches, Z., & Lichtenburg, F. (1984). Inter-industry technology flows and productivity growth: A reexamination. *The Review of Economics and Statistics, 66*(2), 324–329. doi:10.2307/1925836

Hayashi, F., & Prescott, E. (2002). The 1990s in Japan: A lost decade. *Review of Economic Dynamics, 5*(1), 206–235. doi:10.1006/redy.2001.0149

Hirota, T. (1994). R&D system in Sumitomo Electric Industries, Ltd. *Kansai Daigaku Shogakuronshu, 38*(6), 917–941.

Hirota, T. (1995). New business creation by corporate venture system: A case of compound semiconductor in Sumitomo Electric Industries, Ltd. *Kansai Daigaku Shogakuronshu, 40*(4-5), 589–610.

Hobjin, B., & Jovanovic, B. (2001). The information-technology revolution and the stock market: Evidence. *The American Economic Review*, *91*(5), 1203–1220.

Hur, K. I., & Watanabe, C. (2001). Unintentional technology spillover between two sectors: Kinetic approach. *Technovation*, *21*, 227–235. doi:10.1016/S0166-4972(00)00037-7

Jaffe, A. B. (1986). Technological opportunity and spillovers of R&D: Evidence from firm's patents, profits and market value. *The American Economic Review*, *76*(5), 984–1001.

Jin, J. Y., Perote-Pena, J., & Troege, M. (2004). Learning by doing, spillovers and shakeouts. *Journal of Evolutionary Economics*, *14*, 85–98. doi:10.1007/s00191-003-0177-5

Kennedy, C. (1964). Induced bias in innovation and the theory of distribution. *The Economic Journal*, *74*(295), 541–547. doi:10.2307/2228295

Larsen, E., & Lomi, A. (2002). Representing change: A system model of organizational inertia and capabilities as dynamic accumulation process. *Simulation Modelling Practice and Theory*, *10*, 271–296. doi:10.1016/S1569-190X(02)00085-0

Lichtenthaler, U. (2008). Open innovation in practice: An analysis of strategic approaches to technology transactions. *IEEE Transactions on Engineering Management*, *55*(1), 148–157. doi:10.1109/TEM.2007.912932

Matsumoto, K., Ouchi, N., Watanabe, C., & Griffy-Brown, C. (2002). Optimal timing of the development of innovative goods with generation – an empirical analysis focusing on Canon's printer series. *Technovation*, *22*, 175–185. doi:10.1016/S0166-4972(01)00003-7

McGahan, A. M., & Silverman, B. S. (2006). Profiting from technological innovation by others: The effect of competitor patenting on firm value. *Research Policy*, *31*, 1222–1242. doi:10.1016/j.respol.2006.09.006

Mizuhara, N. (2007). Meeting the challenge of creating a new material for producing environmentally friendly, small, and high-efficiency devices. *Sumitomo Electric Industries, R&D Professionals*, from http://www.sei.co.jp/RandD_e/researcher/012/index.html

Motohashi, K. (2005). University-industry collaborations in Japan: The role of new technology-based firms in transforming the national innovation system. *Research Policy*, *34*, 583–594. doi:10.1016/j.respol.2005.03.001

Nakagawa, M. (2008). *An empirical analysis in nonferrous metal technology: Change in technology spillover structure in Japan's nonferrous metal industry.* Unpublished doctoral dissertation, Tokyo Institute of Technology, Tokyo, Japan.

Nakagawa, M., & Watanabe, C. (2007). Moving beyond organizational inertia as a survival strategy for resource-based industry in a service-oriented economy: Lessons from cross-sector technology spillover in the nonferrous metal industry. *Journal of Service Research*, *7*(1), 7–35.

Nakagawa, M., & Watanabe, C. (in press). A revival of compound semiconductor material innovation: Strategic technology spillovers in Japan's nonferrous metal industry in the 2000s. *Journal of Service Research.*

Nakagawa, M., Watanabe, C., & Griffy-Brown, C. (2009). Changes in the technology spillover structure due to economic paradigm shifts: A driver of the economic revival in Japan's material industry beyond the year 2000. *Technovation*, *29*, 5–22. doi:10.1016/j.technovation.2008.08.003

Nakanishi, Y. (2002). Empirical evidence of inter-industry R&D spillover in Japan. *Journal of Economic Research, 7*, 91–104.

National Academy of Sciences. (2005). *Rising above the gathering storm: Engineering and employing America for a brighter economic future*. Washington, DC: National Academy Press.

Ornaghi, C. (2006). Spillovers in product and process innovation: Evidence from manufacturing firms. *International Journal of Industrial Organization, 24*, 349–380. doi:10.1016/j.ijindorg.2005.07.002

Roller, L.-H., & Sinclair-Desgagne, B. (1996). Industrial organization and business strategy on the heterogeneity of firms. *European Economic Review, 40*, 531–539. doi:10.1016/0014-2921(95)00066-6

Solow, R. M. (1957). Technical change and the aggregate production function. *The Review of Economics and Statistics, 39*(3), 312–320. doi:10.2307/1926047

Watanabe, C. (1999). Systems option for sustainable development. *Research Policy, 28*(7), 719–749. doi:10.1016/S0048-7333(99)00018-9

Watanabe, C., & Ane, B. K. (2003). Coevolution of manufacturing and service industry functions. *Journal of Service Research, 3*(1), 101–118.

Watanabe, C., Takayama, M., Nagamatsu, A., & Tagami, T. (2002). Technology spillover as a complement for high-level R&D intensity in the pharmaceutical industry. *Technovation, 22*, 245–258. doi:10.1016/S0166-4972(01)00004-9

Watanabe, C., & Tokumasu, S. (2003). Optimal timing of R&D for effective utilization of potential resources in innovation. *Journal of Advances in Management Research, 1*(1), 11–27.

Watanabe, C., & Wakabayashi, K. (1996). The perspective of techno-metabolism and its insight into national strategies. *Research Evaluation, 6*(2), 69–76.

Watanabe, C., Zhu, B., Griffy-Brown, C., & Asgari, B. (2001). Global technology spillover and its impact on industry's R&D strategies. *Technovation, 21*, 281–291. doi:10.1016/S0166-4972(00)00048-1

ADDITIONAL READING

A special report on businesses in Japan. (2007, December1). *The Economist,* pp. 3-20.

Audretsch, D. B., & Lehmann, E. E. (2005). Does the knowledge spillover theory of entrepreneurship hole for regions? *Research Policy, 34*, 1191–1201. doi:10.1016/j.respol.2005.03.012

Barzel, Y. (1968). Optimal timing of innovations. *The Review of Economics and Statistics, 50*(3), 348–355. doi:10.2307/1937928

Cameron, G. (2005). The Sun also rises: Productivity convergence between Japan and USA. *Journal of Economic Growth, 10*, 387–408. doi:10.1007/s10887-005-4735-6

Christensen, C. M. (1997). *The innovator's dilemma: When new technologies cause great firms to fail*. Boston, MA: Harvard Business School Press.

Christensen, C. M., Anthony, S. D., & Roth, E. A. (2004). *Seeing what's next: Using the theories of innovation to predict industry change*. Boston, MA: Harvard Business School Press.

Christensen, C. M., & Raynor, M. E. (2003). *The innovator's solution: Creating and sustaining successful growth.* Boston, MA: Harvard Business School Press.

Council of Competitiveness. (2004). *Innovate America: Thriving a world of challenge and change*. Washington, D.C.: COC.

de Man, A., & Duysters, G. (2005). Collaboration and innovation: A review of the effects of mergers, acquisitions and alliances in innovation. *Technovation, 25*, 1377–1387. doi:10.1016/j.technovation.2004.07.021

Dertouzos, M. L., Lester, R. K., & Solow, R. M. (1989). *Made in America: Regaining the productive edge*. Cambridge, MA: MIT Press.

Hammer, M., & Champy, J. (1993). *Reengineering the corporation: A manifesto for business revolution*. New York, NY: Harper Business.

Innovation 25 Strategy Council. (2007). *Innovation 25*. Innovation 25 Strategy Council, Tokyo, Japan.

Kelly, D. J., & Nakosteen, R. A. (2005). Technology resources, alliances, and sustained growth in new, technology-based firms. *IEEE Transactions on Engineering Management, 52*(3), 292–300. doi:10.1109/TEM.2005.851272

Lichtenthaler, E. (2005). Corpoerate diversification: Identifying new business systematically in the diversified firm. *Technovation, 25*, 697–709. doi:10.1016/j.technovation.2003.12.005

Lichtenthaler, U., & Ernest, H. (2006). Attitudes to externally organising knowledge management tasks: A review, reconsideration and extension of the NIH syndrome. *R & D Management, 36*(4), 367–386. doi:10.1111/j.1467-9310.2006.00443.x

Lui, S. S., & Ngo, H. (2005). An action pattern model of inter-firm cooperation. *Journal of Management Studies, 42*(6), 1123–1153. doi:10.1111/j.1467-6486.2005.00536.x

Milson, M. R., & Wielmon, D. (2006). Driving new product success in the electrical equipment manufacturing industry. *Technovation, 26*, 1268–1286. doi:10.1016/j.technovation.2005.08.008

Nieto, M., & Quevedo, P. (2005). Absorptive capacity, technological opportunity, knowledge spillovers, and innovative effort. *Technovation, 25*, 1141–1157. doi:10.1016/j.technovation.2004.05.001

Ohmura, A., Ouchi, N., Morisaki, S., & Watanabe, C. (2003). Functionality development as a survival strategy for fine ceramics. *Technovation, 23*(10), 833–842. doi:10.1016/S0166-4972(02)00028-7

Ohmura, A., & Watanabe, C. (2005). Inside the black box of cross-functional spillover: A lesson from the functionality development of the ceramics. *Journal of Advances in Management Research, 2*(2), 7–23.

Ohmura, A., & Watanabe, C. (2006). Cross-products technology spillover in inducing a self-propagating dynamism for the shift to a service oriented economy: Lessons from high-performance fine ceramics. *Journal of Service Research, 6*(2), 145–149.

Soh, P., & Roberts, E. B. (2005). Technology alliances and networks: An external link to research capability. *IEEE Transactions on Engineering Management, 52*(3), 419–428. doi:10.1109/TEM.2005.850727

Takayama, M., Watanabe, C., & Griffy-Brown, C. (2002). Remaining innovative without sacrificing stability: An analysis of strategies in the Japanese pharmaceutical industry that enable firms to overcome inertia resulting from successful market penetration of new product development. *Technovation, 22*, 747–759. doi:10.1016/S0166-4972(01)00067-0

Tanabe, K., & Watanabe, C. (2005). Soft policy instruments for inducing industrial innovation in a service-oriented economy: A comparative analysis of the vision system and university system. *Journal of Service Research, 5*(1), 123–154.

Watanabe, C. (2006, October). *Innovation and development: Japan's co-evolutionary dynamism between innovation and industrial system.* Paper presented at The 10th East Asia Round Table Meeting of Academies of Engineering, Tokyo, Japan.

Watson, S., & Hewett, K. (2006). A multi-theoretical model of knowledge transfer in organizations: Determinants of knowledge contribution and knowledge reuse. *Journal of Management Studies*, *43*(12), 141–173. doi:10.1111/j.1467-6486.2006.00586.x

KEY TERMS AND DEFINITIONS

Substrates, Thin Films: Compound semiconductor devices such as lasers, light emitting diodes, and microwave ICs, have their specified structures made on thin films of compound semiconductors. Thin films are made on substrates. Substrates are normally made of similar compound semiconductor crystals, however, in some cases, made of silicon, or sapphire crystals.

Compound Semiconductor: Semiconductor is classified as "single element semiconductor" and "compound semiconductor". The former includes silicon (Si) and germanium (Ge); the latter includes gallium arsenide (GaAs), indium phosphide (InP), gallium nitride (GaN), etc. Si is mainly used as a material of ICs for computer, or photovoltaic battery. Compound semiconductors are materials for light emitting, and light receiving devices for optical communications, compact disc, DVD, and Blu-ray players. Thin film manufacturing technology and substrate manufacturing technology are process technologies on compound semiconductor materials. Laser fabrication technology is a technology to fabricate lasers from compound semiconductor materials.

VCZ Technology: (Vapor pressure controlled Czochralski Method). A manufacturing method developed at SEI. A method that compound semiconductor crystal is manufactured in conditions of high vapor pressure using specific gases.

GaAs: The chemical formula for gallium arsenide. A chemical formula consists of symbols of constituent elements. Gallium arsenide (GaAs), indium phosphide (InP), gallium phosphide (GaP), indium arsenide (InAs), cadmium telluride (CdTe), zinc serenade (ZnSe), gallium nitride (GaN), aluminum nitride (AlN), aluminum gallium nitride (AlGaN), and aluminum gallium indium nitride (AlGaInN) are presented in this chapter.

Surface Acoustic Wave (SAW) Filter: A surface acoustic wave filter is a small device that reduces noise in communication devices. It is used in cellular phones, optical telecommunication devices, and so on.

ENDNOTES

1 Technological capital: As the phrase "technological capital" has a technology-specific meaning, this chapter uses this phrase instead of "intellectual capital." This chapter focuses on the use of intellectual capital in the research and development process.

2 Sumitomo Electric Industries, Ltd.: We conducted an empirical analysis on the nonferrous metal industry in Japan as an example, because it supplies high quality materials, such as power and communication cables, that support Japan's industrial society. In addition, it has also developed new materials used for IT infrastructure, such as fiber-optics and wireless communication devices. The contribution of the nonferrous metal industry to the wider economy can largely be attributed to efforts made in technological innovation. In addition, as this industry has continuously launched a plenty of new products in various fields of technology, this industry appears to have adopted technology - oriented strategy.

3 While Jaffe (1986) used patents granted, we use patent applications in this study. How-

ever, because not every patent application is granted, applications mark shorter intervals between technology footprints than patents granted.

4 We added 15 patent applications left off the list: JP2001-315703, 315704, 315705, 330068, 330181, JP2002-22987, 27982, 27983, 27984, 27985, JP2003-1255, 119334, 120130, 123180, and 153621.

5 The Industrial Property Digital Library (http://www.ipdl.inpit.go.jp/homepg.ipdl) served by the National Center for Industrial Information and Trading. This chapter examines patent applications filed with the Japan Patent Office (JPO). As Sumitomo Electric Industries, Ltd. is a Japanese firm, it can reasonably be assumed that SEI first files patent applications with the JPO. Therefore, the patent application IDs presented in this chapter are JPO patent application IDs.

6 JP1998-4857: The patent application ID consists of a Japan Patent Office abbreviation, JP, the year of application, 1998, and a serial number for acceptance, 4857.

Chapter 7
Absorptive Capacity, R&D Intensity and Innovation Results

Jorge Cruz-González
Universidad Complutense de Madrid, Spain

Miriam Delgado-Verde
Universidad Complutense de Madrid, Spain

ABSTRACT

The dynamic capabilities Perspective has emerged in the last decade to improve the explanatory power of the Resource-Based View of the firm with regard to obtaining sustainable competitive advantage by firms operating in rapid change environments. However, some aspects of its conceptualization, especially in relation to the factors that make up them, do not appear to be entirely clear. Accordingly, the aim of the present chapter is to provide a conceptual synthesis of the construct, trying to identify the key factors involved in the development of dynamic capabilities by firms. Based on a sample of 177 companies around the world operating in the software industry, the authors derive an empirical study. Their findings suggest that firms that jointly develop absorptive capacity and innovative capacity achieve better results.

INTRODUCTION

Generally, the main paradigms that try to explain how companies could obtain higher revenues derived from a sustainable competitive advantage are the Competitive Forces Approach (Porter, 1981) and the Resource-Based View (Barney, 1991).

The first perspective, which is based on Industrial Economy, argues that the structure of the industry within which firms operate, fundamental unit of analysis of this theory, determines the possibility of obtaining higher economic returns (Teece, Pisano & Shuen, 1997). The character and the action of five competitive forces determine the degree of attractiveness of the industry (Porter, 1981). According to this perspective, called Industry Effect or Model based on the Market Power, the strategic alternatives of the firms are limited to the research, selection and taking advantage of the most attractive industries, that is to say, in those industries where there is some imperfection in the product market.

The main criticism of the Competitive Forces Approach comes from the study conducted by Rumelt (1991). The author empirically supports

DOI: 10.4018/978-1-61520-875-3.ch007

that the differences between the profitability of firms within the same industry are greater than the differences between the profitability of those operating in different industrial sectors, supporting the reasoning of the studies that view the firm as a set of heterogeneous resources and capabilities.

On the other hand, the second perspective, so-called Resource-Based View, argues that the "VRIN" resources (Valuable, Rare, Inimitable and Non sustainable: Barney, 1991) of the firms enable them to obtain higher economic returns (Teece et al., 1997; Eisenhardt & Martin, 2000). Therefore, the resources and capabilities that generate heterogeneity allow firms to obtain a sustainable competitive advantage (Barney, 1991; Nelson, 1991). In addition, if the complementariness between resources and related systems of activity occurs, its potential to generate competitive advantage is increased (Porter, 1996), rising the probability to obtain quasi-rents. The company is considered as a set of resources and capabilities and, therefore, according to this theory, the possible strategic alternatives for the company consist on obtaining the resources and developing the key capabilities that will allow to deploy value-creation strategies at a lower cost than their contribution to the above mentioned creation of value (Makadok, 2001), therefore, taking advantage of the possible imperfections in the factor market.

In this sense, we must differentiate the concept of resources, the basic unit of analysis of this theory, understood as the set of factors or assets which the company possesses and controls (Barney, 1991), and capabilities, defined as the abilities and organizational competences that allow to suitably develop an activity from the combination and coordination of the individual available resources (Grant, 1996). We assume, therefore, a hierarchical conception based on the level of aggregation and coordination of both terms.

The criticism regarding to the Resource-Based View, as well as the Competitive Forces Approach, are focused on its static character, and the sustainability of the competitive advantage, which derives from the assumptions raised by both theories, is considered improbable in dynamic environments (Eisenhardt & Martin, 2000). Indeed, resources accumulation, or to operate in an "attractive" industry, is not sufficient to achieve and, especially, maintain sustainable competitive advantage in the face of changes that could affect either factor or product markets.

In this sense, and following the reasoning of Priem & Butler (2001), the value of the resources (and capabilities) varies as the competitive environment changes. These authors claim that the value of a resource is determined by external factors which are not considered in the Resource-Based View (according to which these external factors remain invariable), since if factors regarding products and clients vary, then, the value of the (strategic) resources that sustain the competitive advantage of the firm also can vary, therefore, eroding this advantage. The reasoning can be extrapolated to the Competitive Forces Approach. The changes that affect the product market so reducing firm market power will make the industry less attractive, eroding its competitive advantage.

These arguments have contributed to the emergence, in the last decade, of a new approach in the strategic management literature that tries to explain how the firms can obtain and maintain competitive advantage in the face of changes affecting product market and/or factor market. This new perspective in the strategic thinking, focal piece of the analysis carried out in the present chapter, has been called dynamic capabilities (Teece et al., 1997; Eisenhardt & Martin, 2000; Makadok, 2001; Winter, 2003; Teece, 2007). The main argument of this new perspective is that neither the Competitive Forces Approach, nor the Resource-Based View are able to explain how firms obtain competitive advantage when operating in rapid change environments (Teece et al., 1997), defending the idea that the competitive advantage in such environment rests in the development of dynamic capabilities more skilfully or more fortuitously than competitors and the creation of resources' configurations that maintain this advantage (Eisenhardt & Martin, 2000). Firms that have succeeded in such competitive environments are those that have demonstrated: i) fast response capacity, and ii) fast and flexible product innovation (Teece et al., 1997).

Therefore, the dynamic capabilities perspective tries to improve the explanatory power of both theoretical approaches regarding with the sustainable character of the competitive advantage in fast changing environment, while considering the evolutionary nature of firm resources and capabilities in relation to the changes in the environment in which it operates (Nelson, 1991; Wang & Ahmed, 2007).

The strategic relevance of dynamic capabilities, as well as the interest that it provokes in the academic literature is, therefore, justified. Nevertheless, in spite of its relevance and wide and increasing number of scientific studies on this phenomenon, the concept of dynamic capabilities suffers from certain terminological deficiencies as its definition is not consistent in the literature (Wang & Amhed, 2007). More important, however, is the lack of consensus with regard to the factors that influence the dynamic capabilities development by firms. In addition, the consequences and implications derived from the dynamic capabilities development must be examined. These three objectives are intended to be achieved in this work, and they are developed in each of the three following sections. First of all, on the basis of the theoretical revision of the main studies that analyse the subject of the dynamic capabilities, we try to offer a conceptual synthesis of the term in order to clarify the definition provided by the theory. Once this objective is achieved, we will try to identify the key factors involved in the development of dynamic capabilities by firms. In order to present a general vision of the antecedents those take part in the construct, we offer a theoretical model in which such antecedents are included, as well as the consequences derived from the dynamic capabilities development. Finally, an empirical study is carried out on a sample of 177 worldwide firms operating in a rapid change sector such as the software development industry.

DYNAMIC CAPABILITIES CONCEPT

With the aim to offer a consistent definition of the dynamic capabilities concept, Table 1 summarizes various definitions adopted by the main authors who have studied the subject (take into account that until the work of Teece et al., 1997, the term had not been coined).

Although the above conceptualizations differ in the form they are presented, it is possibly to identify two common elements: i) adaptive capability, and ii) to rapid change environment.

In a dynamic (Schumpeterian, evolutionary, rapid change or of high speed, according to authors) context, if the firm intends to maintain its competitive advantage, it must be able to alter (to adapt, to evolve, to renew, to reshape, according to authors) its base of resources and capabilities so that they allow it to compete with competitive advantage in the new environment.

Therefore, we can define the concept of dynamic capabilities as the firm ability to continuously modify its base of resources and capabilities so that it is always adapted to fast changing environment, adopting, therefore, the seminal definition of Tecce, Pisano & Shuen (1997). The following section will explain what a dynamic environment consists of.

According to Wang & Amhed (2007), there is a hierarchy of capabilities integrated by three levels: level zero consists on firm resources, fundamental to the development of their capabilities, standing on the basis of the competitive advantage when they demonstrate VRIN characteristics. Nevertheless, and as it has already been argued in the previous section, such characteristics do not persist in the face of changes affecting factor and/or product markets. At the first level of the

Table 1. Dynamic capabilities definitions.

AUTHORS	CONCEPTUALIZATION
SCHUMPETER (1934)	*Innovative capacity:* Architect of the process of "creative destruction". Characteristics that reunite the entrepreneurs as key actors in the continuous change process.
NELSON (1991)	*Core competences evolution*: Constructed on a hierarchic base of organizational routines defined based on a Schumpeterian or evolutionary context.
COLLINS (1994)	He argues that there are certain organizational capabilities that guide the ratio of change of ordinary capabilities.
PISANO (1994)	*Altering resources capabilities*: The organizational antecedents and the strategic routines by means of which the managers alter the firm resource base (they acquire and remove resources, integrate them jointly and recombine them) to generate new value creation strategies.
HENDERSON & COCKBURN (1994)	*Architectural competencies:* They provoke the creation, evolution and recombination of the resources in search of new sources of competitive advantage.
TEECE, PISANO & SHUEN (1997)	*Dynamic capabilities*: Firm's ability to integrate, build, and reconfigure internal and external competences to address rapidly changing environments. Therefore, they reflect an organization's ability to achieve new and innovative forms of competitive advantage.
EISENHARDT & MARTIN (2000)	They consist on specific strategic and organizational processes (such as product development, alliances, or strategic decision making) that create value in firms facing dynamic markets by means of the manipulation of resources towards new value creation strategies.
MAKADOK (2001)	Processes through which firms generate economic income by being more effective than their rivals deploying resources.
ZOLLO & WINTER (2002)	Learned pattern of collective activity by means of which the organization systematically generates and modifies its operational routines.
WINTER (2003)	Organizational capabilities (high level routines or set of routines) affected by change. They are investments on firm capabilities renewal.
AUGIER & TEECE (2007)	They refer to the (inimitable) capacity of the firm to form, reform, shape and reshape its resource base in order to be able to respond external changes in markets and technologies.
TEECE (2007)	Difficult to imitate firm capabilities necessary to adapt to changing customers needs and new technological opportunities. Also, they include the firm ability to influence the environment in which it operates, to develop new products and processes and to design and implement new viable business models.
WANG & AHMED (2007)	Form orientation towards continuous integration, reconfiguration, renovation and recreation of its resources and capabilities and, more importantly, to the increase and rebuilding of its core competences in response to changing environment in order to maintain the competitive advantage.

hierarchy we can find firm capabilities, which will result in a greater performance when it gains the ability to unfold resources to achieve a desired objective. The core or central capabilities of the firm are classified at the second level of the hierarchy, defined as the set of strategically relevant capabilities to its actual competitive advantage. Nevertheless, the core capabilities can become irrelevant or even become "core rigidities" in rapid change environment. At the top of the capabilities hierarchical pyramid are situated the previously defined dynamic capabilities. These capabilities do not only allow achieving a competitive advantage in the present context, but they also drive the change in dynamic environments. Therefore, they can be understood as a capability to create or to develop new capabilities based on the evolutionary character of the environment in which the company operates.

Accordingly to the above discussion, we propose that the mechanism by which firms can be continuously adapted to the changes that take place in a dynamic environment requires two key components. First of all, that the company will be able to learn from the changes that occur in its environment, that is to say, it must be capable of absorb the new knowledge coming from outside its boundaries. Secondly, that the firm induces such changes in the context, capitalizing the new knowledge, staying ahead of its competitors when responding to the new demands. Both aspects are dealt with in greater depth in the following section.

DYNAMIC CAPABILITIES DEVELOPMENT

In this section, an analysis of the necessary factors to develop dynamic capabilities by firms will be carried out. First of all, we discuss the firm environmental dynamism as a necessary condition for the impulse for this type of capabilities, since if the environment in which the company operates is not characterized by the occurrence of changes, or if the changes are easily predictable, it does not seem relevant to speak about any kind of adaptation to the changes that take place in the context. Secondly, the necessary characteristics of the firm or the mechanisms which enable it to be continuously adapted to the changes that take place in the previously defined environment will be addressed.

Dynamic Environments

Dynamism implies change (Winter 2003). Environmental dynamism can be caused by one main factor or by a combination of different factors, which include the technological innovation, regulatory changes, economic cycle, competitive change and changes in demand patterns (Wang & Ahmed, 2007).

If the environment in which firm operates is static, perfectly comprehensible, stable or its changes are predictable, the strategic value of resources that configure the competitive capabilities of the firm will also be stable, not diminishing, and so will be the competitive advantage (Priem & Butler, 2001). In this type of environment it is not necessary to speak about dynamic capabilities, maintaining the traditional concept of organizational routines, that is to say, complicated and very detailed, following an analytical process that rests on the existing knowledge and a linear execution with predictable results (Eisenhardt & Martin, 2000).

Nevertheless, when markets are dynamic or 'high speed', characterised by unclear boundaries, uncertain successful business models and their participants (buyers, suppliers, competitors and complementary sectors) are ambiguous and changing, the general structure of the industry is not clear, therefore the uncertainty cannot be modelled through probabilities, because it is not possible to specify a priori the possible future states. This aspect poses important problems, as much for the detection of external threats and opportunities in Industrial Economy, as for the identification of the internal strengths and weaknesses from the Resource-Based View. In this type of environments, dynamic capabilities necessarily rest much less on the existing knowledge and more on the fast creation of new knowledge (Eisenhardt & Martin, 2000). In this type of context the success depends on the discovery and the development of opportunities, the efficient combination of the internally and externally generated innovations, the efficient transfer of technology in the company and between companies, the protection of intellectual property, the improvement of the 'best practice' in business processes, the invention of new business models, making impartial decisions and preventing imitation and other forms of replication by rivals (Teece, 2007).

Absorptive Capacity and Innovative Capacity

As it has been mentioned before, we propose two key components of the mechanism which allows the firm to be adapted to external changes. The first one consists on firm's ability to learn from the environment, absorbing and internalising new knowledge coming from outside. The second one consists on producing these changes in the context, capitalising this knowledge. As we will present in the following subsections, both mechanisms refer to absorptive capacity and innovative capacity, respectively.

Some authors have also included adaptive capability as another component factor of dynamic capabilities (Wang & Ahmed, 2007), defined as the firm ability to identify and capitalize on new emerging market. This academic stream considers than the focus of adaptive capability resides on the correct alignment of firm internal factors with the external context factors.

Nevertheless, it seems that including adaptive capacity, as defined, as a component factor of dynamic capabilities would be a tautological error, because of its definition is implicit in the dynamic capabilities concept, which is, the ability to adapt to fast changing environments (Teece et al., 1997).

Absorptive Capacity

In order to be adapted to changes occurring in a dynamic environment, firms need to demonstrate absorptive capacity. Cohen and Levinthat (1990) define it as firm ability to recognize the value of new external information, assimilate it, and apply it with commercial ends. Nevertheless, this absorptive capacity definition includes the concept of dynamic capabilities (absorption and innovation capacities).

On the other hand, Zahra & George (2002) claim that absorptive capacity is critical for the success in the face of outside technological changes, stating that this capacity is an important factor for the dynamic capabilities development in multiple industries, affecting the nature and sustainability of firms competitive advantage. These authors argue that absorptive capacity is a multidivisional concept and they propose four component factors: acquisition of knowledge, assimilation, transformation and exploitation. Nevertheless, as in the previous case, if absorptive capacity agglutinates these four elements, the absorptive capacity concept is put on the same level as that of dynamic capabilities, including again the innovative capacity concept. In both studies, it is argued that absorptive capacity is critical for the development of both innovative capacity (Cohen & Levinthal, 1990) and dynamic capabilities (Zahara & George, 2002). Therefore, it does not seem right to raise an absorptive capacity definition including both dimensions.

According to Arbussá & Coenders (2007), we suppose a less broad definition of absorptive capacity. They assume that there are two types of absorptive capacity: (i) the capability to scan the external environment for new knowledge and (ii) the capability to integrate new external knowledge into firm innovation process.

Linking this definition with both previous works, the concept of absorptive capacity, as it has been defined, would be formed exclusively by the two first components of Zahra & George (2002) proposed definition, that is to say, the potential absorptive capacity (acquisition and assimilation of new external knowledge). This is coherent with the first part of Cohen & Levinthal (1990) absorptive capacity definition, that is, firm's ability to recognize the value of new external information and assimilate it.

In the theoretical contribution of Teece (2007), an explanative model of dynamic capabilities concept is presented, with a series of factors that affect their development. Based on the previous discussion, some elements composing that model can be characterised in the context of absorptive capacity. These are: processes to exploit the complementary and suppliers innovation; processes to exploit exogenous developments in science and technology; processes to identify new market segments and customer necessities; decentralisation and 'near decomposability'; learning and knowledge management.

Therefore, based on the previous arguments we propose that:

Hypothesis 1: In rapid change environments, the higher the firm absorptive capacity, the higher will be the firm performance.

Innovative Capacity

Again, we propose that firms facing rapid change environments need to demonstrate innovative capacity to be continuously adapted to the changes occurring in these kind of contexts. According to Schumpeter (1934), the innovative capacity is characterised by the development of new products and services, the development of new production methods, the identification of new markets, the discovery of new provisioning sources and the development of new organizational forms. On the other hand, Wang & Ahmed (2007) define it as the ability of the company to develop new products and/or markets, aligning a strategic innovative orientation with innovative behaviours and processes. Therefore, the innovative capacity explains the link between the firm resources and capabilities and its products and markets portfolio when it faces rapid change environments.

Obviously, in order to maximize its innovative capacity output, firm needs to develop absorptive capacity as it has been defined in the previous paragraph (if the company is a completely closed system, the its innovation success would be due to coincidence, since it would not obtain fed back by the evolution of context in which it operates). This argument is put forward by Cohen & Levinthal (1990) when they claim that absorptive capacity is critical to firm innovative capabilities. Therefore, completing the argument discussed in the previous section, we suggest that innovative capacity is equivalent to the second component of absorptive capacity proposed in Zahra & George (2002) theoretical model, that is, the realized absorptive capacity (transformation and exploitation of knowledge). Although this knowledge could be partly generated from external sources, after being absorbed by the organization, it is already integrated into firm knowledge base and need to be transformed and utilized with commercial ends. Again, as in the previous case, the innovative capacity defined in this way is coherent with the second part of Cohen & Levinthal (1990) absorptive capacity definition, that is, applying the new knowledge (already absorbed) with commercial objectives.

Also, and returning to the dynamic capabilities model proposed by Teece (2007), several component factors of dynamic capabilities could be included within innovative capacity, eg: processes of directing the internal R&D and selecting new technologies; design of the business model; selection of protocols in the decision making; selection of the limits of the company; direction of the complementary assets; corporative government.

Therefore, following the previous reasoning, the second hypothesis is proposed:

Hypothesis 2: In rapid change environments, the higher the firm innovative capacity, the higher will be the firm performance.

Therefore, we can consider that both aspects are the most important factors composing the mechanism that allows the dynamic capabilities development by firms, reinforcing together its ability to integrate, reconfigure and renew its resources and capabilities along with external changes. As it can be seen, both factors are correlated, although as it has been discussed, they are conceptually different.

The followed reasoning supposes that in order to develop dynamic capabilities, firms need to develop both absorptive capacity and innovative capacity, as they have been defined. On the other hand, if the organization demonstrates a high level of only one of these components, it does not guarantee firm success in rapid change environments. Indeed, the suggestion is that absorptive capacity acts as essential input of the innovative capacity efficiency, so the absorptive capacity that does not become effective in this process, as well as the innovative capacity which is not sustained on an ample base of absorptive capacity, may involve an additional cost to the firm. Consequently, the efforts of the firms that operate in dynamic environments need to be focused on the maximization of the absorptive capacity output, maximising, in this way, the efficiency of the innovative capacity.

Therefore, according to the argumentation presented throughout the present study, we propose the third hypothesis:

Hypothesis 3: In rapid change environments, if a firm develops dynamic capabilities, it will achieve higher performance than if it only develops absorptive capacity or innovative capacity separately.

EMPIRICAL STUDY

The following section offers the basic information regarding the empirical study. Firstly, we present information about data sources and sample. Next, the variables and the measurement methods are discussed, as well as the methodology used for carrying out the statistical analysis. Finally, the main results of the study are presented.

Data and Sample

The "Monitoring Industrial Research: The 2006 EU Industrial R&D Investment Scoreboard" report, prepared for the European Commission, provides a classification of the first thousand companies based on their investment in R&D during the year 2005. This report provides data divided in groups in six different ways: i) classification of the first thousand companies of the European Union based on their investment in R&D; ii) the same companies classified by state members; iii) the same companies classified by industrial sector; vi) classification of the first thousand companies outside the European Union based on their investment in R&D; v) the same companies classified by countries; and vi) the same companies classified by industrial sector.

The report provides the following information for each firm (for years 2004 and 2005): the name of the company; the position in the corresponding ranking (EU or non EU); country; industrial sector; investment in R&D; net sales; number of employees; operational profit; market capitalization; and capital investment.

Therefore, this report is an important source of data for the objectives intended in this work, as much for the available information that it contains as for the fact that it includes companies from all over the world, which allows to overcome one of the most common limitations in the social science research, that is, the possibility of results generalization.

Our empirical test is focused on companies operating in the software industry and which were included either in the classification of the first thousand European companies by R&D investment, or in the analogous classification for the companies outside the European Union. There were one hundred companies devoted to the software development in the European classification, that is, 10% of most innovative companies in Europe operate in the software industry. In the case of non European classification, this percentage declines slightly to 9%, with 90 companies devoted to this activity (most of these companies from United States). Thus, it is an innovation-intensive industry, being a propitious environment for our analysis.

Therefore, the study started with 190 firms worldwide dedicated to software development. After eliminating the cases without data in those magnitudes that compose some of the variables used in the analysis, the final sample to carry out the study were 177 firms.

Variables

There are four variables necessary to carry out the empirical test. The first three, that is, absorptive capacity, innovative capacity and dynamic capabilities, are independent variables. On the other hand, the fourth variable, that is, firm performance, is the dependent variable. Next, the measures used for each variable are described.

Absorptive capacity: in order to carry out the measurement of this variable we have used R&D intensity (Investment in R&D/ Net Sales). This ratio has been used as a traditional measure of absorptive capacity in the specialist literature since Cohen & Levinthal (1990) put it forward in their work. By dividing the investment in R&D by the volume of sales, we eliminate the possible effects derived from firm size, therefore using variables that control those effects is unnecessary.

Innovative capacity: we propose an indicator of the efficiency in the application of knowledge to measure the innovative capacity of firms (Increase of the Net Sales/ Increase of R&D Intensity). In their recent study, Leskovar-Spacapan & Bastic (2007) use in one of their questionnaire items the increase of market quota derived from innovation in order to measure the so called "innovative advantage". Again, as in the previous case, variables controlling size effects are unnecessary, as the focus is on variations.

Dynamic capabilities: According to the theoretical argumentation discussed above, we use iteration term of the two previous variables to measure dynamic capabilities. This measurement provides information about the extent to which a firm counts both on absorptive capacity and innovative capacity.

Firm performance: we have used the operative profit as percentage of the net sales in 2005 to measure the dependent variable.

Methodology

The proposed methodology for the study is composed by three regressions estimated by Ordinary Least Squares in which the dependent variable is firm performance and the independent variables are, in each case, absorptive capacity (H1), innovative capacity (H2) and dynamic capabilities (H3). The statistic software used for the estimations of the three models is SPSS 15.0 (Statistical Package for the Social Sciences) for Windows.

RESULTS

Table 2 presents the results of the three regressions.

As it can be seen in the estimation results of Model 1, the standardized coefficient corresponding to absorptive capacity is negatively related to firm performance and statistically significant, therefore rejecting our first hypothesis. When we presented this concept, we argued that a high absorptive capacity not effectively realised implies an additional cost to the firm. This reasoning is supported by the results of Model 1, especially if, as in this case, absorptive capacity is measured based on countable data.

According to the estimation results of Model 2, the standardized coefficient corresponding to innovative capacity is positively related to firm performance and statistically significant, supporting our second hypothesis. The higher the firm innovative capacity, the higher is the firm performance.

Finally, as it can be seen in Model 3, the data support our third hypothesis. When a company develops both absorptive capacity and innovative capacity it achieves a higher performance. The standardized coefficient corresponding to dynamic capabilities is positively related to firm performance. This coefficient is considerably higher than Model 1 (only considering absorptive capacity) and Model 2 (only considering innovative capacity), and statistically highly significant. Moreover, the explained variance in this model clearly overcomes the explained variance of the two previous estimated models. Therefore, the empirical evidence supports the third hypothesis. When a company develops dynamic capabilities, it achieves higher performance than only developing absorptive capacity or innovative capacity separately.

Table 2. Empirical results.

MODEL 1.- ABSORPTIVE CAPACITY			
VARIABLE	**β**	**$β_{STANDARDIZED}$**	**p-value**
Constant	25,209	-	,000
ACAP	-1,136	**-0,719**	**,000**
F = 187,297	-	-	,000
$R^2 = 0{,}517$ **$R^2_A = 0{,}514$**			
MODEL 2.- INNOVATIVE CAPACITY			
VARIABLE	**B**	**$β_{STANDARDIZED}$**	**p-value**
Constant	-1,000	-	0,731
INNCAP	0,289	**0,191**	**,011**
F = 6,618	-	-	,011
$R^2 = 0{,}036$ **$R^2_A = 0{,}031$**			
MODEL 3.- DYNAMIC CAPABILITIES			
VARIABLE	**B**	**$β_{STANDARDIZED}$**	**p-value**
Constant	-0,682	-	,775
ACAP x INNCAP	0,017	**0,358**	**,000**
F = 25,710	-	-	,000
$R^2 = 0{,}128$ **$R^2_A = 0{,}123$**			

CONCLUSION, LIMITATIONS AND FUTURE RESEARCH

First of all, this work intended to offer a consensus description of dynamic capabilities according to the main definitions of the concept raised in previous studies. Based on literature review, it seems that the definition of the seminal work of Teece et al. (1997) dominates in the literature, although the presented definitions differed in the way they were expressed. Adaptive capacity to rapid change environments constitutes, therefore, a suitable synthesis of the concept. The second objective was to identify the component factors of the mechanism by which firms achieve a state of adaptation to the changes affecting the environment in which it operates. With such changes as a necessary condition, there seem to be two key components of dynamic capabilities: absorptive capacity, which allows the firm to identify and internalize the new external knowledge, and innovative capacity, which allows the firm to capitalize its knowledge-base by adequately responding to such changes.

Finally, an empirical study has been carried out based on a sample of firms operating in a dynamic industrial sector, such as the software development industry. In order to support our theoretical arguments with real data, we have obtained a considerable support for the proposed reasoning. The data sustain the argument that if firms only invest in the development of absorptive capacity, without commercially exploiting the new externally acquired knowledge, it will incur an additional cost without positive impact on the revenues, therefore worsening its performance.

On the other hand, it has been proved that when a firm develops innovative capacity, it achieve better results, although the explanatory power of this variable on firm performance is considerably small. Finally, the data supports the argument that if the company develops its dynamic capabilities (impulse of its absorptive and innovative capacities together), it will achieve higher performance than if only develops absorptive capacity (negative impact) or innovative capacity (low impact) separately.

The fact that our sample does not belong only to one region enriches the obtained results, allowing for their generalisation. Nevertheless, as we have focused only on one industrial sector, this possibility of generalization is limited to the studied industry. This fact supposes the firs limitation in our study, as well as the definition of a future research direction, that is, to replicate the study in other and multiple industrial sectors.

An additional limitation, as well as, again, an interesting future research direction, is the simplification of absorptive capacity and innovative capacity measurements. The development of adequate and complete measurements of both concepts is necessary for literature advancement. In the future, it will be necessary to develop more complex measuring instrument, such as compound measures, also conjugating primary source data, that include the mechanisms by means of which firms acquire information from the environment (i.e.: customers, competitors, providers, public authorities and institutions, research centres), their ability to internalize it (i.e.: qualified personnel, adequate technology, directive capacities, learning mechanisms) and the efficiency of the company to capitalize its knowledge (i.e.: new products and markets, new production processes, innovations in marketing).

Another limitation is the use of financial performance as dependent variable. It would be more adequate to use market data. Moreover, it was not possible to include delays in the dependent variable. In this sense, and according to Kor & Mahoney (2005), in technological-based sectors the investment in R&D becomes into products capable to generate revenues in three years. This fact needs to be taken into consideration in future empirical research.

Finally, another interesting future research direction is the identification and understanding of the process through firms manage theirs resources to maximise the efficiency between their absorptive and innovative capacities, so that, in the optimal situation all the absorbed knowledge is transformed in responses to external changes or in changes that the company induces in its environment like development of new products, processes, organizational forms or business models.

REFERENCES

Arbussá, A., & Coenders, G. (2007). Innovation Activities, Use of Appropriation Instruments and Absorptive Capacity: Evidence from Spanish Firms. *Research Policy, 36*(10), 1545–1558. doi:10.1016/j.respol.2007.04.013

Augier, M., & Teece, D. J. (2007). dynamic capabilities and Multinational Enterprise: Penrosean Insights and Omissions. *Management International Review, 47*(2), 175–192. doi:10.1007/s11575-007-0010-8

Barney, J. B. (1991). Firm Resources and Sustained Competitive Advantage. *Journal of Management, 17*(1), 99–120. doi:10.1177/014920639101700108

Cohen, M. D., & Levinthal, D. A. (1990). Absorptive Capacity: A New Perspective of Learning and Innovation. *Administrative Science Quarterly, 35*(1), 128–152. doi:10.2307/2393553

Collins, D. J. (1994). Research Note: How Valuable are Organizational Capabilities? *Strategic Management Journal, 15*(Winter Special Issue), 143-152.

Eisenhardt, K. M., & Martin, J. A. (2000). dynamic capabilities. What are They? *Strategic Management Journal, 21*(10/11), 1105–1121. doi:10.1002/1097-0266(200010/11)21:10/11<1105::AID-SMJ133>3.0.CO;2-E

Grant, R. M. (1996). Prospering in Dynamically Competitive Environments. *Organization Science, 7*(4), 375–387. doi:10.1287/orsc.7.4.375

Henderson, R. & Cockburn, I. (1994). Measuring Competence? Exploring Firm Effects in Pharmaceutical Research. *Strategic Management Journal, 15*(Winter Special Issue), 63-84.

Kor, Y., & Mahoney, J. T. (2005). How Dynamics, Management, and Governance of Resource Deployments influence Firm-level Performance. *Strategic Management Journal, 26*(5), 489–496. doi:10.1002/smj.459

Leskovar-Spacapan, G., & Bastic, M. (2007). Differences in Innovation Capability in Transition Economy: Internal Aspect of the Organization's Strategic Orientation. *Technovation, 27*(9), 533–546. doi:10.1016/j.technovation.2007.05.012

Makadok, R. (2001). Toward a Synthesis of Resource-Based and Dynamic-Capability Views of Rent Creation. *Strategic Management Journal, 22*(5), 387–401. doi:10.1002/smj.158

Nelson, R.R. (1991). Why Do Firms Differ, and How Does it Matter? *Strategic Management Journal, 12*(Winter Special Issue), 61-74.

Pisano, G.P. (1994). Knowledge, Integration, and the Locus of Learning: An Empirical Analysis of Process Development. *Strategic Management Journal, 15*(Winter Special Issue), 85-100.

Porter, M. E. (1981). The Contribution of Industrial Organization to Strategic Management. *Academy of Management Review, 6*(4), 609–620. doi:10.2307/257639

Porter, M. E. (1996). What is Strategy? *Harvard Business Review, 74*(6), 61–78.

Priem, R. L., & Butler, J. E. (2001). Is the Resource-based 'View' a Useful Perspective for Strategic Management Research? *Academy of Management Review, 26*(1), 22–40. doi:10.2307/259392

Rumelt, R. P. (1991). How much does Industry Matter? *Strategic Management Journal, 12*(3), 167–185. doi:10.1002/smj.4250120302

Schumpeter, J. (1934). *The Theory of Economic Development*. Cambridge, MA: Harvard University Press.

Teece, D. J. (2007). Explicating dynamic capabilities. The Nature and Microfoundations of (Sustainable) Enterprise Performance. *Strategic Management Journal, 28*(13), 1319–1350. doi:10.1002/smj.640

Teece, D. J., Pisano, G., & Shuen, A. (1997). Dynamic capabilities and Strategic Management. *Strategic Management Journal, 18*(7), 509–533. doi:10.1002/(SICI)1097-0266(199708)18:7<509::AID-SMJ882>3.0.CO;2-Z

Wang, C. L., & Ahmed, P. K. (2007). dynamic capabilities: A Review and Research Agenda. *International Journal of Management Reviews, 9*(1), 31–51. doi:10.1111/j.1468-2370.2007.00201.x

Winter, S. G. (2003). Understanding dynamic capabilities. *Strategic Management Journal, 24*(10), 991–995. doi:10.1002/smj.318

Zahra, S. A., & George, G. (2002). Absorptive Capacity: A Review, Reconceptualization, and Extension. *Academy of Management Review, 27*(2), 185–203. doi:10.2307/4134351

Zollo, M., & Winter, S. G. (2002). Deliberate Learning and the Evolution of dynamic capabilities. *Organization Science, 13*(3), 339–351. doi:10.1287/orsc.13.3.339.2780

KEY TERMS AND DEFINITIONS

Resources: Tangible or intangible assets.

Capabilities: Intangible configurations of resources.

Core Competences: Capabilities in which the firm basis its actual competitive advantage.

Dynamic Environment: Competitive context characterized by rapid, continuous and unpredictable changes in product and/or factor markets.

Dynamic Capabilities: Firm's ability to adapt to changing environmental conditions by continuously acquiring and applying new knowledge.

Innovative Capacity: Firm's ability to develop new products, services or new methods of production, as well as the identification of new markets, the discovery of new sources of provision and the development of new organizational forms.

Absorptive Capacity: Firm's ability to explore the environment in search of new knowledge and to integrate that new external knowledge in its organizational knowledge.

Section 3
Intellectual Capital and Technological Innovation

Chapter 8
An Intellectual Capital-Based View of Technological Innovation

Miriam Delgado-Verde
Universidad Complutense de Madrid, Spain

Jorge Cruz-González
Universidad Complutense de Madrid, Spain

ABSTRACT

The growing importance of knowledge as a key productive factor in today's society requires a change in thinking about innovation in general terms as well as in specific terms, such as technological innovation, product innovation and organizational innovation (Nonaka, 1994). Also, companies are aware that knowledge is the most valuable and strategic resource to face the current environment (Chen et al., 2004). In this sense, intellectual capital is increasingly interesting to companies that get their benefits through innovation and knowledge. For this reason, it is important to study the relationship between intellectual capital and innovation (Edvinsson and Sullivan, 1996). In fact, it is widely accepted that an organization capability to innovate is closely tied to it is intellectual capital, or its ability to utilize its knowledge resources (Subramaniam and Youndt, 2005). So, this research proposes a theoretical model to identify and classify the different elements of intellectual capital (human capital, organizational capital, technological capital, relational capital and social capital) as a source of various types of technological innovation (product, process, radical and incremental innovation), analyzing the relationship between each of those components and the four types of innovation.

INTRODUCTION

The present situation is causing a new competitive dynamic (Johnson et al. 2002; Leitner, 2005), in which firms give increasingly greater importance to the intangible resources and capabilities when they face competitors. Therefore, it is widely recognized that the new knowledge and its implementation as a key factor in achieving and maintaining competitive advantage (Schumpeter, 1942; Itami and Roehl, 1987; Galende, 2006).

In this sense, in the current competitive and dynamic environment, technological innovation is becoming, ever more, in a key aspect of the business

DOI: 10.4018/978-1-61520-875-3.ch008

competition (Alegre et al., 2005). For this reason, studies focused in improving the firm processes are a major concern.

The changing environment, globalization phenomenon and accelerated product cycles are prompting pressures in the business environment, so that it is necessary to accumulate a large volume of knowledge in order to develop subsequently different types of innovations.

In addition, in the professional field, managers need tools to evaluate the management and results of innovation activities. Thus, the reasons for the importance of technological innovation may be summarized in (Alegre et al., 2005):

1. The development of new technologies and major understanding of existing technologies increase the possibility of creating new products and processes.
2. Facing a number of companies capable of competing at international level, they have risen considerably.
3. Satisfying consumers who have become more sophisticated and discerning, so that the range of products has increased significantly.

Owing to the importance of a study that not addresses only to consider the relationship between intellectual capital and innovation in general, but that takes into account each component of intellectual capital and various types of innovation, it seems necessary to conduct an analysis as this to cover the gap. Besides, as it can see in the literature review, many empirical researches examine the innovation and its connection with several issues about intellectual capital, but they are not covered by these works within the intellectual capital. Therefore, since it is a question little discussed in intellectual capital literature, this work tries to study technological innovation in depth, differentiating four types.

Thus, our study proposes a classification of intellectual capital, differentiating five elements (human capital, organizational capital, technological capital, relational capital and social capital) and linking each of them with four types of technological innovation (product, process, radical and incremental), trying to analyze a research issue that has not been well treated from the Intellectual Capital-Based View.

THEORETICAL BACKGROUND

Given the competitive paradigm presented in the introduction section, new theoretical approaches have appeared from the academic literature, trying to explain the nature and sources of business success from internal factors owned by the firms.

Thus, in recent decades, the explanation of sustained competitive advantage has been supported primarily on the possession and/or control of endogenous factors to the company (Wernelfelt, 1984; Rumelt, 1991). This view, so-called Resource-Based View (Amit and Schoemaker, 1993; Barney, 1991), attaches special importance to the intangible factors (Itami and Roehl, 1987; Hall, 1993), which are based on knowledge and information and constitute cornerstones of intellectual capital studies.

Besides, previous empirical analysis carried out in several researches leads to conclude that the factors associated with such resources owned by the company, basically it is intellectual capital, have a more significant influence on innovation, market and financial performance (Bowman and Helfat, 2001; Reed et al., 2006).

However, the Resource-Based View has some weakness due to its static view of the competitive analysis or the difficulty in making operative its principles (Priem and Butler, 2001). Therefore, different approaches have appeared in order to overcome these problems, improving the internal perspective of Business Management.

Thus, Capabilities Dynamics Perspective (Eisenhardt and Martin, 2000; Carpenter et al., 2001) appears as an extension of the Resources-Based View, explaining the firms survival and competitive success from their ability to develop knowledge and skills in order to adapt to the environment requirements.

On the other hand, Knowledge-Based Theory of the firm (Nonaka, 1991; Kogut and Zander, 1993; Hedlund, 1994) examines the role of firms in creating, storing and applying knowledge (Grant and Banden-Fuller, 1995), considering the evolutionary point of view.

With regards to the difficulty in making operational the resources and capabilities, Intellectual Capital-Based View (Reed et al., 2006) appears for identifying and measuring the organizational intangible resources and capabilities that may be a advantage competitive for the companies. This approach takes a static perspective of the intangible factors in order to carry out an inventory of intangible resources, trying to identify, evaluate and classify the different stocks of knowledge possesses or controlled by organizations.

INTELLECTUAL CAPITAL AND TECNOLOGICAL INNOVATION

Different works have examined firm intellectual capital and its elements (Edvinsson and Sullivan, 1996; Bontis, 1998; CIC, 2003b; Subramaniam and Youndt, 2005; among others). In this sense, this study explains a comparative analysis about them, developing a comprehensive definition and classification in order to fit the intellectual capital with Resource-Based View.

In this sense, intellectual capital can be defined as a set of intangible resources and capabilities related to different categories of knowledge, whether individual, organizational or inter-organizational, which could achieve a competitive advantage.

Once the classification and the terminology used by different authors have been revised, it can assert that the intellectual capital includes three main elements: human capital, structural capital and relations-based capital. However, taking into account that this study aims for a more detailed and deeper analysis of intellectual capital, we propose to firm intellectual capital into five components. Therefore, the structural capital is divided into organizational and technological capital, and the relations-based capital is divided into relational and social capital.

Among the different reasons for taking into account five components, there are the following ones: 1) On the one hand, this research is focused to technological innovation, so it's essential to take out from structural capital the idea offered by the CIC (2003a) about the technological capital, since it plays an important role to innovate, as a basis of knowledge and technological infrastructure. In addition, the organizational capital and the technological capital (parts in which is divided the structural capital) are uneven nature, different strategic implications (CIC, 2001) and require different strategies for their accumulation and explotation (Tseng and Goo, 2005).

On the other hand, this work considers the social capital as an independent element because it has a different meaning to relational capital. It refers to the informal and personal relationships among employees inside and outside the firm that are not predetermined by the company. Those personal relationships are not included in the relational capital or human capital due to (i) the nature of the relations between a company as social entity and its stakeholders is substantially different from the nature of informal relationships among different individuals who belong to the firm, and among them and individuals belong to other companies; and (ii) the individual nature of human capital.

In answer to the first reason, the relations between a firm and its stakeholders have a different nature from the relationships among people who work together day by day or have a friendship for their job because in these latter may be a greater closeness and trust. In addition, the relations maintained by the company as a social entity are at the institutional level.

With respect to the second reason, relationships between people are considered at group level, while human capital is at individual level. Furthermore, if such relationships were taken into account within the human capital, it would not be reflected the importance of these informal networks. Also, despite the fact that social capital is linked to human capital, social relationships are different aspects to education or training, which are common measures presented by Blyler and Coff (2003).

Regarding innovation, many studies have analysed this concept (Schumpeter, 1912; Myers and Marquis, 1969; Damanpour and Evan, 1984; Nonaka, 1994; CE, 1995; Damanpour and Gopalakrishnan, 1998; Subramaniam and Youndt, 2005; OECD, 2006; among others). Moreover, taking into account that the current economic environment is characterized by its dynamism, innovations can be seen as cornerstone in adapting firms to their markets (Stieglitz and Heine, 2007).

It is also interesting to highlight that Swart (2006) notes that innovation (among other concepts such as intangible assets, embedding tacit routines, core competencies and knowledge creation) has an important account for explaining those assets that create value continually over the physical and financial resources.

Thus, after reviewing previous studies, this research proposes innovation (based fundamentally on firm intellectual capital) as a process by which manages to create a new idea that will be marketed later.

There are different typologies of innovation (product and process innovation; incremental and radical innovation; architectural and modular innovation, market-pull and technology-push innovation; competence-enhancing and competence-destroying innovation). However, this study will be based on the two most used and accepted by the literature, so we take into account four types of innovation.

On the one hand, if the focus is on the output or the process of carrying out innovation, we can refer to product and process innovation respectively (CE, 1995; Tidd, 2001; Adner, 2002; OECD, 2006). On the other hand, if we consider the degree of originality and novelty of the innovation process, then the types proposed are incremental or radical innovation (Henderson and Clark, 1990; Damanpour and Gopalakrishnan, 1998; Tidd, 2001; Koberg et al., 2003; Subramanian and Youndt, 2005; Stieglitz and Heine, 2007; among others).

Finally, it must highlight the work of Subramaniam and Youndt (2005), since it links the two constructs or complex concepts of this study, intellectual capital and technological innovation. They develop an empirical analysis about the influence of three elements of firm's intellectual capital (human capital, social capital and organizational capital) on the incremental and radical innovation, analyzing how different aspects of intellectual capital and certain relations among them affect different types of innovation capability. Thus, they enhance the understanding of knowledge-innovation link, finding that organizational capital positively influence incremental innovative capability; human capital has a negative influence on radical innovative capability; the interaction of human and social capital positively influence on radical innovative capability; social capital positively influence both incremental and radical innovative capabilities; but they did not find social capital to further enhance organizational capital's influence on incremental innovative capability.

INTELLECTUAL CAPITAL AS SOURCE OF TECHNOLOGICAL INNOVATION

When we analyze researches related to firm intellectual capital and innovation, it can observe that many authors (Van de Ven, 1986; Edvinsson and Sullivan, 1996; Damanpour and Gopalakrishnan, 1998; Bontis, 1998; Almeida and Phene, 2004; Leitner, 2005; Swart, 2006; Wu et al., 2007;

among others) postulate the existence of a positive relationship between the two concepts.

According to Edvinsson and Sullivan (1996) and Sullivan (2001), firms which returns are obtained from innovation and knowledge are increasingly more interested in the management of intellectual capital. Therefore, it's important analyze the relationship between intellectual capital and innovation. Indeed, Edvinsson and Sullivan (1996) argue that innovations create value within the firm and these innovations mainly come from the intellectual capital.

Human Capital as Source of Technological Innovation

One of the key determinants of value creation for knowledge-intensive firms is the innovation created by their human capital (Edvinsson and Sullivan, 1996). Likewise, Wang and Chang (2005) assert that human capital is the most important component of intellectual capital and knowledge as well as capabilities of employees are sources of innovation.

The review about the measurement of innovation management by Adams et al. (2006) refers to the importance of the human factor, taking into account the number of people assigned to the innovation tasks relating to propensity to innovate, skills, experience and education. In the same sense, Amabile et al. (2007) argue that the responses of creative employees, that are, having original ideas for changing products, services and processes, it is relevance to carry out innovations.

Moreover, some authors (Baldridge and Burnham, 1975) study demographic characteristics (gender, age and cosmopolitan condition), concluding that don't influence on the innovative behaviour, while others authors (Amabile, 1998) show that innovative groups should be formed by individuals with several of those characteristics.

Finally, in the Damanpour's (1991) metha-analysis it studied different sources of innovation, including professionalism. This element relates to education and experience of employees is positive and significantly associated with technical innovation.

So, taking into account the above it can formulate the following proposition:

Proposition 1: human capital is a source of technological innovation

Human Capital as Source of Product and Process Innovation

In this sub-section, it presents the influence of human capital or its dimensions on product and process innovation.

In the study on the effect of intellectual capital in the entrepreneurial characteristic of the firm, Hayton (2005) shows empirically the positive relationship between the diversity of human capital, both educational (considering areas of science, business and humanities) and functional (functional experience), and innovation; since the diversity promotes the search for knowledge and organizational learning processes and the different cognitive approaches promote creativity. In addition, this study measured innovation, considering (among others) the number of new products and services. However, it can't be showed that the stock of senior executives has a positive influence on that innovation. The latter may be due to the fact that it isn't enough a higher education and experience, but it is important to motivate employees for reaching innovations.

In her analysis on knowledge management for service innovation from the epistemological classification of knowledge (tacit and explicit knowledge), Leiponen (2006) finds that higher education of service expert supports weakly development of new services (this study incorporates new services as product innovation), possibly due to it facilitates communication within the firm by providing shared knowledge base for knowledge integration and exchange and across organizational boundaries by enhancing the firm's capacity to acquire external knowledge for innovation activities. While, the expert skills of employees have a negative influence on the likelihood of new services introductions. Regarding the latter, Díaz et al. (2006) study the impact of technological organizational knowledge on innovation and they are based on professional experience when they analyze the influence of the qualified staff recruitment on product innovation, obtaining a positive relationship, since the knowledge developed outside the firm adds an important value for carrying out innovations.

On the other hand, the negative relationship between experience and the likelihood of developing a product innovation may be due to the fact that experience is referred to the knowledge accumulated by an individual over time. Since this type of knowledge is likely to be the most difficult to combine with other knowledge assets in innovation projects, because it is deeply embedded in individual skills and practices and is difficult to communicate to other employees (Leiponen, 2006).

Pizarro et al. (2007) found that human capital, mainly the specificity of employees rather than the value of their knowledge, has a positive influence on product innovation, in terms of this research. These authors declare that individuals with valuable knowledge improve information process, learning and its implementation. Moreover, they assert that innovations require individuals with a unique and exclusive knowledge of the firm, relating to specific knowledge of employees.

Regarding support, leadership and coordination provided by managers, Damanpour (1987) and Lloréns at al. (2005) find a positive relationship on technological innovations. In the same sense, Akgün et al. (2007) also take into account the managerial commitment, which indicates developing and facilitating managerial support and leadership commitment for the innovation process and employee motivation, showing that there is a positive relationship between managerial commitment and product innovation. Therefore, commitment is an important issue because top managers can play a key role in the innovation process through the adaptation and internal dissemination within organisational (Boer and During, 2001).

So, based on the previous arguments, it can be enunciated the first sub-proposition:

Proposition 1a: human capital is a source of product innovation

With respect to process innovation, Damanpour (1991) also points out that professional knowledge, including education and experience of employees, is positively related to this type of innovation.

In the same sense, Hayton (2005) indicates the positive relationship between education and functional diversity of employees and "innovation" dependent variable, but highlighting new processes. Thus, just like in the previous sub-proposition, it argues that the diversity and heterogeneity of human capital represents a wide cognitive range which facilitates the acquisition of new knowledge. On the other hand, following the same logic explained on the dependent variable, it presents the impossibility of finding a positive relationship between the senior executives stock and new processes.

Regarding the professional experience included in the "qualified personnel recruitment", Díaz et al. (2006) found a positive influence on the likelihood of having innovative capabilities in the firm, including new processes, among other innovative aspects. Thus, just as explained above, it is represented the importance of knowledge owned by employees when they start to work in a new firm.

Thus, according to the literature review, it formulated the second sub-proposition:

Proposition 1b: human capital is a source of process innovation

Human Capital as Source of Radical and Incremental Innovation

As for radical innovation, Subramaniam and Youndt (2005) show that human capital negatively influences radical innovative capability. This result is also supported by Leiponen (2006) in her study on the determinants of new services. The latter author considers the effect of individuals experience on the likelihood of new service introductions, which require that the firm seek or create new knowledge and capabilities. In this sense, it can be understood it relates to radical innovation, as well as since this author considers independently the improvements of the new services, that is, incremental innovation.

Personal experience can be understood as individual tacit knowledge accumulated by each individual over time (Leiponen, 2006). Therefore, and based on the previous arguments about the relationship between human capital and the possibility of carrying out a product innovation, this kind of knowledge is difficult to communicate to other employees and it can not be a source of innovation. However, we propose that difficulties in communicating this type of knowledge is not a sufficient condition to point out that the experience negatively influences radical innovation; simply it may not be a source of radical innovation.

Continuing with the arguments of Leiponen (2006), Subramaniam and Youndt (2005) the negative relationship may be due to the resistance of independent experts to share their ideas with other employees. Hence, this negative influence suggests that individual experience in itself does not lead to such innovation, but to achieve a radical innovation it's necessary that people share their knowledge with others.

Nevertheless, Damanpour (1991) indicates a positive and significant relationship between the professionalism of workers,- gathering education and experience-, and firm radical innovation. Likewise, and providing a extended empirical data, Díaz et al. (2006) show a positive influence between professional experience and the likelihood of carrying out new products and processes, which can be considered radical innovation since the authors differentiate between new and improved significantly.

On the other hand, Hayton (2005) indicates that the educational and functional diversity of the employees have a positive influence on radical innovations, including the number of radical products, services and processes. However, they can not find a positive relationship between senior managers stock and such innovation.

Taking into account the first concept presented by Hayton (2005), Leiponen (2006) also indicates a positive relationship between a higher level of education and the likelihood of introducing radical new services, since a higher education level may facilitate internal communication to share knowledge and knowledge acquisition from external environment for its integration into the organization.

In addition, from a quantitative view on human resources, Leiponen (2006) shows that the number of employees has a positive impact on the likelihood of carrying out a radical innovation.

Finally, it is important highlight that among the critical success factors responsible of radical innovations, it can be found the support, involvement, and commitment of CEOs and senior managers. The role of this kind of innovation in achieving the long-term strategy and organizational aims must be clearly specified and reinforced at all levels (Simon et al., 2003).

Thus, and based on the previous arguments, it can formulate the third sub-proposition:

Proposition 1c: human capital is a source of radical innovation

Respecting incremental innovation, Hayton (2005) points out the existence of a positive relationship between the educational and functional diversity of employees, and the number of new products, processes and services produced by an organization. These innovations can be labelled as incremental innovations because the author distinguishes them from radical innovations when he defines innovation dependent variable. However, he did not find empirical support about the role of stocks of senior management on incremental innovation. Likewise, Leiponen (2006) can not show that upper level education as well as the number of employees have a positive influence on the likelihood of carrying out improvements in the new services. Among the possible reasons of, it can be found that knowledge must be codified in order to develop improvements in existing services, besides having a good level of education.

As in previous sub-proposition, Diaz et al. (2006) show that professional experience, incorporated in qualified staff recruitment, has a positive relationship on the likelihood of develop products and processes significantly improved, this is, incremental innovation.

Therefore, it raises the fourth sub-proposition:

Proposition 1d: human capital is a source of incremental innovation

Organizational Capital as Source of Technological Innovation

Van de Ven (1986) analyzed the innovation process and showed the importance of different factors. Within them he highlighted the implementation and institutionalization of innovative ideas as key determinants in the innovation process. In the same sense, Salman and Saives (2005) suggest that the accumulation of knowledge enhances the organizational capacity for recognizing and assimilating new ideas, as well as its ability to convert this knowledge into innovations.

On the other hand, there are several works that study the organizational culture as another important organizational factor (Aiman-Smith et al., 2005; Adams et al., 2006; Swart, 2006). According to them, innovations depend on the type of culture implemented in the organization. On that account, those firms that encourage risk taking, creative behaviour or freedom to explore will facilitate obtaining innovations. But it is necessary to take sufficient control to manage innovation in an effective and efficient way.

Also, its interesting pay attention to organizational flexibility and responsibility for change, organizational centralization, organizational formalization, internal communication, and so on, because they may encourage or hamper the innovation process (Boer and During, 2001; Benito and Varela, 2002; Adams et al., 2006). For example, low formalization and centralization is suitable at the beginning of the innovation process. In addition, it is necessary to take in account that the formalization should not be understood as an inflexible process, but as a flexible planning.

Furthermore, based on Chesbrough and Teece (2003) and Adams et al. (2006), the present study points out that different organizational forms will lead to different types of innovations.

Lastly, Damanpour (1991) showed that specialization (number of types of categories within firm) and functional differentiation (number of units below the level of senior management) have a positive and significant relationship with technical innovation, while centralization has a negative effect.

Thus, the arguments presented lead us to formulate the following proposition:

Proposition 2: organizational capital is a source of technological innovation

Organizational Capital as Source of Product and Process Innovation

Damanpour (1991), and Muñoz and Cordón (2002) study decentralization and specialization/complexity, asserting that have a positive and significant influence on technical and organizational innovation, respectively, including new products and services, among others. It may be due to the fact that, on the one hand, decentralization allows greater employee participation in decision-making, which facilitate organizational adaptations to changes because not is necessary that all the decisions being taken by senior management and, on the other hand, complexity contributes different organizational skills.

Also, Damanpour (1991), and Muñoz and Cordón (2002) analyze formalization, finding that it is not relevance on innovation. According to the authors, the organizational procedures facilitate innovation in dynamic and uncertain environments, since it may be suitable taking into account innovation within formal definition about jobs in the firm, achieving flexibility through reorganization of tasks to carry out.

Leiponen (2006) performed a research about the determinants of innovation, trying to test empirically the positive influence of team skills and knowledge on the likelihood of carrying out a product innovation. Therefore, she considers the importance of routines and processes of teams belonging to the firm. Likewise, Kyriakopoulos and Ruyter (2004) indicate that these organizational routines have a positive effect on the creativity of new product (referring to changes in the new product), but they add that such link has an inverted U-shaped; hence, if the organization is too routine its effect on innovation will be negative.

In reply to these results, it seems that too routine level within firm may hamper creativity and capability to perceive changes in the environment, although some structure level allows defusing potential situations of uncertainty and improving the exchange and understanding of knowledge flowing by the organization. This explanation is in the same direction to the arguments on the formalization presented by Muñoz and Cordón (2002).

With respect to competitors and taking into account various theoretical perspectives (Transaction Costs Theory, Resources-Based View, Evolutionary Theory, and so on.) Martínez et al. (2007) analyzed the organizational flexibility, both functional and strategic, finding a positive influence on product innovation. In this sense, these flexibilities are considered within the organizational capital, because the former refers to how the firm distributes the different practices that exist in the organization (multipurpose teams, rotation tasks or telecommuting) and the latter refers to the quickly response of the firm before changes in the environment. The authors argue that it is necessary that employees possess the ability for adapting to different tasks, that capacity is achieved by the rotation of tasks and versatile teams. So that such practices allow a greater knowledge flow that helps improve the innovation process performance. In addition, firms involve in dynamic environments should adapt to market and develop products that satisfy customers, hence the important of strategic flexibility.

On the other hand, Un and Cuervo-Cazurra (2004) show that organizational-level integrative socialization, routine communication, and reward positively influence the product innovation. It is due to the fact that socialization and routine communication facilitate the development of the capability to create knowledge by promoting understanding among individuals within different knowledge sets, and rewards knowledge creation by influencing the willingness of individuals to interact and create knowledge.

With regard to communication, Akgün et al. (2007) also studied the learning capability, taking into account the knowledge transfer and integration. They refer to internal spreading of knowledge through verbal and non-verbal communications, including conversations, dialogues, debates and interactions among individuals, both formal and informal, indicating empirically a positive relationship with product innovation.

Another organizational capital dimension considered in this study is the organizational culture. In this sense, Pizarro et al. (2007) found that the entrepreneurial culture has not a direct influence on product innovation. Furthermore, Prajogo and Ahmed (2006) can not verify that the incentives to innovate have a positive effect on product innovation. In this sense, the former suggest that the entrepreneurial culture is a contextual variable and not directly affects innovation, but it has a moderating role on the relationship between the knowledge value of employees and innovation. Therefore, it is necessary an entrepreneurial culture in order to promote the exchange of knowledge among employees. Likewise, the latter indicate that human resources should be managed (incentives to innovate) to get the ability to innovate, achieving innovations.

Nevertheless, taking into account the learning capability, Akgün et al. (2007) observe that a common identity and shared vision, interconnecting the activities of employees, and a climate of accepting new ideas and points of view, allowing individual knowledge to be constantly renewed, widened and improved, have a positive effect on product innovation. Also, Hegde and Shapira (2007) show that practices of knowledge management, promoting shared information and knowledge and encouraging stay in the firm, have a positive effect on product innovation.

Finally, Wu et al. (2007) point out that the structural capital, including organizational processes, among others, has a positive effect on innovative performance, referring to new products introduced by the firm in the market. Therefore, structural capital will be a source of product innovation if employees are encouraged and stimulated to create new ideas and to innovate in new products, since it will increase their intrinsic motivation, enhancing innovative performance. Also, they are considered evaluation system designed and organisational culture related to arguments processed by Un and Cuervo-Cazurra (2004), who analyzed organisational strategy, and Prajogo and Ahmed (2006) and Pizarro et al. (2007), who studied culture to innovate.

Thus, based on the previous arguments, we can enunciate the first sub-proposition of organizational capital:

Proposition 2a: organizational capital is a source of product innovation

Respect to process innovation, and in general terms, the work of Damanpour (1991) shows a positive association between organizational specialization and formalization, and that type of innovation.

In this same sense, Muñoz and Cordón (2002) analyze decentralization and complexity, finding a positive relationship with organizational innovation, which they are included new technologies, among others, which are considered process innovation. On the other hand, they can not show the negative influence of formalization on this kind of innovation. As it has been stated previously, decision-making by different employees speed up adaptation to changes and the diversity of skills helps to achieve innovation. Moreover, in dynamic and uncertain environments may be useful to include innovation within formal description of jobs, so formalization not adversely affects innovation.

In the same way that it has explained about product innovation, since these authors consider both types of innovation in innovation performance, Martínez et al. (2007) indicate that functional and strategic flexibility positively influence process innovation.

Finally, with regard to organizational culture, Hegde and Shapira (2007) show that knowledge management practices, such as promoting a culture of information and knowledge sharing, motivating employees and executives to permanence in the firm, positively influence process innovation. While Prajogo and Ahmed (2006) can not find that incentives to innovate have a positive effect on innovation, including process innovation. It is due to the fact that it is necessary to possess the capacity to innovate for giving effect to these incentives to innovate.

Then, based on these studies, it makes the second sub-proposition:

Proposition 2b: organizational capital is source of process innovation

Organizational Capital as Source of Radical and Incremental Innovation

In parallel to previous issues, researchers have tried to identify different organizational capital aspects related to radical innovation. In this sense, Kyriakopoulos and Ruyter (2004) found a positive influence of organizational routines on the creativity of the new product. Therefore, as the dependent variable is referred to changes in new product with regard to ideas and practices on the current market, it can be equated with the degree of novelty, so if it is greater, it will be a radical innovation. We shouldn't forget that these authors argue that such link has an inverted U-shaped, for this reason firm should be balanced it in order to not become negative.

Furthermore, Leiponen (2006) adds that processes and routines made up of complementary competences of individuals belonging to a team help radical innovation. It may due to the need for different skills in order to create new knowledge.

Koberg et al. (2003) indicate that intra-organizational structural links, referring to aspects such as communication and coordination within the firm, have a positive relationship with radical innovation. In addition, Un and Cuervo-Cazurra (2004) show that organizational-level integrative socialization and routine communication positively influence new product development (differentiating it from changes in product), so it can be assimilated to radical innovation. In reply to these results, it seems like that the greater degree of these three factors exists between employees of different functions, the greater diffusion of heterogeneous knowledge, which will facilitate the creation of radical products.

On the other hand, Damanpour (1991) finds a positive and significant relationship between specialization and radical innovation, because such specialization means a broader and more heterogeneous knowledge base, which facilitates the creation of new ideas.

Finally, Un and Cuervo-Cazurra (2004) explained that organizational-level integrative reward positively influences radical innovation. Thus, rewards derived from the creation of knowledge lead to better predisposition of employees towards new knowledge. Indeed, according to Simon et al. (2003), reward systems should take into account the differences between radical and incremental innovations, so that individuals of the projects about radical innovations are rewarded equitably regarding those employees who are engaged in incremental innovations.

Therefore, based on previous arguments, it is made the third sub-proposition:

Proposition 2c: organizational capital is a source of radical innovation

With respect to incremental innovation, Subramaniam and Youndt (2005) found that organizational capital, including institutionalized knowledge and codified experience, has a positive impact on incremental innovative capability, since it enhances organizational capability to exploit knowledge existing in the firm, reinforcing organizational dominant knowledge.

The work of Kyriakopoulos and Ruyter (2004) shows the positive relationship between organisational routines and new product creativity, that if it has a lower degree of change, as the authors say, it can be assimilated the idea of incremental product innovation. In addition, they point out that this relationship has an inverted U-shaped, so if the degree of organizational routine is too high, its effect on innovation will change, converting into negative. Furthermore, Leiponen (2006) specifies that processes and team routines positively influence in improvements of existing services, but this influence is lower than on radical innovation, since it suppose that it's most important diversity of knowledge in radical innovation, obtaining such diversity combining different individuals competences.

Taking into account specialization issues, Damanpour (1991) shows a positive relationship with incremental innovation, adding that a management attitude towards change, -aspect that can be understood as "culture to innovate"-, leads to an internal climate in order to innovate, so that managerial support assumes incentives for employees to innovate.

Cabello-Medina et al. (2006) point out that firms, which produce response to its customers needs (strategic flexibility), are forced to continually introduce modifications in its products, or innovations is an incremental sense, because such modifications usually not represent radical changes.

Koberg et al. (2003) explained that the intra-organizational structural links, referring to communication and coordination within the firm, have a positive relationship with incremental innovation. Furthermore, Un and Cuervo-Cazurra (2004) showed that organizational-level integrative socialization and routine communication positively influence product innovation, considering the modification of the product, so that it is assimilated by incremental innovation.

Finally, Un and Cuervo-Cazurra (2004) stated a positive influence of rewards on product innovation, including modifications of product. Thus, organizational-level integrative rewards help incremental product innovation, as it has been presented in the previous sub-propositions.

Therefore, after reviewing the literature, it can enunciate the fourth sub-proposition:

Proposition 2d: organizational capital is source of incremental innovation

Technological Capital as Source of Technological Innovation

In our proposal, technological capital suposses a cornerstone in the model of innovation and intellectual capital presented (McEvily and Chakravarthy, 2002). Among the possible reasons, technological capital is intimately linked to the technological innovation capability, since it is included both input measures related to innovation and prior technological knowledge historically possessed by firm, which is labelled as path dependence (Adner, 2002; Joia, 2004; Nerkar and Roberts, 2004).

Grant (1991) points out that innovations offer a temporary competitive advantage in emerging technological firms, where the speed of technological change is high. Therefore, in such cases, it is important to establish technological capabilities to carry out a direct current of innovations.

Specifically, Alegre-Vidal et al. (2004) argue that R&D is one of activities of the technological innovation process that traditionally has received more attention by the literature. In this sense, the effort of a firm can be represented by its R&D expenditures, R&D intensity (ratio between R&D expenditure and turnover), participation in R&D projects with research institutions, or the formal existence of R&D department. In this sense, many studies show a relationship between R&D expenditure and innovation (Achilladellis and Antonakis, 2001; Adams et al., 2006; García and Mulero, 2007; among others). In addition, García and Mulero (2007) suggest that R&D expenditure may represent the innovative capability of the firm because current R&D expenditure is often the result of prior R&D expenditure that obtained successful performance.

On the other hand, researches have analyzed R&D personnel (Deward and Dutton, 1986; Boer and During, 2001; Becheikh et al., 2005; Blumentritt and Danis, 2006; Díaz and De Saá, 2007). Specifically, Díaz and De Saá (2007) pointed out that innovative firms will develop their R&D human resources, training them in order to reach an internal generation of knowledge and making possible interaction among individuals for expanding and generating new knowledge. Moreover, investments in human capital, specifically in technical specialists, involve an advantage to adoption of technological innovation process (Deward and Dutton, 1986).

Then, based on previous arguments, it makes the following proposition:

Proposition 3: technological capital is source of technological innovation

Technological Capital as Source of Product and Process Innovation

Regarding R&D expenditure or intensity, Tsai (2001), Yam et al. (2004), Huergo (2006), Gallego and Casanueva (2007), and Hegde and Shapira (2007) found that there is a positive effect of these technological capital dimensions on product innovation. However, practically not exist studies on this topic from the intellectual capital-based view framework.

As for employees who are engaged in tasks related to R&D or concepts related to it, there are several studies that observe the relationship between R&D personnel and product innovation (Damanpour, 1991; Huergo, 2006; Hegde and Shapira, 2007; Gallego and Casanueva, 2007; De Saá and Díaz, 2007). All of them point out that this relationship is positive because R&D human capital has scientific and technical knowledge and experience and training in R&D activities (De Saá and Díaz, 2007).

Besides, Hayton (2005) can not prove that intellectual property positively influences the number of new product and services. According to this author, it may be due to the correlation between the three dimensions of intellectual capital (human capital, intellectual property and reputation capital), since it might have reduced the power of statistical analysis; and also it may be due to different levels of importance on intellectual property for each industry.

In this sense, the decisions and activities related to R&D, technological or R&D activities management has a positive relationship with product innovation (Yam et al., 2004; Prajogo and Ahmed, 2006; Huergo, 2006). Specifically, Yam et al. (2004) studied the R&D capability of the firm, referring to the ability to integrate R&D strategies, implement projects and manage the project portfolio, and Prajogo and Ahmed (2006) considered R&D and technology skills and strengths to analyze the capability to innovate. Therefore, those firms that possess such capabilities will have a greater potential to carry out technological innovations because of their greater knowledge base, referring to past situations that made them increase their expertise in this field.

Thus, considering above review, it is made the first sub-proposition:

Proposition 3a: technological capital is a source of product innovation

With regard to process innovation, as in product innovation, Huergo (2006), Gallego and Casanueva (2007), and Hegde and Shapira (2007) indicate that there is a positive effect between R&D expenditure or intensity and process innovation.

As for employees who are engaged in tasks related to R&D, Damanpour (1991), De Saá and Díaz (2007), Hegde and Shapira (2007), Gallego and Casanueva (2007) indicate that the resources of technical knowledge, the R&D personnel, training and proportion of scientific and technical employees and the average number of R&D employees, respectively, have a positive influence on process innovation.

Hayton (2005) can not show again that intellectual property have a positive influence on innovation, which appearing "number of new processes", probably it is due to the correlation among the three dimensions of intellectual capital and industrial differences.

On the other hand, as previous sub-proposition, Prajogo and Ahmed (2006) point out that the ability to innovate has a positive effect on innovation variable, which incorporates process innovation. This is because the technological and R&D

management involves knowledge key as source of technological innovations by the necessity of starting from a previous experience in matters relating to the research and development.

Therefore, from the previous arguments, it is described the second sub-proposition:

Proposition 3b: technological capital is a source of process innovation

Technological Capital as Source of Radical and Incremental Innovation

With respect to types of innovation concerning the degree of novelty, the academic literature presents few studies relating to technological capital.

Regarding radical innovation, Leiponen (2006) can not show a positive influence between R&D permanent activities and R&D intensity, and the likelihood of carrying out a radical product innovation. The explanation for this finding may be the industry where the empirical study was tested, as R&D institutionalized is important only to improve existing services, but it is not important to create new services; unlike manufacturing sectors. In addition, innovative firms need not be intensive in R&D, this is, not require substantial monetary investment.

Nevertheless, R&D personnel item does have a positive effect on the likelihood of carrying out a radical innovation (De Saá and Díaz, 2007). So, generation of knowledge through the development of R&D internal human resources may have a positive impact on innovation, including new product and/or process independently of improvement of them. Thus, if firms have R&D personnel with abilities for absorbing and assimilating external knowledge, probably they will reach radical innovations by the diversity of knowledge internalized within the organization. Also, Damanpour (1991) showed the positive association between technical knowledge resources, referring mainly to technical staff, and radical innovation, since a greater level of technical knowledge mean that new ideas are achieved more easily.

On the other hand, for the same reason as it was explained in the previous two types of innovation, Hayton (2005) can not show that intellectual property have a positive influence on the number of radical products, services and processes.

Therefore, according to studies analyzed, the relationship between technological capital and radical innovation is not clear, so that they are formulated two sub-propositions:

Proposition 3c[1]: technological capital is a source of radical innovation

Proposition 3c[2]: technological capital is not a source of radical innovation

As for incremental innovation, Leiponen (2006) points out that there is a positive influence between R&D intensity and R&D permanent activities, and the likelihood of carrying out an incremental product innovation. However, this influence has significance limited, suggesting that innovative service firms require less R&D investment than the manufacturing sector, and that many of the service innovations are conducted through ad-hoc organizational arrangements.

In addition, as it has been presented in radical innovation, De Saá and Díaz (2007) found a positive relationship between R&D personnel and the likelihood of making a significant improvement in current products and/or processes. Therefore, development of R&D internal human resources through progress in its ability to innovate will mean a greater amount of incremental innovation. Also, Damanpour (1991) indicated a positive association between technical knowledge resources and incremental innovation.

Finally, as it has been indicated in the above propositions, Hayton (2005) can not show either that intellectual property will have a positive influence on innovation, including the number of new products, services or processes, which are considered as incremental innovation, as the authors also refer specifically to those which are radical.

Then, on based these arguments, it can be enunciated the fourth sub-proposition:

Proposition 3d: technological capital is source of incremental innovation

Relational Capital as Source of Technological Innovation

With regard to relational capital, it is interesting to highlight the importance that is assuming the role of networks, communities and ties in researches on innovation performance.

Nowadays, many innovative firms spend little on R&D, and yet they achieve success innovations because of knowledge and experience of a wide range of external sources (Laursen and Salter, 2006). Thus, many studies suggest that knowledge from beyond the organizational boundaries of firm is useful for innovation (Bossink, 2002; Chang, 2003; Phene et al., 2006). More specifically, the latter work shows that external sources of knowledge are vital for a better product and process innovation. In addition, the authors state that the creation of radical innovations is a common function of external knowledge and firms can access it.

Likewise, Damanpour (1991) and Gallego and Casanueva (2007) assert that external communication and cooperation, respectively, have a positive association with technological innovation. It's due to the fact that involvement and participation of members in inter-organizational activities may lead to innovative ideas development.

Salman and Saives (2005), based on Powell et al. (1996), underline clearly that organizational capability to innovate can't be studied without considering external organizational relationships. In this sense, the authors go further and argue that it's essential that the firm is located in a central position for creating indirect links with other firms, and thereby it has access to a wider range of activities and a greater chance of locating in positions of most interest in order to obtain relevant information. Therefore, it is assumed that usually indirect networks foster the conditions for innovating and enable knowledge sharing and transferring. Likewise, Almeida and Phene (2004) state that networks with much knowledge offer greater opportunities for accessing to such knowledge and thus greater opportunities to innovate.

When Damanpour and Gopalakrishnan (1998) relate to sources of innovation, they highlight mergers, acquisitions, joint ventures and strategic alliances, considering them alternative sources in an innovation introduction. Owing to the fact that such sources are relationships that firm maintains with external agents, they are part of relational capital.

In the same sense, Dyer and Singh (1998) indicate that alliances partners of the firm are, in many cases, the most important source of new ideas and information leading to innovations, because the inter-organizational relationships facilitate the exchange of knowledge. On the other hand, and based on other authors, McEvily and Eisenhardt (2004) argue that research into alliances has explored effective mechanisms for learning and innovation. Specifically, King et al. (2003) suggest that alliances between small and large organizations lead to technological innovations because both organizations have complementary resources that may facilitate the success of technological innovation if they are combined.

Also, Swart (2006) points out that knowledge diversity makes possible to create the innovative capability, and for achieving innovations it may be necessary to maintain long-term relationships with major knowledge and information exchanges. Moreover, innovation can also be attributed to collaboration among organizations, this is relational capital.

Regarding agents, Adams et al. (2006) found widespread recognition in the literature about that collaboration with suppliers and customers have a significant impact on the innovation process. Also, Chang (2003) observed the different relationships with allies, universities and R&D institutions or participation in government projects and their importance in technological innovation. Specifically, Blumentritt and Danis (2006) suggest that the demands of customers and competitive pressures influence innovation, so if the firm interacts with customers, understanding their needs and satisfying their aims, it will achieve a positive impact on technical innovation (Han et al., 1998).

Then, once it has exposed this literature review, it state the forth proposition:

Proposition 4: relational capital is a source of technological innovation

Relational Capital as Source of Product and Process Innovation

In the present and next section, the literature review is presented and classified according to the typology of innovation, as in previous components of intellectual capital, and considering various types of agents with whom relations are maintained.

Various studies show a positive influence of relationships maintained with customers and suppliers on product innovation, when they analyze collaboration or cooperation alliances (Chang, 2003; Díaz et al., 2006; Gallego and Casanueva, 2007; Martínez et al., 2007), because of the opportunity to access resources that enable firms to face more complex challenges

Likewise, Wu et al. (2007) find a positive relationship between relational capital, appearing customers and suppliers interactions, and product innovation, since a firm can acquire complementary knowledge and skills and achieve a greater number of innovations or it can acquire key information and knowledge (Yli-Renko et al., 2001) about technologies, markets and needs of customers (Díaz and De Saá, 2007). However, Huergo (2006) can not show that vertical cooperation in R&D, which also considers both relationships with customers and suppliers, is a relevant source for such innovation. It may be because the latter author base exclusively on R&D cooperation, an important dimension to process innovation rather than product innovation, since their main purpose is to improve procedures among different actors in the value chain.

With respect to relationships with competitors, Chang (2003) and Díaz et al. (2006) show that they have a positive effect on product innovation, although a minor effect that relationships with customers and suppliers, according to Díaz et al. (2006). However, Gallego and Casanueva (2007) indicate that product innovation is derived largely from use of knowledge and information gathered from collaborations with other organizations which aren't direct competitors, that is, they excluded potential competitors by opportunistic behaviour.

Also, various studies analyze relationships with universities, research institutes or organizations, participation in European Union and/or government projects. In this sense, Chang (2003), Díaz et al. (2006) and Gallego and Casanueva (2007) show a positive relationship between universities and/or technology centres and product innovation. In addition, Díaz et al. (2006) suggest that participation in European Union projects has a positive impact on the likelihood of carrying out product innovations and that this relationship has a minor effect than relationships with customers or suppliers. This finding may be due to the fact that vertical relationships are more common and closer to the business and, therefore, they may mean a greater value for the firm at the time of carrying out innovations.

With regard to relations with partners, Stuart (2000) and Sampson (2007) indicate that have a positive effect on product innovation. Furthermore, considering intellectual capital, Wu et al. (2007) find a positive relationship between the relational capital, which includes partners interactions, and product innovation. Thereby, external knowledge acquisition helps achievement of innovations because complementary skills are shared. Moreover, Stuart (2000) suggested that innovative organizations possess key technological capabilities, so it's expected that the know-how gained from those partners should support to the firm for developing new technology in the next period.

Likewise, Hagedoorn and Duysters (2002) showed that strategic technology alliances (cooperation between firms in which at least one innovative activity is combined or a technology exchange is carrying out) are the main mechanism of firms, which operate in technology-intensive sectors and want to acquire external innovative capabilities, which are considered new products.

Specifically, Li and Atuahene-Gima (2002) analyzed the relationship between different types of alliances and product innovation. They state that relationship is positive if alliance is for products development and negative if it is for marketing alliance. The former has a positive relationship owing to the existence of technical skills transferences from partners; and the latter has a negative relationship because of this type of alliance can divert limited resources, which are necessary for technological innovation, toward sales activities for supporting to the partner, instead of learning technical skills, hampering the achievement of product innovation.

Hayton (2005) argues that there is a positive relationship between corporate reputation and innovation, including new products and services. This is because a positive reputation reduces the risk perceived by potential agents to establish a new interrelationship or alliance.

Finally, Tsai (2001) shows the positive relationship between central network position and product innovation in relative terms. The present study finds that such position is valuable at inter-organizational level, so it is assumed that the better position in relationships with other firms or agents, the greater product innovation by the possible exceptional information and by higher and better contacts are accomplished.

Therefore, according to the arguments presented, it is formulated the first sub-proposition:

Proposition 4a: relational capital is a source of product innovation

With regard to process innovation, Chang (2003), Díaz et al. (2006), Gallego and Casanueva (2007) and Martínez et al. (2007) found a positive effect between relationships with customers and suppliers, and process innovation, when they were considering collaboration or cooperation alliances. It may be due to the importance of vertical links throughout productive process for achieving process innovations.

More specifically, as it has been explained in the previous sub-propositions, Huergo (2006) showed how R&D cooperation with customers and suppliers is relevant source to the likelihood of carrying out process innovation, since this kind of cooperation brings knowledge about procedures to develop among different stakeholders.

On the other hand, Chang (2003), and Díaz et al. (2006) analyzed the relationship maintained with competitors, pointing out that it has a positive effect on process innovation, although a minor effect that relationship with customers and suppliers, according to Díaz et al. (2006). However, Gallego and Casanueva (2007) maintain again that process innovation comes from collaboration with other organizations which are not direct competitors, that is, they excluded potential competitors by opportunistic behaviour.

As for universities, research institutes or organizations, participation in European Union and/ or government projects, Chang (2003), and Díaz et al. (2006) come back to that issue. But in this case, the relationship has a minor effect on the likelihood of carrying out process innovative that maintained with customers or suppliers (Díaz et al., 2006). And, more specifically, Barañano et al. (2005) suggest that the most significant relationships are with the university or research institutes. Regarding this, it may state that relationships with agents, who are part of the production process of the organization, usually lead theirs links to achievement of process innovation for improving or creating a new one.

Therefore, Chang (2003), Díaz et al. (2006), and Gallego and Casanueva (2007) indicate the existence of a positive relationship between universities and/or research institutes and process innovation, which also occurs at a level lower than in relationships with customers and suppliers (Díaz et al., 2006), for the same reason that the earlier argument.

On the other hand, as it has been presented in the previous sub-proposition, Hagedoorn and Duysters (2002) refer to strategic technological alliances, indicating that the main mechanism for firms, which operate in technology-intensive sectors and want to acquire external innovative capabilities, which are considered new processes.

In the same way as regards product innovation, Hayton (2005) shows the existence of a positive relationship between corporate reputation and the number of new products, collected within "innovation" variable.

Thus, after arguments submitted, it is made the second sub-proposition:

Proposition 4b: relational capital is a source of process innovation

Relational Capital as Source of Radical and Incremental Innovation

In general, Damanpour (1991) found that external communication is positively associated with radical innovation. This is because the exchange of inter-organizational information enriches organizational knowledge and promotes the creation of new ideas.

Thus, considering radical innovation, Santamaría and Nieto (2007) point out that ties with customers, suppliers and research organizations positively influence the likelihood of carrying out radical product innovation, although less than on incremental product innovation, being the largest and the most significant relationship maintained with suppliers, as it helps new products development and improves theirs quality. In this sense, and as the "innovative capability" dependent variable refers to new products and processes or significantly improved, we can say that Díaz et al. (2006) also find that relationships with customers and suppliers (collaboration alliances), universities, participation in technological development organizations, joint venture, and European Union projects have a positive effect on the likelihood of carrying out radical innovation, showing further that those relationships maintained with customers and suppliers have a greater importance on innovation capability. In addition, Simon et al. (2003) state that relationships with universities, government laboratories and research organizations enable complement internal resources and make more feasible innovation process.

Moreover, Nieto and Santamaría (2007) found that relationships with competitors have a negative effect on the likelihood of carrying out a radical product innovation owing to possible opportunistic behaviour, reason presented by Gallego and Casanueva (2007) regarding product and process innovation. So, it may be more reasonable to collaborate with competitors when the aim is the establishment of standards or at the stage of basic research. However, Díaz et al. (2006) show the opposite when considering capital alliances (alliances for acquiring technological knowledge), which they are included competitors.

With regard to this duality, this study agrees with the arguments of Nieto and Santamaría (2007), because it seems clear that in the launching of new product, organizations were seeking to reach a competitive advantage and therefore they should avoid any contact with competitors in that process.

Furthermore, firms involved in various kinds of cooperative agreements may introduce a greater degree of innovation, in other words, radical innovation (Chang, 2003). Thus, collaboration with more than one agent has a positive effect on the likelihood of carrying out radical product innovation, which is higher than on incremental product innovation (Nieto and Santamaría, 2007). So, the greater diversity of knowledge, derived from heterogeneous relationships with different agents, the better position will acquire to achieve more radical innovations.

On the other hand, Hayton (2005) shows the existence of a positive relationship between corporate reputation and innovation, which are considered radical innovations. In this sense, reputation may reduce doubts arisen before a possible agreement with other organizations or agents. This situation promotes achievement of a radical innovation, since it may be necessary to have other resources or skills not possessed by the organization, which can be achieved maintaining inter-organizational relations.

Then, considering above arguments, it can be enunciated the third sub-proposition:

Proposition 4c: relational capital is a source of radical innovation

Regarding incremental innovation, we discuss again the investigation conducted by Damanpour (1991), which indicates that external communication is positively associated with incremental innovation, since a greater inter-organizational exchange of information promotes the achieving of innovations owing to heterogeneous knowledge.

Nieto and Santamaría (2007) argue that relationships with customers, suppliers and research organizations have a positive effect on the likelihood of achieving incremental product innovation, being a greater effect than on carrying out radical process innovation. Furthermore, suppliers come back to be the agents who have a greater and more significant effect on this kind of innovation, as the collaboration with them helps to develop new processes and improve productivity. Likewise, if we highlight the aspect considered by Díaz et al. (2006) in innovative capability, that is, products and processes improved significantly, we also find that collaboration and capital alliances positively influence the likelihood of carrying out incremental innovation, being more relevant collaboration alliances.

Moreover, Nieto and Santamaría (2007) found that collaboration with more than one agent has a positive influence on the likelihood of achieving incremental product innovation, although a lesser degree than on radical product innovation. Therefore, the greater diversity of knowledge, derived from heterogeneous relationships with different agents, the better position will acquire to achieve more incremental innovations.

Hayton (2005) shows again that the corporate reputation has a positive relationship with incremental innovation. In this case, it may not be as decisive as in the case of radical innovation, but surely these linkages provide interesting and useful knowledge for conducting an incremental innovation.

Thus, based on these authors, it formulates the fourth sub-propositions:

Proposition 4d: relational capital is a source of incremental innovation

Social Capital as Source of Technological Innovation

With regard to social capital, Tsai and Ghoshal (1998) think that social capital contributes to a firm ability to create value in the form of innovations. In this sense, Nahapiet and Ghoshal (1998) connect innovation to social capital, pointing out that the former is achieved by combining different knowledge and experience, diversity of opinion is a way to expand knowledge, and condition of sharing is basic in order to exchange knowledge. However, it is necessary to underline that these authors do not understand social capital just like us. So, we are only considering social and personal relationships among individuals belonging to the same or different organization.

On the other hand, Madhavan and Grover (1998) defended the idea that individuals who are part of a group and act together have the potential to create new knowledge. Such knowledge is the combination of explicit and tacit knowledge generated in labour relations. These individuals will use their own knowledge system to build relationships with the group, creating knowledge or innovative capability (Swart, 2006).

Bouty (2000), more specifically, states that scientists from different organizations can meet at conferences or annual meetings, informally exchanging information and services. Thus, through this interesting learning process, it may have important implications for innovation.

Lastly, based on scientific literature, it points out that there are arguments and evidence that close networks act as facilitators of the tasks aimed at innovation (Moran, 2005). Thereby, mutual confidence and free favours mean strength in form of social norms that lead to exchange of knowledge. In addition, the cohesion created by that proximity increases the degree and speed of transferring information between individuals and guarantees how this information will be used.

Then, based on this literature review, it is formulated the following propositions:

Proposition 5: social capital is source of technological innovation

Before sub-propositions are justified, it is necessary to refer to important gap in the scientific literature on the relationship between social capital and the four types of innovation analyzed in this research.

Social Capital as Source of Product and Process Innovation

With regard to product innovation, Tsai and Ghoshal (1998) indicate that some aspects of social capital, as social interaction ties, which refer to social relations as well as close social events; trust and trustworthiness, at business unit level, positively influence product innovation. However, those relationships show an indirect effect, since social interaction ties, trust and trustworthiness have positive effect on resource exchange and combination, which have positive effect on product innovations.

Despite the fact that their study does not take into account social capital, they examine issues relating to this concept from our point of view, so they are going to be analyzed in this section. Thus, Un and Cuervo-Cazurra (2004), as Tsai and Ghoshal (1998), indicate that organizational-level integrative routine communication, including meetings between management personnel or non-management personnel from different functions during personal time to discuss work-related issues as well as non-work-related issues, positively influence product innovation.

Therefore, from the ideas above, it can formulate the first sub-proposition:

Proposition 5a: social capital is a source of product innovation

Even though in the review have not been found studies that justify the relationship between social capital (or dimensions associated with it) and process innovation; this work suggests, as in the previous propositions, that there is a positive relationship between social capital and such type of innovation.

So, we state the following sub-proposition:

Proposition 5b: social capital is source of process innovation

Social Capital as Source of Radical and Incremental Innovation

Regarding radical innovation, Subramaniam and Youndt (2005) show that social capital has a positive impact on radical innovative capability, finding that social capital not only acts as a moderator, as they thought initially, but it also has a direct effect on innovation. These findings underscore the significance of interrelationships, partnerships, and collaborative networks to an organization innovation versatility, since they also found, as it is described in the following sub-proposition, that social capital has a positive impact on incremental innovation.

We refer again to study of Un and Cuervo-Cazurra (2004), who argue that organizational-level integrative routine communication, which includes personal relationships, have a positive effect on product innovation. They take into account within product innovation both new product development and product modification, so we equate new product development with radical innovation. Therefore, since we are analysing social capital, we highlight meetings between management personnel or non-management personnel from different functions during personal time to discuss work-related issues as well as non-work-related issues in order to emphasize the importance of this capital on technological innovation.

Then, based on these arguments, it is presented the third sub-proposition:

Proposition 5c: social capital is source of radical innovation

With respect to incremental innovation, Subramaniam and Youndt (2005) also indicate that social capital positively affects the incremental innovative capability, so that this capital not only acts as a moderator, as they have proposed, but it has a direct effect on innovation. Moreover, as shown above, this capital has a greater impact on incremental innovation than on radical innovation. This may be due to the fact that personal relationships between individuals who interact regularly strengthen existing knowledge, while for carrying out a radical innovation may be necessary to interact with individuals who are not so close in order to achieve higher levels of Knowledge diversity.

Also, as discussed above, Un and Cuervo-Cazurra (2004) suggest that organizational-level integrative routine communication have a positive effect on product modification. Therefore, we consider the existence of a positive relationship between social capital and incremental innovation.

Thus, according to these two researches, it can enunciate the fourth sub-proposition:

Proposition 5d: social capital is a source of incremental innovation

CONCLUSION

This theoretical proposal highlights the important role it may play the intellectual capital in the technological innovation process, due to the rapid changes that were being appeared in the current competitive environment, which affecting some factors required to achieve business success.

Therefore, it developed a important analysis of each of the variables, so that intellectual capital has been divided into five elements (human capital, organizational capital, technological capital, relational capital, and social capital) and it has been considered four types of innovation (product, process, radical, and incremental innovation) to carry out the examination of the relationship between intellectual capital and technological innovation. In this way, from a theoretical perspective, possible relationships between each component of intellectual capital and the types of innovation have been analyzed, concluding that the most of them have a positive influence on the different types of technology innovation.

As for directions for future research, it is necessary to empirically analyze the findings of this study about the influence of knowledge or intangible resources on the types of innovation from intellectual capital-based view, providing value to the existing gap in the practice literature.

REFERENCES

Achilladelis, B., & Antonakis, N. (2001). The Dynamics of Technological Innovation: The Case of the Pharmaceutical Industry. *Research Policy, 30*(4), 535–588. doi:10.1016/S0048-7333(00)00093-7

Adams, R., Bessant, J., & Phelps, R. (2006). Innovation Management Measurement: A Review. *International Journal of Management Reviews, 8*, 21–47. doi:10.1111/j.1468-2370.2006.00119.x

Adner, R. (2002). When are Technologies Disruptive? A Demand-Based View of the Emergence of Competition. *Strategic Management Journal, 23*, 667–688. doi:10.1002/smj.246

Aiman-Smith, L., Goodrich, N., Roberts, D., & Scinta, J. (2005). Assessing Your Organization's Potential for Value Innovation. *Research Technology Management, 48*(2), 37–42.

Akgun, A. E., Keskin, H., Byrne, J. C., & Aren, S. (2007). Emotional and Learning Capability and their Impact on Product Innovativeness and Firm Performance. *Technovation, 27*, 501–513. doi:10.1016/j.technovation.2007.03.001

Alegre, J., Lapiedra, R., & Chiva, R. (2005). *Propuesta y Validación de una Escala de Medida del Desempeño Innovador de la Empresa.* ACEDE Conference, La Laguna.

Alegre-Vidal, J., Lapiedra-Alcamí, R., & Chiva-Gómez, R. (2004). Linking Operations Strategy and Product Innovation: An Empirical Study of Spanish Ceramic Tile Producers. *Research Policy, 33*, 829–839. doi:10.1016/j.respol.2004.01.003

Almeida, P., & Phene, A. (2004). Subsidiaries and Knowledge Creation: The Influence of the MNC and Host Country on Innovation. *Strategic Management Journal, 25*, 847–864. doi:10.1002/smj.388

Amabile, T. M. (1998). How to Kill Creativity. *Harvard Business Review, 76*, 76–87.

Amabile, T. M., Barsade, S., Mueller, J., & Staw, B. (2007). La conexión entre las emociones y la creatividad en el trabajo. *Harvard Deusto Business Review, 159*, 36–44.

Amit, R., & Schoemaker, P. J. H. (1993). Strategic Assets and Organizacional Rent. *Strategic Management Journal, 14*, 33–46. doi:10.1002/smj.4250140105

Baldridge, J. V., & Burnham, R. A. (1975). Organizational Innovation - Individual, Organizational and Environmental Impacts. *Administrative Science Quarterly, 20*, 165–176. doi:10.2307/2391692

Barañano, A. M., Bommer, M., & Jalajas, D. S. (2005). Sources of Innovation for High-Tech SMEs: A Comparison of USA, Canada, and Portugal. *International Journal of Technology Management, 30*(1/2), 205–219. doi:10.1504/IJTM.2005.006351

Barney, J. B. (1991). Firm Resources and Sustained Competitive Advantage. *Journal of Management, 17*, 99–120. doi:10.1177/014920639101700108

Becheikh, N., Landry, R., & Amara, N. (2005). Lessons from Innovation Empirical Studies in the Manufacturing Sector: A Systematic Review of the Literature from 1993-2003. *Technovation, 26*(5/6), 644–664.

Benito Torres, J. L., & Varela González, J. A. (2002). Influencia del Tipo de Proceso y del Grado de Novedad sobre las Actividades Ejecutadas durante el Desarrollo de Nuevos Productos. *Revista Europea de Dirección y Economía de la Empresa, 11*(2), 173–187.

Blumentritt, T., & Danis, W. M. (2006). Business Strategy Types and Innovative Practices. *Journal of Managerial Issues*, *18*(2), 274–291.

Blyler, M., & Coff, R. W. (2003). Dynamic Capabilities, Social Capital, and Rent Appropriation: Ties that Split Pies. *Strategic Management Journal*, *24*, 677–686. doi:10.1002/smj.327

Boer, H. & During, W.E. (2001). Innovation, What Innovation? A Comparison between Product, Process and Organizational Innovation. *International Journal of Technology Management, 22* (1/2/3), 83-107.

Bontis, N. (1998). Intellectual Capital: an Exploratory Study that Develops Measures and Models. *Management Decision*, *36*, 63–76. doi:10.1108/00251749810204142

Bossink, B. A. G. (2002). The Development of Co-Innovation Strategies: Stages and Interaction Patterns in Interfirm Innovation. *R & D Management*, *32*(4), 311–320. doi:10.1111/1467-9310.00263

Bouty, I. (2000). Interpersonal and Interaction Influences on Informal Resource Exchanges between R&D Researches across Organizational Boundaries. *Academy of Management Journal*, *43*, 50–65. doi:10.2307/1556385

Bowman, E. H., & Helfat, C. E. (2001). Does Corporate Strategy Matter? *Strategic Management Journal*, *22*, 1–23. doi:10.1002/1097-0266(200101)22:1<1::AID-SMJ143>3.0.CO;2-T

Cabello-Medina, C., Carmona-Lavado, A., & Valle-Cabrera, R. (2006). Identifying the Variables Associated with Types of Innovation, Radical or Incremental: Strategic Flexibility, Organisation and Context. *International Journal of Technology Management*, *35*, 80–106. doi:10.1504/IJTM.2006.009230

Carpenter, M. A., Sanders, G., & Gregersen, H. B. (2001). Bundling Human Capital with Organizational Context: the Impact of International Assignment Experience on Multinational Firm Performance and CEO Pay. *Academy of Management Journal*, *44*, 493–511. doi:10.2307/3069366

CE. (1995). *Libro Verde de la Innovación*. Bruselas.

Chang, Y. (2003). Benefits of Co-operation on Innovative Performance: Evidence from Integrated circuits and Biotecnology Firms in the UK and Taiwan. *R & D Management*, *33*, 425–437. doi:10.1111/1467-9310.00308

Chen, J., Zhu, Z., & Xie, H. Y. (2004). Measuring Intellectual Capital: a New Model and Empirical Study. *Journal of Intellectual Capital*, *5*, 195–212. doi:10.1108/14691930410513003

Chesbrough, H. W., & Teece, D. J. (2003). Organizarse para Innovar: ¿Cuándo es Virtuoso lo Virtual? *Harvard Deusto Business Review*, *112*, 22–30.

CIC. (2001). Identificación y Medición del Capital Tecnológico de la Empresa, ed. CIC-IADE, Madrid.

CIC. (2003a). Modelo Intellectus: Medición y Gestión del Capital Intelectual, ed. CIC-IADE, Madrid.

CIC. (2003b). Metodología para Elaboración de Indicadores de Capital Intelectual, ed. CIC-IADE, Madrid.

Damanpour, F. (1987). The Adoption of Technological, Administrative, and Ancillary Innovations: Impact of Organizational Factors. *Journal of Management*, *13*(4), 675–688. doi:10.1177/014920638701300408

Damanpour, F. (1991). Organizational Innovation: A Meta-Analysis of Effects of Determinants and Moderators. *Academy of Management Journal*, *34*(3), 555–590. doi:10.2307/256406

Damanpour, F., & Evan, W. M. (1984). Organizational Innovation and Performance: The Problem of 'Organizational Lag. *Administrative Science Quarterly*, *29*, 392–409. doi:10.2307/2393031

Damanpour, F., & Gopalakrishnan, S. (1998). Theories of Organizational Structure and Innovation Adoption: the Role of Environmental Change. *Journal of Engineering and Technology Management*, *15*, 1–24. doi:10.1016/S0923-4748(97)00029-5

De Saá, P., & Díaz, N. L. (2007). Incidencia de los Recursos Humanos de I+D Internos y Contratados en la Innovación. *Cuadernos de Economía y Dirección de la Empresa*, *33*, 7–30.

Deward, R. D., & Dutton, J. E. (1986). The Adoption of Radical and Incremental Innovations: an Empirical Analisys. *Management Science*, *32*(11), 1422–1433. doi:10.1287/mnsc.32.11.1422

Díaz-Díaz, N. L., Aguiar-Díaz, I., & De Saá-Pérez, P. (2006). Technological Knowledge Assets and Innovation. *International Journal of Technology Management*, *35*(1-4), 29–51. doi:10.1504/IJTM.2006.009228

Díaz-Díaz, N. L., & De Saá-Pérez, P. (2007). *El Papel de los Recursos Humanos de I+D en la Absorcio□n del Conocimiento Adquirido mediante Alianzas*. ACEDE Conference, Sevilla.

Dyer, J. H., & Singh, H. (1998). The Relational View: Cooperative Strategy and Sources of Interorganizational Competitive Advantage. *Academy of Management Review*, *23*, 660–679. doi:10.2307/259056

Edvinsson, L., & Sullivan, P. (1996). Developing a Model for Managing Intellectual Capital. *European Management Journal*, *14*, 356–364. doi:10.1016/0263-2373(96)00022-9

Eisenhardt, K. M., & Martin, J. A. (2000). Dynamic Capabilities: What are They? *Strategic Management Journal*, *21*, 1105–1121. doi:10.1002/1097-0266(200010/11)21:10/11<1105::AID-SMJ133>3.0.CO;2-E

Galende, J. (2006). Analysis of Technological Innovation from Business Economics and Management. *Technovation*, *26*, 300–311. doi:10.1016/j.technovation.2005.04.006

Gallego, A., & Casanueva, C. (2007). *El Peso de la Cooperación en la Innovación de la Empresa Industrial Española*. ACEDE Conference, Sevilla.

García, T., & Mulero, E. (2007). Medida de los Factores Claves del Éxito de la I+D: El Constructo y sus Dimensiones. *Cuadernos de Economía y Dirección de la Empresa*, *32*, 15–48.

Grant, R. M. (1991). The Resource-Based Theory of Competitive Advantage: Implications for Strategy Formulation. *California Management Review*, *33*(3), 114–135.

Grant, R. M. & Baden-Fuller, C. (1995). A Knowledge-Based Theory of Inter.-Firm Collaboration. *Academy of Management Journal*, Best Paper Proceedings, 17-21.

Hagedoorn, J., & Duysters, G. (2002). Learning in Dynamic Inter-Firm Networks: The Efficacy of Multiple Contracts. *Organization Studies*, *23*(4), 525–548. doi:10.1177/0170840602234002

Hall, R. (1993). A Framework Linking Intangible Resources and Capabilities to Sustainable Competitive Advantage. *Strategic Management Journal*, *14*, 607–618. doi:10.1002/smj.4250140804

Han, J. K., Kim, N., & Srivastava, K. (1998). Market Orientation and Organizational Performance: Is Innovation a Missing Link? *Journal of Marketing*, *62*(4), 30–45. doi:10.2307/1252285

Hayton, J. C. (2005). Competing in the New Economy: The Effect of Intellectual Capital on Corporate Entrepreneurship in High-Technology New Ventures. *R & D Management*, *35*, 137–155. doi:10.1111/j.1467-9310.2005.00379.x

Hedlund, G. (1994). A Model of Knowledge Management and the N-Form Corporation. *Strategic Management Journal*, *15*, 73–90.

Hegde, D., & Shapira, P. (2007). Knowledge, Technology Trajectories, and Innovation in a Developing Country Context: Evidence from a Survey of Malaysian Firms. *International Journal of Technology Management, 40*, 349–370. doi:10.1504/IJTM.2007.015757

Henderson, R. M., & Clark, K. B. (1990). Architectural Innovation: the Reconfiguration of Existing Product Technologies and the Failure of Established Firms. *Administrative Science Quarterly, 35*, 9–30. doi:10.2307/2393549

Huergo, E. (2006). The Role of Technological Management as a Source of Innovation: Evidence from Spanish Manufacturing Firms. *Research Policy, 35*, 1377–1388. doi:10.1016/j.respol.2006.07.005

Itami, H., & Roehl, T. (1987). *Mobilizing Invisible Assets*. Cambridge, MA: Harvard University Press.

Johnson, L. D., Neave, E. H., & Pazderka, B. (2002). Knowledge, Innovation and Share Value. *International Journal of Management Reviews, 4*, 101–134. doi:10.1111/1468-2370.00080

Joia, A. (2004). Are Frecuent Customer Always a Company's Intangible Asset?: Some Findings Drawn from an Exploratory Case Study. *Journal of Intellectual Capital, 5*(4), 586–601. doi:10.1108/14691930410567031

King, D. R., Covin, J. G., & Hegarty, W. H. (2003). Complementary Resources and the Exploitation of Technological Innovations. *Journal of Management, 29*(4), 589–606.

Koberg, C. S., Detienne, D. R., & Heppard, K. A. (2003). An Empirical Test of Environmental, Organizational, and Process Factors Affecting Incremental and Radical Innovation. *The Journal of High Technology Management Research, 14*, 21–45. doi:10.1016/S1047-8310(03)00003-8

Kogut, B., & Zander, U. (1993). Knowledge of the Firm and the Evolutionary Theory of the Multinacional Corporation. *Journal of International Business Studies, 24*, 625–645. doi:10.1057/palgrave.jibs.8490248

Kyriakopoulos, K., & De Ruyter, K. (2004). Knowledge Stocks and Information Flows in New Product Development. *Journal of Management Studies, 41*, 1469–1498. doi:10.1111/j.1467-6486.2004.00482.x

Laursen, K., & Salter, A. (2006). Open for Innovation: The Role of Openness in Explaining Innovation Performance among U.K. Manufacturing Firms. *Strategic Management Journal, 27*(2), 131–150. doi:10.1002/smj.507

Leiponen, A. (2006). Managing Knowledge for Innovation: The Case of Business-to-Business Services. *Journal of Product Innovation Management, 23*, 238–258. doi:10.1111/j.1540-5885.2006.00196.x

Leitner, K. (2005). Managing and Reporting Intangible Assets in Research Technology Organisations. *R & D Management, 35*, 125–136. doi:10.1111/j.1467-9310.2005.00378.x

Li, H., & Atuahene-Gima, K. (2002). The Adoption of Agency Business Activity, Product Innovation, and Performance in Chinese Technology Ventures. *Strategic Management Journal, 23*, 469–490. doi:10.1002/smj.233

Madhavan, R., & Grover, R. (1998). From Embedded Knowledge to Embodied Knowledge: New Product Development as Knowledge Management. *Journal of Marketing, 62*, 1–12. doi:10.2307/1252283

Martínez, A., Vela, M. J., Pérez, M., & De Luis, P. (2007). Flexibilidad e Innovación: El Efecto Moderador de la Cooperación. *Revista Europea de Dirección y Economía de la Empresa, 16*, 69–88.

McEvily, S. K., & Chakravarthy, B. (2002). The Persistence of Knowledge-Based Advantage: An Empirical Test for Product Performance and Technological Knowledge. *Strategic Management Journal*, *23*, 285–305. doi:10.1002/smj.223

McEvily, S. K., Eisenhardt, K. M., & Prescott, J. E. (2004). The Global Acquisition, Leverage, and Protection of Technological Competencies. *Strategic Management Journal*, *25*(8/9), 713–722. doi:10.1002/smj.425

Montes, J. L., Moreno, A. R., & Morales, V. G. (2005). Influence of Support Leadership and Teamwork Cohesion on Organizational Learning, Innovation and Performance: An Empirical Examination. *Technovation*, *25*(10), 1159–1172. doi:10.1016/j.technovation.2004.05.002

Moran, P. (2005). Structural vs. Relational Embeddedness: Social Capital and Managerial Performance. *Strategic Management Journal*, *26*(12), 1129–1151. doi:10.1002/smj.486

Muñoz, A., & Cordón, E. (2002). Tamaño, Estructura e Innovación Organizacional. *Revista Europea de Dirección y Economía de la Empresa*, *11*, 103–120.

Myers, S., & Marquis, D. G. (1969). *Successful Industrial Innovation*. Washington, DC: National Science Foundation.

Nahapiet, J., & Ghoshal, S. (1998). Social Capital, Intellectual Capital, and the Organizational Advantage. *Academy of Management Review*, *23*, 242–266. doi:10.2307/259373

Nerkar, A., & Roberts, P. W. (2004). Technological and Product-Market Experience and the Success of New Product Introductions in the Pharmaceutical Industry. *Strategic Management Journal*, *25*, 779–800. doi:10.1002/smj.417

Nieto, M. J., & Santamaria, L. (2007). The Importance of Diverse Collaborative Networks for the Novelty of Product Innovation. *Technovation*, *27*, 367–377. doi:10.1016/j.technovation.2006.10.001

Nonaka, I. (1991). The Knowledge-Creating Company. *Harvard Business Review*, *69*, 96–104.

Nonaka, I. (1994). A Dynamic Theory of organizacional Knowledge Creation. *Organization Science*, *5*, 14–37. doi:10.1287/orsc.5.1.14

OECD. (2006). *Manual de Oslo. Guía para la recogida e interpretación de datos sobre innovación* (3rd ed.). París: OECD and Eurostat.

Phene, A., Fladmoe-Lindquist, K., & Marsh, L. (2006). Breakthrough Innovations in the U.S. Biotechnology Industry: The Effects of Technological Space and Geographic Origin. *Strategic Management Journal*, *27*, 369–388. doi:10.1002/smj.522

Pizarro, I., Real, J. C., & De la Rosa, M. D. (2007). *El Papel del Capital Humano y la Cultura Emprendedora en la Innovación*. ACEDE Conference, Sevilla.

Powell, W. W., Koput, K. W., & Smith-Doerr, L. (1996). Inter-Organizational Collaboration and the Locus of Innovation: Networks of Learning in Biotechnology. *Administrative Science Quarterly*, *41*(1), 116–145. doi:10.2307/2393988

Prajogo, D. I., & Ahmed, P. K. (2006). Relationships between Innovation Stimulus, Innovation Capacity, and Innovation Performance. *R & D Management*, *36*, 499–515. doi:10.1111/j.1467-9310.2006.00450.x

Priem, R. L., & Butler, J. E. (2001). Is the Resource-Based "View" a useful perspective for Strategic Management Research? *Academy of Management Review*, *26*, 22–40. doi:10.2307/259392

Reed, K. K., Lubatkin, M., & Srinivasan, N. (2006). Proposing and Testing an Intellectual Capital-Based View of the Firm. *Journal of Management Studies*, *43*, 867–893. doi:10.1111/j.1467-6486.2006.00614.x

Rumelt, R. (1991). How Much Does Industry Matter? *Strategic Management Journal*, *12*, 167–185. doi:10.1002/smj.4250120302

Salman, N., & Saives, A. (2005). Indirect Networks: an Intangible Resource for Biotechnology Innovation. *R & D Management*, *35*, 203–215. doi:10.1111/j.1467-9310.2005.00383.x

Sampson, R. C. (2007). R & D Alliances and Firm Performance: The Impact of Technological Diversity and Alliance Organization on Innovation. *Academy of Management Journal*, *50*, 364–386. doi:10.2307/20159859

Schumpeter, J. A. (1912). Teoría del Desenvolvimiento Económico, ed. Fondo de Cultura Económica, México, tercera edición.

Schumpeter, J. A. (1942). *Capitalism, Socialism and Democracy*. New York: Harper and Brothers.

Simon, E. S., McKeough, D. T., Ayers, A. D., Rinehart, E., & Alexia, B. (2003). How do You Best Organize for Radical Innovation? *Research Technology Management*, *46*(5), 17–20.

Stieglitz, N., & Heine, K. (2007). Innovations and the Role of Complementarities in a Strategic Theory of the Firm. *Strategic Management Journal*, *28*, 1–15. doi:10.1002/smj.565

Stuart, T. (2000). Interorganizational Alliances and the Performance of Firms: A Study of Growth and Innovation Rates in a High-Technology Industry. *Strategic Management Journal*, *21*, 791–811. doi:10.1002/1097-0266(200008)21:8<791::AID-SMJ121>3.0.CO;2-K

Subramaniam, M., & Youndt, M. A. (2005). The Influence of Intellectual Capital on the Types of Innovative Capabilities. *Academy of Management Journal*, *48*, 450–463. doi:10.2307/20159670

Sullivan, P. H. (2001). *Rentabilizar el Capital Intelectual. Técnicas para Optimizar el Valor de la Innovación*. Buenos Aires: Paidós.

Swart, J. (2006). Intellectual Capital: Disentangling an Enigmatic Concept. *Journal of Intellectual Capital*, *7*, 136–159. doi:10.1108/14691930610661827

Tidd, J. (2001). Innovation Management in Context: Environment, Organiztion and Performance. *International Journal of Management Reviews*, *3*, 169–183. doi:10.1111/1468-2370.00062

Tsai, W. (2001). Knowledge Transfer in Intraorganizational Networks: Effects of Network Position and Absorptive Capacity on Business Unit Innovation and Performance. *Academy of Management Journal*, *44*, 996–1004. doi:10.2307/3069443

Tsai, W., & Ghoshal, S. (1998). Social Capital and Value Creation: the Role of Intrafirm Networks. *Academy of Management Journal*, *41*, 464–476. doi:10.2307/257085

Tseng, C., & Goo, Y. J. (2005). Intellectual Capital and Corporate Value in an Emerging Economy: Empirical Study of Taiwanese Manufacturers. *R & D Management*, *35*, 187–201. doi:10.1111/j.1467-9310.2005.00382.x

Un, C. A., & Cuervo-Cazurra, A. (2004). Strategies for Knowledge Creation in Firms. *British Journal of Management*, *15*, S27–S41. doi:10.1111/j.1467-8551.2004.00404.x

Van de Ven, A. H. (1986). Central Problems in the Management of Innovation. *Management Science*, *32*, 590–607. doi:10.1287/mnsc.32.5.590

Wang, W., & Chang, C. (2005). Intellectual Capital and Performance in Causal Models: Evidence from the Information Technology Industry in Taiwan. *Journal of Intellectual Capital*, *6*(2), 222–236. doi:10.1108/14691930510592816

Wernelfelt, B. (1984). A Resource-Based View of the Firm. *Strategic Management Journal*, *5*, 171–180. doi:10.1002/smj.4250050207

Wu, S., Lin, L., & Hsu, M. (2007). Intellectual Capital, Dynamic Capabilities and Innovative Performance of Organisations. *International Journal of Technology Management*, *39*, 279–296. doi:10.1504/IJTM.2007.013496

Yam, R. C. M., Cheng, J., Fai, K., & Tang, E. P. Y. (2004). An Audit of Technological Innovation Capabilities in Chinese Firms: Some Empirical Findings in Beijing, China. *Research Policy*, *33*, 1123–1140. doi:10.1016/j.respol.2004.05.004

Yli-Renko, H., Autio, E., & Sapienza, H. J. (2001). Social Capital, Knowledge Acquisitions, and Knowledge Exploitation in Young Technology-Based Firms. *Strategic Management Journal*, *22*, 587–613. doi:10.1002/smj.183

KEY TERMS AND DEFINITIONS

Intellectual Capital: A set of intangible resources and capabilities related to different categories of knowledge (individual, organisational or inter-organisational), which provides a firm with a competitive advantage.

Human Capital: A set of individual knowledge and skills possessed by employees and that are useful for carrying out activities within the firm.

Organizational Capital: Methods, structures or processes developed by the firm and necessary to carry out the job of employees.

Technological Capital: Processes, activities and know-how necessary to create or improve goods or services as well as productive processes.

Relational Capital: Knowledge coming from relationships between the firm (as a social entity) and its stakeholders.

Social Capital: Knowledge derived from informal and personal relationships among employees that are not predetermined by the company.

Technological Innovation: A result of a process by which a firm gets to create a new idea that will be marketed later.

Chapter 9
Structural Capital and Innovation Capabilities:
Theory and Empirical Evidence from Spain

Elsa Mercedes Alama Salazar
Universidad de Piura, Perú

ABSTRACT

Organizational knowledge, as well as their different manifestations, may play a critical role for effective development of the firm processes and products (Kogut and Zander, 1996; Teece, 1998; Alegre and Lapiedra, 2005). The present work explores the influence that the different manifestations of organizational knowledge, or structural capital, show on the innovation capability of professional service firms. To achieve this purpose, in a first section an exploratory factor analysis is carried out. From it, the main blocks of organizational capital are obtained. In the next step, the basic hypotheses about the positive influence of these types of structural assets, on firm innovation are presented.

INTRODUCTION

During the last years, research about the role of knowledge in firm activities and value creation has been one of the most fruitful within the business management field. The Knowledge based View has highlighted the relevance of knowledge as a key organizational factor (Zack, 1999), considering that it is one of the main determinants for the existence of the firm, its growth and development, its internal organization, and business success.

DOI: 10.4018/978-1-61520-875-3.ch009

Nevertheless, previous approaches as the Resource based View (Wernerfelt, 1984; Barney, 1991, 2001; Peteraf, 1993) or parallel research streams as the Dynamic Capabilities approach (Teece, Pisano and Shuen, 1997) have also emphasized the role that intangible resources and organizational capabilities, which are based on information and knowledge, play in building robust competitive advantages.

As Galende and De la Fuente (2003) point out, a good piece of research is devoted to address the innovation processes from an external perspective, leaving aside the internal complexity that characterizes innovation dynamics. Besides, it can be said that the innovative capability of a certain firm depends very closely on the intellectual assets and knowledge that it possesses, as well as on its ability to deploy them (Alegre and Lapiedra, 2005; Subramaniam and Youndt, 2005).

Nevertheless, these approaches pose certain problems (Priem and Butler, 2001) like the lack of a clear definition for the notion of competitive advantage, or the challenges and difficulties for the identification and measurement of resources and capabilities that become critical for that competitive advantage. This kind of problems hinders the empirical test of the main axioms of these theoretical streams, making difficult to advance towards a consolidated theory. In this sense, and following the recommendations outlined by Reed, Lubatkin and Srinavasan (2006), a feasible and mainly pragmatic solution is using the so called Intellectual Capital based View.

This theoretical-pragmatic approach is a narrowly focused or specialization of the Resource based View around those resources or factors with intangible nature that may become the main responsible for firm success. This approach arisen from practitioners in the decade of the nineties (Brooking, 1996; Edvinsson and Malone, 1997; CIC, 2003) and it distinguishes different blocks of intellectual capital or types of organizational knowledge stocks. There is some general agreement when depicting the following main kinds of intellectual capital: a) human capital, or those knowledge, abilities, experiences and attitudes possessed by the organizational members (CIC, 2003; Subramaniam and Yound, 2005); b) structural capital, which includes those pieces of knowledge that provide coherence and guidance for the whole organization (Edvinsson and Malone, 1997); and c) relational capital, that arises form the relationship processes that the firm maintains with the external agents that surround it (CIC, 2003; Reed et al., 2006).

This work tries to throw some more light on the relations between organizational knowledge stocks and the processes of firm innovation. To carry out this task, in a first section a review of some of the main works that have tried to link intellectual capital or knowledge and technological innovation is carried out, taking into account that in this research technological innovation is understood as the successful development of new projects, products, services and ideas. Later on, based on the literature review the research hypotheses are discussed, addressing their empirical test in the results section. Conclusions, limitations and future research suggestions complete the structure of the paper.

BACKGROUND

The Intellectual and Structural Capitals of the Firm

The literature contains several definitions of intellectual capital, but while to some extent there is certain convergence among these definitions, we can also find substantial differences (Hayton, 2005). In general, intellectual capital is a broad-based term considered synonymous with firm intangible assets (Tseng and Goo, (2005). Nevertheless, taking into account that intangible resources and firm capabilities have a common essence because they are based on knowledge, or they are simply a form of knowledge, intellectual capital constitutes a representation of the endowment, domain or stocks of firm knowledge.

Therefore, if according to the Resource based View these are the ultimate responsible for the competitive advantage of the firm, analyzing knowledge stocks or intellectual capital should allow us to determine the possibilities of the organization to generate a sustainable competitive advantage.

Both academics and practitioners have embraced the intellectual capital perspective as a useful framework for describing knowledge stocks, and its relation with value creation. The intellectual capital perspective focuses on how to obtain the maximum value from intangible assets endowment, so this perspective provides an interesting bridge between resources and corporate value (Tseng and Goo, 2005), which makes operative the resource based view for the case of intangible assets.

To be able to better cope with the challenge of measuring and valuing intellectual capital, the most comprehensive approaches are models based on a structured differentiation (Leitner, 2005). Although several taxonomies of intangibles have been suggested over the past decade, taking into account an overview of the internationally proposed metrics for intellectual capital, it can be argued that a certain general agreement exists when establishing a classification of intangible assets according to a framework for grouping and identification using the social character of knowledge (Brown and Duguid, 1991). This way, intellectual capital is usually classified in three basic components (Hermans and Kauranen, 2005): human capital, organizational or structural capital, and relational capital, which represent the most commonly, accepted categories of intangibles (Pike, Roos, and Marr, 2005).

The knowledge tied to the people that form the firm is critical for business activities. In fact, people are hired and fired according to the assessment of their knowledge and its usefulness for the firm tasks. The main characteristic of human capital is the impossibility of separating this kind of intangible assets from the people that develop them. When speaking about human capital we are considering an individual capital that comes to the firm and remains at it by means of employment contracts, which link the employees with the organization.

The second block of intangible elements presented in the general classification of intellectual capital is named structural capital. Most organizations form a conglomerate of different communities of practice or overlapped and interdependent work groups (Brown and Duguid, 1991). The relationships among these groups that integrate the organization allow it to develop, starting from independent and partial contributions, a series of knowledge based assets or intangible elements that present an outstanding systemic, integrative, synthetic and synergistic character. This way, structural capital 'knits' jointly the technical dexterity of the organizational members (Edvinsson and Malone, 1997), and it includes all the intangible assets that shape the real firm structure, culture, as well as its internal organizing principles. This kind of intellectual capital fosters the knowledge flow, and it brings a relevant improvement in organizational effectiveness through coordination, because it integrates, in an appropriate way, the different functions of the firm.

Lastly, relational capital gathers those intangible assets that the firm obtains when it maintains successful relationships with agents of its environment as clients, suppliers or allies.

It is widely recognized, in particular from the Resource Based View, that Intellectual capital affects organizational performance and competitive advantage (Wu, Tsai, Cheng, and Lai, 2006). Nonetheless, whereas in management literature, firm value is often explained by intellectual capital, empirical contributions on these matters are scarce (Hermans and Kauranen, 2005). Thus, an improved understanding of how intellectual capital provides sustained competitive advantage is required, deepen into what is known about what occurs in the 'black box' between intellectual capital, and specially organizational capital, and corporate value (Tseng and Goo, 2005).

Products and Service Innovation

Although the value of tangible assets in generally recognized, managers need to know how their investments in intangible assets are associated to firm returns (Wu et al., 2006). The challenge for the users of intellectual capital, whatever it will be managers or investors, is to make this kind of assets meaningful, connecting them to corporate goals in order to understand their impacts on financial returns (Leitner, 2005). Here is where our model tries to provide the missing link between intellectual capital and value creation, taking into account firm innovation.

This way, the second part of the model to be studied refers to innovation, which has been a frequently addressed topic for specialized scientific literature, due to its relevance for firm survival and success. However, the analysis and understanding of this phenomenon continues raising debates (Subramanian and Youndt, 2005). According to these authors, innovation can be defined as the identification and seizure of opportunities to create new products, services or processes.

Although the literature (Tushman and Nadler, 1986; Van de Ven, 1986; Grandson, 2001; Hill and Rothaermel, 2003; Stieglitz and Heine, 2007, among others) recognizes a wide range of innovation types within the firm (product / process, radical / incremental, technological / managerial, market push / market pull, or competition fostering / competition lowering), most of the empirical works use the product-process typology of innovation. This research will also employ this kind of typology that classifies innovation types according to the results or outputs of the innovation process. This way, according to Nieto (2001), when new technological knowledge is materialized in the development of new products or in the improvement of those already existent we are talking about a product innovation, while when it is materialized in new productive processes and practices or when these are improved, we are speaking about process innovation.

MAIN FOCUS OF THE CHAPTER

Hypotheses

The role that organizational knowledge stocks or structural capital play on the firm capabilities to innovate through its products, projects or services, constitutes the main argument of this paper. In this sense, once the theoretical background has been presented, it is necessary to move towards a specific model to be tested. This model is shaped by three basic hypotheses to be tested that will be commented along this section. Later on, the variables and measurement tools employed are included, as well as the research resume of the empirical research.

As it has been pointed out previously, from a theoretical point of view there is a very close relation between knowledge stocks and flows, and with the innovation process (Nonaka and Takeuchi, 1995; Díaz, Díaz and De Saá, 2006). Nevertheless, from an empirical point of view and excepting pieces of research like that of Subramaniam and Youndt (2005) or Smith and Sharif (2007), the works that have addressed this topic from the Knowledge Based View and the Intellectual Capital Based View are quite scarce.

Having a brilliant, motivated and experienced human capital should be the base of all innovation process in the company. Thus, this kind of intellectual capital provides the main source for developing new ideas and knowledge (Snell and Dean, 1992). Highly motivated and trained employees workers may question the already established organizational routines, so this kind of human capital become critical to push the firm into its technological limits, constituting the best incentive towards obtaining new knowledge and innovations (Nonaka and Takeuchi, 1995; Hill and Rothaermel, 2003).

Beyond human capital, or individual knowledge, an important part of the knowledge, abilities, experiences and behaviors required for the successful development of new products and services lie inside the own organization. As Van de Ven (1986) points out, the innovation process –in general terms- is a collective achievement of the organizational members, where structural support becomes a key element. Institutionalization acts as a mean for preserving organizational knowledge and routines, which in turn fosters the accumulation, preservation and improvement of collective knowledge.

Following Tseng and Goo (2005), a good structural capital will translate the human dimension innovation into company property. For doing so, firms must support and nurture the brightest individuals to share their innovation, knowledge, and abilities through organization learning. Tacit issues like managerial commitment, a common identity and shared vision or a climate of openness and experimentation compose the learning capability of the firm (Akgün, Keskin, Byrne, and Aren, 2007). Nevertheless, although organized information cannot substitute tacit knowledge, it can significantly enhance it to fill existing knowledge gaps, so information technologies can support the transformation of information into organizational knowledge (Adamides and Karacapilidis, 2006)., In this vein, operational processes, information systems, organization culture, internal organization structure and administrative systems, will have a positive influence on the innovative capabilities of the firm. Thus, having databases, proceeding manuals, effective information systems, or cultural values devoted to innovation promotion can constitute important sources for innovative success. This relation appears in the following main hypothesis:

H1: *The higher structuralcapital endowments will be in the firm, the stronger will be its innovation capabilities.*

Research Design

Measuring Variables

Bearing in mind the purpose of identifying and measuring the different components of structural capital, as well as capturing technological innovation, as well as their main dimensions, a literature review was carried out. From it, a series of dimensions and variables were designed, and they were used as the basis for the elaboration of a questionnaire with Likert 1-7 scales, which were employed for gathering primary data.

This way, structural capital is the platform or supporting structure that allows the coordinated execution of tasks inside the firm. From the literature review eight dimensions were obtained, like IT competences (Tippins and Sohi, 2003; Adamides and Karacapilidis, 2006), licenses, manuals and databases to store knowledge (Wilcox King et al., 2001; Youndt and Snell, 2004), structures, systems and processes (Wilcox King et al., 2001; Tippins and Sohi, 2003), funds and efforts devoted to innovation (Chen et al., 2004; Youndt, Subramanian, and Snell, 2004; Alegre et al., 2006;) or organizational culture (Carmeli, 2004; Carmeli and Tishler, 2004). The questions employed for measuring organizational capital can be found on Additional Readings.

Innovation outcomes were measured through three indicators. The first one was devoted to the number of new products, services, and/or projects implemented by the firm in a certain time (Tsai and Ghosal, 1998) because this was the most direct way of knowing the success of the innovation efforts carried out by the firm. The second indicator was based on the satisfaction of managers with the success of firm innovative projects, assuming a positive relation between perceived success and effective or real success like it has been done in other studies (Akgün et al., 2007). The third indicator followed the same logic than the first one, asking about the number of new technologies developed recently by the company (Chen et al., 2004). As for the case of independent variables, the items included in the questionnaire for measuring innovation outcomes can be found in Annex.

The choice of these items for measuring the innovation capabilities of services firms follows a synthesis approach, which suggests that manufacturing and service innovation share many similarities, although being cautious not to include metrics which are standard in manufacturing industries, but may be less meaningful in the case of services, as R&D and patents (Mansury and Love, 2008).

Population, Sample and Information Sources

Bearing in mind the hypothesis obtained from the literature review, they were empirically tested in a set of firms where the intangible resources represent an important determinant of business competitiveness. Population had to be selected picking up firms whose distinctive capabilities will be based on the knowledge of their employees, internal organizing principles, and the relations with the environmental agents and mainly the customers.

Finally, the professional service firms of Spain found on the SABI[1] database were chosen for the fieldwork, following the classification system provided by the NAICS 2002 codes. Their belonging to the professional service sector grants the homogeneity required for this kind of research (Rouse and Daellenbach, 1999). Only those firms with at least 50 employees were selected to assure the existence of the different types of intellectual capital, due to the fact that for very small business the different "capitals" may not be distinguished clearly.

RESULTS

The gathered information was processes through two stages. In the first one, an exploratory factor analysis was carried out in order to determine the basic components that can be found in the final firm sample. Next, on a second stage, a multiple regression model was designed to test the effects of the different assets of structural capital on the firm innovation capabilities.

With the aim of testing the main dimensions included in each of the components of structural capital, a factor analyses was carried out, because this technique allows us to synthesize the information contained in wide data set into a limited number of components that will represent the original variables with a minimum information loss. Figure 2 shows the results of this test for the whole set of questionnaire items about organizational capital.

The measurement scale for structural capital was composed taken items from several works devoted to measuring this concept in a whole (Youndt and Snell, 2004; Carmeli and Tishler, 2004) and from others interested in capturing some of its dimensions (Tippins and Sohi, 2003; Wilcox King et al., 2001). Before carrying out the exploratory factor analysis for this scale, the preliminary tests were run: determinant of the

Figure 1. Research resume

Research aim	Intellectual capital and innovation
Criteria defining firm population	Professional service firms: - Included in the SABI database - Located in Spain - Within NAICS 5412, 5413, 5415, 5416, 5417, 5418, 5419 y 5511 - With at least 50 employees
Population	981 firms
Sample and response rate	120 firms (12.23%)
Method for data gathering	Survey (questionnaire)
	Answered by the CEO, Human Resource Top Manager, Intellectual Capital Top or equivalent top managers
Statistical software used	SPSS 13.0 and AMOS 7.0

Figure 2. Exploratory factor analysis of structural capital

Questionnaire items	F1	F2	F3	F4	F5
SC18: Employees and management in our firm make sincere efforts to solve common problems	.840				
SC20: There is complete trust between management and employees	.813				
SC17: There fluid communication between management and employees	.804				
SC15: The employees are committed and hold a high sense of responsability to the organization	.780				
SC16: Our employees get enough work-related information to perform their jobs effectively	.743				
SC19: Employees generally view the conditions of their employment here as fair	.715				
SC14: The organizational goals are clear and agreeable to all members	.689				
SC13: Our employees share the values, beliefs and symbols of the organization	.618				
SC7: Relative to our firm´s largest competitor we spend a much larger of R&D investment		.846			
SC8: Regarding our nearest competitor, the quality and quantity of employees in I+D is top		.837			
SC9: Our organization possesses knowledge and ability to link operational goals and objectives with compensation plans		.544			
SC5: Our company grants incentives to the innovative employees		.540			
SC6: The average cost for project of innovation is minor that of the nearest competitor		.465			
SC12: Our company has developed quality programes during the last 3 years			.841		
SC11: We have a standard procedure for effectively dealing with customer complaints			.663		
SC10: Our company has documented procedures that help carry out routine			.481		
SC3: Much of our organization´s knowledge is contained in manuals, databases, etc.				.802	
SC2: Our organizations uses patents and licenses as a way to store knowledge				.733	
SC4: The average time for the project				.523	

Figure 3. Exploratory factor analysis of innovation outcomes

Questionnaire items	F1
IN2: Regarding our nearest competitor, the number of new technologies developed over the past 3 years is higher	0.808
IN0: The number of products / services / projects that we have launched over the past 3 years is higher than the nearest competitor	0.801
IN1: The satisfaction of our company for the efficiency of innovation projects is high	0.736
% of Standard deviation	69.775
Determinant of the correlation matrix = 0.045 Kaiser-Meyer-Olkin index = 0.776 Barlett test (Chi-squared, gf) = 355.512 (45) Signification level = 0.000	

*Figure 4. Regression results (structural capital and innovation outcomes). Signification level ***p<0,01 **p<0,05 *p<0,1*

Variables	Innovation Outcomes SC and INN Hypoth. 1
OCU	0.392***
INE	0.107
SSP	0.001
KSP	-0.006
ITC	0.259***
Age	0.168*
Size	-0.084
R	0.519ª
R2	0.269
Adjusted R2	0.224
Standard Error	0.8810
Durbin-Watson	1.876
F	5.898***

matrix, KMO test, and Bartlett test. The results of these diagnostic tests can be found on figure 4, and they advice to carry on with the factor analysis.

Following with this technique, the first iteration for extracting the main components provided a total of six relevant factors for structural capital. Nevertheless, one of these components contained only one item, so we proceed to run a second factor analysis, limiting the results to only five components. The results of this second test are shown in figure 2.

This way, the first component of structural capital included eight questionnaire items regarding organizational culture (OCU) which was quite coherent among them, because they address the way of working and understanding within the organization, as well as its core values. In this sense, issues like the efforts of the employees and managers for solving problems, the communication among organizational members, the access to information, or the sense of commitment and adhesion to the core values of the firm, explain a 23.204% of the variance.

The second component of structural capital is focused around the innovation efforts (INE) of the firm, including a set of five items devoted to the R&D budget, the quantity and quality of the R&D personnel, the innovation related to management activities, and the incentives designed for innovative employees. This component represents a 12.674% of the explained variance. The third component obtained from the factor analysis gathers three items related to the formal structure, systems, and processes (SSP) of the firm. It makes reference to the routines that the firm uses for performing its day by day tasks which are widely known by the employees, and that represent a 8.911% of the explained variance.

The fourth component obtained for structural capital is related to the ways in which knowledge is stored and preserved (KSP), containing three items, and explaining a 8.055% of the explained variance. The fifth and last component was labelled "IT competences" (ITC). These competences are show through the frequent use of information technologies, and the level that the firm has reached in its use. This component includes two items, and it represents 8.027% of the explained variance. This way, it is necessary to highlight that the percentage of explained variance, taking into account the five components of the organizational capital measurement scale becomes 31.321.

Then, reliability tests for the components of the structural capital scale are performed through the Cronbach's Alpha coefficients. The analysis shows a very low reliability level for the KSP dimension. Eliminating the item SC04 raises this Alpha level, although it remains under the 0.6 level. Nevertheless, we have decided not to eliminate the rest of items because we are performing an exploratory analysis. The item SC05 was also eliminated from the INE component, and doing so, the Cronbach's Alpha raises to 0.754. The coefficients of the dimensions OCU, INE and SSP were 0.902, 0.704, and 0.620.

The results of this preliminary analysis point out that the five components found for structural capital provide a coherent grouping for the questionnaire items. Nevertheless, some authors, like Subramanian and Youndt (2005), do not distinguish between "structure, systems, and processes" and "knowledge storage and preservation". Instead, they argue that organizational abilities for knowledge storage and appropriating can be found equally in databases, manuals and tutorials, organizational processes, cultural systems and business models. This notion of structural capital is quite monolithic an one-dimensional, representing all the abilities, knowledge and information that remains at the firm when the employees go home (Bontis, 1996).

Finally, the dependent variable of the model was measured through the scale devoted to innovation outcomes, which included three items. From another exploratory factor analysis we can argue the existence of a sole component which was labelled it as "innovation" (INN). It shows a Cronbach's Alpha of 0.730, demonstrating the reliability of the measurement tool.

Once the components of the different dimensions of firm structural capital have been obtained, and trying to answer the second question that guides this research, a causal model is proposed, according to the multiple lineal regression rules, in order to analyze the influence of structural capital on the innovative capabilities of the firm.

As it can be seen in the regression model resume and in the variance analysis, both Durbin-Watson coefficient and Snedecor´s F show values above the critical values, so we can conclude that the proposed model is statistically significant as a whole.

In a more detailed comment, the empirical test of hypothesis 1 was addressed, trying to demonstrate the influence of structural capital on firm innovation. The resulting model becomes statistically significant. It shows a higher adjusted R^2 coefficient (22.4%) and it corroborates a positive relation with innovation outcomes for those components of structural capital dealing with organizational culture and IT competences.

FUTURE RESEARCH DIRECTIONS

Further research is needed in order to analyze the different components of structural capital, and in a wider framework, intellectual capital, in other industrial settings, and also for determining their influence on product or service innovation. Besides, longitudinal studies about this topic would provide a better understanding of how organizational learning takes place, and how it is related to the innovation process. This systematic evidence may be very useful for both academics and practitioners interested in using an 'Intellectual Capital based View of Innovation'. This way, there is a long way to be walk, because as Park and Kim (2006) suggest, firm knowledge assets and innovation must be combined into new terms, demonstrating empirically that managing knowledge, organizational capital, and intellectual capital is crucial for succeeding in the implementation of innovation based strategies (Liu, Chen, and Tsai, 2005).

CONCLUSION

As Hayton (2005) points out, there has been limited research into intellectual capital and its outcomes. The main contribution of the empirical findings of this research is precisely providing evidence that supports that structural capital is one of the main sources for firm innovation. Although our results corroborate those of previous works, there are also some differentiated and valuable characteristics.

First, the theoretical framework adopted is the Intellectual Capital based View (Reed et al., 2006), which until the present date has little empirical support for its main arguments. In this sense, as the Intellectual Capital based View proposes intangible assets as the source of sustainable competitive advantage, several specific kinds of competitive have been considered by the previous literature, as corporate value (Tseng and Goo, 2005), anticipated future sales (Hermans and Kauranen, 2005), corporate entrepreneurship (Hayton, 2005), or directly firm performance (Wu et al., 2006). In our case, the competitive advantage has been considered in the shape of innovation outcomes. Thus, we have tried to provide additional empirical evidence in testing this research stream, moving from 'Resource Based View of the innovation process' (Pike, Roos, and Marr, 2005) towards an 'Intellectual Capital based View of Innovation'.

Second, exploratory factor analysis takes the existence of different dimensions of structural capital. This follows the somewhat general international agreement, but in this research the main dimensions of each of these capitals are shown for the case of professional service firms. This context, although is clearly knowledge-intensive, is an understudied one, if we compared it with high-technology industries, which have receive much more attention from scholars.

Third, the causal analysis provides support for the main thesis of this research for the selected sample of professional service firms. Thus, findings show a statistically positive and significant influence of organizational capital assets on innovation capability. From these findings it is necessary to highlight the key role of firm competences focused on information technologies.

In spite of the contributions of this paper, it is necessary to point out its main limitation, which advice taking our findings with care, and consider them as tracks, directions or partial evidence for the main hypotheses tested. We must also take into account that the conducted empirical research has a cross-sectional nature and it has been designed specially for a certain industry. This implies that our findings may not be generalized in industrial or geographically without some caution. It makes sense to think that other kind of industries or different countries could show other critical elements of intellectual capital as main determinants of the firm innovation activities.

REFERENCES

Adamides, E. D., & Karacapilidis, N. (2006). Information Technology Support for the Knowledge and Social Process of Innovation Management. *Technovation, 26*, 50–59. doi:10.1016/j.technovation.2004.07.019

Akgün, A. E., Keskin, H., & Aren, S. (2007). Emotional and Learning Capabilities and their Impact on Product Innovativeness and Firm Performance. *Technovation, 27*, 501–513. doi:10.1016/j.technovation.2007.03.001

Alegre, J. & Lapiedra, R. (2005). Gestión del Conocimiento y Desempeño Innovador: un Estudio del Papel Mediador del Repertorio de Competencias Distintivas. *Cuadernos de Economía y Dirección de la empresa, 23*, 117-138.

Barney, J. (1991). Firm Resources and Sustained Competitive Advantage. *Journal of Management, 17*, 99–120. doi:10.1177/014920639101700108

Barney, J. (2001). Is the Resource-Based View a Useful Perspective for Strategic Management Research? Yes. *Academy of Management Review, 26*, 41–57. doi:10.2307/259393

Bontis, N. (1996). There's a Price on your Head: Managing Intellectual Capital Strategically. *Business Quarterly, 60*, 41–47.

Brooking, A. (1996). *Intellectual Capital. Core Asset for the Third Millennium Enterprise*. London: International Thomson Business Press.

Brown, J., & Duguid, P. (1991). Organizational Learning and Communities of Practice: Towards a Unified View of Working, Learning and Innovation. *Organization Science, 2*, 40–57. doi:10.1287/orsc.2.1.40

Carmeli, A. (2004). The link between organizational elements, perceived external prestige and performance. *Corporate Reputation Review, 6*, 314–331. doi:10.1057/palgrave.crr.1540002

Carmeli, A., & Tishler, A. (2004). The relationships between intangible organizational elements and organizational performance. *Strategic Management Journal, 25*, 1257–1278. doi:10.1002/smj.428

Chen, J., Zhu, Z., & Yuan, H. (2004). Measuring intellectual capital: a new model and empirical study. *Journal of Intellectual Capital, 5*, 195–212. doi:10.1108/14691930410513003

CIC (2003). *Modelo Intellectus: Medición y Gestión del Capital Intelectual* (Serie Documentos Intellectus No. 5). Madrid: Centro de Investigación sobre la Sociedad del Conocimiento (CIC).

Cohen, W., & Levinthal, D. (1990). Absorptive Capacity: A New Perspective on Learning and Innovation. *Administrative Science Quarterly, 35*, 128–152. doi:10.2307/2393553

Coombs, J. E., & Bierly, P. E. (2006). Measuring Technological Capability and Performance. *R & D Management, 36*, 421–438. doi:10.1111/j.1467-9310.2006.00444.x

Díaz, N., Aguiar, I., & De Saá, P. (2006). El Conocimiento Organizativo Tecnológico y la Capacidad de Innovación. Evidencia para la Empresa Industrial Española. *Cuadernos de Economía y Dirección de la Empresa, 27*, 33–60.

Edvinsson, L., & Malone, M. (1997). Intellectual Capital. Realizing your Company's True Value by Findings its Hidden Brainpower. New Cork: Harper Collins Publishers, Inc.

Fernandez, E., Montes, J., & Vazquez, C. (1998). Los Recursos Intangibles como Factores de Competitividad en la Empresa. *Dirección y Organización*, *20*, 83–98.

Galbreath, J. (2005). Which Resources Matter the Most to Firm Success? An Exploratory Study of Resource-Based Theory. *Technovation*, *25*, 979–987. doi:10.1016/j.technovation.2004.02.008

Galende, J. (2006). Analysis of Technological Innovation from Business Economics and Management. *Technovation*, *26*, 300–311. doi:10.1016/j.technovation.2005.04.006

Galende, J., & De la Fuente, J. (2003). El Proceso de Innovación de la Empresa Española: Identificación de Patrones de Innovación. *Cuadernos de Economía y Dirección de la Empresa*, *16*, 145–171.

Gallego, I., & Rodríguez, L. (2005). Situation of Intangibles Assets in Spanish Firms: An Empirical Análisis. *Journal of Intellectual Capital*, *6*, 105–126. doi:10.1108/14691930510574690

Hatck, N., & Dyer, J. (2004). Human Capital and Learning as a Source of Sustainable Competitive Advantage. *Strategic Management Journal*, *25*, 1155–1178. doi:10.1002/smj.421

Hayton, J. C. (2005). Competing in the New Economy: the Effect of Intellectual Capital on Corporate Entrepreneurship in High-Technology New Ventures. *R & D Management*, *35*, 137–155. doi:10.1111/j.1467-9310.2005.00379.x

Hermans, R., & Kauranen, I. (2005). Value Creation Potential of Intellectual Capital in Biotechnology – Empirical Evidence from Finland. *R & D Management*, *35*, 171–185. doi:10.1111/j.1467-9310.2005.00381.x

Hill, C., & Rothaermel, F. (2003). The Performance of Incumbent Firms in the Face of Radical Technological Innovation. *Academy of Management Review*, *28*, 257–274. doi:10.2307/30040712

Hitt, M., Bierman, L., Shimizu, K., & Kochhar, R. (2001). Direct and Moderating Effects of Human Capital on Strategy and Performance in Professional Service Firms: A Resource-Based Perspective. *Academy of Management Journal*, *44*, 13–28. doi:10.2307/3069334

Huselid, M. (1995). The Impact of Human Resource Management Practices on Turnover, Productivity, and Corporate Financial Performance. *Academy of Management Journal*, *38*, 635–672. doi:10.2307/256741

Joia, A. (2004). Are Frequent Customer Always a Company's Intangible Asset? Some Findings drawn from an Exploratory case Study. *Journal of Intellectual Capital*, *5*, 586–601. doi:10.1108/14691930410567031

Kogut, B. (2000). The Network as Knowledge: Generative Rules and Emergence of Structure. *Strategic Management Journal*, *21*, 405–425. doi:10.1002/(SICI)1097-0266(200003)21:3<405::AID-SMJ103>3.0.CO;2-5

Kogut, B., & Zander, U. (1996). What Firms Do? Coordination, Identity, and Learning. *Organization Science*, *7*, 502–518. doi:10.1287/orsc.7.5.502

Leitner, K. (2005). Managing and Reporting Intangible Assets in Research Technology Organisations. *R & D Management*, *35*, 125–136. doi:10.1111/j.1467-9310.2005.00378.x

Liu, P. L., Chen, W. C., & Tsai, C. H. (2005). An Empirical Study on the Correlation Between the Knowledge Management Method and the New Product Development Strategy on Product Performance in Taiwan's Industries. *Technovation*, *25*, 637–644. doi:10.1016/j.technovation.2003.11.001

Mansury, M. A., & Love, J. H. (2008). Innovation, Productivity and Growth in US Business Services: A Firm-Level Analysis. *Technovation*, *28*, 52–62. doi:10.1016/j.technovation.2007.06.002

Moran, P. (2005). Structural vs. Relational Embeddedness: Social Capital and Managerial Performance. *Strategic Management Journal, 26*, 1129–1151. doi:10.1002/smj.486

Nieto, M. (2001). *Bases para el Estudio del Proceso de Innovación Tecnológica en la Empresa.* León, Spain: Universidad de León.

Nonaka, I. (1994). A Dynamic Theory of Organizational Knowledge Creation. *Organization Science, 5*, 14–37. doi:10.1287/orsc.5.1.14

Nonaka, I., & Takeuchi, H. (1995). *The Knowledge Creating Company: How Japanese Companies Create the Dynamics of Innovation.* New York: Oxford University Press.

Park, Y., & Kim, S. (2006). Knowledge Management System for Fourth Generation R&D. *Technovation, 26*, 595–602. doi:10.1016/j.technovation.2004.10.008

Peteraf, M. (1993). The Cornerstones of Competitive Advantage: a Resource Based View. *Strategic Management Journal, 14*, 179–191. doi:10.1002/smj.4250140303

Peteraf, M., & Barney, J. B. (2003). Unrevailing the Resource-Based Tangle. *Managerial and Decision Economics, 24*, 309–323. doi:10.1002/mde.1126

Pike, S., Roos, G., & Marr, B. (2005). Strategic Management of Intangible Assets and Value Drivers in R&D Organizations. *R & D Management, 35*, 111–124. doi:10.1111/j.1467-9310.2005.00377.x

Prahalad, C., & Hamel, G. (1990). The Core Competence of the Corporation. *Harvard Business Review, 90*, 79–91.

Priem, R., & Butler, J. (2001). Is the Resource-Based View a Useful Perspective for Strategic Management Research? *Academy of Management Review, 26*, 22–40. doi:10.2307/259392

Reed, K., Lubatkin, M., & Srinivasan, N. (2006). Proposing and Testing an Intellectual Capital-Based View of the Firm. *Journal of Management Studies, 43*, 867–893. doi:10.1111/j.1467-6486.2006.00614.x

Rouse, M. J., & Daellenbach, U. S. (1999). Rethinking Research Methods for the Resource-Based Perspective: Isolating Sources of Sustainable Competitive Advantage. *Strategic Management Journal, 20*, 487–494. doi:10.1002/(SICI)1097-0266(199905)20:5<487::AID-SMJ26>3.0.CO;2-K

Smith, R., & Sharif, N. (2007). Understanding and Acquiring Technology Assets for Global Competition. *Technovation, 27*, 643–649. doi:10.1016/j.technovation.2007.04.001

Stieglitz, N., & Heine, K. (2007). Innovations and the Role of Complementarities in a Strategic Theory of the Firm. *Strategic Management Journal, 28*, 1–15. doi:10.1002/smj.565

Subramaniam, M., & Youndt, M. (2005). The Influence of Intellectual Capital on the Types of Innovative Capabilities. *Academy of Management Journal, 48*, 450–463. doi:10.2307/20159670

Teece, D. (1998). Capturing Value from Knowledge Assets: The New Economy, Markets for Know-How and Intangible Assets. *California Management Review, 40*, 55–79.

Tippins, M., & Sohi, R. (2003). IT Competency and Firm Performance: Is Organizational Learning a Missing Link? *Strategic Management Journal, 24*, 745–761. doi:10.1002/smj.337

Tsai, W., & Ghosal, S. (1998). Social Capital and Value Creation: The Role of Intrafirm Networks. *Academy of Management Journal, 41*, 464–476. doi:10.2307/257085

Tseng, C., & Goo, Y. J. (2005). Intellectual Capital and Corporate Value in an Emerging Economy: Empirical Study of Taiwanese Manufacturers. *R & D Management, 35*, 187–201. doi:10.1111/j.1467-9310.2005.00382.x

Tushman, M., & Nadler, D. (1986). Organizing for Innovation. *California Management Review, 28*, 74–92.

Van de Ven, A. (1986). Central Problems in the Management of Innovation. *Management Science, 32*, 590–607. doi:10.1287/mnsc.32.5.590

Wernerfelt, B. (1984). A Resource-Based View of the Firm. *Strategic Management Journal, 5*, 171–180. doi:10.1002/smj.4250050207

Wilcox, A., Fowler, S., & Zeithaml, C. (2001). Managing Orgnizational Competencies for Competitive Advantage: The Middle-Management Edge. *The Academy of Management Executive, 15*, 95–106.

Wu, W., Tsai, H., Cheng, K., & Lai, M. (2006). Assessment of Intellectual Capital Management in Taiwanese IC Design Companies: Using DEA and the Malmquist Productivity Index. *R & D Management, 36*, 531–545. doi:10.1111/j.1467-9310.2006.00452.x

Youndt, M., & Snell, S. (2004). Human Resource Configurations, Intellectual Capital and Organization Performance. *Journal of Managerial Issues, 16*, 337–360.

Youndt, M., Subramanian, M., & Snell, S. (2004). Intellectual Capital Profiles: An Examination of Investments and Returns. *Journal of Management Studies, 42*, 335–361. doi:10.1111/j.1467-6486.2004.00435.x

Zack, M. (1999). Developing a Knowledge Strategy. *California Management Review, 41*, 125–145.

ADDITIONAL READINGS

Questionnaire Items

Figure 5. Scale used for structural capital

Questionnaire items
SC0: Our firm possesses a high degree of computer-based technical expertise
SC1: We routinely utilize computer-based systems to work
SC2: Our organizations uses patents and licenses as a way to store knowledge
SC3: Much of our organization's knowledge is contained in manuals, databases, etc.
SC4: The average time for the project development of innovation is major that of the most nearby competitor
SC5: Our company grants incentives to the innovative employees
SC6: The average cost for project of innovation is minor that of the nearest competitor
SC7: Relative to our firm´s largest competitor we spend a much larger of R&D investment
SC8: Regarding our nearest competitor, the quality and quantity of employees in I+D is top
SC9: Our organization possesses knowledge and ability to link operational goals and objectives with compensation plans
SC10: Our company has documented procedures that help carry out routine
SC11: We have a standard procedure for effectively dealing with customer complaints
SC12: Our company has developed quality programmes during the last 3 years
SC13: Our employees share the values, beliefs and symbols of the organization
SC14: The organizational goals are clear and agreeable to all members
SC15: The employees are committed and hold a high sense of responsability to the organization
SC16: Our employees get enough work-related information to perform their jobs effectively
SC17: There fluid communication between management and employees
SC18: Employees and management in our firm make sincere efforts to solve common problems
SC19: Employees generally view the conditions of their employment here as fair
SC20: There is complete trust between management and employees

Figure 6. Scale used for innovation outcomes

Questionnaire items
The number of products / services / projects that we have launched over the past 3 years is higher than the nearest competitor
The satisfaction of our company for the efficiency of innovation projects is high
Regarding our nearest competitor, the number of new technologies developed over the past 3 years is higher

Chapter 10
Prioritizing Corporate R&D Capabilities:
The Intellectual Capital Perspective

Yuan-Chieh Chang
National Tsing Hua University, Taiwan

Pei-Ju Yu
Chunghua Institution for Economic Research, Taiwan

Hui-Ru Chi
National Changhua University of Education, Taiwan

ABSTRACT

This chapter examines a comprehensive list of intellectual capital (IC)-related indicators for developing corporate R&D capabilities along the input-process-result (IPR) processes. Via factor analysis, 43 R&D related IC indicators were abstracted into 11 factors. Corporate R&D managers prioritized these IC factors by completing analytical hierarchy process (AHP) questionnaires. The results of AHP are as follows: (1) the result phase is the pivotal of developing corporate R&D capabilities in three phases, (2) the top three weighting factors are the relational and process capitals (cost effectiveness to customers) in the result phase, followed by organizational capital (strategy fitness) in the input phase, and human capital (competency of R&D personnel) in the input phase; (3) strategy fitness in the input phase, project execution capability in the process phase; and cost effectiveness to customers in the result phase is the most crucial IC capabilities. Some discussions and conclusions were drawn.

INTRODUCTION

As Drucker (1993, p. 42) suggested, "The traditional factors of production— land, labor, and capital— have not disappeared. But they become secondary. Knowledge is becoming the only meaningful resource". However, although the paradigm has shifted from manual work towards knowledge work, managerial awareness of the importance of "intangible assets" is quite low (Litschka et al., 2006). Guthrie (2000, p. 11) argued that "the limitation of the existing financial reports for capital markets and other stakeholders have led to a search for new ways to measure and report a company's intellectual capital", especially in the rising importance of

DOI: 10.4018/978-1-61520-875-3.ch010

the knowledge-based economy and information society. Intellectual capital is concerned with how better to manage and measure knowledge and other intangibles in the company (Mouritsen, 2002). Obviously, technological innovation has become the main source of competitive advantages for firms (Burgelman et al., 2004; Teece, 1987; Rothwell, 1992). It is well documented that R&D activities form the central element of technological innovation of firms (Cohen and Levinthal, 1990; Evangelista et al. 1997; Yam et al., 2004). However, the R&D processes are increasingly more expensive, costly and complex since the 1990s. How to manage corporate R&D capabilities in the intellectual capital view is a great challenge to many firms.

Previous research on IC proposed its structure and candidate indicators (Edvinsson, 1997; Stewart, 1997; Sullivan, 2000; Sveiby, 1997). Carlucci, Marr, and Schiuma (2004) showed that the management of IC directly impacts business performance. Furthermore, measuring intellectual capital can be used to help formulate business strategies and provide evaluation bases for venture capitalists (Liu, 2006). Yet, managers and researchers have experienced difficulties in conducting IC measurement model which includes measuring priority among candidate IC indicators (Mouritsen, Bukh, Larsen, & Johansen, 2002). The chapter aims to measure R&D specific IC indicators to bridge this gap.

There are increasing studies to develop and evaluate weights of different dimensions and indicators of IC by AHP (analytical hierarchy process) in different firms, industries and nations (Bozbura et al., 2007; Chen, 2009; Han and Han 2004; Liu, 2009). Han and Han (2006) identified key IC dimensions and indicators that were better presented IC in the Korean mobile phone industry. Bozbura et al. (2007) identified the priority of human capital measurements by a group of IC academics and professionals in Turkey. Chen (2009) evaluated the weights of IC dimensions in the four Taiwanese high-tech firms. Finally, Liu (2009) mainly identified the importance of IC dimensions and indicators in the Taiwanese e-learning platform industry. However, few studies have empirically linked intangibles and corporate R&D capabilities. We argue that the weights of IC dimensions and indicators are R&D context-specific. This prioritizing of R&D capabilities provides a better compass to manage intangibles more effectively in the increasingly uncertain and complicated R&D activities.

The organization of this chapter is as follows. This Chapter reviews the definition and taxonomy of intellectual capital, categorizing R&D related IC indicators and proposing a R&D input-process-result framework. We explain the exercise of analytical hierarchy process. The results of AHP were shown. We compare the research results to previous literature. Finally, some managerial implications are drawn for R&D managers in firms and organizations.

BACKGROUND

Definition of Intellectual Capital

Teece (1986) argued that when established firms possessed some specific assets, they might prevent themselves from new entrants' attack. He used the term complementary assets to illustrate factors such as specialized manufacturing capability, access to distribution channels, service network and complementary technologies. Stewart (1991, p.44) mentioned: "Every company depends increasingly on knowledge – patents, processes, management skills, technologies, information about customers and suppliers, and old-fashioned experience... added together, this knowledge is intellectual capital." He defines IC as "the sum of everything everybody in your company knows that gives you a competitive edge in your marketplace". Then, Stewart (1997) defined IC as "the intellectual material – knowledge, information, intellectual property, experience –that can be put to use to

Table 1. Definition of intellectual capital: A summary

Scholar	*Definition*
Stewart (1991)	The sum of everything everybody in your company knows that gives you a competitive edge in your marketplace.
Edvinsson &Sullivan (1996)	IC as knowledge that can be converted into value.
Stewart (1997)	The intellectual material – knowledge, information, intellectual property, experience –that can be put to use to create wealth.
Edvinsson &Malone(1997)	IC is the procession of the knowledge, applied experience, organizational technology, customer relationships and professional skills that provide Skandia with a competitive edge in the market.
Roos et al. (1998)	IC includes all the processes and the assets which are not normally shown on the balance-sheet and all the intangible assets (trademarks, patents and brands) which modern accounting methods consider... it includes the sum of the knowledge of its members and the practical translation of his/her knowledge.
Bontis (1998)	IC is the pursuit of effective use of knowledge (the finished product) as opposed to information (the raw material).
McElroy (2002)	A major modification to Edvinsson's model which includes the addition of both social capital and the underlying notion of social innovation capital

create wealth. Edvinsson and Sullivan (1996) defined IC as knowledge that can be converted into value. Edvinsson and Malone (1997) defined IC as: "the possession of the knowledge, applied experience, organizational technology, customer relationships and professional skills that provide a company with a competitive edge in the market."

Roos, Roos, Dragonetti and Edvinsson, (1998) claimed that IC includes all the processes and the assets which are not normally shown on a balance-sheet and all the intangible assets (trademarks, patents and brands) which modern accounting methods consider. It includes the sum of the knowledge of its members and the practical translation of their knowledge. Bontis et al. (2000) defined IC as "elusive, but once it is discovered and exploited, it may provide an organization with a new resource-base from which to compete and win." in 1996. McElroy (2002) proposed a major modification to Edvinsson's model to include the addition of both social capital and the underlying notion of social innovation capital. It is found that some studies tend to treat intellectual capital equivalent to intangible assets that are not normally shown in the financial report (Roos et al., 1998; Kauffmann & Schneider, 2004). However, intellectual property is one of intellectual capital (Stewart, 1991, 1997; Roos et al., 1998). In summary, intellectual capital represents the value gap between market value between financial one. The definitions of IC are summarized in Table 1.

Corporate R&D-Related Intellectual Capital

The literature review presents a great number of classification schemes for intellectual capital. Edvinsson (1997) defined IC as human capital and structural capital. Bontis (1998) categorized into three types: human capital, structural capital and customer capital. Bozbura and Beskese (2005) prefer three dimensions of intellectual capital that includes human, relational, and organizational components. However, a convergent taxonomy of IC emerged, categorizing IC into: (1) human capital, (2) structural capital including: organizational capital, innovational capital, and process capital; (3) relational capital (do Rosario Cabrita & Vaz, 2005).

Before the chapter elaborated corporate R&D related IC capabilities, R&D and innovation are interchangeable throughout the chapter.

Therefore, the chapter reviewed corporate R&D related intellectual capital from the literature of R&D management and innovation management. Rothwell (1992) provided seven critical factors in managing successful industrial R&D including: (1) good internal and external communication, (2) treating innovation as a corporate wide task, (3) implementing careful planning and project control, (4) strong market orientation, (5) providing good technical services to customers, (6) presence of certain key individuals as technical gatekeepers, and (7) high quality management. Burgelman et al. (2004) proposed corporate innovative capabilities can be measured from five major dimensions: (1) resource availability and allocation, (2) capabilities to understand competitor's innovative strategy and industrial evolution, (3) capabilities to understand technological development, (4) structural and cultural context and (5) strategic management capability. Yam et al. (2004) examined technological and innovative capabilities of Chinese firms, which include: (1) learning capability, (2) R&D capability (e.g., R&D strategy, project implementation, portfolio management, and R&D expenditure), (3) resource allocation capability, (4) manufacturing capability, (5) marketing capability, (6) organizing capability, and (7) strategic planning capability. Recently, Adams et al. (2009) proposed a holistic framework to measure innovation management at the level of the firm with seven dimensions, namely: input management, knowledge management, innovation strategy, organizational culture and structure, portfolio management, project management and commercialization. The chapter interweaves types of IC and factors influencing R&D effectiveness and innovation management (Adams et al., 2009; Burgelman et al., 2004; Chiesa et al., 1996; ERIAM, 2004; Lee et al., 1997; Rothwell, 1992; Yam et al., 2004). We elaborated these five types of IC in the R&D context in the following.

R&D Human Capital

Human capital is the primary component of IC (Edvinsson & Malone, 1997; Stewart, 1997; Bontis et al., 2000) because human interaction is the critical source of intangible value in an intellectual economy (O'Donnell, 2003). Edvinsson (1997) defined human capital as the combined knowledge, skill, innovation, and ability of a company's individual employees. It is also includes the company's values, culture, and philosophy. Roos et al. (1998) argued that employees generate IC through their competence, their attitude and their intellectual agility. Bontis (1998) described human capital as the firm's collective capability to extract the best solutions from the knowledge of its individuals. Unfortunately, loss of personnel can result in the loss of corporate memory and hence become a threat to the organization. Bontis et al. (2000) defined human capital as the individual knowledge stock of an organization as represented by its employees.

Based on Edvinsson and Malone's (1997) definition, human capital is combined knowledge, skill, and ability of employees. Lee and Lee (1996) identified three key HC elements influencing the R&D effectiveness of Korean firms. These three elements are: (1) degree of R&D professionalization, (2) skill of R&D personnel, and (3) adequate education and training. Moreover, the knowledge of R&D employees can be embedded in their knowledge and how know (EIRAM, 2004) and their experiences (Edvinsson and Malone, 1997). As Roos et al. (1998) and Edvinsson and Malone (1997) mentioned that the attitude, motivation and satisfaction of R&D employees act key elements of HC. As Bontis (1998) illuminated that the loss of employees can result in the loss of corporate memory. The chapter argues that R&D employee retention is one of key elements in HC.

R&D Organizational Capital

Edvinsson (1977) regarded organizational capital that is the company's investment in systems, tools, and operating philosophy. Organizational capital speeds the flow of knowledge through the organization, as well as out to the supply and distribution channels. It is the systemized, packed, and codified competence of the organization as well as the systems for leveraging that capability. Damanpour (1991, P. 556) defined "organizational innovation" as the adoption of innovations in organizations, that encompasses the generation, development and implementation of new ideas and behavior. Consequently, organizational capital is the organizational properties (e.g., specialization, formalization, centralization) that enhance or hinder R&D capabilities.

European Industrial Research Management Association (EIRMA) (2004) concluded that R&D organizational capital can be reflective on the organizational mission, vision and leadership. Nokia claimed their company mission is to connect people. Sony has "like no other" and claims a leader in the world home entertainment business. Once a firm's mission and vision are set, enough R&D budgeting is necessary to support the mission and vision (EIRAM, 2004) or leadership (Rothwell, 1992). However, integrating R&D strategy and business strategy is crucial to increase R&D productivity and effectiveness (Lee et al., 1996; EIRMA, 2004). Otherwise, new products will not reach the market and make profit. There are different kinds of R&D derived businesses in terms of risk, time horizon and complexity. A good strategic business portfolio can minimize risk and maximize profit among different businesses (EIRMA, 2004). Finally, the organizational capital can be measured by the degree of goal achievement and general management improvement (Lee at al., 1996).

R&D Innovational Capital

Innovation capital refers to the renewal capability and the results of innovation in the form of protected commercial rights, intellectual property, and other intangible assets and talents used to create and rapidly bring to market new products and services (Edvinsson, 1997). Lee et al. (1996) suggested that innovation capital can be measured by: (1) R&D investment, (2) R&D facilities, and (3) expansion and diversification of research areas. ERIAM (2004) developed some innovation capital indicators that influence R&D effectiveness: (1) intellectual property management, (2) ideas management, (3) technology leadership, (4) innovativeness. Adams et al. (2009) regarded idea generation and information flows that are good measurements for managing knowledge during the innovation process.

R&D Process Capital

Process capital is work processes (such as ISO 9000), technologies, and employee programs that augment and enhance the efficiency of manufacturing or the delivery of services (Edvinsson, 1997). It is the kind of practical knowledge used in continuous value creation. Chen (2009) proposed that process capital can be measured by: (1) quality of decision; (2) using IT capacity, (3) process expense and (4) IT equipment invested. In measuring R&D effectiveness for the Korean firms, Lee et al. (1997) developed four process capital indicators: (1) feasibility of R&D plans, (2) validity of selected R&D topics, (3) efforts to strictly follow plans, and (4) collaboration between R&D, manufacturing and marketing. EIRMA (2004) proposed that corporate R&D can be implemented more effectively if the firms have: (1) process intelligence, (2) project portfolio management, (3) knowledge management, and (4) cost efficiency in place. Adams et al. (2009)

proposed four good indexes: tools, project efficiency, communications and collaboration to measure project management.

R&D Relational Capital

The size and structure of the customer base is crucial to a firm's future value because customer relationships are the origin of the financial flows (Edvinsson, 1997). Bonits (1998) identified customer capital as the potential an organization has due to ex-firm intangibles. These intangibles include the knowledge embedded in customers, suppliers, the government or related industry associations. Roos et al. (1998) identified relational capital as the relationships with internal and external stakeholders. The relational capital contains the relations with customers, shareholders, suppliers, rivals, the state, governmental institutions and society (Bozbura, 2004). Relational capital can be regarded as social capital that the knowledge embedded within, available through and utilized by interactions among individuals and their networks of interrelationships (Subramaniam and Youndt, 2005). Todtling et al. (2009) found that radical innovation tends to draw knowledge links with nearby universities and research institutes. However, incremental innovation seems to occur often in interactions with business partners widely. Egbetokun et al. (2009) revealed that Nigerian firms most frequently work with their customers and suppliers as their innovation partners. In summary, the relational capital is the knowledge embedded in organizational relationships with customers, suppliers, stakeholders, strategic alliance partners, universities, research institutes etc.

For enhancing R&D effectiveness, EIRMA think that R&D should be managed in the stakeholder approach by including: (1) market intelligence, (2) social responsiveness and investor relationship, (3) external partnerships with customers, suppliers, universities and research institutes, (4) technical service, (5) customer satisfaction of products/services, and (6) branding. The relational capital be measured by market share and customer retention or lost (Edvinsson and Malone, 1997). Finally, Adams et al. (2009) suggested that three indexes are useful to measure innovation commercialization: market research, marketing testing, and marketing and sales.

RESEARCH FRAMEWORK AND PRIORITIZING INTELLECTUAL CAPITAL

The R&D performance measurement methods described in the literature are so varied and uniquely designed for particular situations that conducting a selection process for R&D performance indicators would be a challenging task (Ellis, 1997). Chiesa et al. (1996) developed a model to audit corporate innovation capability. The process audit model includes: (1) concept generation process, (2) product development process, (3) production process, (4) technology acquisition process, (5) leadership process, (6) resource provision process, (7) systems and tools provision process. Brown and Svenson (1998) regarded R&D as a processing system that includes a number of phases that contain several subjects for the measurement of performance. The system includes contents of input, processing system, output of the processing systems, receiving systems and outcome (input-process-output, IPO), as well as their linkages to each other. A similar IPO concept of the measurement method, so-called Input-Process-Result (IPR) Phase (EIRMA, 2004; Lee et al., 1996) was adopted to construct the research framework of measuring R&D capabilities from the perspective of intellectual capital. The chapter adopted the IPR framework to examine R&D phase-specific. The framework can help R&D managers to develop IC along different R&D phases. Sometimes, the R&D intellectual capital can be overlapping between input, process, and result phases. However, based on references and expert interviews, the chapter was able to delineate the IC items in Figure 1.

Figure 1. Corporate R&D capabilities and intellectual capital: input-process-result framework

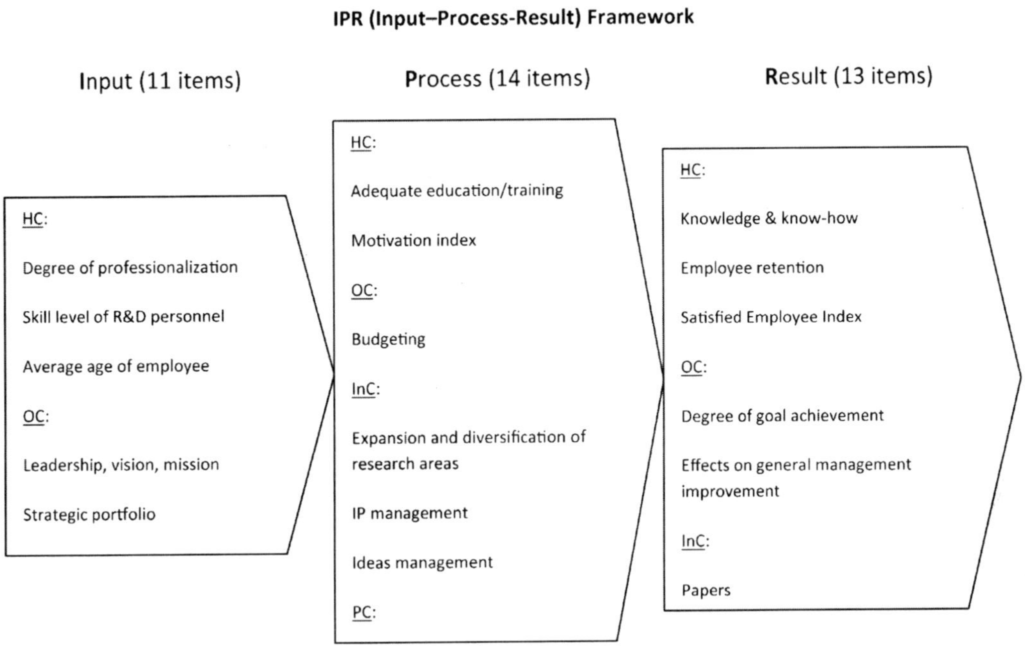

Based on the list of R&D IC indicators (Table 2), we further re-allocated them into: 11 indicators in the input phase, 14 indicators in the process phase and 13 items in the result phase (Figure 1).

Yam et al. (2004) developed seven capabilities for auditing technological innovation capabilities in mainland China. They developed 17 items to measure R&D capability. These 17 items can be categorized into R&D input related indicators: (1) relevance of R&D plan to corporate plan, (2) R&D investment in products, (3) R&D investment in processes, (4) number of new product per year, (5) R&D personnel ratio. These R&D process related indicators include: (1) mechanisms to encourage and reward inventiveness and creativity; (2) range of functions involved in concept development and screening, (3) clear project target and management, (4) presence of cross-functional teamwork, (5) quality and availability of product champions, (6) efficiency of R&D personnel communication, (7) application of advanced design tools, (8) functional integration in the innovation process, (9) technology transfer mechanisms from research to development, and (10) presence of established protocols. The R&D result related indicators include: (1) feedback from manufacturing, and (2) feedback from customers. Yam et al. (2004) mainly developed R&D process related indicators and focused on project implementation, communication, coordination and integration.

However, it is difficult to measure the value of such intangible assets and intellectual capital (Marr, Gray, and Neely, 2003; Hung and Chang, 2006; Chen, 2009). Most of the previous papers related to IC only proposed indicators for evaluating IC (Liu, 2006). The indicators of IC are complex and require decision makers to consider multiple criteria and their relative importance at the same time (Bukh et al., 2001). It is, to some extent, difficult for them to consider all of the factors involved in a complex IC indicator appropriately (Han & Han, 2004). The selection of IC statement indicators is a multi-criteria decision problem that requires resolutions involved with various stakeholders' interests (Bozbura, Beskese

Table 2. R&D related intellectual capital

Intellectual capital	R&D capabilities
Human capital	• Degree of professionalization (Lee et al., 1996) • Skill level of R&D personnel (Lee et al., 1996; Rothwell, 1992; Adams et al., 2009) • Adequate education/training (Lee et al., 1996) • Knowledge & Know-How (EIRMA, 2004) • Employee motivation (Edvinsson & Malone, 1997) • Employee retention (Edvinsson & Malone, 1997) • Employee satisfaction (Edvinsson & Malone, 1997) • Employee experience (Edvinsson & Malone, 1997)
Organizational capital	• Degree of goal achievement (Lee et al. 1996) • Effects on general management improvement (Lee et al., 1996) • Leadership, vision, mission (EIRMA, 2004; Rothwell, 1992; Adams et al., 2009) • Budgeting (EIRMA, 2004) • Strategic Portfolio (EIRMA, 2004; Burgelman et al., 2004; Yam et al., 2004) • Business Alignment (EIRMA, 2004; Burgelman, et al., 2004)
Innovational capital	• Enough R&D investment (Lee et al., 1996; Burgelman et al., 2004; Yam et al., 2004) • Enough R&D facilities (Lee et al., 1996) • Expansion and diversification of research areas (Lee et al., 1996) • IP management (EIRMA, 2004) • Ideas management (EIRMA, 2004; Adams et al., 2009) • Technology leadership (EIRMA, 2004Burgelman et al., 2004; Yam et al., 2004) • Innovativeness (EIRMA, 2004)
Process capital	• Feasibility of R&D plans (Lee et al., 1996) • Validity of selected R&D topics (Lee et al., 1996) • Effort to strictly follow plans (Lee et al., 1996; Yam et al., 2004) • Collaboration between R&D and production/marketing (Lee et al., 1996) • Process Intelligence (EIRMA, 2004; Rothwell, 1992) • Portfolio Management (EIRMA, 2004; Adams et al., 2009) • Knowledge Management (EIRMA, 2004; Adams et al., 2009) • Cost Efficiency (EIRMA, 2004)
Relational capital	• Market intelligence (EIRMA, 2004; Burgelman et al., 2004) • Societal responsiveness (EIRMA, 2004) • External partnerships (EIRMA, 2004) • Usefulness of developed technology (EIRMA, 2004; Adams et al., 2009) • Technical service (EIRMA, 2004; Rothwell, 1992) • Customer satisfaction of products/services(EIRMA, 2004; Rothwell, 1992) • Image, brand & stock value (EIRMA, 2004) • Market share (Edvinsson & Malone, 1997) • Customer lost (Edvinsson & Malone, 1997)

& Kahraman, 2007). Bozbura et al. (2007) identified five attributes of human capital, namely: talent, strategic integration, cultural relevance, knowledge management and leadership. There are twenty indicators were developed to measure these five attributes of human capital. By using fuzzy AHP, a group of experts ranked "creating results by using knowledge" as the most important indicator, followed by " employees' skills index", "sharing and reporting knowledge", "succession rate of training program" and employee's satisfaction index" in 20 indicators (Bozbura et al., 2007). Thus, Analytical Hierarchy Process (further discussed in the following part) makes the decision process more rational by taking all the data given by the managers and synthesizes them in a way that the managers might not be able to do due to the limited cognitive ability of the human mind (Calantone et al., 1999).

Based on fuzzy AHP results of four Taiwanese high-tech firms' evaluation among four dimensions of IC (human capital, customer capital, innovation capital and process capital), human capital was ranked the most important dimension, followed process capital, customer capital and innovation capital across four firms (Chen, 2009). Chen (2009) concludes that fuzzy AHP is a useful evaluation method for deciding what their objectives are and discover which dimensions of IC can be improved. By developing measurements of IC in the e-learning platform industry, Liu (2006) identifies five dimensions of IC (human capital, customer capital, innovation capital, service capital and structure capital) and fifteen indicators. The Liu's (2006) IC assessment can be applied to formulate business strategy, evaluate firms by venture capitalists, and allocate resource.

Notes: HC: Human Capital, OC: Organizational Capital, InC: Innovational Capital; PC: Process capital, RC: Relational Capital

METHODS

Exploratory Factor Analysis

According to SPSS software, the five steps of exploratory factor analysis (EFA) are as follows: First there is the Bartlett Test of Sphericity and the Kaiser-Meyer-Olkin (KMO). The Bartlett's test of sphericity is a widely programmed statistical test of appropriateness and it should be applied prior to factor analysis. In addition, a test of the appropriateness of a matrix for factoring is the KMO measure of sampling adequacy (MSA) (Kaiser, 1970). The KMO provides a measure of the extent to which the variables belong and whether they are appropriate for factor analysis or not. Hair et al. (2009) show KMO should be at least above 0.6.

The second step is the Extraction Method. Since principal factor analysis is the most widely used mathematical method for such decomposition (Gorsuch, 1983), it was chosen as the extraction method in this study. The third step is eigenvalues, although factor selection should not be based on this criterion alone. In 1960, Kaiser recommended holding the factors with eigenvalues greater than 1, and this is adopted here. Fourth is rotation. Here, the characteristics for which most variables with high loading on the most important factor and low loading on all other factors make the interpretation difficult; therefore, factor rotation is used to discriminate between factors (Field, 2005). There are two basic types of rotation: orthogonal and oblique. Orthogonal means the factors are assumed to be uncorrelated with one another, while oblique rotation, which is the case for this study, derives factor loadings based on the assumption that the factors are correlated. Hence, Promax, which is one of the oblique rotations, adopted as the rotation method. Fifth is the reliability test and reliability means that a scale should be consistent. Cronbach's α is the most common measure of scale reliability, it achieves high reliability when alpha is greater than or equal to.7.

In the study, forty-three indicators were collected from the literature on IC research, R&D management and innovation management. Before drafting the factor-analysis questionnaire, two R&D managers were interviewed in order to test the validity of question items. There were three parts of the factor analysis questionnaire. The first part is the basic information about the respondent; the second part is the basic company information; and the third part is the importance of IC-related capabilities along the 3-phase R&D process. All items can be rated Saaty 1~9 scale. The data collection used an on-line questionnaire tool created by Chunghwa Telecom Co., Ltd. Telecommunication Laboratories.

The survey subjects are targeted at the companies from the Inc. 500 list with R&D Centers in Taiwan. Therefore, under the above constraint, there were eighty-eight candidate companies. However, instead of sampling, we inquired by email whether the top managements were willing to answer the questionnaire. Consequently, 35 respondents had shown their willingness.

The results of the first questionnaire, designed for factor analysis, are presented as below. There were thirty-five respondents in total. The value of KMO in the input phase was 0.603 and the four factors can explain 74.056% of variance. The value of KMO in the process phase was 0.722 and 70.452% of variance can be explained by the three factors. The value of KMO in the result phase was 0.689 and the four factors can explain 73.732% of variance, showing the sampling adequacy (Table 3). The factor naming is referred to previous literature as well. Cronbach's α was exploited to test the reliability. The overall reliability is 0.813, 0.899 and 0.885 respectively in the input, process, and result phases, which represent high reliability. Each factor of the reliability from 0.576 to 0.920, showing acceptable reliability.

ANALYTICAL HIERARCHY PROCESS

The world is a complex system of interacting elements. Therefore, in order to deal with unstructured social, economic, and political issues, there is a need to order our priorities, to agree that one objective outweighs another in the short term, and to make tradeoffs to serve the greatest common interest (Saaty, 1995). AHP enables us to make effective decisions on complex issues by simplifying and expediting our natural decision-making processes. It also provides an effective structure for group decision making by imposing discipline on the groups' thought processes (Saaty, 1995). In short, AHP was created by Thomas Saaty to structure complex judgments (Saaty, 1980; Saaty & Kearns, 1985).

A brief description of AHP process is follows. First, a hierarchy is structured from the top-down, much like relevance trees (see Figure 2). This stage allows a complex decision to be structured into a hierarchy descending from an overall objective to various 'criteria', 'sub-criteria', and so on until the lowest level. According to Saaty (2000), a hierarchy can be constructed by creative thinking, recollection and using people's perspectives.

Second, people can judge between two items more easily than they can make composite judgments of multiple items all at once. Therefore, AHP uses pairwise comparison among each relevant pair of items as the basic judgments. The preferences are quantified by using a nine-point scale (Saaty, 2000). Third, the stage is to synthesize the judgments within a given matrix (for local priorities) and then across matrices (global priorities). The method of calculating the eigenvalue is usually used by AHP to evaluate the vectors of priorities of parameters. The vector of priorities of the parameters in the lower level in the hierarchy is first calculated and then it progresses to get the overall priority vector.

There are four methods for approximating the solution: (a) NRA (normalization of row average) method sums the elements in each row and normalize by dividing each sum by the total of all the sums. (b) NRC (normalization of the reciprocal sum of columns) method is to take the sum of the elements in each column and form the reciprocals of these sums. Then normalize by dividing each reciprocal by the sum of the reciprocals. (c) ANC (average of normalized columns) method is to divide the elements of each column by the sum of that column (i.e., normalize the column) and then add the elements in each resulting row and divide this sum by the number of elements in the row (n). This is a process of averaging over the normalized columns. (d) NGM (normalization of the geometric mean of the rows) method is to multiply the n elements in each row and take the nth root. In practice, NGM is the most common method to be applied, which is adopted in the study. λmax can be calculated as Equation 1:

$$\lambda_{\max} = \frac{1}{n}\left(\frac{w_1'}{w_1} + \frac{w_2'}{w_2} + \Lambda + \frac{w_n'}{w_n}\right) \quad (1)$$

Table 3. Denomination of each factor in IPR phase

Factor	Item	Factor Loading	Eigenvalue	% of Variance	Cronbach's Alpha
F1: Competence of R&D Personnel (Roos et al., 1997)	Intellectual agility of personnel	.850	4.323	33.252	.812
	Skill level of R&D personnel	.820			
	Degree of professionalization	.806			
	The attitude of personnel	.682			
F2: Enough R&D Investment & Forecasting (Lee et al., 1996)	Enough R&D investment	.876	2.252	17.322	.867
	Enough R&D facilities	.861			
	Process intelligence	.854			
F3: Strategy Fitness (Edvinsson, 1997)	Strategic portfolio	.921	1.776	13.663	.755
	Business alignment	.845			
	Leadership, vision, mission	.596			
F4: Market intelligence (Roos et al., 1998)	Average age of employee	.851	1.277	9.820	.699
	Societal Responsiveness	.789			
	Market Intelligence	.645			
F5: Project Execution (Edvinsson, 1997)	Effort to strictly follow plans	.878	6.343	45.307	.920
	Feasibility of R&D plans	.878			
	Portfolio management	.876			
	Collaboration between R&D and production/marketing	.862			
	Validity of selected R&D topics	.780			
F6: Managing Creative People (Edvinsson, 1997)	Ideas Management	.852	1.944	13.883	.835
	Expansion and diversification of research areas	.832			
	Motivation Index	.772			
	Adequate education/training	.720			
	Knowledge Management	.630			
	IP Management	.598			
F7: Strategic External Partnership (Roos et al., 1998)	Budgeting	.826	1.577	11.262	.777
	Technical service	.820			
	External Partnerships	.732			
F8: R&D yield (Edvinsson, 1997)	Outside acknowledge of R&D activities —patents and royalty	.912	6.261	39.130	.870
	Innovativeness	.871			
	Outside acknowledge of R&D activities —papers	.727			
	Leadership	.723			
	Effects on general management improvement	.687			
	Satisfied Employee Index	.632			
	Degree of goal achievement	.613			

continued on following page

Table 3. continued

F9: Customer Relationship management (Sveiby, 1997)	Image/Brand	.885	2.964	18.527	.867
	Market share	.844			
	Customer lost	.818			
	Product leader customer satisfaction	.742			
	Stock Value	.709			
F10: Cost Effectiveness to Customers (Lee et al., 1996)	Cost efficiency	.864	1.386	8.664	.744
	Usefulness of product or technique	.768			
F11: Accumulation of R&D personnel knowledge (EIRMA, 2004)	Knowledge & know-how	.815	1.186	7.411	.567
	Employee retention	.737			

Figure 2. Selecting the most desirable alternative

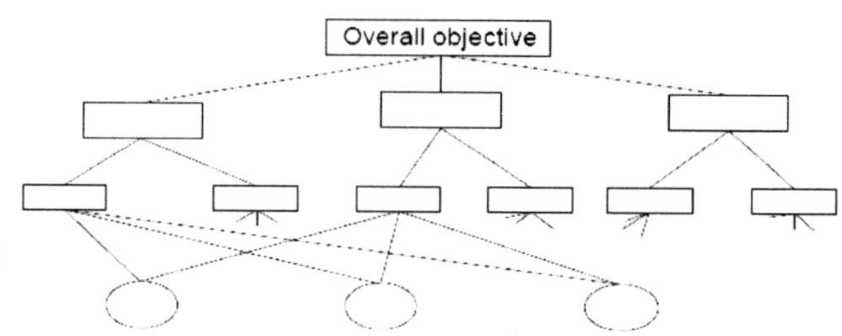

Fourth, checking that the judgments combined are reasonably consistent with each other, collections of pairwise judgments are apt to show inconsistencies. These may reflect crude scaling. Besides, raters may just be flagrantly inconsistent. AHP provides a helpful indicator to signal the degree of inconsistency in a matrix of judgments: Consistency Ratio (CR), defined as Equation 2:

$$CR = \frac{CI}{RI} \qquad (2)$$

where CI is called the consistency index and RI is the Random Index.

Furthermore, Saaty (1980, 2000) provided average consistencies (RI values) of randomly generated matrices. CI for a matrix of order n is defined as Equation 3:

$$CI = \frac{\lambda_{\max} - n}{n - 1} \qquad (3)$$

Saaty (1980) suggests that the acceptable value of CI should be less than or equal to 0.1. Consequently, this procedure provides an overall numerical ranking of available alternatives (Calantone et al., 1999; Forgionne et al., 2002; Tam & Tummala, 2001; Han & Han, 2004).

AHP QUESTIONNAIRE DESIGN

The chapter categorized four factors in input phase, three factors in the process phase, and four factors in the result phase (Table 4). In order to weigh factors among phases, input human capital (I_HC): competence of R&D personnel, input innovation capital and process capital (I_InC+PC): enough R&D investment and forecasting, input-

Table 4. Factor naming

R&D phase	IC factor naming
Input	I_HC: Competences of R&D personnel
	I_InC+PC: Enough R&D investment & forecasting
	I_OC: Strategy Fitness
	I_RC: Market intelligence
Process	P_PC: Project Execution
	P_InC+HC: Managing creative people
	P_RC+OC: Strategic external partnership
Result	R_InC+OC: R&D yield
	R_RC1: Customer relationship management
	R_RC2+PC: Cost effectiveness to customers
	R_HC: Accumulation of R&D personnel knowledge

organizational capital: (I_PC): strategy fitness and input relational capital (I_RC): market intelligence are four factors in the input phase. The reason to combine InC and PC is due to the R&D investment is based on the results of R&D forecasting. So R&D investment has to tighten with R&D forecasting as a process capital.

Process capital: project execution, Innovation and human capital: managing creative people, relational and organizational capital: and strategic external partnership is three IC factors in the process phase. Finally, result innovation and organizational capital: R&D yield, result relational capital: customer relationship management, result relational and process capital: cost effectiveness to customers, and result human capital: accumulation of R&D personnel knowledge is included in the result phase. AHP questionnaire was applied. However, the main challenge here was to denominate the name of each factor based on the literature. To sum up, the second questionnaire was based on the results of the first questionnaire.

Based on AHP, Figure 3 shows the structure of the second questionnaire. Its goal is to determine which factor outweighs the others in the case of assessing R&D capabilities. Therefore, the objective (level 1) is to assess R&D capabilities; the criteria (level 2) are the input phase, the process

Figure 3. Corporate R&D capabilities: The IC framework

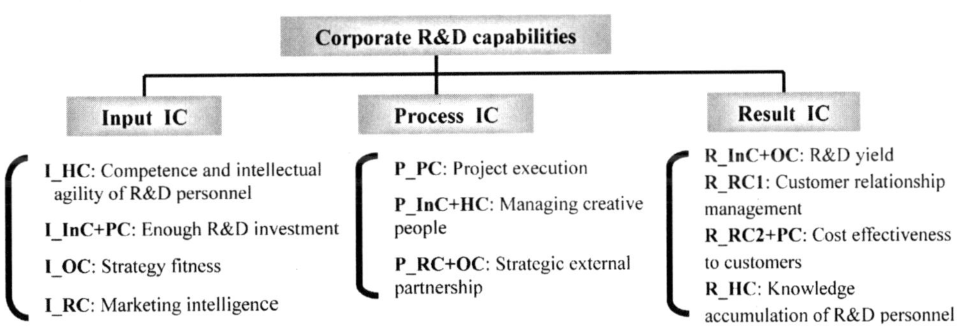

Table 5. Demographics of respondents

Demographics	Details	Number	Percentage
Industry	Semi-Conductor	3	12%
	3C	9	36%
	Optotronics	1	4%
	Biopharmaceutical	2	8%
	Material and Chemical	3	12%
	Others	7	28%
	Total	25	100%
Position	Chief Technology Officer(CTO)	4	16%
	Vice President (VP)	6	24%
	R&D Manager	6	24%
	Project Manager	2	8%
	Others	7	28%
	Total	25	100%
Working years	0-5 year	2	8%
	5-10 year	5	20%
	10-15 year	6	24%
	above 15 year	12	48%
	Total	25	100%
Tenure in this department	0-5 year	9	36%
	5-10 year	12	48%
	10-15 year	1	4%
	above 15 year	3	12%
	Total	25	100%

phase, and the result phase; the sub-criteria (level 3) under input phase, process phase, and result phase.

The first part of the questionnaire is the main content, pair-wise comparisons between factors (Saaty 1980, Saaty 1995); It includes four sections: 1) the relative importance of the input phase and there are six questions of comparing ($C_2^4 = 6$); 2) the relative importance of the process phase, and there are four questions for comparing ($C_2^3 = 4$); 3) the relative importance of the result phase and there are six questions of comparing ($C_2^4 = 6$); 4) the relative importance of the three phases and there are four questions of comparing ($C_2^3 = 4$). The second part is the basic information about the interviewee. The targeted companies were also from the top 500 company list especially the ones with R&D Centers. Nevertheless, instead of sampling, we inquired by email whether the high-level managers were willing to answer the questionnaire. The twenty-five respondents were mainly CTOs, VPs, and R&D Managers in R&D centers.

RESULTS

There were twenty-five respondents in AHP survey. In term of industry, the major respondents were from computation, communication and consumer electronics, occupying 36%, followed by semiconductor (12%), and material and chemicals (12%) (Table 5). In term of position, the major respondents were in position of vice presidents of R&D (24%), R&D Managers (24%), and chief technology officers (16%). The top managers' opinions were well represented and occupied more than 60% respondents in the AHP survey. Finally, more than 70% of respondents had more than 10 years working experience and more than 60% of

Table 6. Matrix of criteria comparison

Phases	*P1*	*P2*	*P3*	*Relative Weight*	*Rank*
P1: Input_IC	1	1.22	0.65	0.30	2
P2: Process_IC	0.82	1	0.63	0.26	3
P3: Result_IC	1.54	1.58	1	0.44	1

λmax=3.0032, CI=0.0016, RI=0.58, CR=0.0028

Table 7. Matrix of sub-criteria comparison under input phase

Input	*I_HC*	*I_InC+PC*	*I_OC*	*I_RC*	*Relative Weight*	*Rank*
I_HC: Competence of R&D personnel	1	1.41	0.32	0.88	0.18	3
I_InC+PC: Enough R&D Investment and forecasting	0.71	1	0.50	0.84	0.17	4
I_OC: Strategy fitness	3.16	2.02	1	1.93	0.43	1
I_RC: Marketing intelligence	1.14	1.19	0.52	1	0.21	2

λ_{max} =4.05, CI=0.0182, RI=0.90, CR=0.0202

respondents had stayed in the existing R&D departments for more than 5 years. The respondents are well represented by top R&D executives, and R&D working experiences.

Weight Analysis

At level 2 analysis, the principal eigenvalue of this matrix is λmax= 3.0032 with the consistency ratio CR=0.0028. The CR is less than 0.1 indicating a high consistent result of the AHP survey (Table 6). The results indicate the highest weight, criteria P3 "Result Phase" (0.44); followed by criteria P2: Process IC (0.30) and criteria P2: Process IC (0.26) (Table 6). The respondents give highest weight to the result IC, then input IC, and least to process capital. It implies that R&D managers emphasize most the result-oriented IC despite the R&D results are uncertain and risky in nature.

At level 3 of weighting four indicators in the input phase, the principal eigenvalue of this matrix is λmax= 4.05 with the consistency ratio CR=0.0202 less than 0.1, suggesting the AHP survey result is highly consistent (Table 7). Among four IC criteria in the input phase, the relative weights are highest input-organizational capital: strategic fitness (0.43), followed by the I-RC: marketing intelligence (0.21), I-HC: competence of R&D personnel (0.18) and I-IC+PC: enough R&D investment and forecasting (0.17). The results suggest that R&D mangers should emphasize strategic integration between R&D strategy and business strategy and external marketing intelligence in planning stage of R&D activities.

At level 3 of weighting 3 IC indictors in the process phase, the principal eigenvalue of this matrix is λmax= 3.0012 with the consistency ratio CR=0.0010 less than 0.1, suggesting the AHP survey result is highly consistent (Table 8). Among three IC criteria, the relative weights are highest P-PC: project execution (0.43), followed by P-IC+HC: managing creative people (0.42) and P-RC+OC: strategic external partnership (0.15). The results suggest that R&D mangers should emphasize on project execution and managing creative people to increase R&D effectiveness in the process phase.

At level 3 of weighting 4 IC indictors in the result phase, the principal eigenvalue of this matrix is λmax= 4.1529 with the consistency ratio

Table 8. Matrix of sub-criteria comparison under process phase

Process	*P_PC*	*P_InC+HC*	*P_RC+OC*	*Relative Weight*	*Rank*
P_PC: Project Execution	1	0.97	2.93	0.43	1
P_InC+HC: Managing creative people	1.03	1	2.71	0.42	2
P_RC+OC: Strategic external partnership	0.34	0.37	1	0.15	3

λ_{max} =3.0012, CI=0.0006, RI=0.58, CR=0.0010

Table 9. Matrix of sub-criteria comparison under result phase

Result phase	*R_IC*	*R_RC1*	*R_RC2+PC*	*R_HC*	*Relative Weight*	*Rank*
R_IC: R&D yield	1	1.55	0.30	0.50	0.16	4
R_RC1: Customer relationship management	0.64	1	0.63	0.85	0.18	3
R_RC2+PC: Cost effectiveness to customers	3.41	1.60	1	1.64	0.40	1
R_HC: Knowledge accumulation of R&D personnel	1.99	1.18	0.61	1	0.26	2

λ_{max} =4.1529, CI=0.0510, RI=0.90, CR=0.0566

CR=0.0010 less than 0.1, suggesting the AHP survey result is highly consistent (Table 9). Among four IC criteria, the relative weights are the highest in result-relational capital+ process capital: cost effectiveness to customers (0.40), followed by R-HC: knowledge accumulation of R&D personnel (0.26), R-RC: customer relationship management (0.18) and R-IC+OC: R&D yield. The results suggest that R&D mangers should emphasize on cost effectiveness to customers and knowledge accumulation of R&D personnel in the process phase.

Global Priority Analysis

The global weight for each phase can be calculated by multiplying the relative weight of each sub-criterion with the relative weight and then summing up those values (Table 10). Under the global priority, the top five priority indicators are: R-RC2+PC: cost effectiveness to customer (0.18), followed by I-OC: strategy fitness (0.13), R-HC: knowledge accumulation of R&D Personnel (0.11), P-PC: project execution (0.11), and P-HC+PC: managing creative people (0.11). The relatively least important indicators are: strategic external partnership (0.04), enough R&D investment and forecasting (0.05) and competence of R&D personnel (0.06) among 11 factors.

DISCUSSIONS

The chapter examined the weighting of 11 IC indicators in three R&D phases. The weight of each phase (level 2) interprets that the results phase (0.44) is the most principal one, followed by the input phase (0.30); however, the process phase (0.26) is less valued. The above results are consistent with the statement by Lee et al. (1996) that although measurement methods cover different aspects of an R&D department or an organization, most attempts to assess effectiveness relied on the goal-oriented approach, which focuses on output and outcomes. In other words, it is shown that the output/results phase is an essential focus when prioritizing R&D capabilities. This is also consistent to Han & Han (2004) study that output indicators (e.g., customer capital-related indicators and corporate performance indicator) gained

Table 10. IC-related R&D capabilities: the local and global priorities

Level	Indicators		Local Priority (Rank)	Global Priority (Rank)
1	Corporate R&D Capabilities		1	1
2	Input_IC		0.3009 (2)	0.3009
2	Process_IC		0.2619 (3)	0.2619
2	Result_IC		0.4373 (1)	0.4373
3	Input_IC	I_OC: Strategy fitness	0.4338 (1)	0.1305 (2)
3		I_RC: Marketing intelligence	0.2121 (2)	0.0638 (8)
3		I_HC: Competence of R&D personnel	0.1832 (3)	0.0551 (9)
3		I_InC+PC: Enough R&D Investment and forecasting	0.1708 (4)	0.0514 (10)
3	Process_IC	P_PC: Project Execution	0.4263 (1)	0.1116 (4)
3		P_InC+HC: Managing creative people	0.4231 (2)	0.1108 (5)
3		P_RC+OC: Strategic external partnership	0.1506 (3)	0.0394 (11)
3	Result_IC	R_RC2+PC: Cost effectiveness to customers	0.4041 (1)	0.1767 (1)
3		R_HC: Knowledge accumulation of R&D personnel	0.2557 (2)	0.1118 (3)
3		R_RC1: Customer relation management	0.1786 (3)	0.0781 (6)
3		R_InC+OC: R&D yield	0.1617 (4)	0.0707 (7)

higher weights than the input- and process- IC indicators in reporting better intangibles of the Korean mobile phone industry. The importance of result-IC was also confirmed by Egbetokun et al. (2009) that found that "satisfying customer's demand" and "improving product quality" were the top reasons for innovation in Nigerian firms. Moreover, it is aligned with the results of Bozbura et al. (2007) that ranked the most important output indicators, e.g., creating results by using knowledge among 20 HC-related indicators. Besides, the importance of output IC, our results also indicated that input IC has higher weights than process indicators. It is the same line with Bozbura et al. (2007) that has shown the most important HC indicators employee's skills (input IC), followed sharing and reporting knowledge (process IC) among 20 HC related indicators in Turkey.

The top five factors (level 3) of global priority are cost-effectiveness to customers (R-RC2+PC), strategy fitness (I_OC), knowledge accumulation of R&D personnel (R_HC), project execution (P-PC), and managing creative teams (P_InC+HC). Existing studies have illuminated that the weights of IC dimensions and indicators are firm-specific (Chen, 2009), industry-specific (Han and Han, 2004, Liu, 2006); and nation-specific (Bozbura et al., 2007). However, the weighting of IC dimensions and indicators can be intra-IC dimension comparison (human capital in Bozbura et al., 2007) and cross-dimension comparison (Chen, 2009; Han and Han, 2004). The results of the chapter were hard to compare to the previous studies because the chapter adopted a R&D-specific, both cross-phase and cross-dimension comparison in nature. Complementarily, the existing studies on key success factors influencing innovation performance tended to find the casual relationships and provided a more dynamic view among these 11 IC factors (Subramaniam and Youndt, 2005; Wu et al., 2008; Yam, et al., 2004). Yam et al.'s (1994) findings of R&D investment (similar to I_InC+PC: enough R&D investment and forecasting) has the most widespread impact on the firms' innovation, sales and product performance. Project portfolio management similar to I_OC: strategy fitness in

Figure 4. Decomposition of IC-related R&D capabilities by phase

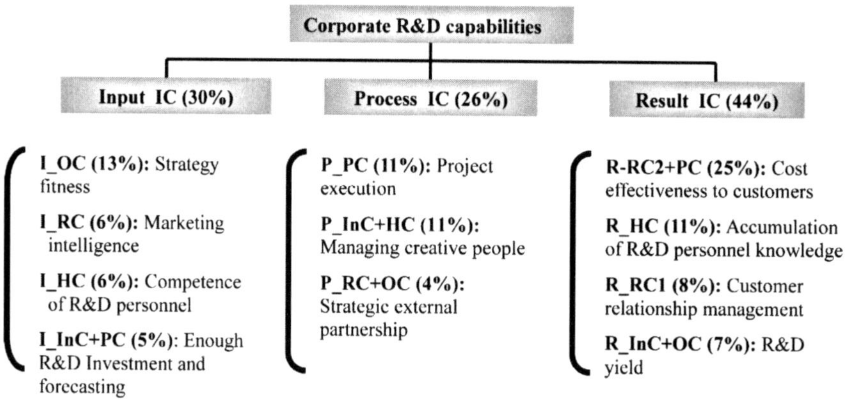

the input phase has a positive impact of the firms' innovation and product performance. Project implementation similar to P_PC: project execution has a positive impact on product performance only. Both Nahapiet and Ghoshal (1998) and Wu et al. (2008) find that dense social capital similar relational capital in the chapter plays a facilitator of creating and sharing other ICs, in turn enhances organizational advantage. Although the chapter provides the weights of IC indicators in different R&D phases, it is misunderstood to treat each R&D phase and each IC indicator independently. Good R&D capabilities must allow enhancing cross-functional coordination and integration such as high quality production, strong market orientation and technical services (Rothwell, 1992; Cooper and Kleinschmidt, 1997).

CONCLUSION

In the rise of the knowledge-based economy, intellectual capital becomes the company's most important asset. Moreover, technological innovation forms one of curial determinants for the firm's competitive advantage. Essentially, corporate R&D capability development has increasingly faced severe challenges since R&D has become more expensive, complicated and uncertain then before. To alleviate this problem, the chapter synthesized the literatures of R&D management and intellectual capital and identified forty-three elements of corporate R&D capabilities. Via exploring factors analysis, these forty-three elements were condensed into eleven factors in a R&D input-process-output framework. Further, the chapter adopted AHP method to weigh IC factors in the IPR framework.

One of major findings suggests that R&D managers regarded IC factors in the result phase are the main concern for assessing corporate R&D capabilities, followed by these factors in the input phase and the process phase (Figure 4). This indicated that the main purpose of corporate R&D capabilities is to turn creative ideas, people, and processes into useful products for customers despite lots of uncertainty and risk prevails. Secondly, the top three weighting factors are cost effectiveness to customers, strategy fitness, and knowledge accumulation of R&D personnel among 11 IC factors. Particularly, strategy fitness places the most significant role in the input phase; project execution is the primary indicator in the process phase; cost effectiveness to customers is valued most in the result phase. We concluded that weighting of IC indicators are quite R&D phase-specific. These relative weights in different phases and different dimensions could guide the emphasis of R&D managers or evaluators while they develop/evaluate corporate R&D capabilities from intellectual capital angle.

The chapter advanced weighting of IC can be R&D phase-specific. Moreover, the chapter bridged the validity gap and omission gap discussed in Adams et al. (2009). We are able to validate the IC R&D related capabilities via the AHP exercise. Moreover, we also provide some potential indicators (omission gap) to measure strategic fitness and customer cost-effectiveness. Therefore, R&D managers can have a holistic view on the IC factors in the input, process, and result phases in order to manage R&D systematically. In addition, the R&D phase-specific IC factors are able to draw right attention of R&D managers when they are situated in different R&D phases. However, it is crucial to know that these 11 IC factors work inter-dependently. Despite good R&D people and project management, a good innovative product never reached to customers without good marketing. The chapter emphasize that corporate R&D capabilities should be a cross-functional and cross-organizational (e.g., customers, suppliers and partners) in nature. Then, the relational capital itself enhances creating other human capital, process capital and innovative capital. Consequently, the future research might examine the inter-relations between these 11 factors and elaborate the dynamics of IC factors in the R&D context.

The chapter mainly surveyed senior R&D managers in high-tech manufacturing companies, especially information and communication technology industry. The Taiwanese manufacturing firms mainly concentrated on development and process innovation rather than research and radical innovation. Moreover, the respondents are mainly from larger firms because most large firms are able to establish corporate R&D centers. Consequently, the results are represented from large, ICT firms who focused incremental and process R&D capabilities. The further research might include respondents from different industries (e.g., traditional or service industries) on different types of R&D activities.

Future research could examine corporate R&D capabilities in the other countries to compare the differences and similarities between Taiwan and other countries. Besides, how these eleven IC factors dynamically interact with each other to generate, develop and maintain corporate R&D capabilities are still unknown. A casualty model to elaborate these interactions is encouraged to clarify the relationship between IC and technological innovation in the rise of knowledge based economy.

ACKNOWLEDGMENT

The earlier version of this chapter was presented in *European Conference of Intellectual Capital,* INHolland University of Applied Sciences, Haarlem, the Netherlands, 28-29, April, 2009. The authors would like to take this opportunity for thanking the financial supports from National Science Council, Taiwan (Project No.: NSC96-2416-H-007-004-MY3) and National Tsing Hua University, Taiwan (Project No.: 98N-2925E1).

REFERENCES

Abeysekera, I., & Guthrie, J. (2004). Human capital reporting in a developing nation. *The British Accounting Review, 36*, 251–268. doi:10.1016/j.bar.2004.03.004

Adams, R., Bessant, J., & Phelps, R. (2006). Innovation Management measurement: a review. *International Journal of Management Reviews, 8*(1), 21–47. doi:10.1111/j.1468-2370.2006.00119.x

Ahonen, G., & Grojer, J.-E. (2005). Social accounting in the Nordic countries: from social accounting towards accounting in a social context. In Jonsson, S., & Mouritsen, J. (Eds.), *Accounting in Scandinavia–The Northern Lights* (pp. 43–60). Copenhagen: Copenhagen Business School Press.

Benson, G., Finegold, D., & Mohrman, S. (2004). You paid for the skills, now keep them: tuition reimbursement and voluntary turnover. *Academy of Management Journal*, *47*(3), 315–333. doi:10.2307/20159584

Bontis, N. (1998). Intellectual capital: An exploratory study that develops measures and models. *Management Decision*, *36*(2), 63–76. doi:10.1108/00251749810204142

Bontis, N., Keow, W. C., & Richardson, S. (2000). Intellectual capital and business performance in Malaysian industries. *Journal of Intellectual Capital*, *1*(1), 85–100. doi:10.1108/14691930010324188

Bozbura, F. T. (2004). Measurement and application of intellectual capital in Turkey. *The Learning Organization*, *2*, 357–367. doi:10.1108/09696470410538251

Bozbura, F. T., & Beskese, A. (2005). Prioritization of organizational capital measurement indicators using fuzzy AHP. In *Proceedings of the 11th international fuzzy systems association world Congress*, (pp. 1756–1760).

Bozbura, F. T., Beskese, A., & Kahraman, C. (2007). Prioritization of human capital measurement indicators using fuzzy AHP. *Expert Systems with Applications*, *32*(4), 1100–1112. doi:10.1016/j.eswa.2006.02.006

Brown, M. G., & Svenson, R. A. (1998). Measuring R&D productivity. *Research Technology Management*, *41*(6), 30–35.

Bukh, P. N., Larsen, H. T., & Mouritsen, J. (2001). Constructing intellectual capital statements. *Scandinavian Journal of Management*, *17*(1), 87–108. doi:10.1016/S0956-5221(00)00034-8

Burgelman, R., Maidique, M. A., & Wheelright, S. C. (2004). *Strategic Management of Technology and Innovation* (pp. 8–12). New York: McGraw-Hill.

Calantone, R. J., Benedetto, C. A., & Schmidt, J. B. (1999). Using the analytic hierarchy process in new product screening. *Journal of Product Innovation Management*, *16*(1), 65–76. doi:10.1016/S0737-6782(98)00036-8

Carlucci, D., Marr, B., & Schiuma, G. (2004). The knowledge value chain—how intellectual capital impacts business performance. *International Journal of Technology Management*, *27*, 575–590. doi:10.1504/IJTM.2004.004903

Chen, H. H. (2009). Measuring intellectual capital using fuzzy analytic hierarchy process. *International Journal of Innovation and Learning*, *6*(1), 51–61. doi:10.1504/IJIL.2009.021682

Chiesa, V., Coughlan, P., & Voss, C. A. (1996). Development of a technical innovation audit. *Journal of Product Innovation Management*, *13*(2), 105–136. doi:10.1016/0737-6782(95)00109-3

Chunghwa Telecom Co. Ltd. Telecommunication Laboratories, from http://qqq.cht.com.tw/webform/index.htm

Cohen, W. M., & Levinthal, D. A. (1990). Absoprtive capability a new perspective on learning and innovation. *Administrative Science Quarterly*, *35*, 128–152. doi:10.2307/2393553

Cooper, R. G., & Kleinschmidt, E. J. (1997). Winning businesses in product development: The critical success factors. *Journal of Product Innovation Management*, *14*(2), 132–132. doi:10.1016/S0737-6782(97)83946-X

Damanpour, F. (1991). Organizational innovation: a meta-analysis of effects of determinants and moderators. *Academy of Management Journal*, *34*(3), 555–590. doi:10.2307/256406

do Rosario Cabrita, M., & Vaz, J. L. (2005). IC and value creation: Evidence from the Portuguese banking Industry. *Electronic Journal of Knowledge Management*, *4*(1), 11–20.

Drucker, P. F. (1993). *Post-capitalist society*. Oxford, UK: Butterworth-Heinemann.

Edvinsson, L. (1997). Developing IC at Skandia. *Long Range Planning, 30*(3), 366–373. doi:10.1016/S0024-6301(97)90248-X

Edvinsson, L., & Malone, M. S. (1997). *Intellectual capital: Realizing your company's true value by finding its hidden brainpower.* New York: Harper Collins.

Edvinsson, L., & Sullivan, P. H. (1996). Developing a model for managing intellectual capital. *European Management Journal, 14*(4), 356–364. doi:10.1016/0263-2373(96)00022-9

Egbetokun, A., Siyanbola, W., Sanni, M., & Olamade, O. (2009). What drive innovation? inferences from an industry wide survey in Nigeria. *International Journal of Technology Management, 45*(1/2), 123–140. doi:10.1504/IJTM.2009.021524

EIRMA. (2004). *Assessing R&D Effectiveness. Working Group Number 62 Report, Europe Industrial Research Management Association.* Paris: EIRMA.

Ellis, A. (1997). *A reason and emotion in psychotherapy.* New York: Lyle Stuart.

Erickson, G., & Jacobson, R. (1992). Gaining comparative advantage through discretionary expenditures: The returns to R&D and advertising. *Management Science, 38*(9), 1264–1279. doi:10.1287/mnsc.38.9.1264

Evangelista, R., Perani, G., Rapiti, F., & Archibugi, D. (1997). Nature and impact of innovation in manufacturing: some evidence from the Italian innovation survey. *Research Policy, 26*, 521–536. doi:10.1016/S0048-7333(97)00028-0

Feiwel, G. R. (1975). *The IC of Michael Kaleck: a study in economic theory and policy.* Knoxville, TN: The University of Tennessee Press.

Fernström, L. (2005). A marketing perspective on intellectual capital. In Marr, B. (Ed.), *Perspectives on Intellectual capital* (pp. 82–95). Burlington, MA: Elsevier Butterworth Heinemann. doi:10.1016/B978-0-7506-7799-8.50011-5

Field, A. (2005). *Discovering statistics using SPSS.* London: Sage.

Flamholtz, E. G., Bullen, M. L., & Hua, W. (2002). Human resource accounting: a historical perspective and future implications. *Management Decision, 40*, 947–954. doi:10.1108/00251740210452818

Galbraith, J. K. (1969). *A theory of price control.* Cambridge, MA: Harvard University Press.

Gorsuch, R. L. (1983). *Factor analysis.* Hillsdale, NJ: L. Erlbaum Associates.

Guthrie, J. (2000). Measuring up to change. *Financial Management*, (Dec): 11.

Hair, J. F., Black, W. C., Babin, B. J., & Anderson, R. E. (2009). *Multivariate data analysis* (7th ed.). New York: Prentice Hall.

Hall, R. (1992). The strategic analysis of intangible resources. *Strategic Management Journal, 13*(2), 135–144. doi:10.1002/smj.4250130205

Han, D., & Han, I. (2004). Prioritization and selection of intellectual capital measurement indicators using analytic hierarchy process for the mobile telecommunications industry. *Expert Systems with Applications, 26*(4), 519–527. doi:10.1016/j.eswa.2003.10.010

Hung, Y. C., & Chang, E. L. (2006). Measurement scale of intellectual capital for the information service industry. *International Journal of Services and Standards, 2*(2), 154–175. doi:10.1504/IJSS.2006.008730

Itami, H., & Roehl, T. W. (1987). *Mobilizing invisible assets.* Boston: Harvard University Press.

Kaiser, H. F. (1970). A second generation little jiffy. *Psychometrika, 35*(4), 401–415. doi:10.1007/BF02291817

Kaiser, H. F., & Rice, J. (1974). Little jiffy, Mark IV. *Educational and Psychological Measurement, 34*(1), 111–117. doi:10.1177/001316447403400115

Kaufmann, L., & Schneider, Y. (2004). Intangibles: a synthesis of current research. *Journal of Intellectual Capital, 5*, 366–388. doi:10.1108/14691930410550354

Kerssens-van Drongelen, I. C., & Cooke, A. (1997). Design principles for the development of measurement systems for research and development processes. *R & D Management, 27*(4), 345–357. doi:10.1111/1467-9310.00070

Klock, M., & Megna, P. (2000). Measuring and valuing intangible capital in the mobile communications industry. *The Quarterly Review of Economics and Finance, 40*(4), 519–532. doi:10.1016/S1062-9769(00)00055-7

Lee, M., Son, B., & Lee, H. (1996). Measuring R&D effectiveness in Korean companies. *Research Technology Management, 39*(6), 28–31.

Lev, B., & Sougiannis, T. (1996). The capitalization, amortization, and value-relevance of R&D. *Journal of Accounting and Economics, 21*(1), 107–138. doi:10.1016/0165-4101(95)00410-6

Litschka, M., Andreas, M., & Susanne, S. (2006). Measuring and analyzing intellectual assets: An integrative approach. *Journal of Intellectual Capital, 7*(2), 160–173. doi:10.1108/14691930610661836

Liu, C.-C. (2006). Developing measurements of intellectual capital in the e-learning platform industry by the analytic hierarchy process. *International Journal of Innovation and Learning, 3*(4), 374–386.

Marr, B. (2005). *Perspectives on intellectual capital: Multidisciplinary insights into management, measurement and reporting*. Boston: Elsevier.

Marr, B., Gray, D., & Neely, A. (2003). Why do firms measure their intellectual capital? *Journal of Intellectual Capital, 4*(4), 441–464. doi:10.1108/14691930310504509

Marr, B., Schiuma, G., & Neely, A. (2002). Assessing strategic knowledge assets in e-business. *International Journal of Business Performance Management, 4*, 279–295. doi:10.1504/IJBPM.2002.000119

McElroy, M. W. (2002). Social innovation capital. *Journal of Intellectual Capital, 3*(1), 30–39. doi:10.1108/14691930210412827

Mouritsen, J. (2002). Developing and managing knowledge through IC statements. *Journal of Intellectual Capital, 3*(1), 10–29. doi:10.1108/14691930210412818

Mouritsen, J., Bukh, P. N., Larsen, H. T., & Johansen, M. R. (2002). Developing and managing knowledge through intellectual capital statements. *Journal of Intellectual Capital, 3*(1), 10–29. doi:10.1108/14691930210412818

Nahapiet, J., & Ghoshal, S. (1998). Social capital, intellectual capital and the organizational advantage. *Academy of Management Review, 23*(2), 242–266. doi:10.2307/259373

O'Donnell, D. (2003). Human interaction: The critical source of intangible value. *Journal of Intellectual Capital, 4*(1), 82–99. doi:10.1108/14691930310455405

Ojanen, V., & Vuola, O. (2006). Coping with the multiple dimensions of R&D performance analysis. *International Journal of Technology Management, 33*(2), 279–290. doi:10.1504/IJTM.2006.008315

Roos, J., Roos, G., Dragonetti, N., & Edvinsson, L. (1998). *Intellectual capital: Navigation in the new business landscape*. New York: New York University Press.

Rothwell, R. (1992). Successful industrial innovation: critical factors for the 1990's. *R & D Management*, *22*(3), 221–239. doi:10.1111/j.1467-9310.1992.tb00812.x

Rousseau, D. M., & Shperling, Z. (2003). Pieces of the action: Ownership and the changing employment relationship. *Academy of Management Review*, *28*(4), 553–570. doi:10.2307/30040747

Saaty, T. L. (1980). *The analytical hierarchy process*. New York: McGraw Hill.

Saaty, T. L. (1995). *Decision making for leaders*. Pittsburgh, PA: RWS Publications.

Seetharaman, A., Low, K. L., & Saravanan, A. S. (2004). Comparative justification on intellectual capital. *Journal of Intellectual Capital*, *5*, 522–539. doi:10.1108/14691930410566997

Stewart, D. W. (1981). The application and misapplication of factor analysis in marketing research. *JMR, Journal of Marketing Research*, *18*(1), 51–62. doi:10.2307/3151313

Stewart, T. A. (1991). Brainpower. *Fortune*, *123*(11), 44–60.

Stewart, T. A. (1997). *Intellectual capital: The new wealth of organizations*. New York: Doubleday.

Subramaniam, M., & Youndt, M. A. (2005). The influence of IC on the types of innovative capabilities. *Academy of Management Journal*, *48*(3), 450–463. doi:10.2307/20159670

Sullivan, P. (2000). *Value-driven intellectual capital—How to convert intangible corporate assets into market value*. New York: Wiley.

Sveiby, K. E. (1997a). *The new organizational wealth: Managing and measuring knowledge-based assets*. San Francisco, CA: Berrett-Koehler.

Sveiby, K. E. (1997b). The intangible assets monitor. *Journal of Human Resource Costing & Accounting*, *1*, 73–97. doi:10.1108/eb029036

Teece, D. J. (1986). Profiting from technological innovation: Implications for integration, collaboration, licensing and public policy. *Research Policy*, *15*(6), 285–305. doi:10.1016/0048-7333(86)90027-2

Todtling, F., Lehner, P., & Kaufmann, A. (2009). Do different types of innovaiton rely on specific kinds of knowledge interactions? *Technovation*, 59–71. doi:10.1016/j.technovation.2008.05.002

Vuontisjarvi, T. (2006). Corporate social reporting in the European context and human resource disclosures: an analysis of Finnish companies. *Journal of Business Ethics*, *68*, 331–354. doi:10.1007/s10551-006-9094-5

Werner, B. M., & Souder, W. E. (1997). Measuring R&D performance: State of the art. *Research Technology Management*, *40*(2), 34–42.

Wu, W.-Y., Chang, M.-L., & Chen, C.-W. (2008). Promoting innovation through the accumulation of intellectual capital, social capital and entrepreneurial orientation. *R & D Management*, *38*(3), 265–277. doi:10.1111/j.1467-9310.2008.00512.x

Yam, R., Guan, J. C., Pun, K. F., & Tang, E. (2004). An audit of technological innovation capabilities in Chinese firms: some empirical findings in Beijing, China. *Research Policy*, *33*, 1123–1140. doi:10.1016/j.respol.2004.05.004

KEY TERMS AND DEFINITIONS

Intellectual Capital: The possession of the knowledge, applied experience and organizational technology can be put to use to create wealth.

Corporate R&D Capabilities: The capabilities of a corporate focus on the activities performing research and development.

Human Capital: The combined knowledge, skill, innovation, and ability of a company's individual employees. It is also includes the company's values, culture, and philosophy.

Structural Capital: The hardware, software, databases, organizational structure, patents, trademarks, and everything else of organizational capability that supports the employees' productivity.

Relational Capital: The relationships between internal and external stakeholders. It is the knowledge embedded in organizational relationships with customers, suppliers, stakeholders, strategic alliance partners, etc.

Organizational Capital: It is the company's investment in systems, tools, and operating philosophy that speeds the flow of knowledge through the organization.

Innovational Capital: This refers to the renewal capability and the results of innovation in the form of protected commercial rights, intellectual property, and other intangible assets and talents used to create and rapidly bring to market new products and services.

Process Capital: This is those work processes, technologies (such as ISO 9000), and employee programs that augment and enhance the efficiency of manufacturing or the delivery of services.

Input-Process-Result Model: The system includes contents of input, processing system, output of the processing systems, receiving systems and outcome, as well as their linkages to each other.

Analytical Hierarchy Process: AHP is especially suitable for complex decisions which involve the comparison of decision elements which are difficult to quantify. It is based on the assumption that when faced with a complex decision the natural human reaction is to cluster the decision elements according to their common characteristics.

Section 4
Social Capital, Intellectual Capital, and Technological Innovation

Chapter 11
Co-Evolving Relationships and Innovation Dynamics

Charles Ehin
Westminster College, USA

ABSTRACT

This chapter takes a critical look at the interplay of three key engagement elements: transactions, conversations, and relationships as they relate to innovation dynamics. It further pinpoints the importance of tracking four key factors—self-organization, tacit knowledge, social capital, and human nature—for the development of innovation rich social relationships. Most significantly, the author identifies a new dynamic organizational component: the shared-access domain or organizational sweet spot, and its significance to the innovative capacity of an enterprise. Research from such diverse fields as anthropology, evolutionary psychology, social neuroscience, and complex adaptive systems are used in an attempt to show commonalities in these disciplines in determining the effects of various organizational contexts/ecologies on the expansion or contraction of the shared-access domain. In essence, the chapter identifies methodologies and interrelated multidisciplinary factors for managing, or rather "unmanaging," knowledge professionals.

INTRODUCTION

A recent Modern Survey (2008) of the US workforce points out that employee engagement in the US is deteriorating even as the financial crisis worsens. The November 2008 survey results were as follows:

- August '07: Disengaged 17%, Under Engaged 41%, Moderately Engaged 30%, Fully Engaged 12%.
- August '08: Disengaged 21%, Under Engaged 40%, Moderately Engaged 27%, Fully Engaged 12%.

Overall, disengaged workers are estimated to cost the US economy more than $300 billion annually.

DOI: 10.4018/978-1-61520-875-3.ch011

The obvious question is, "Why are so many workers in the United States actively disengaged from their jobs?" Is it caused by the decreasing relative wages of the middle class workers, promises of bonuses that go unpaid, results of employee surveys that produce no results, antagonistic work environments or simply a lack of purpose? All these issues certainly have an impact. I suggest, however, that the problem is much more fundamental than that.

Doc Searls (2007), the Senior Editor of the *Linux Journal*, I believe best describes the intricacies and dynamics of the predicament. He states:

Think of markets as three overlapping circles: Transaction, Conversation and Relationship. Our financial system is Transaction run amok. Metastasized. Optimized at all costs. Impoverished in the Conversation department, and dismissive of Relationship entirely. We've been systematically eliminating Relationship for decades, excluding, devaluing and controlling human interaction wherever possible, to maximize efficiency and mechanization.

What Searls makes clear is that we have and continue to ignore the human factors in our organizations and business networks. Thus, for practical purposes, we are ignoring the suggestions and innovative ideas of almost 25 million people in the US alone. That is an unimaginable waste of any nation's resources. What needs to be fully recognized is that the "hidden assets" of social systems are deeply embedded within the relationships of the people we work and live with.

Therefore, we need to look beyond transactions and conversations within and external to an organization's constituents. That is because we are dealing with the talents and skills of human beings who are "intrinsically social creatures." Once we are able to reach that conclusion it becomes easier to comprehend that relationships are the essence of life, work and innovation dynamics.

BACKGROUND

For the better part of the past two decades I have been investigating the fundamental interplay of a number of individual and group associated behavioral factors and how they relate to the expansion or contraction of the innovative capacity of an organization or a social network. I began by first taking a more thorough look at anthropology and paleontology.

What I discovered was that our hunter-gatherer ancestors had gotten along quite well for 99 percent of the roughly 200,000 year existence of our species (Homo sapiens) without the use of any rigid hierarchical social structures (Ehin, 1995). Thus, top-down administrative frameworks are "relative" newcomers to the societal scene. Hierarchical organizational structures essentially began to evolve around 12,000 B.C. with the advent of agriculture and the domestication of animals. The irony is that this significant fact is seldom talked about in business schools or in company boardrooms.

Next I delved into sociobiology and evolutionary psychology (Ehin, 2000). That was followed by the examination of many of the latest findings in molecular biology, social neuroscience, and complex adaptive systems (Ehin, 2005 and 2009). In general what has become quite clear to me and many others is that biological entities, including humans, cannot and should not be fully controlled. They are emergent and constantly evolving complex adaptive systems. In fact, we now have solid evidence that attempts to control people stifle innovation (Lehrer, 2008).

Hence, for innovation to thrive among people they need to be immersed in flexible biophysical and social environments. When it comes to developing productive social systems we need to be less dependent on Industrial Age cause-and-affect thinking and, instead, begin to rely more on the dynamic principles of self-organizing.

Consequently, it should not come as a great surprise that traditional management concepts seldom work any longer especially when it comes to knowledge workers. That is mainly due to the continued use of cause-and-affect theoretical constructs. People are not machines. They are organic self-organizing entities from their DNA molecules to their interactions with the external world.

More specifically, studies of breakthroughs in neuroscience by Rock and Schwartz (2006) lead them to the following conclusion related to organizational transformations:

- Change is pain. Organizational change is unexpectedly difficult because it provokes sensations of physiological discomfort.
- Behaviorism doesn't work. Change efforts based on incentive and threat (the carrot and the stick) rarely succeed in the long run.
- Humanism is overrated. In practice, the conventional empathic approach of connection and persuasion doesn't sufficiently engage people.
- Focus is power. The act of paying attention creates chemical and physical changes in the brain.
- Expectation shapes reality. People's preconceptions have a significant impact on what they perceive.
- Attention density shapes identity. Repeated, purposeful, and focused attention can lead to long-lasting personal evolution.

Evolutionary psychology and social neuroscience are converging (Cacioppo and Patrick, 2008). Thus, if we want to expand the innovative capacities of our organizations we need to pay much closer attention to human nature. Reinventing traditional methodologies will not help us advance any further even if they may have given us some success in the past. The new science of the brain and DNA is helping to rewrite not only the origins but also the innate behavior of our kind. That's where our attention should also be from a social systems perspective.

With that in mind I will now present some of the theoretical constructs that I have developed. My primary focus will be on the importance of co-evolving relationships and the "shared-access domain" (Ehin, 2009) of an organization in the realm of intellectual capital generation.

THE GRAND ORGANIZATIONAL MYTH

The main question that needs to be answered is, "How do we begin to solve the huge worker disengagement crisis?" This problem is not only plaguing the United States but also every other industrialized nation. I suggest the best place to start is to first identify the greatest management myth ever conceived and perpetuated on humans. It started with the advent of large societies and their respective armies. The myth, however, gained peak support during the Industrial Revolution and still is high on the "things to do list" of many administrators and management consultants. Ironically, as I pointed out above, there is no anthropological evidence that our kind used such structures prior to the beginning of the Agricultural Revolution.

The Grand Organizational Myth, as depicted by Figure 1, is the misnomer that organizations can actually be managed by such rigid mechanical top-down structures. Experience and research is slowly beginning to peel back the facade and show how social systems really function. Social systems are organic. They are composed of very vibrant constantly evolving self-regulating networks. They don't look or perform like "machines." In fact, social systems work around mechanical structures whether devised by heads of prison administrators or managers set on running very "strictly disciplined" organizations. Most sage individuals have known that for ages. Only recently, however, have we begun to accurately recognize how our social systems actually function (Allee, 2003). Figure 2 is a realistic portrayal of how work gets accomplished in organizations.

The obvious question is, "Why don't social aggregations conform to formal structures, directives or policies precisely as demanded?" The simple answer is that it's physiologically and psychologically impossible for any organic entity (person or group) to respond to external demands (such as a directive or a certain work context) exactly as intended. The reason for that is quite straight forward. Of all the billions of people that have lived or continue to live on this planet no two people have ever been genetically identical. Even identical twins are not completely identical. Therefore, no individual or group can look at the same directive (or anything else) and see it exactly the same way as another individual or group. In addition, people innately hook up with people and social networks they believe have the same general goals and interests that they are intent on pursuing. Hence, no matter how determined some managers are in trying to control people it's, in reality, an impossible task.

SOCIAL ENGAGEMENT DYNAMICS

As Doc Searls has indicated transactions, conversations and relationships are the mainstays of social engagement. In Figure 3 I illustrate the close dynamic relationships of the three engagement components and suggest the model can be used to analyze social engagement dynamics at any level.

Figure 1. Organizational myth

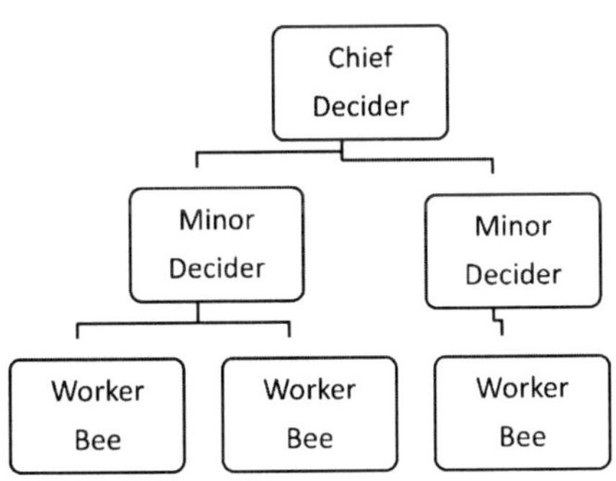

That is, the model is appropriate for not only examining the interactions of two individuals, but also small teams, entire organizations, and extended worldwide networks. You can even apply this model to show how individual identities develop and are maintained. My focus will primarily be on the relationship aspect of the model, the most widely ignored engagement component in the majority of organizational schemes.

Transactions are exchanges of tangible or intangible items between two or more parties. They can include monetary attributes such as stocks, tangibles such as machinery, intangibles such as patents, or all three factors.

The point to remember is that no matter what a transaction includes, it's always explicit whether there's a signed agreement or not. The process is easily traced because an event or several events must take place before a transaction is completed.

The same is true for conversations. At least two people, by whatever means, have to exchange ideas or stories for a conversation to take place. Transactions and conversations are both explicit, even under highly informal circumstances. In other words, they are specific, definable, and fully developed. They also have an explicit value whether economic, intellectual, emotional, or spiritual.

Not so with relationships. Because relationships are based on spontaneity and intimacy, two commodities that cannot be traced, relationships are implicit. Unlike transactions and conversations, they are not solid or definable. By their very nature they exist because of unexpressed

Figure 2. Organizational reality

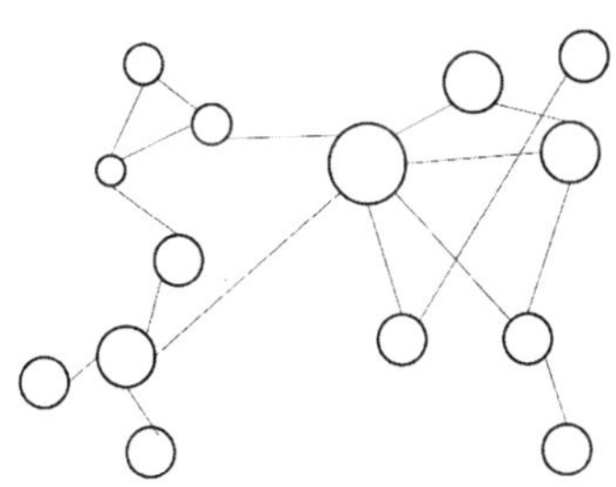

Figure 3. Engagement Dynamics (Adopted from Ehin, 2009).

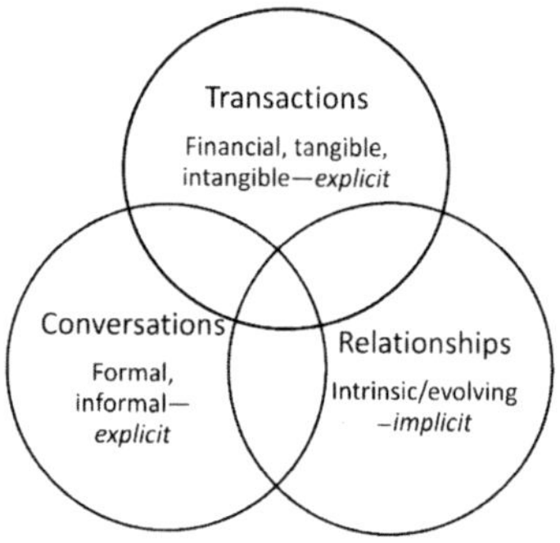

agreement or affinity which is difficult, if not impossible, to define. This implicit nature is a major difference between relationships and the other two features.

Relationships also can have intrinsic value in the same ways as transactions and conversations. However, since relationships are implicit, the values derived from them develop over time and are not necessarily the initial foundations for the associations.

Individual and Group Identity Dynamics

Relationships and individual identities are constantly evolving for better or for worse, depending on the biophysical and social contexts. It's an unplanned, self-organizing process between two or more parties where the outcomes are unpredictable. Given their implicit, unpredictable nature, relationshipscan be influenced, but not controlled, by third parties or varying environmental contexts.

Most significantly, people either individually or in groups are constantly striving to both uphold and maintain their identities in every social situation. Let's first take a closer look at the constant struggle for identity at the individual level. The most well known authority on this subject, Harrison White (2008), suggests that we pay close attention to the following concerning identity dynamics:

- Humans have an intrinsic need to express our identities and differences in social settings.
- Interactions trigger identity control efforts.
- Identity control is seldom about domination.
- Finding a footing in social situation is the aim of our identity control efforts.
- Established footings bring orientation among people's identities.

What this means is that in any social situation we are usually most concerned about our individual identities above everything else. Also, our actions and reactions are not always made consciously. They are emergent and serendipitous depending on the circumstances. For example, our efforts to maintain our identities are quite different in a family, work or a country club setting.

Similar dynamics come into play at a group or organizational level. Extensive research by Stacey, Griffin and Shaw (2000) indicates the following about group identity dynamics:

- We have an inherent need to express our identities and differences as a group.
- A group's identity emerges from the relationships of its members, not an edict from management.
- Identities and differences emerge through self-organization or reciprocalinteractions.

I'm going to define self-organization shortly but first I'll make two critical observations about group identity dynamics. As pointed out above, group identities emerge like individual identities. This process cannot be controlled by managers or third parties. In addition, group members are usually more concerned about their own and the group's identity than formal organizational goals and objectives. What that tells us very clearly is that if people are not given sufficient opportunity by an organization to develop a firm footing for their identities they will become disengaged and find support elsewhere. Thus, it's very critical that people are immersed in a supportive environment. I'll talk about that later in the chapter.

The key point to remember is that because relationships and identities arise naturally, they are *emergent*. Since organizations are generally populated by at least two or more individuals, relationships will arise with all their delightful volatility and variations, no matter what type of organization you consider.

Relationships are the informal social fabric of every organization and network whether we are dealing with a neighborhood book club, the office grapevine, or the United Nations. Thus, creative enterprises will learn how to openly support the development of these constantly evolving emergent systems rather than ignore them or, worse, push them underground.

THE TWO SIDES OF ORGANIZATIONS

Having peeled back the organizational façade it's now fitting to make a broad and fundamental statement about all organic systems. "All life forms are self-organizing systems by design, down to their individual cells and molecules" (Ehin, 2005). Essentially, what this means is that the self-organizing process acts as a catalyst in all the mental and physical activities that we are involved in. If that was not the case our kind probably would not have advanced as far as we have.

This leads us to the two sides of organizations. As portrayed by Figure 4, organizational life can typically be separated into two spheres of influence: management and unmanagement.

The domain of management is an artificial construct usually devised by one or a handful of top executives or founders of an enterprise. Its intent is to control and "explicitly coordinate" the activities of people in the pursuit of organizational goals and objectives. However, as stipulated before, all life forms are self-organizing systems by design, down to their individual cells and molecules. That is, people can be influenced but not controlled. Therefore, unless people are eliminated completely from a formal system (it's fully automated) members will seldom interact with one another exactly as planned.

This brings into play the dynamics of the "unmanagement" realm. Here emergent order and "implicit coordination" rule since there are no bosses or formal rules on this side of the organization. In fact, if management attempts to meddle with the informal domain in any way the informal network(s) will immediately morph into a different configuration like an amoeba.

What this means is that each member of the organization will interpret, from their own perspective, as to how best to work and survive within the parameters of the prescribed structure. Further, people will also discuss their options with other individuals and link up with members they believe have the same general ideas on how to prosper within the boundaries of the formal organization. Simply put, it's an organic unmanageable process.

One can now begin to discern the primary interfaces that come into play on each side of the organization. On the management side the emphasis is primarily on explicit coordination via official directives and policies. Also, here transactions and conversations are generally formal especially among administrators and employees. Additionally, relationships are primarily confined to the people managing the enterprise.

Conversely, in the realm of "unmanagement" self-organization is the catalyst bringing about implicit coordination among the emergent networks situated both within and external to the organization. Further, transactions and conversations are mostly informal. What really stands out,

Figure 4. Two sides of organizations

Management Realm	*UnManagement Realm*
•Artificial/Formal	•Organic/Informal
•Designed Order	•Emergent Order
•Explicit Coordination	•Implicit Coordination

however, on this side of the organization are the intimate and constantly evolving relationships. These relationships are the hidden assets that can make or break an enterprise. They are mostly dependent on the organizational context within which people have to work. This will be covered later. The end result is that on the formal side of the organization "compliance" is given primary emphasis whereas in the unmanagement domain "commitment" is the key binding element.

INNOVATION DYNAMICS

In order to explore and discover the intricacies of innovation dynamics and how to enhance its effectiveness we need to drill even deeper into the relationship component. Thus, I will next identify three critical factors and a catalyst that make constructive relationships so powerful in developing and advancing new ideas. I will then suggest what types of organizational contexts boost relationships by increasing people's attention and influence. Such context will also help narrow the worker disengagement gap.

Self-Organization

Fundamentally, all life forms are not only self-organizing systems by design, but self-organization also constitutes the primary process by which all organic entities interact with one another. Typical examples of this dynamic, from a relationship perspective, are the informal social systems that are a vital component of any organizational framework. Fundamentally, self-organization includes the following features:

- An entity's intrinsic ability to change itself as it interacts with its environment and strives to maintain its identity.
- Interactions that produce self-referential patterns without the need to be designed or managed.
- Evolving patterns that are both sustained and transformed by spontaneous interactions.
- Creativity and destruction are part of the emergent process as are attraction and repulsion.
- All life forms are self-organizing systems by design down to their DNA.

Human Nature

Human nature is important from the standpoint of our innate behavioral tendencies. Humans are not born with a blank slate for a mind. Instead, we arrive with all the basic rudiments of our mental circuitry in place ready to act in response to our immediate environment. At the same time, we are able to learn from our experiences. Thus, humans are equipped not only with instincts, but also with much broader innate drives or predisposed genetic tendencies such as concern for status and for affiliation. This means that our behavior is "influenced" by our genes rather than genetically determined and that we do have free will.

Innate drives fall into two fundamental categories (Stevens and Price, 1996; Ehin, 2000): *a set of self-centered drives* (e.g. concern for control, rank, status, territory, possessions, savagery and bloodlust) and *a set of other-centered drives* (e.g. concern for attachments, affiliation, altruism, care giving, care receiving, morality and empathy). People seem to function best in a constructive social context where both categories of drives can be expressed in a balanced manner. Such an environment consists of individuals who express a mix of moderate self-interest and outward-reaching altruism. This context promotes implicit coordination and is vital for overt sharing of tacit knowledge (discussed next).

Further, ample research indicated that humans are physiologically limited in developing and maintaining mutually beneficial *voluntary* collaborative relationships in groups of more than 150 people (Dunbar, 1996). In larger collectives, relationships become fragmented, ties of common interest cannot be properly sustained, and hierarchical structures begin to take shape. Thus, from a human nature perspective, small size is vital for the development of supportive environmental contexts where emergent relationships can flourish most effectively.

This does not mean, however, that meaningful collaborative relationships cannot be nurtured within large organizations and extended virtual networks. Enterprises consisting of more than 150 people can be segmented into small mutually interdependent groups that are also well connected to other relatively autonomous parts of an organization.

Within extended virtual networks there is a key factor that enhances collaborative relationships—ample opportunities for periodic face-to-face meetings among the participants. Such occasions provide a constructive social context where relationships become more intimate and high levels of mutual trust are developed. As Cacioppo and Patrick (2008) point out, "...most face-to-face encounters in real life allow us to communicate through even more subliminal cues—body chemistry, body language, action semantics, mimicry—in addition to words and gestures. Once again, the mind that seeks to connect is first about the body, and leaving the body behind can make human connections less satisfying."

Tacit Knowledge

Tacit knowledge is the second key factor of innovation dynamics. Michael Polanyi (1958) is credited with originating the concept. Essentially, tacit knowledge encompasses ideas and abstractions at the individual level. More specifically, it is implicit knowledge that is:

- Grounded in personal experience and innate predispositions.
- Carried by people in their minds that is difficult to access or share.
- Difficult to transfer to others without extensive personal contact and trust.
- Based on habits and culture that we do not recognize in ourselves.
- Stored in a different area of the brain than explicit knowledge.
- The wellspring of new codified or explicit knowledge.

Tacit or unrelated knowledge comes to the fore serendipitously and becomes explicit as individuals or small groups confront new or unanticipated situations. Consequently, tacit knowledge is a dynamic resource. Hence, although relatively stable, implicit knowledge continues to be shaped by our interactions with our immediate surroundings and other people.

Thus, tacit knowledge must be allowed to "emerge" through voluntary collaboration or self-organization. People are seldom aware of exactly what unrelated knowledge they possess until confronted with a problem or an opportunity. Therefore, in order for tacit knowledge to properly emerge, people must first be surrounded by a supportive environment.

Threats, for example, create negative emotions that, by necessity, narrow thought patterns. People threatened by the loss of their jobs, a bullying boss, unreasonable deadlines and so on, innately narrow their thought patterns to avoid or eliminate these negative emotions. As a result, such individuals devote little or no time to engage their minds more expansively and resourcefully in search of new ideas or opportunities. They are usually part of the 25 million workers in the US who are disengaged from their jobs.

Social Capital

Social capital is the third major dynamic in the idea generation process. Robert Putnam (2000) the author of the book, *Bowling Alone: The Collapse and Revival of American Community*, is credited with popularizing the concept. More recently published research has further refined the theory (Adler and Seok-Woo, 2002).

Without high levels of social capital, not much in the way of productive work, and especially new knowledge creation, can take place in an organization. Social capital has the following general attributes:

- It is a self-initiated drive by people to promote collective social interests.
- It is the goodwill available to individuals and groups within informal social networks.
- It provides valuable information, influence, and cohesion to the emergent network members.
- It increases in value with use.
- It cannot be bought, sold, traded or managed.

A social entity, if it is to function as a closely-knit group, should support the development of an array of positive personal relationships among its members. Therefore, an organization should first cultivate a distinctive work environment wherein its constituents are able to interact comfortably with each other with minimal personal reservations. We should bear in mind that few things in organizations are accomplished by strictly following formal directives. Hence, without sufficient social capital, an enterprise is, for all intents and purposes, merely a collection of hired hands waiting for instructions from the bosses.

In essence, social capital is based on altruistic and interdependent relationships formed by people over time. It is the lifeblood of tightly knit informal networks which are a key source for information, influential connections, and friendships needed by all of us for meaningful existence.

Figure 5 portrays the four key elements of the innovation process from a comprehensive interactive perspective. What should be immediately apparent is that the dynamics of neither the individual elements nor the system as a whole can be managed in the traditional sense.

What needs to take place in order to breathe life into such a complex vibrant structure is three-fold: selection and assembling of a group of no more than 150 people with the right talents, knowledge and social skills for a particular venture; allowing these folks to select worthy and challenging problems to solve or opportunities to pursue; and help the group develop a constructive social context where they can practice a mix of moderate self-interests and outward-reaching unselfishness leading to high levels of implicit coordination.

THE ORGANIZATIONAL SWEET SPOT

We should remember that the more people are given a voice and implicit control in the management realm, the more they will understand and respond positively to formal organizational goals and initiatives. Also in the process, the organizational informal networks will begin to function more in the open and start making appropriate connections with other emergent groups.

Most importantly, under the right conditions the informal components will begin to overlap more and more with the formal elements of an

Figure 5. Co-evolving relationships

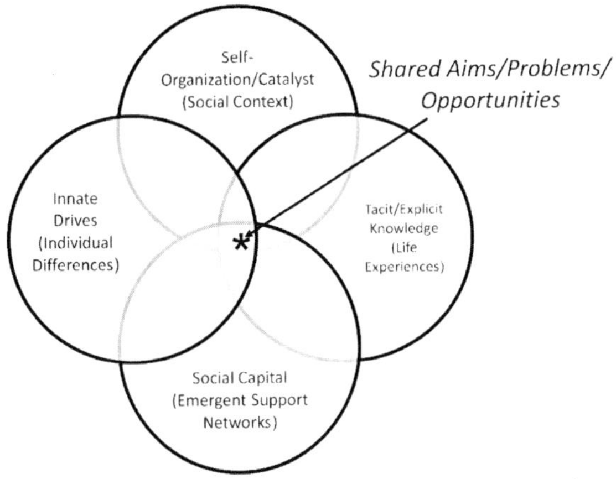

organization. This overlap is a very desirable state for any enterprise. That spot, in essence, represents the area where the formal and informal systems of an organization have reached "a meeting of the minds" over the fundamental goals, policies and processes of an organization.

What is particularly noteworthy about this agreement is that it is not reached through any sort of formal negotiations. Rather, it is emergent. It is a natural outgrowth of day-to-day interactions, or self-organization by the people representing both the management and the "un-management" realms of a venture.

In my new book, *The Organizational Sweet Spot: Engaging the Innovative Dynamics of Your Social Networks* (2009), I have labeled this place of common agreement as the "shared-access domain," as shown in Figure 6. More fittingly, the overlapping area in the model is best referred to as theSweet spot:definedorganizational sweet spot."

It's important to remember that the two systems do not completely merge and become one, even though members of both the formal and informal camps participate. Instead, in the shared-access domain, implicit coordination based on self-organization is the predominant operating mode.

Ideally the two spheres should overlap totally making the entire organization the shared-access domain. In reality, a complete overlap is not possible, even in smaller organizations of fewer than 150 members for two reasons.

First, all organizations, large and small, need some type of a formally recognized framework for internal and especially external communications to effectively coordinate their overall activities.

These communication frameworks do not need to be hierarchical or linear. All that's needed is for everyone involved to understand and adhere to agreed upon communication rules. The bottom line is that all ventures need uniformly recognized policies and procedures to facilitate effective transactions and conversations. This is why the formal side of an organization can never be entirely abolished and, therefore, due to its nature, can never be wholly incorporated into the shared-access domain.

Figure 6. Organizational sweet spot (Adapted from Ehin, 2009)

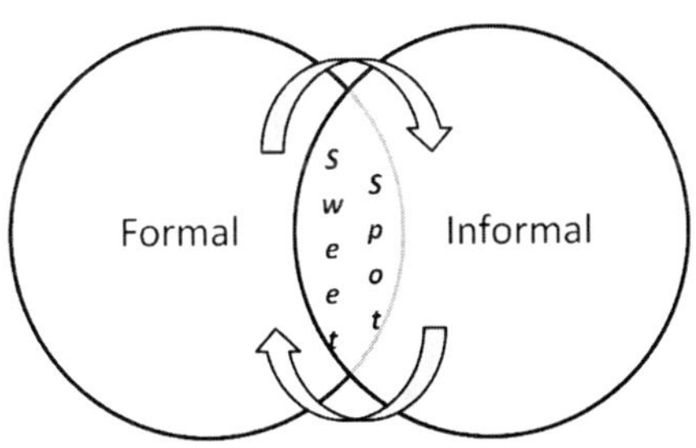

As mentioned earlier, when possible, it's best to encourage and facilitate face-to-face interactions since they better promote co-dependent long-term relationships. Organizations or sub-divisions within organizations that have memberships of fewer than 150 command the advantage in this respect, since face-to-face interactions are more feasible in smaller groups.

Second, every organization, whether it's a book club or a city government, has an informal social system of various emergent networks. However, it is also true that not all informal network members will choose to participate in the shared-access domain activities all the time. This means that at any given moment there will always be some members on the informal side of a venture who are outside the sweet spot.

By taking another look at Figure 6 we can now begin to visualize why the formal and informal circles can never fully overlap. The part of the formal organization which holds all the operationally needed rules and procedures will always remain outside the shared-access domain (to the left in the figure). No matter how flexible or accommodating the organization may be, the formal framework simply cannot be completely incorporated into the sweet spot.

Similarly, a part of the informal organization will also linger outside and to the right of the shared-access domain as depicted in the model. That is, no matter how supportive a venture's social context may be in facilitating self-organization throughout an enterprise, not all emergent network members will choose to participate in the productive activities of the sweet spot.

A certain number of people will always be disengaged from work, busying themselves with such things as honing mutually supportive relationships, establishing a firm footing for their identities, or simply having a bad day. That, of course, prevents the entire informal side of an organization from being totally incorporated into the shared-access domain.

ORGANIZATIONAL CONTEXT AND THE SWEET SPOT

I will next describe two very broad organizational context categories and suggest which option is best suited for supporting the expansion of the organizational sweet spot. Several years ago, out of frustration with the standard definitions, I decided to be more definitive by placing formal organizational structures into just two general categories: *controlled access systems* and *shared-access systems* (Ehin, 2000).

In this classification scheme, a controlled-access system, whether tall or flat, is an organizational framework wherein one individual or a very limited number of people exclusively control access to all major resources including the workforce. All other members of the organization must first get approval from these executives before any of the assets can be used or invested.

In a controlled-access system position power is the predominant force behind all key decisions; hence, "open" self-organizing arrangements are not encouraged or valued, limiting the development of high levels of social capital and the sharing of tacit and explicit knowledge. In other words, compliance instead of commitment is prized in such organizations.

Conversely, in a shared-access system all organizational members have considerable autonomy in decision-making and in resource allocations including hiring and firing of people. In a shared-access system, expert power instead of position power dominates. Thus, major emphasis is placed on situational leadership, open book management, and self-organization in solving problems or in pursuing opportunities. Such an arrangement clearly benefits from expanding social capital and the continuous sharing of tacit and explicit knowledge. Here, personal commitment rather than compliance is the dominant factor for success. That, of course, helps to expand the shared-access domain of the organization.

Basically, a shared-access system minimizes or avoids many of the knowledge sharing "killers" that are prevalent in a traditional organization. For example, knowledge workers in a controlled-access setting are continuously faced with some of the following or similar problems:

- I would like to download a free Web resource which will help me perform my job better, but the IT Department will not allow me to do that.
- I would like to work cross-domain, in an interdisciplinary way, with colleagues in another department, but my manager refuses to give me permission to do so.
- I would like to have access to cost information that pertains to the resources that I am currently using so that I can make more judicious use of them. The Accounting Department, however, has informed me that such information is available on a need-to-know basis only to managers and I am not a manager.

In a controlled-access system, the upshot, from a human nature perspective, is that people are more concerned with their own welfare instead of their co-workers or the organization as a whole. In such an environment, more of the self-centered innate drives are continuously expressed than the other-centered drives. In a shared-access system, the opposite is true. Concern for self and others as well as the organization as a whole is well balanced. In addition, there is great stress placed on interdependence rather than competition. Hence, the genetic predispositions are expressed in a balanced fashion.

Figure 7 recaps the major attributes of a controlled-access system. Leadership, for instance, is based on position or resource holding potential. This means that resource management is not widely distributed among people within such an enterprise. Consequently, most members of the organization need to justify their resource requirements to management prior to their access and use. This can be a very cumbersome and time consuming process delaying suitable responses in high-speed transactions.

Further, the dynamics of the system are driven by self-interest, domination over others, compliance and explicit coordination. As a result there are high levels of psychological tension throughout the organization as members compete for position power and favorable personal assessment.

Also, since self-interests are openly or covertly emphasized less effort is placed on advancing the goals and objectives of the enterprise as a whole. Finally, since informal networks are not recognized and forced to function clandestinely they may not only give little support to the formal organizational programs but even actively obstruct their progress. Clearly, in such a context a considerable segment of the workforce becomes actively disengaged and the organizational sweet spot has little chance of expanding.

A shared-access system epitomizes an engaged organizational context as illustrated by Figure 8. Here leadership is based on talent and expertise or social attention holding potential. It's the complete opposite of leadership grounded in resource holding power, domination and intimidation (actual or perceived). A great deal of value is placed on constructive relationships enhancing collaboration

Figure 7. Controlled-access system attributes

- Leadership based on position power (resource holding power)
- Self-interest
- Domination
- Compliance (explicit coordination)
- High psychological tension
- Unbalanced expression of innate drives
- Informal networks not recognized (pushed underground)
- No limit on size

and commitment leading to sustained high levels of implicit coordination. In addition, informal networks are not only recognized but are actively supported. As a result, psychological tensions are also reduced to minimal levels. It's a constructive context allowing members to express a mix of moderate self-interest and outward-reaching unselfishness and in the process the organizational sweet spot continues to expand.

In the final analysis, shared-access success factors can best be summarized as follows:

- Autonomy to experiment with various options and to join multiple teams.
- Freedom to access and manage resources.
- Resolve to share information.
- Emphasis on expanding knowledge.
- Willingness to take accountable risk.
- Embracing constant and discontinuous change.
- Practicing situational leadership

FUTURE RESEARCH DIRECTIONS

The future of organizational and social network research needs to be firmly founded on the latest discoveries about human nature rather than the further refinement of Industrial Age "mechanistic" processes. There is nothing wrong with applying "refined Taylorism" to non-human systems. However, continuing to do the same when it comes to social entities makes no sense at all.

Figure 8. Shared-access system attributes

- Leadership based on expertise (social attention holding potential)
- Collaboration (attention/influence)
- Commitment (implicit coordination)
- Low psychological tension
- Balanced expression of innate drives
- Informal networks recognized and supported
- Limited size

From this point on our research and organizational change efforts needs to be based on the new knowledge emanating from evolutionary psychology and social neuroscience especially now that both fields are merging. Organizations and social networks are complex adaptive systems founded on self-organizing principles not "machine logic." We need to become comfortable with visualizing and emphasizing the dynamics of social systems from a new perspective that places major emphasis on the self-referential and reflective nature of humans. That includes our innate predispositions to relate to one another in all sorts of unpredictable ways.

We should finally admit that change literally causes pain and is not just something ephemeral or purely psychological that can be accepted almost instantly if only we wanted to. Before change can be brought about we first need time and considerable effort to allow some of the wiring of our neural networks to reconfigure themselves. Hence, we desperately need to leave behaviorism and conventional humanistic approaches of connection and persuasion behind. Instead, we should focus on developing organizational and network ecologies founded on self-organizing principles that give people ample opportunities to find their own way through co-evolving mutually supportive relationships with people around them. Cacioppo and Patrick astutely suggest the types of organizational frameworks we should pursue:

When we feel safe and secure within our social connections, we can move along free of bias and unwarranted expectations. Relaxed and attentive, we can be in sync with the movements of others. With no expectations that we'll be excluded our defensive, fight-or-flight mechanisms are not on the ready alert. Free from all this distraction, we are able to detect more reliably whether any budding connection is promising or an invitation to the blues. Living more calmly in the moment, we can make better choices. Which has the added benefit of helping improve our larger social environment over time. (2008, p. 222)

The bottom line is that we need to develop organizational constructs that are in tune with human nature. Swimming against the stream makes very little sense. After all we have left the Industrial Age and entered the Knowledge Age. Knowledge workers are a different breed who cannot be managed in the traditional sense. Thus, our emphasis should be on learning how to "unmanage" rather than manage these professionals.

CONCLUSION

Whether we like to admit it or not, all activities and interactions between people are governed by the principles of self-organization. Therefore, we need to place greater emphasis on developing constructive social contexts that support the dynamics that allow people to establish meaningful relationships and in the process expand an organization's sweet spot. Continuing to almost exclusively focus on financial transactions while limiting conversations and ignoring relationships makes little sense.

Complex adaptive systems' thinking is vital today in order to properly address never before anticipated problems or new opportunities. Thus, where conventional approaches consistently fail to bring success, more pragmatic approaches need to be found based on the latest multi-disciplinary research in such fields as evolutionary psychology and social neuroscience. Therefore, it's prudent that we place greater emphasis in our organizations on the shared-access ecology success factors such as individual autonomy for responsible experimentation, a culture of sharing information and practicing situational leadership.

Multiple perspectives consistently lead to fresh and more insightful solutions than simply putting new faces on old failed concepts. In the end, it all comes down to dealing realistically with the unavoidable realm of "social emergence" and supporting the expansion of "organizational sweet spots." We certainly do not want to continue to perpetuate the disengagement affliction that keeps depriving us, individually and as a society, from vast reservoirs of innovative and positive relationships.

REFERENCES

Adler, P. S., & Seok-Woo, K. (2002). Social Capital: Prospects for a New Concept. *Academy of Management Review*, *27*, 17–40. doi:10.2307/4134367

Allee, V. (2003). *The Future of Knowledge*. New York: Butterworth-Heinemann.

Baker, S. (2008). *The Numerati*. New York: Houghton Mifflin Company.

Cacioppo, J. T., & Patrick, W. (2008). *Loneliness*. New York: W. W. Norton and Company.

Dunbar, R. (1996). *Grooming, Gossip, and the Evolution of Language*. Cambridge: Harvard University Press.

Ehin, C. (1995). *The Quest for Empowering Organizations: Some Lessons From our Foraging Past. Organization Science*. November-December.

Ehin, C. (2000). *Unleashing Intellectual Capital*. Boston: Butterworth-Heinemann.

Ehin, C. (2005). *Hidden Assets: Harnessing the Power of Informal Networks*. Boston: Springer.

Ehin, C. (2009). *The Organizational Sweet Spot: Engaging the Innovative Dynamics of Your Social Networks*. New York: Springer.

Halstead, J. (2008). U.S. Employee Engagement Deteriorates as Financial Crisis Worsens. *Modern Survey Website*, November 12. Retrieved March 2, 2009 from http://modernsurvey.com/news/?p=124

Lehrer, J. (2008). The Eureka Moment. *New Yorker (New York, N.Y.)*, (July): 28, 40–45.

Polanyi, M. (1958). *Personal Knowledge: Towards a Post-Critical Philosophy*. Chicago: University of Chicago Press.

Putnam, R. D. (2000). *Bowling Alone: The Collapse and Revival of American Community*. New York: Simon & Schuster.

Rock, D. & Schwartz, J. (2006). The Neuroscience of Leadership. *Strategy+Business*, (43), 06207.

Searls, D. (2007). It's the Relationship, smarty. *Weblog*, December 26. Retrieved January 3, 2008.

Stacey, R. D., Griffin, G., & Shaw, P. (2000). *Complexity and Management: Fad or Radical Challenge to Systems Thinking*. New York: Routledge.

Stevens, A., & Price, T. (1996). *Evolutionary Psychiatry: A New Beginning*. New York: Routledge.

White, H. C. (2008). *Identity and Control*. Princeton, NJ: Princeton University Press.

ADDITIONAL READING

Allen, S., Deragon, J. T., Orem, M. G., & Smith, C. F. (2008). *The Emergence of the Relationship Economy*. Cupertino, CA: HappyAbout.Info.

Baker, S. (2008). *The Numerati*. New York: Houghton Mifflin Company.

Ehin, C. (2008). Un-Managing Knowledge Workers. *Journal of Intellectual Capital*, *9*(3). doi:10.1108/14691930810891965

Gigerenzer, G. (2007). *Gut Feelings: The Intelligence of the Unconscious*. New York: Viking.

Gloor, P. A. (2006). *Swarm Creativity: Competitive Advantage through Collaborative Innovation Networks*. New York: Oxford University Press.

Kelso, J. A. S. (1995). *Dynamic Patterns: The Self-Organization of Brain and Behavior*. Cambridge, MA: The MIT Press.

Kluger, J. (2007). What Makes Us Moral. Time. December 3, pp. 54-60.

Kluger, J. (2008). *Simplexity*. New York: Hyperion.

Levine, R., Locke, C., Searls, D., & Weinberger, D. (2000). *The Cluetrain Manifesto*. New York: Basic Books.

Linden, D. (2007). *The Accidental Mind*. Cambridge: Harvard University Press.

Lynch, G., & Granger, R. (2008). *Big Brain: The Origin and Future of Human Intelligence*. New York: Palgrave Macmillan.

Marcus, G. (2008). *Kluge: The Haphazard Construction of the Human Mind*. New York: Houghton Mifflin Company.

Medina, J. (2008). *Brain Rules*. Seattle: Pear Press.

Montague, R. (2006). *Why Choose This Book? How We Make Decisions*. New York: Dutton.

Pinker, S. (2002). *The Blank Slate: The Modern Denial of Human Nature*. New York: Viking.

Ridley, M. (2003). *Nature via Nurture*. New York: Harper Collins Publishers.

Shubin, N. (2008). *Your Inner Fish*. New York: Pantheon Books.

Taleb, N. N. (2007). *The Black Swan*. New York: Random House.

Wheatley, M. J. (2009). *Turning to One Another*. San Francisco: Berrett-Koehler Publishing.

Chapter 12
Social Capital:
Definition, Dimensions and Measurement

Pablo Blasco Bocigas
Universidad Complutense de Madrid, Spain

ABSTRACT

The competitive dynamics of a firm take place through the interaction with different agents inside as well as outside of its organizational limits. Firms have elements of social character that are visible through the interactions between individuals, groups and organizations at different levels. The social capital would allow a firm to connect diverse agents to obtain and exchange resources and transfer knowledge. For this reason, the purpose of this chapter is to focus on finding an adequate defintion for social capital and in the identification of the forms that the research proposed to make this topic operative: networks and relations.

INTRODUCTION

According to the academics and the business world, the importance of the intangible elements as creators of value in the firm is becoming the center of attention in the last years. All the contributions made emphasize the need to value and quantify the intangible resources since they are the "result of the integration of knowledge to the different productive activities of the firm" (Bueno, 1998, p. 215) and given that they are not reflected in the accounts. Within this topic are included diverse aspects such as the knowledge of the environment, the market reputation, the experience and knowledge acquired inside as well as outside of the organization, the innovation and technological diffusion in the firm and the firm's relations with customers, suppliers and other agents (social and economic).

Various elements mentioned above (ways of knowledge transmission in the firm, relations with agents, etc.) have a common component: they all belong to the social capital of the firm. This explanation is based on the fact that inside the firm, the social capital does not include a unique dimension rather it can be included within the three dimensions in which the intellectual capital was divided:

DOI: 10.4018/978-1-61520-875-3.ch012

human capital, structural capital and relational capital (CIC, 2003).

Therefore, the concept of social capital has acquired an important relevance in a wide range of disciplines. An increasing number of sociologists, political scientists, economists and theoreticians of the firm have used the concept to look for answers to multiple questions within their respective fields of investigation (Adler & Kwon, 2002). In this sense, an increasing interest exists among academics in trying to explain how the social context, in which the firm is immersed, significantly influences its behavior and results (Gulati, Nohria & Zaheer, 2000). Any organization possesses elements of social character, visible through the interactions between its individuals and groups, changing the social capital into an observable phenomenon at different levels. For that reason, and if we examine a firm's potential for development, we should analyze the interaction between the different agents (internal and external) that compose it.

Social capital allows a firm to connect diverse groups (in different contexts) in order to obtain and exchange information and transfer knowledge. In this sense, the firm would generate value by establishing strategic relations with other agents in its environment, developing and establishing social networks (flexible and dynamic) which allow the creation and the exchange of information and knowledge and the development of learning and innovation processes for the improvement of its competitiveness and therefore, of its performance.

Our main interest will be focused in defining what is social capital, what are the dimensions that compose it and how can it be effective through two primary measures: networks and relations. In both cases, the options could be used internally as well as externally involving this way the three dimensions of the intellectual capital. Moreover, if we understand the role of social capital, the elements that comprise it and its interrelations, we will be able to obtain an instrument that will allow managers, academics and consultants to explain the results of any organization in its socioeconomic context.

BACKGROUND

Social Capital Definition

Social scientists have proposed a wide number of social capital definitions that vary depending on whether they focus on the substance, the source or the effects of social capital (Adler & Kwon, 2002). In an attempt to collate all the existing meanings, these authors indicate that in order to define the social capital that an agent possesses, academics have focused on three perspectives: (1) the relations that an agent maintains with other agents (external relations) (2) the structure of relations between the agents within a group (internal relations and) (3) both types of perspectives simultaneously. This classification takes into consideration only the fact that the perspective of the ties of an agent in its network is internal or external.

This chapter, in an attempt to expand this concept, tries to establish a *social capital classification* based on the proposed definition. This new classification is suggested by the definition's heterogeneity of the academic investigation and by the belief that more centers of interest exist within the social capital than the location of the ties which connect agents. Thus, we can affirm that there exist definitions of social capital that focus on:

1. Social networks in which agents are located. Definitions that refer to the number and nature of the contacts linked to a network.
2. The resources that an agent could mobilize in its network. Definitions associated with the possession of a determined resource and benefits which could be obtained from it.
3. The interaction measures that promote relations between agents of a same social

structure. We have defined, under this term, the whole set of rules, expectations, capacities, characteristics, perceptions, etc. that act as mechanisms of collective action within a network.

4. The integration of social networks and mobilized resources. In this section we have included the definitions that put together these two visions, when emphasizing the importance of the network as well as of the resources that can be obtained by the agents of the network.
5. The integration of social networks and the rules that govern them. Definitions that try to integrate these two visions, when defining the importance of the network as well as the norms that constitute it.

In Table 1 we have gathered the main definitions of social capital proposed by academics grouped according to the type of definition and author.

This variety of meanings makes us think that social capital is related to structural aspects of the social network of an agent that facilitate determined actions and that are related to the resources located in his/her network as well as those that are accessible to him/her, through that same network. For the purposes of this study, the closest definition to the idea previously mentioned is the one proposed by Nahapiet and Ghoshal (1998) who stated that social capital is "The sum of the actual and potential resources embedded within, available through, and derived from the network of relationships possessed by an individual or social unit. Social capital thus comprises both the network and the assets that may be mobilized through that network."(p. 243).

For that reason, the study of the social capital of a firm requires specifically:

- A structural analysis oriented to identify and to quantify the configuration of its network and,
- a relational analysis that determines the relations to be maintained with other agents of the network.

Social Capital Dimensions

In order to identify the dimensions that allow us to analyze this type of capital, we must take care of the distinction made by Granovetter (1985, 1992), who determined that social capital consists of structural and relational embeddedness. This author stated that there has been failure to recognize the importance of both concrete personal relations and social networks. His work on social embeddedness in the economic exchange is an important precursor of the concept of social capital and this was considered to be an attempt to introduce the idea into the analysis of the social organization of the economic system and of the social relations that motivate it (Burt, 1992; Coleman, 1990). Based on this definition, Nahapiet & Ghoshal (1998) distinguished between the structural dimension or embeddedness and the relational dimension or embeddedness. The structural dimension of social capital includes social interaction, that is to say, the location of the agent's contacts in social structures of interactions. The relational dimension, on the contrary, refers to assets that are embedded in these relations (such as confidence). Nahapiet & Ghoshal (1998) adopted this definition when considering social capital as the sum of actual and potential resources embedded within, available through, and derived from the networks of relationships possessed by a social unit, in this case, the firm.

Structural Dimension of Social Capital

This dimension represents the global group of relations that an agent has (Granovetter, 1992; Nahapiet & Ghoshal, 1998), and is defined by the properties of the social system that define the global patterns of the connections between

Table 1. Social capital definitions (self elaboration based on Adler & Kwon, 2002)

Object	Author	Field	Definition
Social Network	Boxman, De Graaf & Flap (1991, p. 52).	Sociology	"The number of people who can be expected to provide support and the resources those people have at their disposal."
	Burt (1992, p. 9).	Sociology	"Friends, colleagues, and more general contacts through whom you receive opportunities to use your financial and human capital"
	Loury (1992, p. 100).	Business Administration	"Naturally occurring social relationships among persons which promote or assist the acquisition of skills and traits valued in the marketplace...an asset which may be as significant as financial bequests in accounting for the maintenance of inequality in our society."
	Belliveau, O'Really & Wade (1996, p. 1572).	Sociology	"An individual's personal network and elite institutional affiliations."
	Lin (2001, p. 7).	Sociology	"Investment in social relations with expected returns".
	Florin, Lubatkin, & Schulze (2003, p. 374).	Business Administration	"Social capital contributes directly to a venture's resource base, by allowing it to better attract human and financial resources, and also contributes indirectly, through its ability to leverage the productivity of the venture's resources".
Object	**Author**	**Field**	**Definition**
Resources	Bourdieu (1985, p. 248).	Sociology	"The aggregate of the actual or potential resources which are linked to possession of a durable network of more or less institutionalized relationships of mutual acquaintance or recognition".
	Baker (1990, p. 619).	Sociology	"A resource that actors derive from specific social structures and the use to pursue their interests; it is created by changes in the relationship among actors".
	Bourdieu & Wacquant (1992, p. 119).	Sociology	"The sum of the resources, actual or virtual, that accrues to an individual or a group by virtue of possessing a durable network of more or less institutionalized relationships of mutual acquaintance and recognition".
	Leana & Van Buren (1999, p. 538).	Business Administration	"A resource reflecting the character of social relations of a firm".
	Oh, Labianca & Chung (2006 p. 570).	Business Administration	"The set of resources made available to a group through group members social relationships within the social structure of the group itself, as well as in the broader formal and informal structure of the organization".
Object	**Author**	**Field**	**Definition**
Interaction measures	Portes & Sensenbrenner (1993, p. 1323).	Sociology	"Those expectations for action within a collectivity that affect the economic goals and goal seeking behavior of its members, even if these expectations are not oriented toward the economic sphere".
	Fukuyama (1995, p. 10).	Political Science	"The ability of people to work together for common purposes in Groups and organizations".
	Putnam (1995, p. 67).	Political Science	"Features of social organization such as networks, norms, and social trust that facilitate coordination and cooperation for mutual benefit".
	Adler & Kwon (2002, p. 23).	Business Administration	"Social capital is the goodwill available to individuals or Groups. Its source lies in the structure and content of the actor's social relations. Its effects flow from the information influence and solidarity it makes available to the actor".
	Kostova & Roth (2003, p. 301).	Business Administration	"The potential value arising from certain psychological states, perceptions, ad behavioral expectations that social actors form as a result of both their being part of social structures and the nature of their relationships in these structures".
Object	**Author**	**Field**	**Definition**

continued on following page

Table 1. continued

Object	Author	Field	Definition
Integration between social network and resources	Nahapiet & Ghoshal (1998, p. 243).	Business Administration	"The sum of the actual and potential resources embedded within, available through, and derived from the network of relationships possessed by an individual or social unit. Social capital thus comprises both the network and the assets that may be mobilized through that network".
	Siebert, Kraimer & Liden (2001, p. 221).	Business Administration	"Different network structures that facilitate (or impede) access to social resources embedded in the network".
Object	**Author**	**Field**	**Definition**
Integration between social network and interaction measures	Coleman (1990, p. 302).	Sociology	"Social capital is defined by its function. It is not a single entity, but a variety of different entities having two characteristics in common: They all consist of some aspect of social structure, and they facilitate certain actions of individuals who are within the structure".
	Portes (1998, p. 6).	Sociology	"The ability of actors to secure benefits by virtue of membership in social networks or other social structures".

all the agents (Burt, 1992). Thus, the location of the contacts of a firm in a social structure of interactions, will provide it with a series of advantages, opportunities or benefits in order to obtain its objectives measured in terms of valuable information or access to specific resources (Burt, 1992). An agent located in a determined position within a network of interactions would have greater potential to exchange resources due to the advantage conferred by its location in the network (Tsai & Ghoshal, 1998). Gulati, Nohria & Zaheer (2000) consider that the agents whose relations allow them to occupy an excellent position within the network to which they belong, will obtain superior returns since they have better access to information and opportunities. Thus, the network's structure can influence the type of resources which may be accessible to a firm.

Therefore, we must define what factors are decisive for the configuration of the network's structure. These factors are associated with the flexibility and the ease of interchanging information with the members of the network (Nahapiet & Ghoshal, 1998):

1. **Number of ties of the network.** The number of ties of an agent will determine the size of its network. For that reason, wide and reduced networks will exist.
2. Frequency and intensity of interactions that actually exist in the network in relation to those that theoretically could occur. The cohesion of a network (Coleman, 1988) will be determined by strong ties between agents. On the other hand, numerous weak ties in the network will involve diverse groups (Burt, 1992). Therefore, we will distinguish between cohesive and diverse networks.
3. **Connectivity.** To establish how the agents of a network are connected will determine the possibility of access to certain resources, which are otherwise difficult to obtain. In this case, we may distinguish between direct and indirect contacts. Direct contacts will be more common in dense and cohesive networks, whereas indirect contacts are typical in diverse networks.
4. **Hierarchy.** The establishement of internal and external limits of the network (Oh, Labianca & Chung, 2006). In cohesive networks, the limit is laid down inside of the same network whereas the limit of diverse networks is located outside of the network.

Thus, we can establish an initial distinction between two configurations of networks if we take into consideration the frequency and intensity of the interactions that take place since each

Table 2. Main structural characteristics for the design of a network

Characteristics of the network	Type of network	
	Cohesive	Diverse
Number of ties	Reduced	Wide
Frequency and intensity	Strong ties	Weak ties
Connectivity	Direct contacts	Indirect contacts
Hierarchy	Inside limit	Outside limit

one of these forms of network addresses specific requirements in relation to: the number of ties, the strength of ties, the connectivity between the different agents and the limit where we can locate the different agents of the network. *Cohesive networks* will be formed by a limited (or reduced) number of strong and direct ties. This will strengthen the appearance of internal control mechanisms that guide collective action in order to coordinate and control the different activities carried out within the group. For that reason, the limit of this type of network is located inside of it. As a result, it will be the same network that will assume the central position, at the expense of the individuals that compose it.

On the other hand, *diverse networks* will be formed by a wide number of weak and indirect ties between the agents of the network. This situation could entail two consequences for the agent: (1) to determine more or less a central position that can condition the resources that he/she can obtain from the network and, (2) the agent that occupies this position has to define and ensure the control mechanisms of the networks by means of instruments different from those used by the members of a cohesive network, where this centrality is the responsibility of the network and not the agent. Due to these conditions, the limit of a network with diverse ties connected to each other in an indirect way is external for this agent.

The characteristics that affect the design of the network's structure are reflected in Table 2.

Relational Dimension of Social Capital

Some academics have affirmed that the value of social capital is affected by more than the structure of the ties of an agent (Adler & Kwon, 2002; Leana & Van Buren, 1999; Nahapiet & Ghoshal, 1998; Uzzi, 1997). Granovetter (1992) related relational embeddedness to the quality of relations and to the assets in these relations, in a specific period of time. From this relational perspective, the social capital, therefore, reflects the potential benefits derived from the content of the relations reflected in the activities carried on by the agents (Kostova & Roth, 2003). These activities could include attributes such as trust and reliability (Fukuyama, 1995; Putnam, 1993; Tsai & Ghoshal, 1998), norms and penalties (Coleman, 1990; Putnam, 1995), identity (Kogut & Zander, 1996) and identification (Nahapiet & Ghoshal, 1998).

These factors and their possible interrelations could influence the manner in which a firm forms its relations to build social capital. This has led some researchers to consider **trust** to be a critical factor of the dimension (Inkpen & Tsang, 2005; Moran, 2005; Nahapiet & Ghoshal, 1998) since it could be both the antecedent and the result when all the implicated agents in the different social systems (networks) find ways to settle agreements with others (and thus to generate social capital), although the uncertainty and risk of appearance of opportunistic conducts could influence them.

This approach made us establish two forms of trust: dyadic trust derived from the relation with a specific agent, and generalized trust in the network to which it belongs. This distinction can make a firm form its relations in a determined way based on the factors that could positively influence the accomplishment of the objectives through the different mechanisms that guide this relation.

Dyadic Trust

Trust is based on social judgments established by individuals (Tsai & Ghoshal, 1998). In addition, it is one of the fundamental components of intense relations and mutual knowledge (Burt, 1992). Trust in relationships is developed from social interactions (Granovetter, 1985; Gulati, 1995; Tsai & Ghoshal, 1998). If two agents interact over time, their relations can be more concrete and both are perceived as trustworthy (Tsai, 2001). We define dyadic trust as the one that emerges between two agents that maintain or have maintained continued and stable relations on time when sharing common interests and that have been gradually fulfilling a series of mutual expectations and obligations.

Thus, relations based on dyadic trust will encourage the transfer of resources between two firms (Inkpen & Tsang, 2005; Nahapiet & Ghoshal, 1998), although it requires investing time to develop agreements (Dyer & Singh, 1998), and stimulate those exchanges that serve to build or reinforce social capital (Coleman, 1988). For that reason, trust plays an important role in the willingness of any agent to exchange resources, and it can act as a control mechanism of the relation (Uzzi, 1996).

The research confirmed that dyadic trust facilitates collective and collaborative behaviors in the absence of explicit mechanisms to promote and reinforce these behaviors (Coleman, 1990). Lane & Lubatkin (1998) developed a theoretical frame in which control mechanisms and informal rules based on trust were effective in promoting the exchange of resources between two companies because (1) they are not dependent on time and, (2) they are positive actions adopted by firms that expect positive benefits.

For that reason and above all, we must ask what are the different attributes that generate trust and how they interact. Based on this approach, our work considers two factors that generate this type of trust: the similarity between agents and the previous history of interactions based on the reciprocal fulfillment of obligations and expectations throughout the relation.

Similarity Between Agents

In order to lessen the uncertainty caused by the absence of a prior relationship and previous knowledge, it is predictable that an agent establishes an identification process to determine the possible similarities and differences with other agents for the relationship to be guided by a shared model of reference. The agents who share the same mental model will be more susceptible of having relations of high quality of interacting and sharing resources regularly (Leana & Pil, 2006). On the contrary, if two agents involved in a relationship demonstrate opposing identities, this could represent a significant barrier to sharing resources in terms of mutual information and learning and creating or sharing knowledge. Therefore, these shared social models will provide the basis for mutual understanding at the beginning of the relationship and would be shaped by the objectives and control mechanisms that guide the behavior of the agents involved.

The identification will act as a specific factor of the relationship that influences the expectations of the likely value obtained in the exchange of resources and in the motivation to transfer these resources between both (Nahapiet & Ghoshal, 1998), thus facilitating the coordination and the communication among them (Kogut & Zander, 1996).

Prior History of Interactions

The history of interactions reflects the perceived utility of the previous relations between two agents and is defined as the degree to which an interaction produces the desired effect for both (Kostova & Roth, 2003). Some academics have argued that relationships are considered to be previous histories since it is the previous ties that determine the creation of future unions (Gulati, 1995; Nahapiet & Ghoshal, 1998; Tsai, 2000).

The effectiveness of the stated interaction would be measured based on the degree of fulfillment or failure in achieving the objective for which it was created and it will entail positive or negative perceptions and relations with the implicated agent based on these results. Thus, past histories of interactions influence the creation of a shared vision of the relationship (Tsai & Ghoshal, 1998) since codes, shared languages and practices based on shared interest and understanding are created (strengthening the identification process previously described, and achieving a better perception to carry out new agreements). This reliability, manifested through the continuous fulfillment of obligations and expectations over time, will display the quality of a relationship (Dyer & Singh, 1998).

For this reason, the development of a positive history of interactions over time would confirm the reliability of the implicated agents in the relationship and would indicate their intentions to obtain short term results (Tsai & Ghoshal, 1998), to facilitate the exchange of resources (Uzzi, 1996) and to establish new relations and consolidate already existing ones. Some authors (Coleman, 1990; Fukuyama, 1995; Tsai & Ghoshal, 1998; Uzzi, 1996) have denoted reputation as the main consequence of reliability of an agent. Therefore, the reputation could be a condition that provides the firm with the necessary reliability to obtain and use the available resources of its contacts (Nahapiet & Ghoshal, 1998). On the other hand, the absence of a previous history of interactions with a potential agent would entail the appearance of uncertainty, since there are no specific established routines that rule the relation (Gulati, 1995). These initial relations will influence time in a positive way whenever: (1) a previous process of identification has been established or, (2) both agents belong to the same social system or network and therefore share a common identity.

In conclusion, numerous repeated interactions and proven reliability by both parties over time will imply reciprocity based on informal relations (trust) more than formal mechanisms (contractual obligations) which will increase trust between both parties and will increase its expectations on future exchanges as reflected in Figure 1.

Generalized Trust

Putnam (1993) named this type as indirect trust, since it is not based on mutual knowledge between individuals, but in generalized norms and conduct in the social system as a whole. This is generalized

Figure 1. The process of dyadic trust generation between agents

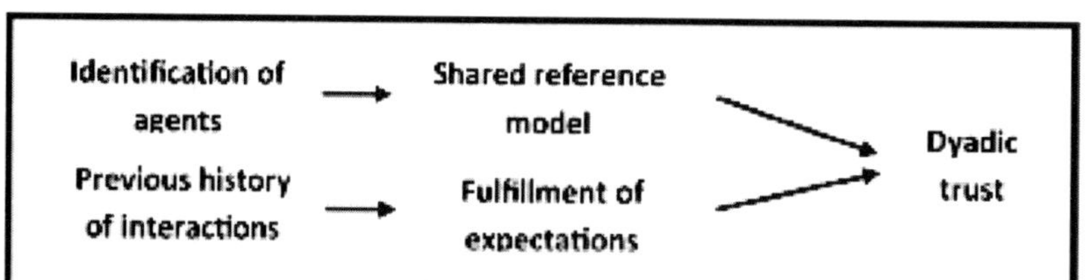

trust and it is a characteristic of systems with strong social capital. An important fact that characterizes this type of trust is the degree by which the agents could trust the others without direct information or previous interaction, simply by virtue of belonging to the same social system (or social network). In this sense, Ring & Van de Ven (1992) developed the concept of resilient trust. This trust is based on experience with other agents and on the belief in moral integrity within the social system, hence the reason why the general norms and common values are the point which maintains unified communities with resilient trust (Granovetter, 1985). The presence of resilient trust could reinforce the social capital within a system because it would increase the interaction between all the members (Granovetter, 1973). Therefore, organizations strong on social capital will show resilient trust, including a situation whereby their members are connected more globally than individually (and vice versa). In this sense, Leana & Van Buren (1999) confirmed that this trust is an antecedent and a result of collective action or success, and is necessary in order that people work jointly on common projects.

On the other hand, we could find a context where trust does not exist or is fragile. Rousseau (1995) defined this kind of trust as transactional. Thus, the transactions realized in a social system based on fragile trust will be ruled by contractual measures since there is no reciprocity within the relations that conform it. Therefore, and in order to have generalized trust, we should take into consideration the fact that the agents belonging to the same social system share a common identity and have generalized norms of reference that guide the collective action and encourage cooperation within the system (or network).

Collective identity is the common understanding of objectives and forms of performance in a social system. The set of common values provides the members of the network with a pattern of coherent norms and rules of acting. Therefore, it comprises collective objectives and aspirations of members of the same social system. Thus, through similar cognitive social processes an agent could generalize his judgments about contacts of the rest of social entities to which they belong and represent (Kostova & Roth, 2003). Therefore, trust in a social system is developed from the reciprocity in the relations in a context where the established social norms are fulfilled (Coleman, 1988, 1990). Thus, in the absence of other type of capital (as economic rewards), the common norms and identity (the feeling of membership) could act as a substitute to obtain collective and collaborative action as it is reflected in Figure 2.

In this same sense, Coleman (1990) states that a generalized norm exists when the socially-defined right to control an action is not possessed by any one of the involved members. Therefore, in consequence, the collective action will be based on norms of general character. Looking into these norms, the agents influence their behavior in order to meet the expectations that others have about them within the system. In these situations, this attribute could be a substitute for formal contracts or for the systems of control of the behavior of its members (Leana & Pil, 2006). The cooperation norms can be a fundamental source for the creation

Figure 2. Process of development of generalized trust in a social system

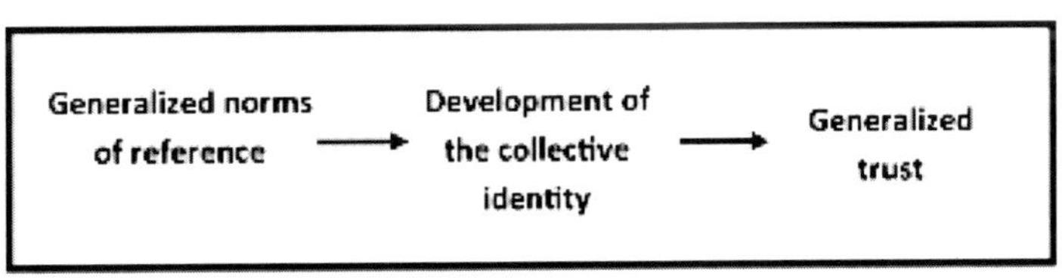

of social capital since in a process of transfer of resources, they would ensure the necessary motivation to establish the exchange engagement (Putnam, 1993). Therefore, the shared norms can be defined as forms of social capital.

Thus, if an agent did not have a previous history of interactions with any member of the group, the existence of shared norms would act as a "promoter function" (or substitute) of a possible relationship between both, previously disconnected. Therefore, community members share strong feelings of trust. For that reason, the generalized norms solve problems of collective action, creating shared interests and a common identity, avoiding opportunistic behaviors between its members. This common interest usually is manifested through cooperative action in the creation of shared projects, facilitating the flow of information and resources or penalizing behaviors that go against the collective identity.

In Table 3, we have collated all the factors that affect the resilient or fragility of the generalized trust.

To summarize, the two dimensions that constitute social capital are structural and relational. The structural dimension makes reference to the patterns by which an agent can shape a concrete type of network: size of the network, frequency and intensity of the interaction, degree of connectivity between members of the network and establishment of the limit of the network. On the other hand, the relational dimension concentrates on the attributes of the relations. In Figure 3 we have summarized these attributes in two categories: dyadic trust or trust between two agents and generalized trust or trust in the system, showing in addition, how they develop over time.

ESSENTIAL FORMS OF SOCIAL CAPITAL: NETWORKS AND RELATIONS

Network as a Primary Form of Social Capital

In spite of the existence of a relative and generalized consensus regarding the definition of the network, our main question must focus on the analysis of what decisions an agent could make at the time of choosing (or designing) a determined *structure of the network* to create social capital. For this reason, it is essential to delineate the structural patterns of an agent's network and at the same time, to study the two proposed issues by the academic research to make it operative.

The main issue indicated by the theoreticians of the network (Bourdieu & Wacquant, 1992; Burt, 1992; Coleman, 1990; Putnam, 1993) allows the analysis of what agents are better connected to obtain higher returns. To hold a concrete position in the structure of the network could be an asset owned by an agent (Lin, 2001). Almost all academics have agreed that a concrete structure could generate an advantage. Therefore, our interest is focused on specifying what network configuration involves the concept "being better connected". It is expected that the diffused information through the networks of all members flows within the groups rather than between groups (Coleman, 1988), which is why not all of them are totally aware of the emerging opportunities in all the groups that shape the network (Leana & Van Buren, 1999). For that reason, we can distinguish two theoretical approaches within the study of the structure

Table 3. Types of generalized trust on the basis of the contextual factors of the relationship

Factors of generalized trust	**Types of generalized trust**	
	Fragile	Resilient
Generalized norms	Oriented to mitigate the risk	Promotes collective actions
Collective identity	Fragile feelings of community	Strong feeling of community

of the network according to the configuration of its ties: the closed network and the existence of structural holes (or diverse network).

Closed Network

According to this approach (Coleman, 1988, 1990) the closed networks (networks in which each one is connected in a way that nobody can escape from the "evidence of others") can acquire the form of social capital. Therefore, this is a form of cohesive network. According to the delimitation in the factors of the structural dimension, a closed or cohesive network is characterized by strong ties and direct contacts between its members. The strength of the ties could serve as a social control mechanism that directs the behavior of the involved members (Rowley, Behrens & Krackhardt, 2000). In the same sense, Uzzi (1996, 1997) shows that the strong ties promote and increase the perspective of the firms in order to develop joint interactions over time, which is why having strong ties in the network reduces the risk of vulnerability (Capaldo, 2006) from outside as well as from inside of the network.

In spite of these considerations, Granovetter (1973) affirms that the strong ties are redundant contacts or contacts that have the same resources (information, knowledge, specific abilities, etc.). Thus, numerous repeated ties between a small number of agents generate an isomorphism within the network, reducing the access of anyone to non-redundant resources (Capaldo, 2006). This author affirms that being located in a narrow group of strong ties and the risk of being unable to access to new opportunities are two of the major risks of a network with strong ties. Therefore, a dense network contains a single non-redundant agent, with the result that the resources that will flow around it will have certain similarity (Burt, 1992). The members involved in a closed or cohesive network structure could obtain certain advantages derived from the amount of the exchanged resources following collective and cooperative norms and behaviors. Therefore, the hierarchy of the organization's ties will be defined inside the network as shown in Figure 4.

For this reason, the direct connectivity between members of the cohesive network can have three consequences:

Figure 3. Summary of the essential components of the social capital dimensions

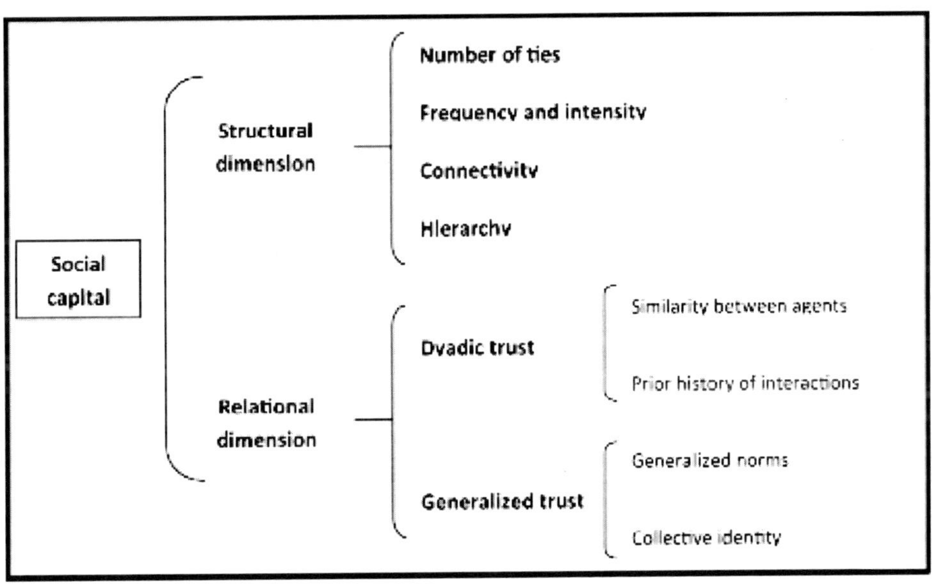

Figure 4. Completely closed network with four agents

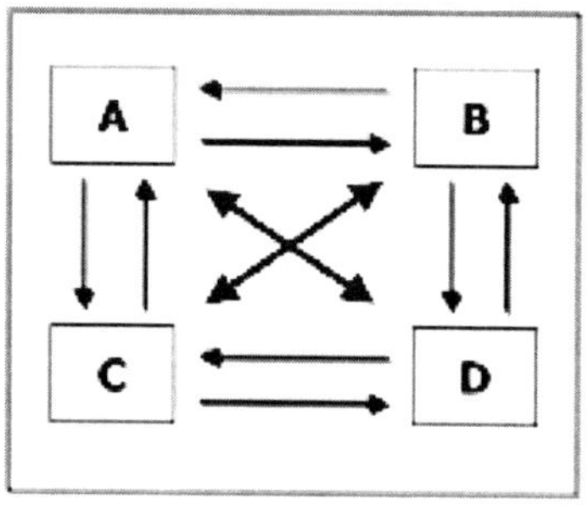

1. Coordinates the collective action and promotes effective incentives to manage the members of the network (Coleman, 1988; Rowley, Behrens & Krackhardt, 2000). Thus, without these two conditions a firm may have difficulty in surviving over time.
2. Implies the creation of norms in the level of the network. Granovetter (1992) argues that "the existence of norms causes the agents with direct contacts to have greater disposition to belong to the same network" (p. 44).
3. Affects the access and the attainment of information. Coleman (1990) asserts that "an important form of social capital is the potential for information that inheres in social relations" (p. 310). Rowley, Behrens & Krackhardt (2000) also state that cohesion of the network is an effective mechanism to obtain precise information.

To summarize, the cohesion of the network is important not only for the appearance of norms but also for other forms of social capital such as reliability of the social structures which allows the proliferation of common expectations (Coleman, 1988; Lin, 2001) between its members.

Structural Holes

This approach was enunciated by Burt (1992) and was named as a perspective of the structural holes.

Burt (1992, 1997) defined social capital on the basis of the resources possessed by an agent and the configuration of the contacts' network structure. This conception focuses not only on the characteristics of the direct ties of this agent but also on the potential pattern of interactions of all the members of its network. The fundamental antecedent of this theory has been Granovetter's work (1973) that focused on defining the strength of the social ties used by an individual in the attainment of his/her particular objectives. This author argues that ties between agents within the same social group will probably be strong (defined in terms of intensity and frequency of its interactions). Nevertheless, the ties that are accessible to a wide social group will be characterized by a weak tie (less intense and less frequent). The weak ties are considered as a bridge between densely-interconnected social groups and as a source of information and unique resources (Burt, 1992; Moran, 2005).

Burt (1992) stated that in order to build a network, the agent who holds a central position must maximize the proportion of non-redundant ties or contacts of the total possible contacts of the network. Therefore, this agent must focus on the diversity of its contacts (direct and indirect) which will be relevant for the group (Capaldo, 2006) since it increases the probability of diversity of the network. In this sense, Burt (1992) ensures that each agent has a network of contacts from which he/she can obtain an advantage. For that reason, this perspective:

1. Describes the network on the basis of the possibility to access to other agents with specific resources, which leads us to consider the network as a conduit that links directly and indirectly all the agents with the one who holds a central position in the network.
2. Essentially describes how the networks are in themselves a form of social capital which are primarily measured by focusing on the analysis of the network's size, measured on the basis of the overall number of ties.

Structural holes exist between two agents connected to a central one, if they are not connected between each other even though they are connected to the central one (non-redundant contacts). Also, structural holes exist because the agents are differentiated on the basis of the position held within the network's structure (Oh, Labianca & Chung, 2006). Thus, different positions in the network could represent different opportunities to access new and diverse specific resources (Tsai, 2000), which is why this position would reveal the agent's ability to locate and access these valuable resources.

As a result, the objective of any firm will be to maximize the number of non-redundant contacts in the network through a large amount of weak (Granovetter, 1973, 1985) and diverse ties (Burt, 1992), which justifies the analysis of the size of the network as the primary means of measurement of social capital. Figure 5 represents the diversity of an agent's network with structural holes composed by twelve contacts: three direct and nine indirect.

Based on what we analyzed earlier, it is clear that the structural holes separate non-redundant resources. In order to be able to measure this redundancy, Burt (1992, 1997) established two fundamental criteria of measurement: the degree

Figure 5. Agent A's network with structural holes

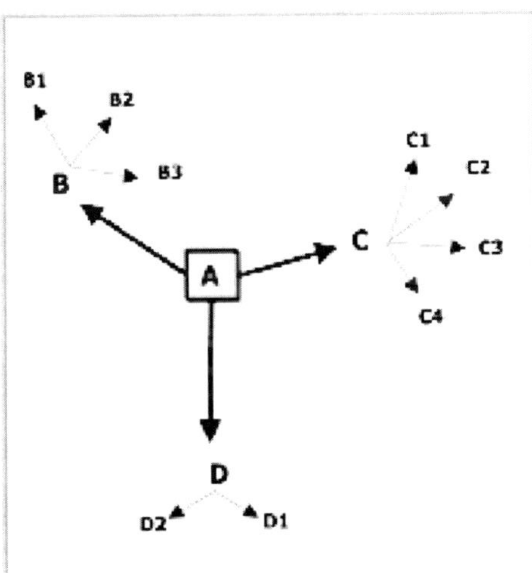

of cohesion of the network and structural equivalence between all its contacts. First of all, the cohesive contacts (strongly connected to each other) are susceptible to having similar information. This viewpoint is the main point of disagreement between this perspective and the one proposed by Coleman (1988, 1990) on the degree of cohesion of a network. Under this criterion, two contacts are redundant for the group because they are tied up through a strong tie. This strength implies the absence of structural holes. Secondly, structurally-equivalent contacts (tie an agent with the same third parts) have the same sources of information (Burt, 1992). Two agents are structurally equivalent for others if they have the same contacts, reaching this way the same group of agents in the network. In Figure 6, we represent an agent's network with a complete degree of cohesion, where the members of agent A's network are connected directly to each other and to a central agent network with redundancy by structural equivalence between its contacts. In this case in particular, the direct contacts are linked to the same third parties.

In addition, this issue allows us to establish the location of the limit for the central agent network taking into consideration the degree of cohesion as well as structural equivalence previously defined. If a central agent possesses two direct contacts with the same two indirect contacts (making them redundant by structural equivalence), this latter could still be connected with different agents beyond this network (making them non-redundant). Therefore, it appears that the limit of the network of this central agent would be located outside the network of the direct contact, trying to look for new sources that provide non-redundant resources to the already owned by its contacts.

To summarize, the vision of social capital arising from this line of research tries to balance the size and diversity of the network with the objective of optimizing the structural holes. We can expect an increase in the number of structural holes under two principles (Burt, 1992):

Figure 6. Network of a central agent with redundancies because of cohesion (left) and because of structural equivalence (right)

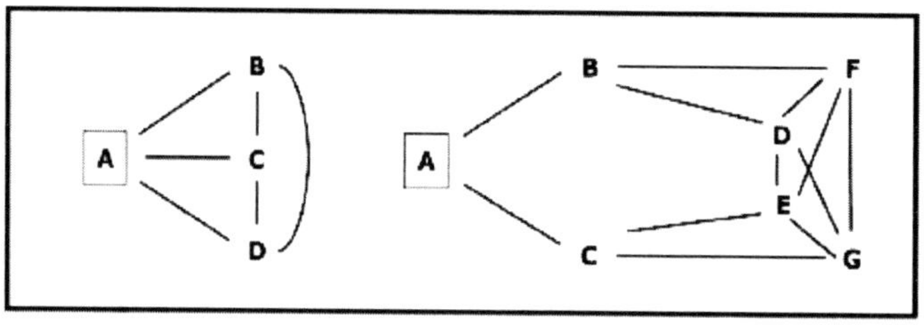

1. Efficiency of the network when maximizing the number of non-redundant contacts in order to maximize the result of the structural holes by contact.
2. Effectiveness of the network when distinguishing primary contacts from secondary ones with the objective of focusing in the resources that preserve the primary contacts. Instead of establishing ties with all contacts, the primary task will be to expand the network in order to access to the highest number of resources.

Based on these two principles, an agent will have to focus his/her efforts in creating social capital through the strengths of weak ties (Burt, 1992; Capaldo, 2006; Granovetter, 1973; Moran, 2005; Rowley, Behrens & Krackhardt, 2000), where few strong ties are connected to different third parties (efficiency) in order to generate an exchange of resources (effectiveness) through high levels of interaction over time. When these interactions increase, the ties become stronger and the agents involve a higher amount of resources.

Our next step will be to detail how the relational perspective defines the nature of the interactions on the basis of its content and the attributes that the agents can use to establish agreements to access and transfer specific resources. Therefore, the configuration of a determined structure influences the accessibility and transfer of the resources through the network. In addition, we need to determine how the degree of interaction of the agents in the network can affect the strength of the relation and therefore create social capital (Granovetter, 1973), since intense and effective interactions can increase the possibility of developing relations with other agents (Kostova & Roth, 2003).

RELATIONS AS A PRIMARY FORM OF SOCIAL CAPITAL

Some researchers focused on the relational aspects of social capital, considering that it is the nature of the resources that leads to the achievement of real benefits for the agents (Kostova & Roth, 2003) more than the structure of the network to which they belong. In order to have social capital, an agent must be connected to others and "these others to others" although not necessarily to the first agent itself. In this same sense, Moran (2005) stated that observing how the network of contacts influences the result of a firm forces us to look beyond the configuration of the network (previously described) and to value the quality of the relations of the embedded agents.

Granovetter (1985) stated that *social relations* are essential in economic activity to generate trust as well as to dissuade opportunistic behavior. An agent could use a wide variety of safeguards to be protected against the risks of opportunism that could appear. The term "safeguard" is a control mechanism with the objective of creating equity perception (or symmetry) between the involved agents (Dyer, 1997; Jones, Hesterly & Borgatti, 1997). The purpose of the safeguards is to provide the necessary control and trust, at the minimum cost, so that the agents consider that engagement in the exchange will be a benefit for them (Williamson, 1985).

Although contracts are considered as the primary measure to formally safeguard the interactions, another type of agreements called "self-reinforced" is also used (Williamson, 1985). Dyer & Singh (1998) distinguished between formal safeguards of self-reinforcing, such as financial penalties for the breach of obligations (Williamson, 1985), and informal ones such as trust (Dyer, 1997; Gulati, 1995; Uzzi, 1997), linked directly to social capital. Thus, any relationship will be composed by interrelations between:

1. Relations based on dyadic trust. Transactions based on the direct contacts between two agents.
2. Relations based on generalized trust. Transactions that form part of the global network where the agent is involved.

Relations Based on Dyadic Trust

Empirical investigation into social capital suggested that dyadic trust (or specific trust between two agents) reduces the risk of appearance of opportunistic behaviors (Gulati, 1995) allowing higher transparency on the exchange. For this reason, an agreement with a high component of dyadic trust would contribute to the increase of the exchange of resources among agents (Kale, Singh & Perlmutter, 2000). Williamson (1985) stated that this reciprocity creates the perception of an objective of "mutual interest" (p. 155), and "transforms a unilateral relation with an agent into a bilateral one" (p. 191) promoted by the similarities between agents and the previous history of interactions, in a way that will allow a gradual appearance of informal mechanisms of control in the exchanges (Jones, Hesterly & Borgatti, 1997).

This mutual interest tries to reduce the risk of opportunism when establishing the terms of the exchange and by consequence, the existence of an asymmetric condition between them (Coleman, 1990) derived from the formal agreements. If an agent in an asymmetric relationship cannot achieve a certain degree of trust and does not have the possibility of increasing the frequency of the transactions, he/she will exclusively establish formal mechanisms to protect himelf/herself from the absence of reliability of the partner. Gulati (1995) showed that trust allows two involved firms in an agreement to reduce their dependency on formal mechanisms to manage their relations.

Coleman (1990) explained that the decisions taken by an agent under the conditions of asymmetry are two: first, to maintain or break off trust and secondly, to invest or not in trust. In the first case, Coleman (1990) assured that the fact that an agent could maintain or break off his/her trust on others depends on the immediate benefits that he/she can obtain (derived from dyadic trust) as opposed to those he/she would have to forego. Therefore, dyadic trust would affect the agents in terms of cost-benefit, when comparing the probability of maintaining trust with the ratio of future, potential losses and earnings (Coleman, 1990). Secondly, Coleman (1990) stated that the additional incentives that exist when investing on trust in a relation lead to the following consequences: (1) the agents involved in an asymmetric transaction will have an incentive to transform agreements into a relation based on dyadic trust (and as a result to convert progressively the asymmetry into symmetry) and, (2) if a relation implies dyadic or mutual trust, both parties would have to be more prone to show mutual reliability, since shared expectations will be created and developed.

For this reason, the objective of an agent will be to change an asymmetric relation into a symmetrical one through the informal control mechanisms according to the increase of the history of interactions and the ratio of future potential earnings. Dyer (1997) stated that the expectation of fulfillment of safeguard increases when the agents involved in a relation: (1) show an engagement to maintain the interaction in the future, (2) increase the volume and variety of shared resources and (3) use informal safeguards to manage the relationship. Therefore, we must consider the reliability of the agents involved in a relationship as an essential attribute in the generation of dyadic trust since it makes easier the exchange of resources (Uzzi, 1996, p. 678), and increases the possibility of carrying out new transactions (Fukuyama, 1995, p. 27).

Relations Based on Generalized Trust

Fukuyama (1995) assured that the existence of generalized trust is extremely important for the economic result since it is a reason to trust others only because they are part of the same social system, exerting an effective role as a coordination mechanism. The agents involved in a social system (and who maintain relations within the global structure) and whose cohesion is based primarily on generalized trust are capable of achieving more effective results.

As the agents involved in a relationship could generate mutual trust through repeated direct contacts or through the proved reliability by the fulfillment of obligations and expectations, trust could also be obtained from the embeddedness of this relationship in a social structure and as a consequence, it would generate good reputation within the network.

For Portes (1998) this embeddedness has two consequences: (1) the donor's return may come not directly from the recipient but from the collectivity as a whole in the form of status honor or approval and, (2) the collectivity itself as a guarantor that whatever debts are incurred will be repaid (p. 8-9).

First, we could define reputation as the reliability of an individual agent or a group. Reputation could be used as a safeguard to foresee a possible deceptive or cooperative behavior. Consequently, the reputation originated by the social structure of the network allows the reduction of uncertainty of collective behavior when providing valuable information about the reliability and goodwill of others (Jones, Hesterly & Borgatti, 1997). Therefore, it serves as a base for the development of control norms to regulate all the exchanges of the network. In addition, the reputation is important under conditions of uncertainty of the environment since when the latter increases, the agents located in the global network of relations used to be more worried to have trustworthy information on the behavior of their potential partners within the social structure.

On the other hand, the information on reputation in a social structures could be inadequate or misinterpreted when it is diffused through large chains of ties in the network. In addition, the information could be distorted when it is filtered by many and diverse parties (Jones, Hesterly & Borgatti, 1997) if they do not share global, determined principles of collective action. For this reason, the global structure as a collective entity has additional mechanisms that could serve to generate relations based on generalized trust with all its members. These sources of social capital are based mainly on all the mechanisms to guide the collective and cooperative action within the group and to penalize deceptive behaviors when some of these members behave in an opportunistic way.

Secondly, the existence of a common identity and shared norms by the members of the network is another essential mechanism to strengthen relations based on generalized trust. Common values and shared norms could generate reliability in the social system for their members. As we stated earlier, the common identity of a social group is a system of fundamental assumptions that specifies the role, general relations and norms to be shared by the members. These assumptions try to coordinate independent tasks between independent entities so that the more complicated tasks can be completed.

Therefore, if a member of the network acts against these common assumptions, the general norms would act as a control mechanism allowing the rest of members to sanction the agent who violates its values and norms (Jones, Hesterly & Borgatti, 1997) being able to exclude him/her from the system for a brief or an indefinite period of time. According to Jones, Hesterly & Borgatti (1997), these collective sanctions act as a safeguard in the exchanges within the system since they define and reinforce the parameters of acceptable behavior when showing the consequences of violating the shared values and norms.

Definitely, when a social structure collectively maintains a group of values and norms, the possibility of appearance of opportunistic problems diminishes. In these situations, the existence of generalized trust (in the global group of relations) could be a substitute for the formal contracts and the systems of control that the collective entities conceive to control the members that could hinder the cooperative action (Leana & Pil, 2006).

FUTURE RESEARCH DIRECTIONS

The social capital construct has a special importance as an explanatory factor of the business outcomes in a global socioeconomic context. Because of this importance, we offer some possible directions for future research that we consider of special interest.

First, we examine in depth the conditions under which social capital is positively related to the firm's outcomes. In order to understand how a firm can obtain and maintain over time a competitive advantage through its network of relations, we will have to consider the conditions under which a firm is better equipped than its competitors to obtain superior benefits derived from its ability to use social capital.

Exploration and exploitation are the two processes associated to this strategic ability (Rowley, Behrens & Krackhardt, 2000). Exploitation requires the use of the existing information and knowledge to improve the efficiency and to increase the derived benefits from the strategy. The exploration involves the search for new information and emerging innovations that will produce future benefits. Following this same line of research, we could enumerate that:

1. Both processes occur in a simultaneous way. Therefore, it would be very interesting to study the type and proportion of the resources and to analyze the necessary capacities for each process according to the environment in which the firm competes.
2. It would be interesting to find out if firms, according to the environment's conditions, favor one type of process or another. In addition, if the condition of the environment changes, whether the social capital of the firm changes too in order to maintain its competitive advantage.

Secondly, we consider the possible relations of the strategic direction of the firm with the different components of social capital. Some of the strategic decisions that the firm can adopt could be formulated in terms related to social capital and in order to obtain additional profits. Thus, directed studies toward the analysis of the implications appear:

1. Formulating a direction of development (such as expansion, related or non related diversification) will involve a specific configuration of social capital (cohesive or diverse network and relations based on a type of specific trust).
2. Choosing the development method also requires a particular configuration of social capital. Thus, if a firm decides to grow internally, externally or by looking for alliances with third parties, it must take into consideration a specific social capital configuration adapted to its decisions.

3. Deciding to compete outside of the firm's country can support the development of determined forms of social capital. The relations with the network of suppliers and customers or the possible ways of entry to a country could be reinforced by an adequate configuration of the different forms of social capital.

Finally, we progress in the identification and measurement of the primary components of social capital of the firms. Therefore, it would be necessary to develop a framework in which all the elements and interrelations are reflected, and that allow the classification of firms on the basis of its social capital configuration. Despite the fact that much of this information is reflected in the system of measurement of intangible assets and of intellectual capital, we need to understand and use all the information that provide the evolution of the "social capital balance" of a firm.

CONCLUSION

Social capital is a construct that includes various scientific disciplines. Because of this diversity, academics have proposed many definitions in recent years. In an attempt to classify all the existing meanings in the social literature, in this chapter we tried to look for common patterns in these meanings on the basis of centers of interest, regardless of the affected discipline. This way, our starting point was the definition proposed by Nahapiet & Ghoshal (1998) who delimited social capital as the resources that an agent can obtain from his/her network of relationships.

Our second point was to analyze this wide and diverse term. Therefore, we considered it of special importance to break down social capital into two dimensions: structural and relational. In addition, we determined the different variables that compose it. For a firm, the structural dimension of social capital relates to the analysis of the essential features of the structure of the networks. Similarly, the relational dimension of social capital comprises two variables: the relational characteristics and the process of having relations with other agents. This classification attempts to bring forward a new orientation on the composition of each one of these dimensions.

If a firm wants to analyze its social capital, it should take into consideration two measures. First, the structural patterns of the different networks in which it is involved. Second, the nature of the relationships. Thus, a firm has two possibilities: relations based on dyadic trust or relations based on generalized trust.

We considered that these options can be applied inside as well as outside of the firm, which is why its design and combination could significantly influence the results inside (through the efficient and coordination of activities) as well as outside (through access to valuable resources or possession of an excellent reputation within its environment) providing it in both cases with a competitive advantage which results in obtaining higher returns over time.

To conclude, we affirm that in order to maximize the potential of its competitive advantage, a firm would have to manage and develop all its networks and relations in each one of the different levels within its socioeconomic context.

REFERENCES

Adler, P. S., & Kwon, S. W. (2002). Social capital: Prospects for a new concept. *Academy of Management Review*, *27*(1), 17–40. doi:10.2307/4134367

Bourdieu, P., & Wacquant, L. (1992). *An invitation to reflexive sociology*. Chicago: University of Chicago Press.

Bueno, E. (1998). El capital intangible como clave estratégica en la competencia actual. *Boletín de Estudios Económicos*, *53*(164), 207–229.

Burt, R. S. (1992). *Structural holes: The social structure of competition*. Cambridge, MA: Harvard University Press.

Burt, R. S. (1997). The contingent value of social capital. *Administrative Science Quarterly*, *42*(2), 339–365. doi:10.2307/2393923

Capaldo, A. (2006). Network structure and innovation: The leveraging of a dual network as a distinctive relational capability. *Strategic Management Journal*, *28*(6), 585–608. doi:10.1002/smj.621

CIC. (2003). *Modelo Intellectus: Medición y gestión del Capital Intelectual. Documento Intellectus nº 5*. Madrid: IADE.

Coleman, J. S. (1988). Social capital in the creation of human capital. *American Journal of Sociology*, *94*(Supplement), S95–S120. doi:10.1086/228943

Coleman, J. S. (1990). *Foundations of social theory*. Cambridge, MA: Harvard University Press.

Dyer, J. H. (1997). Effective interfirm collaboration: How firms minimize transaction costs and maximize transaction value. *Strategic Management Journal*, *18*(7), 535–556. doi:10.1002/(SICI)1097-0266(199708)18:7<535::AID-SMJ885>3.0.CO;2-Z

Dyer, J. H., & Singh, H. (1998). The relational view: Cooperative strategy and sources of interorganizational competitive advantage. *Academy of Management Review*, *23*(4), 660–679. doi:10.2307/259056

Florin, J., Lubatkin, M., & Schulze, W. (2003). A social capital model of high-growth ventures. *Academy of Management Journal*, *46*(3), 374–384. doi:10.2307/30040630

Fukuyama, F. (1995). *Trust: The social virtues and the creation of prosperity*. New York: Free Press.

Granovetter, M. S. (1973). The strength of weak ties. *American Journal of Sociology*, *78*, 1360–1380. doi:10.1086/225469

Granovetter, M. S. (1985). Economic action and social structure: The problem of embeddedness. *American Journal of Sociology*, *91*, 481–510. doi:10.1086/228311

Granovetter, M. S. (1992). Problems of explanation in economy sociology. In Nohria, N., & Eccles, R. (Eds.), *Networks and organizations: Structure, form and action* (pp. 25–56). Boston: Harvard Business School Press.

Gulati, R. (1995). Social structure and alliance formation patterns: A longitudinal analysis. *Administrative Science Quarterly*, *40*(4), 619–652. doi:10.2307/2393756

Gulati, R., Nohria, N., & Zaheer, A. (2000). Strategic networks. *Strategic Management Journal*, *21*(3), 203–215. doi:10.1002/(SICI)1097-0266(200003)21:3<203::AID-SMJ102>3.0.CO;2-K

Inkpen, A. C., & Tsang, E. W. K. (2005). Social capital, networks, and knowledge transfer. *Academy of Management Review*, *30*(1), 146–165. doi:10.2307/20159100

Jones, C., Hesterly, W. S., & Borgatti, S. P. (1997). A general theory of network governance: Exchange conditions and social mechanisms. *Academy of Management Review*, *22*(4), 911–945. doi:10.2307/259249

Kale, P., Singh, H., & Perlmutter, H. (2000). Learning and protection of proprietary assets in strategic alliances: building relational capital. *Strategic Management Journal*, *21*(3), 217–237. doi:10.1002/(SICI)1097-0266(200003)21:3<217::AID-SMJ95>3.0.CO;2-Y

Kogut, B., & Zander, U. (1996). What firms do? Coordination, identity and learning. *Organization Science*, *7*(5), 502–518. doi:10.1287/orsc.7.5.502

Kostova, T., & Roth, K. (2003). Social capital in multinational corporations and a micro-macro model of its formation. *Academy of Management Review, 28*(2), 297–317. doi:10.2307/30040714

Labianca, G., Brass, D. J., & Gray, B. (1998). Social networks and perceptions of intergroup conflict: The role of negative relationships and third parties. *Academy of Management Journal, 41*(1), 55–67. doi:10.2307/256897

Lane, P. J., & Lubatkin, M. (1998). Relative absorptive capacity and interorganizational learning. *Strategic Management Journal, 19*(5), 461–477. doi:10.1002/(SICI)1097-0266(199805)19:5<461::AID-SMJ953>3.0.CO;2-L

Leana, C. R., & Pil, F. K. (2006). Social capital and organizational performance: Evidence from urban public schools. *Organization Science, 17*(3), 353–368. doi:10.1287/orsc.1060.0191

Leana, C. R., & Van Buren, H. J. III. (1999). Organizational social capital and employment practices. *Academy of Management Review, 24*(3), 538–545. doi:10.2307/259141

Lin, N. (2001). Building a network theory of social capital. In Lin, N., Cook, K., & Burt, R. (Eds.), *Social capital: Theory and research* (pp. 3–29). Piscataway, NJ: Aldine Transaction.

Moran, P. (2005). Structural vs. relational embeddedness: Social capital and managerial performance. *Strategic Management Journal, 26*(12), 1129–1151. doi:10.1002/smj.486

Nahapiet, J., & Ghoshal, S. (1998). Social capital, intellectual capital and the organizational advantage. *Academy of Management Review, 23*(2), 242–266. doi:10.2307/259373

Oh, H., Labianca, G., & Chung, M. H. (2006). A multilevel model of group social capital. *Academy of Management Review, 31*(3), 569–582. doi:10.2307/20159229

Portes, A. (1998). Social capital: Its origins and applications in modern sociology. *Annual Review of Sociology, 24*, 1–12. doi:10.1146/annurev.soc.24.1.1

Putnam, R. D. (1993). *Making Democracy Work: Civic Traditions in Modern Italy*. Princeton, NJ: Princeton University Press.

Putnam, R. D. (1995). Bowling alone: America's declining social capital. *Journal of Democracy, 6*(1), 65–78. doi:10.1353/jod.1995.0002

Ring, P. S., & Van de Ven, A. H. (1992). Structuring cooperative relationships between organizations. *Strategic Management Journal, 13*(7), 483–498. doi:10.1002/smj.4250130702

Rousseau, D. (1995). *Psychological Contracts in Organizations*. Thousand Oaks, CA: Sage.

Rowley, T., Behrens, D., & Krackhardt, D. (2000). Redundant governance structures: An analysis of structural and relational embeddedness in the steel and semiconductor industries. *Strategic Management Journal, 21*(3), 369–386. doi:10.1002/(SICI)1097-0266(200003)21:3<369::AID-SMJ93>3.0.CO;2-M

Tsai, W. (2000). Social capital, strategic relatedness and the formation of intraorganizational linkages. *Strategic Management Journal, 21*(9), 925–939. doi:10.1002/1097-0266(200009)21:9<925::AID-SMJ129>3.0.CO;2-I

Tsai, W. (2001). Knowledge transfer in intraorganizational networks: Effects of network position and absorptive capacity on business unit innovation and performance. *Academy of Management Journal, 44*(5), 996–1004. doi:10.2307/3069443

Tsai, W., & Ghoshal, S. (1998). Social capital and value creation: The role of intrafirm networks. *Academy of Management Journal, 41*(4), 464–176. doi:10.2307/257085

Uzzi, B. (1996). The sources and consequences of embeddedness for the economic performance of organizations: The network effect. *American Sociological Review*, *61*(4), 674–698. doi:10.2307/2096399

Uzzi, B. (1997). Social structure and competition in interfirm networks: The paradox of embeddedness. *Administrative Science Quarterly*, *42*(1), 35–67. doi:10.2307/2393808

ADDITIONAL READING

Adler, P. S. (2001). Market, hierarchy and trust: The knowledge economy and the future of capitalism. *Organization Science*, *12*(2), 215–234. doi:10.1287/orsc.12.2.215.10117

Baker, W. (1990). Market networks and corporate behaviour. *American Journal of Sociology*, *96*(3), 589–625. doi:10.1086/229573

Belliveau, M. A., O`Really, C. A. I. I. I., & Wade, J. B. (1996). Social capital at the top: Effects of social similarity and status on CEO compensation. *Academy of Management Journal*, *39*(6), 1568–1593. doi:10.2307/257069

Black, J.A., & Boal, K.B. (1994). Strategic resources: Traits, configurations and paths to sustainable competitive advantage. *Strategic Management Journal, 15*(Winter special number), 131-148.

Borgatti, S. P., & Foster, P. C. (2003). The network paradigm in organizational research: A review of typology. *Journal of Management*, *29*(6), 991–1013.

Boxman, E. A. W., De Graaf, P. M., & Flap, H. D. (1991). The impact of social and human capital on the income attainment of ducth managers. *Social Networks*, *13*, 51–73. doi:10.1016/0378-8733(91)90013-J

Brass, D. J., Galaskiewicz, J., Greve, H. R., & Tsai, W. (2004). Taking stock of networks and organizations: A multilevel perspective. *Academy of Management Journal*, *47*(6), 795–817. doi:10.2307/20159624

Ghoshal, S., & Moran, P. (1996). Bad for practice: A critique of the transaction cost theory. *Academy of Management Review*, *21*(1), 13–47. doi:10.2307/258627

Gulati, R. (1999). Network location and learning: The influence of network resources and firm capabilities on alliance formation. *Strategic Management Journal*, *20*(5), 397–420. doi:10.1002/(SICI)1097-0266(199905)20:5<397::AID-SMJ35>3.0.CO;2-K

Gulati, R., & Gargiulo, M. (1999). Where do interorganizational networks come from? *American Journal of Sociology*, *104*(5), 1439–1493. doi:10.1086/210179

Hansen, M. T. (1999). The search-transfer problem: The role of weak ties in sharing knowledge across organization subunits. *Administrative Science Quarterly*, *44*(1), 82–111. doi:10.2307/2667032

Hansen, M. T., Mors, M. L., & Lovas, B. (2005). Knowledge sharing in organizations: multiple networks, multiple phases. *Academy of Management Journal*, *48*(5), 776–793. doi:10.2307/20159697

Ibarra, H. (1993). Network centrality, power, and innovation involvement: determinants of technical and administrative roles. *Academy of Management Journal*, *36*(3), 471–501. doi:10.2307/256589

Koka, B. J., & Prescott, J. E. (2002). Strategic alliances as social capital: A multidimensional view. *Strategic Management Journal*, *23*(9), 795–816. doi:10.1002/smj.252

Larson, A. (1992). 'Network dyads in entrepreneurial settings: A study of the governance of exchange processes. *Administrative Science Quarterly*, *37*, 76–104. doi:10.2307/2393534

Lin, N. (1999). Social networks and status attainment. *Annual Review of Sociology*, *25*, 467–487. doi:10.1146/annurev.soc.25.1.467

Lin, N., Ensel, W. M., & Vaughn, J. C. (1981). Social resources and strength of ties: structural factors in occupational status attainment. *American Sociological Review*, *46*(X), 393–405. doi:10.2307/2095260

Ostrom, E. (1998). A behavioural approach to the relational choice theory of collective action. *The American Political Science Review*, *92*, 1–22. doi:10.2307/2585925

Podolny, J. M., & Baron, J. N. (1997). Resources and relationships: Social networks and mobility in the workplace. *American Sociological Review*, *62*(5), 673–693. doi:10.2307/2657354

Portes, A., & Sensenbrenner, J. (1993). Embeddedness and immigration: Notes on the social determinants of economic action. *American Journal of Sociology*, *98*(6), 1320–1350. doi:10.1086/230191

Robison, L., Schmid, A., & Siles, E. (2002). Is social capital really capital? *Review of Social Economy*, *60*(1), 1–21. doi:10.1080/00346760110127074

Seibert, S. E., Kraimer, M. L., & Liden, R. C. (2001). A social capital theory of career success. *Academy of Management Journal*, *44*(2), 219–237. doi:10.2307/3069452

Walker, G., Kogut, B., & Shan, W. (1997). Social capital, structural holes and the formation of an industry network. *Organization Science*, *8*(2), 109–125. doi:10.1287/orsc.8.2.109

Yli-Renko, H., Autio, E., & Sapienza, H. J. (2001). Social capital, knowledge acquisitions, and knowledge exploitation in young technology-based firms. *Strategic Management Journal*, *22*(6/7), 587–613. doi:10.1002/smj.183

Zaheer, A., McEvily, B., & Perrone, V. (1998). Does trust matter? Exploring the effects of interorganizational and interpersonal trust on performance. *Organization Science*, *9*(2), 141–159. doi:10.1287/orsc.9.2.141

KEY TERMS AND DEFINITIONS

Social Capital: The sum of the actual and potential resources embedded within, available through, and derived from the network of relationships possessed by an individual or social unit. Social capital thus comprises both the network and the assets that may be mobilized through that network.

Structural Embeddedness: The properties of the social system that defines the global patterns of the connections between all the agents.

Relational Embeddedness: The quality of relations and assets embedded in these relations in a specific period of time.

Cohesive Network: Network in which all of the members are connected by strong ties and direct contact between them.

Structural Hole: Exists between two agents connected to a central one if they are not connected between each other even though they are connected to the central one.

Dyadic Trust: Trust between two parties who have direct knowledge of one another manifested through repeated interactions in the exchanges over time.

Generalized Trust: Trust based on norms and behaviors that are generalized to others in the social unit as a whole.

Chapter 13
Social Capital, Knowledge and Technological Innovation

Salma Alguezaui
Universita' del Salento, Italy

Raffaele Filieri
University of Limerick, Ireland

ABSTRACT

Firm's environment is becoming more complex and dynamic, consequently continuous innovation is considered a key element to gain competitive advantage. According to the knowledge-based view, knowledge is considered the central element in the innovation process and its effective management is the key to achieve higher innovation performance. Since social networks involve knowledge and information embedded into inter-organizational network, social capital is considered the engine fuelling the innovation activities. Through an intensive review of the literature, the chapter seeks to uncover the effect of social capital on innovation through enhancing its fundamental knowledge-based activities. The social capital literature provided controversial approaches about its contribution to the innovation performance. The chapter contributes to the literature by mapping the different views, trying to clarify its contribution to the enhancement of knowledge-based activities, and revealing the effect of different social configurations on innovation outcomes. To do so, the chapter first analyzes the different views of social capital and explains its contribution within the organization studies. Further, the key role of the knowledge and its management to the innovation outcomes is discussed. Finally, the chapter clarifies the implications of social capital to the innovation performance and its typologies as radical or incremental ones, as well as designs a theoretical framework for recognizing both its beneficial and constraining effect. Further, the chapter proposes a balanced approach to the structural social capital reconciling two controversial camps in the literature.

DOI: 10.4018/978-1-61520-875-3.ch013

INTRODUCTION

Several transformations in the business environment have changed the competitive environment to be more dynamic and complex. These environmental features are continuously expanding due to the increasing number of the factors influencing the daily operations of the firm, their interdependency, and their dynamicity. Accordingly, the modern firm is urged to cope and adapt to its environmental needs in order to gain competitive advantage, or even to survive. A McKinsey study of the life expectancy of firms have shown a drastic drop of its average value from 30 to 15 years over the period 1975-2005 (Foster and Kaplan, 2001). Another research that examined 6,772 firms from different industries over a period lasting 25 years, have shown that only a small percentage of these firms have recorded high economic performance (Wiggins and Ruefli, 2002). Within this realm, firms are growingly exposed to environmental threats and eventually to market failure, and only few examples succeeded to adapt to changes in the environment and prosper. Being successful today does not guarantee to survive tomorrow. Continuous innovation is considered a key driver to firms' competitiveness and growth. Thus, as it has been widely taken for granted in the literature, technology, and its underlying knowledge, plays an important role in building firms' sustainable competitive advantage and guaranteeing their success through developing new products, discovering new processes, altering competitive rules, or readjusting the boundaries of their industry (Utterback, 1994).

A stream of research has emerged recognizing the centrality of knowledge and the capability to create such knowledge as the most important source for sustainable competitive advantage (Toffler, 1990; Prahalad and Hamel, 1990; Drucker, 1993; Nonaka, 1994; Leonard-Barton, 1995; Grant 1996; Zack, 1999; Teece et al. 1990). The literature perceives knowledge as the key element of a firm's strategy and emphasizes the importance of strategic activities to manage knowledge (Grant, 1996; Huber, 1991) and to translate dispersed, tacit, and explicit competencies into an aggregated body of organizational knowledge (Nonaka, 1994), giving rise to the knowledge-based view of the firm (Grant, 1996; Nonaka and Takeuchi, 1995). Knowledge is the most valuable intangible asset and is considered the core element for innovation (Clark and Fujimoto, 1990). Firms are increasingly drawing on knowledge beyond their boundaries to perform their innovation activities giving rise to the open innovation model (Chesbrough, 2003). As a matter of fact, innovation is the outcome of effective and efficient production, diffusion and exploitation of novel knowledge.

Although overlooked in the literature over a long period of time, social networks are perceived as the main drivers to learn and exploit competencies, especially those lying outside the firm's boundaries. The social capital of innovators provide information, knowledge and different resources necessary to perform their innovation activities. Though, the literature on social capital and its contribution to the innovation performance is still contentious. Different approaches to social capital has emerged explaining the higher performance of innovation. Focusing on the structural approach, two main perspectives are considered along this chapter: sparse network based on Burt's structural hole theory and cohesive network based on Coleman's argument. Both approaches received empirical support by numerous research studies (Hargadon and Sutton, 1997; Burt, 2004; Obstfeld, 2005; McEvily and Reagans, 2005; Uzzi and Spiro, 2005, Fleming et al., 2007). It is still then ambiguous which is the optimal structural configuration of social capital that better contributes to the innovation performance of a focal firm. Therefore, the chapter discusses about the followers of Burt's and Coleman's approach by displaying the benefits and risks of both structural configurations to the innovation performance and proposing a balanced approach reconciling both camps.

Drawing on the social network theory and the knowledge-based view of the firm, the chapter will seek to clarify the effect of social capital on innovation and its different forms, highlighting its both edges as enabling factor as well as limiting or even detrimental factor. To achieve that, the chapter has been organized into three main sections. The first section will analyze the different views of social capital and explains its contribution within the organization studies. The second section, relying on the knowledge based view of the firm, will highlight the key role of knowledge and its management in the emerging framework of the innovation process. The last section makes in evidence the relationship between social capital and innovation as either an enabler or a limiting factor focusing on a structural perspective. This section will also identifies the different forms of a firm's social network (e.g. cohesive *vs.* sparse networks) and its distinct influence on typologies of innovation (e.g.: radical *vs.* incremental), proposing a more balanced approach.

SOCIAL CAPITAL

Firms are increasingly faced with several daily challenges to grow, and even to survive. Nowadays, different environmental characteristics urge firms to continuously learn and innovate, either incrementally or radically. Firms need to develop and exploit their internal knowledge bases and to draw on and explore the knowledge across their boundaries to innovate. Therefore, firms rely on their external network or their social embeddedness as an essential factor to create, develop, and exploit new opportunities. Thus, social networking activities, and in specific terms social capital, represents a source of competitive advantage and a prerequisite for survival as it provides the focal firm with access to idiosyncratic resources and a network of relationships, enhancing then its capacity for innovation.

The social capital concept is increasingly gaining a considerable interest among academic researchers illustrated through the growing number of articles in the field during the last twenty years (for general review Adler and Kwon, 2002; Borgatti and Foster, 2003). The concept has been analyzed at different levels of analysis (individual, network, community levels) and used in different academic disciplines such as sociology, anthropology, political sciences, economy, and lately organizational studies, etc. The interdisciplinary and inter-level use of the concept has created a huge confusion. Even more, the growing number of scientific articles addressing the social capital issue within the organizational studies led to controversy results. Through reviewing the social capital literature, Adler and Kwon (2002) tried to clarify the concept through developing a conceptual framework identifying the sources, the benefits, the risks, and the contingencies of social capital. The authors highlighted the notion of social capital as an important explanatory factor for various organizational issues such as professional development, ability to find a job, resource sharing among business units, product innovation and the reinforcement of relations with suppliers (Adler and Kwon, 2002). Borgatti and Foster (2003) also recognized the importance of the concept to explain a number of dilemmas in the organizational studies including: mobility, employment, leadership, power, entrepreneurship, team performance, and creativity. Therefore, for the aim of clarity, the next section will provide a review of the historical roots of the concept explaining its various definitions over time, reveal its contribution within the organization studies highlighting both its beneficial and detrimental effects, and identify its different dimensions.

Social Capital: A Revisited Concept

The concept of 'Social Capital' has been officially first articulated by the progressive educator L.

Judson Hanifan in 1916, who noted the need for and importance of renewed community involvement to sustain democracy and development. However, the contemporary literature on social capital recognized Jacobs' (1961) and Loury's (1977) work in the origin of the modern concept of social capital. In fact, the term was coined in the community studies by Jane Jacobs (1961) to emphasize the centrality of strong and interpersonal relationships. These kind of ties lay out the foundations for developing trust, cooperation and collective actions within communities guaranteeing the survival and well-functioning of city neighbourhoods. Later on, the economist Glen Loury (1977) brought the term to support his critiques concerning the neoclassic theories of racial income inequality and their policy implications. However, neither of those authors provided detailed analysis concerning the concept. In fact, the term social capital is extensively traced back to three fundamental authors who laid out the foundations of the concept: Pierre Bourdieu (1986), James Coleman (1988; 1990), and Robert Putnam (1993; 1995; 2000). Additionally, the literature on social capital refers to other major contributors such as Granovetter (1973, 1985), Burt (1992; 1997; 2005), Lin (2001), Lin et al. (2001), Portes and Sensenbrenner (1993), Portes (1998) and Flap and De Graaf (1986).

Bourdieu's work (1980) came to refine the concept and to systematically develop the theory. His focus was oriented towards explaining the deliberate development of social capital and the benefits derived from constructing such resource (Bourdieu, 1985). As a matter of fact, individuals need to invest in order to acquire a certain social capital. Further, he claims out that any form of capital including social one is convertible to economic capital. Bourdieu then defined social capital as "*the aggregate of the actual or potential resources which are linked to possession of a durable network of more or less institutionalized relationships of mutual acquaintance or recognition*" (Bourdieu 1985: 248). This definition reveals two major aspects of the notion: the social ties enabling the focal actor to access resources embedded in his network as well as the quantity and quality of those resources (Portes, 1998). After that, Coleman relied on the studies held by two sociologists, Lin and Granovetter, and two economists, Loury and Ben-Porath, to analyze the role of social capital in creating human capital. Parallel to Bourdieu's definition, social capital in Coleman terms referred to as "*a variety of entities with two elements in common: They all consist of some aspect of social structures, and they facilitate certain action of actors - whether persons or corporate actors- within the structure.*" (Coleman 1988:98; 1990: 302). Accordingly, social capital is defined in terms of structure and the functions it serves to the focal actor. Further, Coleman underlined the relations between social capital and access to resources.

Portes (1998) noted that Bourdieu, in contrast to Coleman, differentiated between the resources embedded into a social setting and the ability of actors within a social structure to obtain those resources. These authors laid on the foundations for further development of the literature and advancement of the concept's definition. Reviewing different definitions, the literature comes to the agreement that social capital refers to the benefits that accrue to a focal actor from being embedded into a social structure. Still, literature defines social capital in different modes such as structure (Coleman, 1988), resources (Bourdieu, 1986; Putnam, 1995), culture of confidence, capacity (Portes, 1998), or an asset within a network (Lin, 1999).

Adler and Kwon (2002) provided a systematic analysis of the literature on social capital. Through examining the previous definitions, the authors argued that the differences among the term's definitions are derived from two main reasons (Adler and Kwon, 2002). First of all, the definition of social capital is related to the researcher's focus on its substance, its sources, or its consequences.

Second, the difference comes from the researchers' focus on the relations tying a focal actor with other actors (external ties), the structure of relations among actors within the collectivity (internal ties), or both. Thereafter, Adler and Kwon (2002) tried to introduce a definition that reconciles the different forms of definitions advanced so far in the literature. Thus, social capital is "*the goodwill available to individuals or groups. Its source lies in the structure and content of the actor's social relations. Its effects flow from the information, influence, and solidarity it makes available to other actor*" (Adler and Kwon, 2002: 23).

The literature on social capital exhibits a general disagreement on defining the concept. The multiplicity of definitions is derived from the numerous studies' approaches, levels of analysis and sources. Social researchers perceive differently the sources of social capital as networks (Coleman, 2000; Putnam, 1995; Burt, 1992; Granovetter, 1973), norms (Portes, 2000; Putnam, 1995), social beliefs (Nahapiet and Ghoshal, 1998) and rules (Adler and Kwon, 2000). Further, there are differences related to the level of analysis undertaken in different researches as individual (Coleman, 2000), a firm (Baker, 1990), a geographic region (Putnam, 1995), a nation (Fukuyama, 1995) or a network (Burt, 1992). Through analyzing the various definitions, there are two different topologies of social capital that were treated throughout the literature. In fact, the social capital was perceived as either a public or private good. The public social capital refers to those assets that flow to and benefit all the individuals making part of a community; whereas, the private one refers to the potential goods that a focal individual may derive from its social network as an exclusive private property. For the aim of the chapter, social capital will be approached as a private good that benefits specific actors in order to attain their specific goals. Next subsection will try to uncover the issue of social capital within the organization studies.

Social Capital in Organization Studies

Recently, management scholars have shifted their view of the firm from the traditional models, whose building blocks are based on opportunism and market transactions (Williamson, 1975), to a more socially-based models where knowledge and its related activities are the key elements to the well-functioning of the firm. One key notion to the social theories of the firm is the social capital concept. Along the same vein, Adler and Kwon (2002) claimed that social capital is 'gaining currency' in numerous research areas related to organization studies including: career success (Burt, 1992), job search (Granovetter, 1995), innovation (Hansen, 1999; Tsai and Ghoshal, 1998), intellectual capital (Hargadon and Sutton, 1997), corporate entrepreneurship (Chung and Gibbons, 1997), new business formation (Walker et al, 1997), supply chain relations (Uzzi, 1997), regional production networks (Romo and Schwartz, 1998) and inter-firm learning (Kraatz, 1998). Within this context, social capital has been generally perceived as the resources (e.g.: norms and sanctions, complementary resources, tacit knowledge and so forth) that can be derived from a personal network or a set of ties to achieve specific goals. In fact, Gabbay and Leenders (1999: 3) define 'corporate social capital' as "*the set of resources, tangible or virtual, that accrue to a corporate player through the player's social relationships, facilitating the attainment of goals.*"

Several recent empirical studies focused on the beneficial effect of social capital to explain a wide range of organizational behaviour and practices at different levels of analysis. Related to the individual-level practices within a firm, authors investigated the impact of social capital on career success, executive rewarding system, recruitment procedure (Belliveau, O'Reilly and Wade, 1996; Fernandez, Castilla and Moore, 2000;

James, 2000; Podolny and Baron, 1997). The social capital has been also researched to explain firm-related practices and behaviours including resource allocation (Bouty, 2000), innovation (Tsai and Ghoshal, 1998) and change management (Gargiulo and Benassi, 2000). Further, social capital approach has been extensively researched in different relationships between organizations (Uzzi, 1997; Kraarz, 1998, Tsai, 2000). In fact, social capital is increasingly being considered as a valuable intangible asset for firms since it provides a network of relationships that can lead to sustained competitive advantage.

Most researchers who addressed the issue of social capital emphasized heavily its positive effect and its productive implications to explain different organizational phenomena and practices. Even more, Burt (2000: 32) refers to *"The social capital metaphor is that people who do better are somehow better connected."* However, Gabbay and Leenders (1999), in their book named 'Corporate Social Capital and Liability,' provided a more balanced and realistic framework analyzing the impact of social networks on the attainment of firms' and leaders' objectives within the context of Small to Medium-sized Enterprises (SMEs). In fact, the authors argued that social capital, in addition to its beneficial effects as enhancing the search for information and other resources, may be an inhibiting factor as impeding the identification of potential opportunities caused by over-embeddedness. Accordingly, Gargiulo and Benassi (1999) described the beneficial and detrimental effects of social capital as the bright and dark sides respectively. In fact, start-ups may benefit from their strong ties (e.g. family ties, friendships, previous working environment) in the early stages of their formation in order to get the needed resources to establish the foundations of the firm (as financial and human resources, emotional support, etc.). At a later stage of the firm's life cycle, the start-up may be over-embedded within its network of strong ties providing a liability instead of a benefit to the firm performance, as it inhibits the start-up from sensing emerging opportunities and realizing potential growth. In addition to the previous authors, Locke (1999) criticized and identified the limitations of the traditional perspective focusing only on the positive side of social capital. He argued that actors may become highly embedded within their networks causing their loss of objectivity, which may result in excluding potentially beneficial ideas or actors. Locke (1999) presented evidence to the downside of social capital through illustrating the Japanese banking crisis, in which the banks put much importance to solidarity than adopting reliable banking practices. Therefore, the perspective claiming that 'the more social capital, the better is' does not hold always true. Within this realm, through an extensive review of the literature, Adler and Kwon (2002) developed an elaborate framework demonstrating both benefits as well as risks embedded in social capital. According to the authors, these risks can be derived from three possible flaws. First, actors may over-invest in specific relationships transforming these assets into liabilities. Second, the strong and localized ties of an actor may result in negative effects due to the exclusionary nature of cliques (e.g.: in-group). Third, the solidarity benefits that may arise from tight relationships may also inhibits the free-flow of novel information or ideas. Conversely to the general view emphasizing the positive and beneficial impact of social capital, there is also evidence for its detrimental implications. Thus, firms need to properly design and adequately invest in its social networking activities to effectively achieve their specific ends.

Dimensions of Social Capital

According to the literature, social capital of a firm represents a potential source of its competitive advantage (Adler and Kwon, 2002; Nahapiet and Ghoshal, 1998). Through analyzing its role in the creation of firm's intellectual capital, Nahapiet and Ghoshal (1998) identified three main dimensions to social capital: structural, relational and cognitive (see Figure 1). The first dimension describes the configuration of the social ties linking different actors, referring to the way the central player gain access to other actors' knowledge and resources. This is represented, for instance, by measures such as density, hierarchy, and connectivity. The second dimension, the relational one, is more concerned with the quality of the relationships that an actor has developed over a history of social interactions. This dimension focuses more on the normative infrastructure underlying a network of relationships such as social norms, obligations and expectation, as well as identity. The cognitive dimension is defined by resources affording shared basis for interpretations and representations among network members. This can be illustrated by shared codes and languages.

Nahapiet and Ghoshal (1998) investigated the effect of social capital dimensions on the creation of new intellectual capital focusing on the intra-firm level. Despite the critical role of social capital for explaining the managerial phenomena within firms (Nahapiet and Ghoshal, 1998; Gargiulo and Benassi, 2000), researchers are increasingly paying more attention to the inter-firm level of social capital (Uzzi, 1997; Kraatz, 1998). In fact, different studies have distinguished between internal and external social capital to recognize the network structural and relational aspects of social capital (Yli-Renko et al., 2000). Firm-internal social capital, on one hand, is captured in the extent and quality of the ties between individuals and units within a firm. On the other hand, firm-external social capital resides in the exchange relationships between firms and their various shareholders. Following the same research path, Inkpen and Tsang (2005) recently extended Nahapiet and Ghoshal's (1998) model to describe the social capital within distinct types of organizational groupings and to explore its facilitating conditions for knowledge transfer. The authors analyzed the structural, relational and cognitive dimensions of social capital across three types of networks: intracorporate networks, industrial districts, and strategic alliances. Inkpen and Tsang defined an intracorporate network as *a group of organizations operating under a unified corporate identity, with the headquarters of the network having controlling ownership interest in its subsidiaries* (2005: 148). The second organizational grouping referred to an industrial district, which consists of firms independently operating though sharing similar goals and geographic areas. Those enterprises rely on similar producers, draw on the same labour pool and aim at similar markets

Figure 1. Dimensions of social capital (adapted from Nahapiet & Ghoshal, 1998)

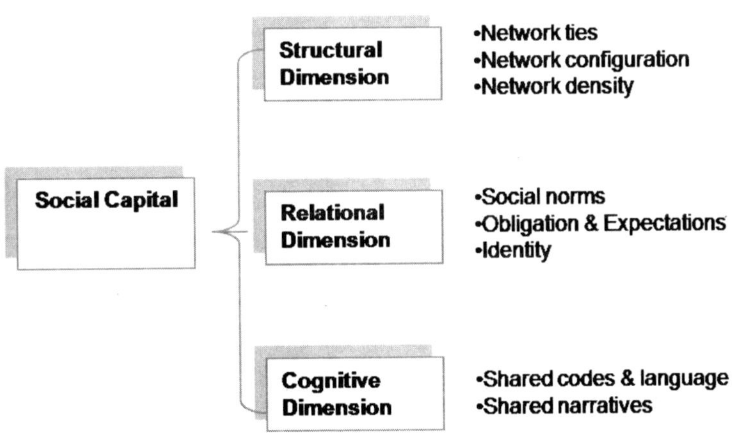

(Inkpen and Tsang, 2005). Instead, strategic alliances differ from the former two organizational groupings as they neither operate within similar geographical areas nor do they share the same formal structure. As a matter of fact, strategic alliances can be represented through horizontal relationships (from same position) or vertical ones (from different positions along the value chain) aiming at deriving common benefits to its participants (ibid). Accordingly, it is in the context of strategic alliance, which constitutes the basis of the emerging innovation framework, that we adopt Inkpen and Tsang's definition of social capital to highlight its contribution to the firm's innovation performance. Accordingly, the next section describes the emerging features of the new innovation paradigm and its implications to the external network structure of the firm. Further, for the sake of simplicity, the chapter concentrates its investigation on the structural dimension of social capital as it is still the most controversial issue in the literature. The authors consider the structural dimension as the source for other dimensions of social capital. Therefore, the subsequent sections try to reveal the effect of the structural dimension of social capital on firm's innovation performance.

A NEW FRAMEWORK FOR THE INNOVATION PROCESS

The Knowledge-Based View

The capacity to continuously innovate and to create business from technology has become the key element for the achievement and maintenance of competitive advantage (Day, 1994; Dickson, 1996; Han, Kim and Srivastava, 1998; Kotler, 1991; Porter, 1990). Product innovation is a complex, cross-functional and dynamic process, which is difficult to manage (Crawford, 1996). It consists of improved or entirely new services or products which are developed to the stage of their commercialization on the market, and whose novelty should be evident to either producers, consumers, suppliers or competitors.

Recently, the knowledge-based view of the firm recognizes the centrality of knowledge and its management as the main factor enhancing firms' ability to successfully innovate and then to compete (Penrose, 1959; Barney, 1991; Grant, 1996). The literature identifies two fundamental types of knowledge: tacit and explicit knowledge (Nonaka and Takeuchi, 1994). The type of knowledge that has more value for innovation is certainly the tacit one. Tacit knowledge potentially has more strategic value, being derived from particular circumstances and events, and thus unique and hard to imitate in other contexts (Zack, 2000). It is also difficult to codify and share in the natural language and is embedded in people's mind and in performing complex tasks (Nonaka and Takeuchi, 1994). Tacit knowledge can be classified into two main categories: technical and market knowledge. Forms of tacit market knowledge are described as brand image, evolving customer needs, emergent latent needs, and customers' feedback and so forth. Tacit market knowledge may enable innovations like the creation of new products, new category entries, additions to product lines, market extension, product improvements, repositioning, etc.. Another type of knowledge refers to the scientific or technical one, such as knowledge for building products, knowledge for minimizing environmental threats of products, knowledge about materials, technologies and the like. This kind of knowledge is fundamental for breakthrough innovations and product improvements. On the other hand, explicit knowledge or information is easy to codify and to communicate. This kind of knowledge, which may be either market or technical knowledge, includes facts, axiomatic propositions, and symbols such as information on size and growth of a market, production schedules, and so forth (Dyer and Nobeoka, 2000).

Knowledge management comprises methods, procedures and tools supporting the core activities: generate, store, codify and apply knowledge

(Heisig and Vorbeck, 1998). Knowledge is an intangible asset, and its management is more complex than managing any other physical asset. Managing knowledge is a very complex task as the benefits of acquiring knowledge are often difficult to recognize and even to measure after the knowledge acquisition. The importance of knowledge management is much more evident in industries characterized by high pace of change. In fact, *"New Product Development (NPD) can be especially costly and complex; when a high degree of component integration is needed; when a nascent technology is being developed and included in a new product; when a company develops a product with a partner; when a company develops a radical innovation"* (Kim and Wilemon, 2003: 15). Today, NPD requires different and cross-functional/industrial knowledge and capabilities and expensive technologies. These additional costs and requirements have further decreased the benefits of innovations (Chesbrough, 2003). Let's think to the knowledge required today to build a vehicle: fuel emissions, traffic safety, style and design, ergonomics, ICTs applications and so on. Managing such complexity requires continuous innovation and then an increased quantity and quality of tacit knowledge and investments in technologies and market knowledge. Within this context, knowledge management has been found to have several beneficial effects as to reduce cycle times of innovation, help firms prevent obsolescence, streamline their business processes, and respond to environmental changes.

Viewing the firm as a bundle of resources, Kogut and Zander (1992) asserted that innovation is created through new combinations of resources, and that knowledge sharing is a strategic mechanism for recombining existing ideas. The history of innovation is full of examples of technologies created by fusing together previously separated knowledge streams (Cradwell, 1995; Matusik and Hill, 1998). Indeed, Schumpeter assumed that innovation 'consists to a substantial extent of a recombination of conceptual and physical materials that were previously in existence' (Nelson and Winter, 1982: 30). Even if ideas or suggestions are not original or new, innovation can emerge through a new combination of these ideas and existing knowledge stocks. Kogut and Zander (1992) argued that the "combinatory capability" to generate new applications from existing knowledge may be a critical asset for sustaining competition. Grant (1996) argued that such a capability is a strategically significant resource to a successful organization. Accordingly, knowledge-building activities are considered as the main bases in determining the combinatory capability of a firm (Leonard-Barton, 1995). Therefore, the innovation process can be regarded as a cycle of knowledge acquisition, storage, codification and application, including the fundamental activity of searching for and sharing new technical or market knowledge (Argote, McEvily, and Reagans, 2003; Carlile and Rebentisch, 2003; Koh and Kim, 2004; March, 1991; March and Simon, 1958). Figure 2 illustrates the fundamental activities for managing the different types of knowledge necessary to achieve effective and continuous innovation.

Distributed Innovation Network

In the last decade, the competition in many industries and in all segments has strongly increased. These changes have pushed all manufacturers to develop complex and highly technical products satisfying the increasing requirements of customers and other actors (government, associations…) in terms of style, quality, safety, comfort and environmental protection, and so on. In order to innovate effectively, businesses have to anticipate changes and acquire the knowledge required for leading the change before competitors. At the same time, businesses have to shorten new product development process time to face international competition. Accordingly, Nonaka stated: *"When markets shift, technologies proliferate, competitors multiply, and/or products become obsolete almost overnight, successful companies are those*

Figure 2. Knowledge-based activities of the innovation process

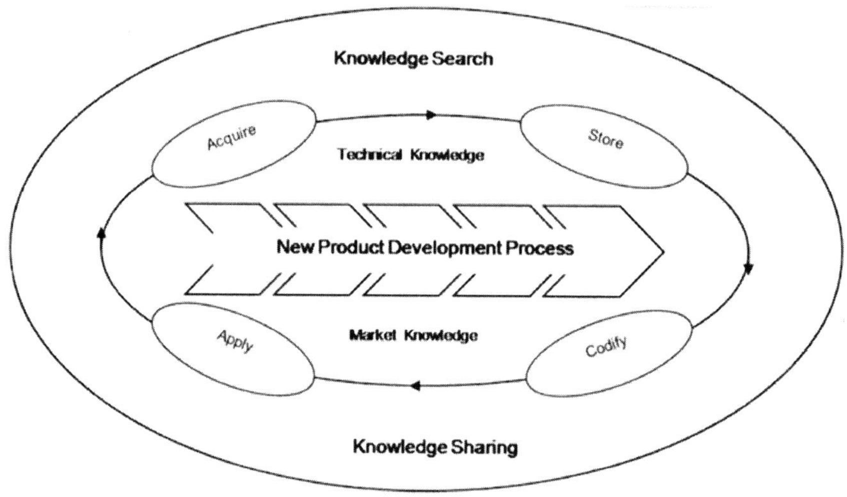

that constantly create new knowledge, disseminate it widely throughout the organization, and quickly embody it in new technologies and products" (Nonaka, 1995:175).

Due to the dynamism of the competitive environment, businesses' in-house generated knowledge and technologies are not enough for firms' survival. As a matter of fact, it has been recognized that an open business framework can prevent environmental threats by effectively exploring and exploiting ideas, knowledge and capabilities lying outside the firm, in collaboration with external players that are formally or informally linked to the central firm (Alguezaui and Filieri, 2008; Filieri 2009). This new framework constitutes the innovation network of the firm (ibid). In the past, it has been proved that innovations can be derived from different players such as lead-users and customers, manufacturers, suppliers (Von Hippel, 1986; Prahalad and Ramaswamy, 2004; Thomke and Von Hippel, 2002; Clark and Fujimoto, 1991), research communities (Henderson and Cockburn, 1994), and even competitors (Gomes-Casseres, 1996). For instance, firms can tap onto the basic and scientific knowledge of universities and their advanced infrastructure to derive profitable businesses. This is especially evident in the technology-oriented industries, such as automotive, biotechnology, aircraft manufacturing and so on. Additionally, customers and lead users have been found to be a source of innovative ideas and valuable knowledge about latent needs, new uses of existing products, novel functionalities, and the like. Firms also collaborate with their competitors and suppliers to enhance products quality, to reduce the cost of operations, and to enter new market segment, etc.. Along the same vein, innovation is becoming distributed across different players, and the NPD process then can be viewed as a process of co-creation (Prahalad and Ramaswamy, 2004). In the traditional conception of the value creation process, consumers were "outside the company", and value creation occurred within firm's boundaries, through its internal R&D activities, and outside markets (Sawhney, Verona, and Prandelli, 2005). Chesbrough (2003) described the distributed nature of the knowledge creation process as the open innovation model.

Consistently, the mission of the today's business is to grasp the right knowledge at the right time. Inter-firm relationship is increasingly considered as the ideal locus of value creation and a way to stimulate firm performances and innova-

Figure 3. New framework of the innovation process

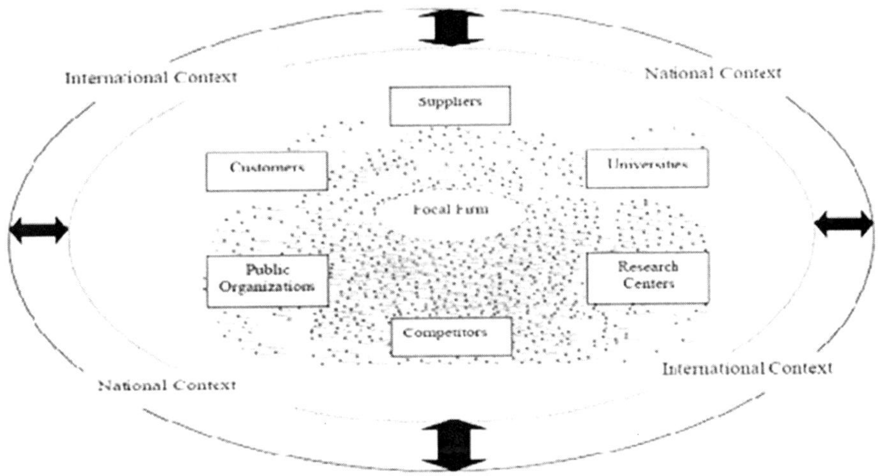

tiveness (Dyer and Singh, 1998; Lorenzoni and Lipparini, 1999). Several scholars have studied the relationship between a firm's strategic alliance and its innovative performance (Shan, Walker, and Kogut, 1994; Kotabe and Swan, 1995; Deeds and Hill, 1996; Baum, Calabrese, and Silverman, 2000; Lerner, Shane, and Tsai, 2003). The literature highlighted three main approaches and theories to explain the reasons behind firm's embeddedness into inter-firm relations (Osland and Yparak, 1993): the Organization Theory (Kogut, 1988); the Competitive Position (Porter, 1986); and the Transaction Cost Economics (Williamson, 1975). Though, these theories were not adequate to include all the motivations behind the current alliances formation; especially, taking into consideration the increasing value of the knowledge and relational resources. Today, firms are dissolving their boundaries for inter-firm relations (Takeishi, 2001) and inter-industry relations (Scott, 1996) at both the national and the international levels, trying to facilitate learning activities (Anand and Khanna, 2000; Dyer and Nobeoka, 2000). Evidence of the importance of knowledge and technologies in alliance formation is the Fiat-Chrysler agreement. Recently, Fiat acquires 35% of Chrysler for sharing its small-car platforms and fuel-efficient engines. The deal is supposed to give the US automaker access to new markets as well as cheaper and more environmentally friendly technologies, which Chrysler can not actually produce due to the delay accumulated in the renovation of knowledge, capabilities and technologies of the manufacturer. Therefore, as illustrated in the following figure, firms build up and develop ties and interaction with diverse external players, at both the national and the international levels, to overcome the risks imposed by their environment and to enhance their innovative capabilities (figure 3).

Firms try to tighten their external relationships in order to improve their collaboration, to achieve organizational efficiency, and to reduce innovation costs. Harrigan and Gulati define strategic alliances as any voluntarily initiated cooperative agreement between firms that involve exchange, sharing, or co-development activities and can include contributions by partners of capital, technology, or firm-specific assets (Harrigan, 1985; Gulati, 1995). Before tightening alliances, firms make foresight and benchmarking about future partnerships, technology, knowledge and capabilities required. Multiple alliances are considered a source of network resources (Gulati, 1995). The strategic alliances become strategic networks as firms exploit and access the resources available among firm's partners. As reducing the product development costs is a critical factor for the innovation process to be effective (Chesbrough, 2007), the strategic value of a network of alliances get focused on that objective. In fact, these networks provide quick and unique access to firm's technology, knowledge, and capabilities required to effectively perform continuous innovation, getting then commitments generally from the more central firm of the network. The growing importance of alliances has increasingly created the need within firms to create business units completely dedicated to the management of alliances, becoming more and more strategic for the business of the firms. Moreover, firms require company wide standards and customized tools for multi-alliance management (Hoffmann, 2005). For instance, Toyota, relying on suppliers for more than 70% of the value of its vehicles, has two major divisions supporting suppliers learning activities, which are the Operation Management and Consulting Division, and if it is necessary other divisions are involved to solve problems with suppliers (Dyer and Nobeoka, 2000). For facilitating the learning of their members, Toyota has developed different associations with different tasks like: on-site consulting, supplier learning teams, problem solving teams, employee transfer, performance feedback, and process monitoring. According to researchers, thanks to its network strategy, Toyota was able to increase worker productivity, to decrease inventories, and to improve product quality at a faster rate than competitors.

Along the network-based innovation process, it is also important to separate multi-player interactions of the New Product Development (NPD) into two distinct stages: "Divergent" and "Convergent thinking". The first stage is an exploratory and open-ended phase and is used for searching novel insights and generating ideas at the ambiguous "fuzzy front end." During the second phase, these ideas are discussed and validated and the new product is co-developed. Within this stage, it is considered fundamental the exploitation of firm's network of alliances, partners, suppliers, and so on. Figure 4 illustrates the different stages of the interactive innovation process.

Perceiving innovation as the product of combining different resources and knowledge streams, two main strategic mechanisms are required to effectively manage the knowledge within firm's network and to enhance the performance of its distributed innovation: knowledge search and sharing (see Figure 1). A key element of the innovation process is to organize search for new ideas and identification of novel knowledge that have commercial potential (Laursen and Salter, 2006), especially at the "fuzzy front end" stage. Indeed, today's firms are urged to search for in-

Figure 4. The different stages of network-based NPD (adapted from Koen et al, 2002)

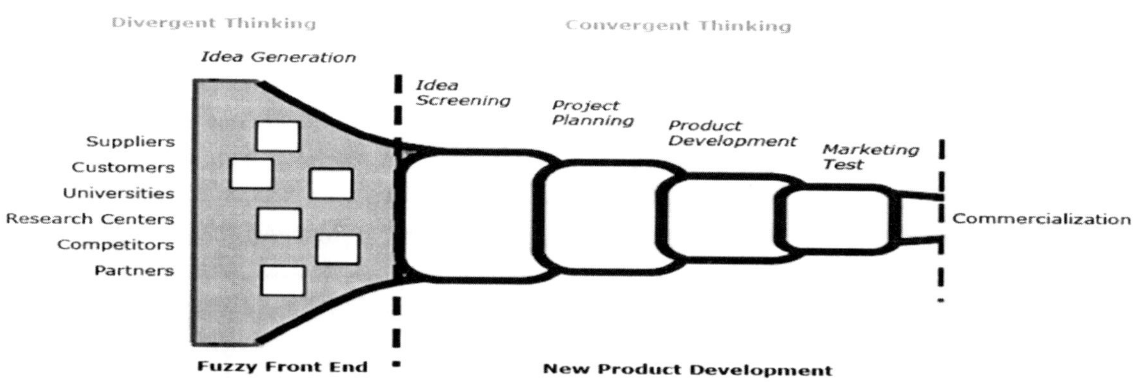

novation opportunities within their innovation networks, crossing their boundaries to keep up with their ever-changing environment and to over-perform its competitors (as illustrated in Figure 3). Following Katila and Ahuja's (2002) distinction of internal search strategies, Laursen and Salter (2006) defined external search strategies in terms of two dimensions: breadth and depth. The first one is defined as "*the number of external sources or search channels that firms rely upon in their innovative activities*" (Laursen and Salter, 2006: 134). The second dimension refers to "*the extent to which firms draw deeply from the different external sources or search channel*" (Laursen and Salter, 2006: 134). Firms need to appropriately design their external search processes according to their innovation strategies in order to find out profitable opportunities.

Knowledge sharing or transfer is considered another prerequisite for resource recombination and new product development and, henceforth, new knowledge creation (Moran and Ghoshal, 1996). The capacity to share such knowledge depends on routines or people' skills in using metaphors, analogies, and models. Star and Griesemer (1989) have emphasized the role of shared artefacts, which convey contextual information about the knowledge being shared, helping to clarify the underlying meaning. Thereafter, another strategic activity is the setting up of routines enabling more effective knowledge sharing and transfer within and across the network of the firm. De Clercq and Sapienza (2001: 114) defined knowledge sharing routines as "*repeated activities that represent the mechanisms for multiple and intensive interactions between parties in developing relational advantages.*" The establishment of different knowledge sharing routines provide several benefits as an efficient transfer of fine-grained and complex knowledge, the enhancing of problem-solving activities, as well as the development and improvement of relationships (Lind and Zmud, 1995). Those beneficial factors contribute heavily to facilitate firm's learning processes and to effectively develop new products. Figure 5 illustrates the network-based innovation process according to the knowledge-based view discussed earlier in the chapter, and highlights two main knowledge management mechanisms according to the different stages of the innovation process.

Additionally, the growing interest in facilitating knowledge sharing activities has led to increased attention being paid to social network analysis as a tool for mapping knowledge sharing, capabilities and the nature of relationships within informal networks or emergent communities of practice (A growing interest towards social network analysis is showed by researchers in the area of knowledge management community (Cross and Parker, 2004; Dahl and Pedersen, 2005; Cross et al., 2006; Allen et al., 2007). Thus, firms can access to a more and extended array of new ideas and knowledge, reducing uncertainty, new product development costs and time. Knowledge and capabilities are shared across informal networks. Informal social networks, or emergent networks, are unsanctioned and ungoverned organic structures connecting a potentially unbounded group of individuals (Mintzberg, 1973; Tichy, 1981). They include the working relationships, collaborations and exchanges of knowledge between individuals, which are not the result of the organizational structures, rather the fruit of the personal initiatives of employees (Cross and Parker, 2004). People during their job continuously ask who knows who, what or how to do things through informal communications. One of the most consistent finding in the social science literature is that who you know often has a great deal to do with what you come to know (Simmel, 1950). However, these networks are not readily available to all people that in a firm need them, their awareness depend on chance conversations, coffee breaks, and other similar events (Davenport and Prusak, 1998). Cross and Parker argued that whether the true extent and membership of an informal network is understood, it may be sup-

Figure 5. Network-based innovation process and its key activities

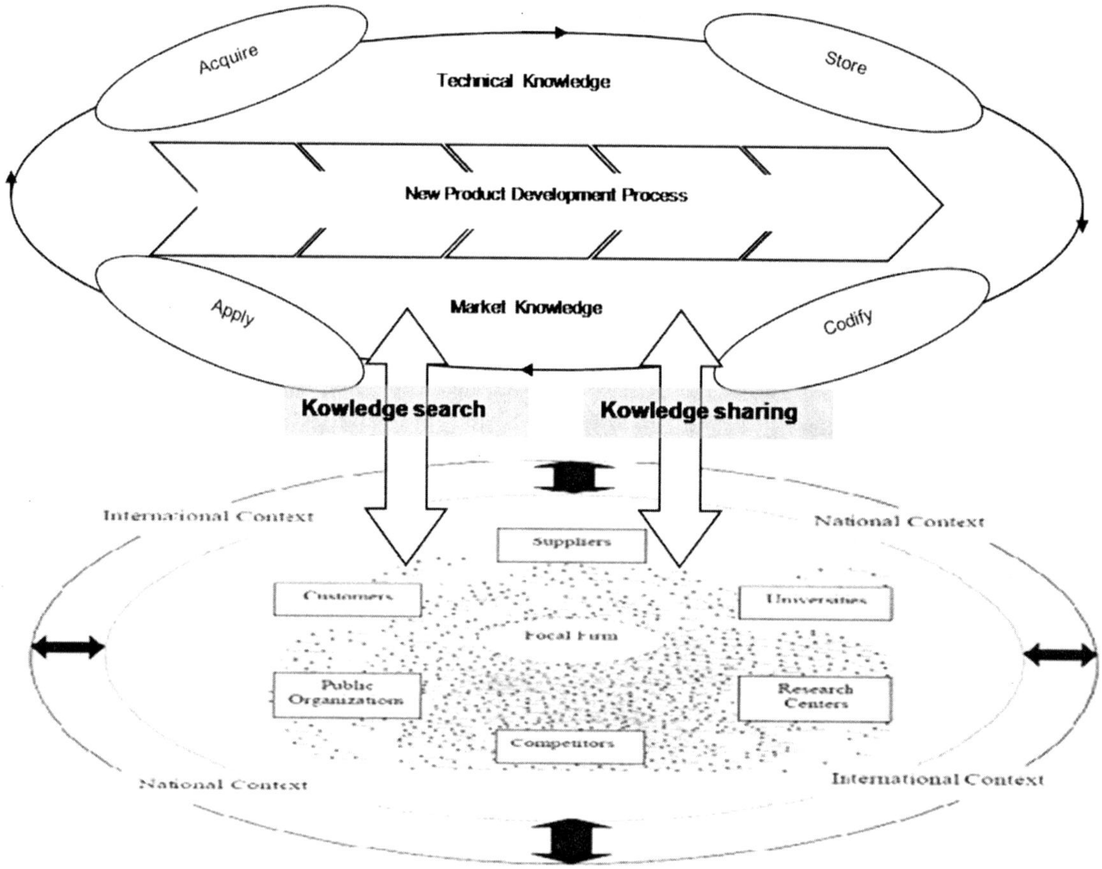

ported and nurtured to increase the wider effectiveness and innovative capacity of the firm (Cross and Parker, 2004).

Evidence of the informal networks' innovation power is the Connect & Develop strategy of Procter & Gamble (Huston and Sakkab, 2006). This strategy is based on the exploitation of the external connections of R&D departments. For every P&G researcher, P&G measured that there were 200 scientists, engineers, in the world; in total just 1,5 million people (Huston and Sakkab, 2006). P&G establishes what is looking for and then collaborates with individuals and organizations throughout the globe searching for proven technologies, packages and products that P&G can improve, scale up or market by their own. Exploiting knowledge and ideas embedded across these ties resulted in *"more than 35% of P&G new products are originated from outside...; 45% of P&G initiatives in product development portfolio have key elements that were discovered externally...and the productivity has increased by nearly 60%"* (Huston and Sakkab, 2006: 61). The positive results obtained by P&G could be increased if all employees of the different departments, from R&D to marketing, set-up routines to make informal relationship management an organizational capability with great potentialities. The Connect & Develop strategy is the evidence that a huge array of innovations and innovators are next to the firm's business environment, however it is critical to search for them and for their knowledge and ideas. Thus, another important element for businesses' success is the firms' awareness about his social capital and the capacity to manage beneficial and productive relationships with an extended typology of players.

SOCIAL CAPITAL AND INNOVATION PERFORMANCE

Past researches highlighted how external actors and the social interactions with them may drive to firm's higher innovation performance and then sustainable competitive advantage. Through empirically investigating the biotechnology start-ups, Shane et al. (1994) found a relationship between firms' cooperation with the external environment and its innovative performance. Along the same vein, Powell and his colleagues (1996) proved that firms embedded in a network of inter-organizational collaborations are more likely to realize higher innovative performance. Indeed, the locus of innovation is mostly laid out in firms' inter-organizational collaborations. Further, Ahuja (2000) empirically demonstrated the benefits of social networks within the framework of R&D's alliances to improve firms' innovative abilities. The author observed that social capital, represented by direct and indirect ties, enhanced the number of R&D alliances during a specific year through providing three main benefits: complementary competences among the partners, knowledge sharing among the partners, and economies of scale in R&D programs (Ahuja, 2000). He had also found that the effect of indirect ties on the innovation performance is moderated by the number of direct ties (Ahuja, 2000).

Firms are significantly urged to collaborate with external partners to continuously innovate in order to keep up with the drastic changes within their markets. This is mainly due to the fact that these firms on their own lack the sufficient resources, capabilities and knowledge needed to perform the innovating activities (Kaufmann and Tödtling, 2002). Consequently, innovation does not come out of a vacuum (Van de Ven, 1986); instead, creating new value is considered the result of a long and complex social process of recombining different and sometimes already existing bodies of knowledge. Thus, interactive learning among organizations plays a critical role for the growth of a firm, and even its survival (Lundvall, 1988). In fact, innovations are viewed as the product of cooperation and continuous interaction between a firm and its external players. Within the realm of a learning economy, firms' competitiveness is then based on innovation, and innovation processes are viewed as socially embedded learning processes. Firms need to develop and exploit their internal knowledge bases and to draw on and to explore the knowledge across their boundaries to innovate. Therefore, modern firms rely on their external networks or their social embeddedness as an essential factor to create, develop, and exploit new opportunities and derive profitable value. Thus, social capital represents a source of competitive advantage and a prerequisite for survival as it provides the focal firm with access to idiosyncratic resources and a network of relationships, enhancing its innovation capacity. In fact, the social network of innovators is considered of great importance as it provides the knowledge and resources obtained from multiple connections or ties.

Benefits and Risks of Social Capital to Innovation Performance

Recently, a controversial debate has arisen among researchers recognizing the beneficial effect of two divergent network structures: dense and sparse networks. While it is widely agreed on that social interactions enhance the innovation performance, it is still ambiguous the optimal structure of social network that better contribute to innovation. Different social network configurations, particularly in terms of both sparse and cohesive structures, present both opportunities and obstacles to the innovation-related activities (Brass, 1995; Hargadon and Sutton, 1997; Burt,

2004; Obstfeld, 2005; McEvily and Reagans, 2005; Uzzi and Spiro, 2005, Fleming et al., 2007). On the one hand, sparse networks, which refers to the foundations of Burt's structural holes theory, implies the absence of connections among the members of the network (Burt 1992). On the other hand, cohesive or dense network, which is the basis of Coleman argument (1988) and others, feature a highly connected web of actors. The literature displays supporting empirical evidence that lay the ground for both camps. Proponents of the cohesive camp argue that dense networks provide numerous benefits to the innovation in terms of trust, redundant information channels, risk sharing attitude, and easy resource mobilization (Uzzi, 1997; Reagans and McEviley, 2003; Uzzi and Spiro, 2005; Obstfeld, 2005). In fact, several authors showed that cohesive social networks facilitate the transfer of information and knowledge, even the more fine-grained one, between individuals (Uzzi, 1997; Hansen, 1999; Reagans and McEviley, 2003). Obstfeld (2005) demonstrated that individuals embedded into cohesive networks are more likely to engage in innovation activities. Conversely, the proponents of structural hole theory claim that actors occupying the position of brokers benefit from information control and privileged access to unique knowledge and opportunities (Burt, 2004). Hargadon and Sutton (1997) provided evidence to the beneficial returns of the brokering position. They investigated the advantages that a product design firm IDEO derives from performing the brokering activities among different industries (Hargadon and Sutton, 1997; Hargadon, 2002). Indeed, the focal firm is proved to rent ideas and technologies from one industry, add some modifications and then apply them to another industrial context.

Cohesiveness, which refers to the degree of tie redundancy and interconnectedness among network members of a focal firm, is the basis of Coleman's arguments. A network is described as cohesive when all actors within that network are connected to each other. Coleman (1988) argues that cohesive network structure provide governance benefits to the focal firm through developing reputation-building mechanism. In fact, cohesiveness in the network develop trust, norms of reciprocity, and a shared identity among the members facilitating then their collaboration through developing informal and trust-based governance mechanisms (Granovetter, 1972; Coleman, 1988; Powell, 1990; Uzzi, 1997; Uzzi and Spiro, 2005). Firms operating within such network are more constrained to perform according to the others' expectations and genuinely cooperate since any deviant behaviour is readily detected and strictly punished. Therefore, cohesive network affords trusting-based context where members exchange fine-grained information (Uzzi, 1997; Hansen, 1999; Reagans and McEvily, 2003), come up with alternative interpretation to the current problems, and develop novel ways to solve these problems (Powell and Smith-Doerr, 1994). Further, cohesive networks enhance intense interactions among firm's partners promoting the effective exchange, combination and use of knowledge (Kogut and Zander, 1992; Nahapiet and Ghoshal, 1998). Tie redundancy characterizing closed networks ensures multiple pathways through which information quickly and reliably achieve all network members (Schillings and Phelps, 2007). In fact, communication and learning between network partners is greatly enhanced through developing shared understanding of problems and solutions (Brown and Duguid, 1991). In addition to that, rich information channels provide quick and easy exchange of feedback and ideas improving then novelty and creativity (Milliken et al., 2003). Consequently, closed networks provide several benefits to the innovating activities through providing governance mechanisms, enhancing information flows and tacit knowledge sharing, and promoting joint problem solving (Krackhardt, 1992; Uzzi, 1996).

Whereas cohesive networks display various advantages to firms' innovation performance, such networks also present a 'dark side,' constraining the effective implementation of the innovation activities. As a matter of fact, closed networks prevent its members from searching new partners and isolate them from the external world, resulting in a 'lock-in' situation (Grabher, 1993; Uzzi, 1997; Johannisson, 2000). Thus, being over-embedded within a closed network stifles the firm's ability to innovate and reduce its economic performance (Uzzi, 1997; Burt, 1992, 1997, 2000, 2004). Indeed, firms embedded within a cohesive networks are compelled to maintain strong ties to their partners that are no more beneficial. This incurs higher cost for maintaining these relationships and impedes the firm as well from developing new relationships. Further, firms are more exposed to redundant knowledge flows. Additionally, closed networks get its members short sighted, leading them to reject outsiders' knowledge namely the 'Not Invented Here' syndrome. Therefore, cohesive networks restrain firm's capability of exploring new knowledge and exploiting novel recombination.

While Coleman's approach to social capital laid the basis for the proponents of cohesive networks, Granovetter (1973) discussed the strength of weak ties providing the foundations for the proponents of brokerage. Burt (1992) further developed the brokerage argument and emphasized the benefits of sparse network structures, proposing the structural hole theory. The main line of reasoning behind that theory claims that firms embedded into sparse networks and occupying positions of brokerage experience a number of benefits. Sparse network supports focal firm's access to novel information and its exploration of new opportunities enhancing the innovation performance. People coming from different backgrounds and having numerous perspectives characterize the open networks providing free flow of information, ideas and knowledge. Firms that perform brokering activities are more exposed to new knowledge and well positioned to generate novel combinations (Brass, 1995; Burt, 1992; Hargadon and Sutton, 1997). While they provide benefits to innovation through promoting access to new knowledge, sparse networks have also detrimental effects to innovation. Sparse networks impede intense interactions necessary to deeply understand knowledge available within the environment and to achieve the realization of innovation. Obstfeld (2005) described the inability of sparse networks to handle the innovation till its realization as an 'action problem.' Therefore, as the following table demonstrates, various forms of a firm's social capital exhibit different benefits and risks to the innovation outcomes.

Social Capital and Novelty of Innovation

The innovation literature recognizes two main distinct categories of innovation outcomes depending on their novelty: incremental and radical innovations (Henderson and Clark, 1990; Utterback, 1996; Garcia and Calantone, 2002; Gatignon et al., 2002). On the one hand, incremental innovation implies minor changes to the firm's current products, services, processes, administrative or technical conditions. These changes do not cause the firm to depart from its current technological paradigm. On the other hand, radical innovation refers to novel combination of different fields of expertise resulting in new technological patterns. Radical innovations are disruptive and are associated with exploration and competence destruction, while incremental innovations have to do with exploitation and competence-enhancing measures (Anderson and Tushman 1990). These two forms of innovation rely on distinct knowledge bases and are performed through different processes. Thus, different configurations of social capital are needed to satisfy the distinct requirements of radical and incremental innovations (Rowley et al. 2000; Rueff, 2002).

Incremental innovations came to light as a result of the small improvements effectuated on the dominant design. Firms then aiming at incremental innovation need to derive knowledge residing in various sources of knowledge (Laursen and Salter,

2006). Therefore, sparse networks contribute greatly to the higher performance of incremental innovation by providing minor changes to the dominant design. In contrast, radical innovation requires deep knowledge of the respective technological domains and develops within a trusting context since it involves more risky investments. Indeed, firms developing radical innovation needs to create shared frames of reference with their partners and common repertoires of communication to solve various problems and receive feedbacks on the emergent designs (Orlikowski and Yates, 1994). By doing so, different participants are able to deploy their expertise and their tacit knowledge to jointly develop radical innovations. Laursen and Salter (2006: 137) stated that "*radical innovators are likely to draw more deeply from external sources of innovation than firms that are not radical innovators.*" Accordingly, cohesive networks provide the focal firm the trusting context to share tacit knowledge and to undertake risky investments, which are key elements to develop radical innovations. In this vein, firms then may opt for sparse or cohesive networks depending on their innovation strategies and objectives.

Balanced Approach of Social Capital to Innovation

Analyzing and weighing up both the beneficial and detrimental effects of both types of structural social capital (see table 1), the benefits and risks of both networks represent the image of each other. Indeed, sparse network provides the benefits of accessing to novel knowledge at the expense of effectively exchanging tacit and rich knowledge. The opposite features are true for cohesive networks. Hence, while the two network structures display contradictory features, their benefits and risks are complementary to one another. Even though structural holes provide unique access and control over novel knowledge as well as efficient search for new opportunities, cohesive networks afford the fundamental communication tools (e.g. tacit knowledge exchange) and governmental mechanisms (e.g. trusting relationships) to effectively exploit the previously identified opportunities and bring them into market. In attempt to reconcile both views, Burt claimed that *"although brokerage across structural holes is the source of added value, closure can be critical to realising the value buried in the holes"* (2000: 410). Therefore, firms succeeding in developing a balanced mixture of both types achieve higher innovation performance. These firms are able to derive and optimize the benefits of both cohesive and sparse networks and overcome then their shortcomings.

There are two strategies to balance both types of social capital depending on firm's industrial dynamics of the specific sector. The first strategy consists of developing and nurturing both open and closed networks simultaneously over time. Within a highly competitive and dynamic environment, this strategy aims at creating sparse network to explore new opportunities in parallel with maintaining cohesive networks that exploit these opportunities. Along the same line of reasoning, Dittrich and Duysters' (2007) analysis of Nokia's

Table 1. Benefits and risks of social capital to innovation

	Benefits	Risks
Closure (Coleman)	• Effective joint problem-solving • Exchange of fine-grained information • Multiple pathways of information	• Costly relationship maintenance • Redundant Knowledge flows • 'Not Invented Here' syndrome
Sparseness (Burt)	• Access to unique knowledge • Control of information exchange • Efficient knowledge search	• Action problem

external ties well illustrates the first strategy. For earlier generations of mobile telephony, the product development of Nokia was kept as internal activities, inside its R&D department. However, recently, firms operating in the telecommunication industry adopt multiple technical trajectories with no general dominant design leading the industry. Within the realm of this dynamic environment, the development of a product or technology can not be carried out by a firm on its own. Instead, interaction with various actors within firm's innovation system is critical to the product development at each stage. Therefore, Nokia shifted its innovation strategy and opened up its processes for developing the third generation telephony. In addition to its earlier and well-established relationships aiming at the exploitation of innovations, the firm now develops less strong ties for the exploration of new technologies. Nokia then keeps a balance of both an exploitation network formed of dense ties and an exploration network adopting a more organic and flexible way of working. In fact, it maintains a relationship with approximately 100 worldwide universities from Boston to Beijing for basic and long-term research. At the same time, Nokia retains and strengthen its relationships with different tiers of suppliers and customers to capitalize on existing knowledge and resources. While keeping both networks independent from each other, Nokia takes advantage from its open network by efficiently searching for novel knowledge and by identifying new technological opportunities. Furthermore, the cohesive network enables Nokia to exploit the previously identified opportunities and to effectively develop new products by enhancing knowledge transfer among its members and facilitating joint problem-solving activities and feedback loops. Consequently, Nokia plays three roles between the two networks: as a broker managing the flows of knowledge, as an administrator governing the set of relationships and managing the activities carried out within the network, and as arbitrator setting appropriately the value to the network members.

The second strategy is perceived as cyclic going from sparse network to a more cohesive one once a new opportunity is identified and needs to be developed. As it has been discussed earlier in the chapter, firms try to search distantly and broadly in the innovation system during the "fuzzy front end" phase. Along that phase, the firm opens up its network's boundaries and encourages new partners bringing fresh knowledge and new opportunities to join its existing network. That stage of the innovation process is characterized by divergent thinking where the firm searches for distinct and various knowledge sources. After the generation of a new idea, the focal firm strengthen its relations with the potential partners and establishes a social system for collaboration among the network's members, facilitating then knowledge exchange within the network. The network members consequently tend to work together into an intimate and interactive manner, to productively develop the innovation and to generate value for the whole system. Therefore, the network's structural configuration enter into a cycle from being sparse and open to new and diverse partners supporting the strategic search of the focal firm, to being a more cohesive system where knowledge is effectively and easily exchanged among the different partners. Gilsing and Nooteboom (2005) illustrate the dynamics of moving between sparse and dense network by analyzing the network of firms operating within the biotechnology sector. The authors described the movement between sparse and dense networks as the cycle of discovery, where firms go from the exploration to the exploitation phase and vice versa (ibid.).

The two strategies to balance both sparse and cohesive networks are critical for firms that aim at enhancing their innovation performance. Each network configuration support specific stage of the distributed innovation process. Sparse networks enhance the 'fuzzy front end' phase where divergent thinking is more critical and search for novel knowledge is highly emphasized. Cohesive networks instead back up more the development phased where the various knowledge streams needs to converge into the development of a single novel artefacts. In highly dynamic environment (e.g. telecommunication industry), firms are urged to continuously explore and exploit new opportunities as well as to develop several technologies in parallel. According to this context, firms need to simultaneously balance both sparse and cohesive networks over time. However, in less dynamic environment where market trends are more predictable, firms may go into a cycle of discovery and development over time. Under this condition, firms open up their network boundaries to new entrants. After recognizing and generating new knowledge, these firms establishes cohesive network for the effective development of the new concepts. Therefore, firms may design their networks configuration according to the dynamics of their environment.

CONCLUSION

The competitive advantage of the firm is increasingly based on its capacity to exploit new opportunities and to innovate. The innovating related activities are no more the mere product of R&D labs. Instead, innovation relies more and more on the external capabilities of the firm's networks. Being described as a knowledge intensive process, firms bridge their boundaries to exploit the knowledge embedded within their external social networks. The social capital derived from the networks has a strong effect on the businesses' innovation performance and their success. The implications of social capital to the innovation performance is controversial and it emphasizes solely its positive side. However, it has been proved that social networks may display also negative consequences.

Researchers have recognised two main social capital's configurations: cohesive and sparse networks. The chapter sought to shed light on the effect of these two forms of social capital on firms' innovation performance. Cohesive networks are argued to have strong and highly interconnected ties to each other. This kind of social capital provides trust, norms, commitment and network of ties. Accordingly, firms may engage in risky investments, share tacit knowledge, and exchange reliably feedbacks and solutions in relation to specific problems promoting the performance of their innovation activities. Yet, cohesive networks may represent also constraining factor to innovation. In fact, cohesive networks may cause the focal firm to be over-embedded creating a situation of lock-in. in this case, firms are unable to explore novel information and knowledge that go beyond their existing networks. In addition to that, these firms may incur the cost of maintaining a set of relationships and ties to other firms that are deemed to be unprofitable. Another form of social capital that has been widely researched is the sparse networks. These networks provide the focal firm the ability to access novel information, benefiting firm's innovation performance. However, sparse networks have limiting effect on innovation as well. Indeed, sparse networks may display an action problem (Obstfeld, 2005), referring to inability of these network to realize the innovation. Taking into consideration the different needs of radical and incremental innovations, the two main forms of social capital may provide different supports to two typologies of innovation. In fact, the chapter highlighted the main contribution of cohesive networks to radical innovation through intense interactions and a trusting social context. The sparse networks support more the incremental innovation as they provide different views and small improvement to the dominant design.

The chapter further highlights the contradictory features of sparse and cohesive networks and the complementarities between their functions. In fact, sparse networks enhance the identification of new opportunities and support more efficient search strategies characterizing the 'fuzzy front end' stage of the innovation process. Contrary to the cohesive networks that improve knowledge sharing and facilitate intensive interactions among the network's members for effective joint-problem solving and development of the new product. Thus, both types of networks complement each other to achieve higher innovation performance. Firms whose aim is to optimize their innovation capacities need to balance both sparse and cohesive networks. The chapter highlights two strategies supporting the balanced approach. The first strategy consists of a simultaneous maintenance of both types of networks over time. The other strategy is perceived as a cycle moving from open to closed networks along the different stages of the innovation process.

Thus, the chapter sought to design a theoretical framework for analyzing the effect of different configurations of the structural social capital on the innovation performance. Accordingly, the chapter suggests several avenues of future research. Future research may go one step further by empirically demonstrating the hypotheses developed in the current chapter within different contexts. Another research stream can be to uncover the network dynamics of developing radical and incremental innovations. The chapter also adopt an approach to balance both types of social capital and described two strategies to achieve that balance. Future research may advance those hypothesis by discovering other strategies of balancing sparse and cohesive networks and evaluating their innovative results under different conditions.

REFERENCES

Adler, P. S., & Kwon, S. W. (2002). Social capital: Prospects for a new concept. *Academy of Management Review*, *27*(1), 17–40. doi:10.2307/4134367

Ahuja, G. (2000). Collaborative networks, structural holes, and innovation: a longitudinal study. *Administrative Science Quarterly*, *45*(3), 425–455. doi:10.2307/2667105

Alguezaui, S., & Filieri, R. (2008, May). *Innovation Across Tech-Firm's Boundaries: a Knowledge-Based View*. Paper presented at the Conference on Offshoring & Outsourcing: The Organizational and Geographical Relocation of High-Value Company Functions, SDA Bocconi, Milan.

Allen, J., James, A. D., & Gamlen, P. (2007). Formal versus informal knowledge networks in R&D: a case study using social network analysis. *R & D Management*, *37*(3), 179–195. doi:10.1111/j.1467-9310.2007.00468.x

Anand, B. N., & Khanna, T. (2000). Do firms learn to create value? The case of alliances. *Strategic Management Journal*, *21*(3), 295–315. doi:10.1002/(SICI)1097-0266(200003)21:3<295::AID-SMJ91>3.0.CO;2-O

Argote, L., & Ingram, P. (2000). Knowledge transfer: A basis for competitive advantage in firms. *Organizational Behavior and Human Decision Processes*, *82*(1), 150–169. doi:10.1006/obhd.2000.2893

Argote, L., McEvily, B., & Reagans, R. (2003). Managing Knowledge in Organizations: An Integrative Framework and Review of Emerging Themes. *Management Science*, *49*(4), 571–582. doi:10.1287/mnsc.49.4.571.14424

Belliveau, M. A., O'Reilly, C. A., & Wade, J. B. (1996). Social capital at the top: effects of social similarity and status on CEO compensation. *Academy of Management Journal*, *39*(6), 1568–1593. doi:10.2307/257069

Borgatti, S., & Cross, R. (2003). A Social Network View of Organizational Learning: Relational and Structural Dimensions of 'Know Who.'. *Management Science*, *49*(4), 432–445. doi:10.1287/mnsc.49.4.432.14428

Borgatti, S. P., & Foster, P. (2003). The network paradigm in organizational research: A review and typology. *Journal of Management, 29*(6), 991–1013.

Bourdieu, P. (1986). The forms of capital. In Richardson, J. G. (Ed.), *Handbook of Theory and Research for the Sociology of Education* (pp. 241–258). New York: Greenwood Press.

Brown, J. S., & Duguid, P. (1991). Organizational learning and communities-of-practice: Toward a unified view of working, learning, and innovation. *Organization Science, 2*(1), 40–57. doi:10.1287/orsc.2.1.40

Burt, R. (1992). *Toward a Structural Theory of Action*. New York: Academic Press.

Burt, R. (1997). The contingent value of social capital. *Administrative Science Quarterly, 42*(2), 339–365. doi:10.2307/2393923

Burt, R. (2000). The network structure of social capital. In Staw, B., & Sutton, R. (Eds.), *Research in organizational behavior* (*Vol. 22*). New York: JAI Press.

Burt, R. (2001). *Social Capital: Theory and Research*. New York: Aldine De Gruyter.

Burt, R. (2004). Structural Holes and Good Ideas. *American Journal of Sociology, 110*(2), 349–399. doi:10.1086/421787

Chen, M. H., & Wang, M. C. (2008). Social Networks and New Venture's Innovative Capability: The Role of Trust within Entrepreneurial Teams. *R & D Management, 38*, 253–264. doi:10.1111/j.1467-9310.2008.00515.x

Chesbrough, H. (2003). *Open Innovation: The New Imperative for Creating and Profiting from Technology*. Boston: Harvard Business School Press.

Chesbrough, H. (2007). Why firms should have open business models. *MIT Sloan Management Review, 48*(2), 21–28.

Clark, K. B., & Fujimoto, T. (1990). The power of product Integrity. *Harvard Business Review, 68*(6), 107–118.

Coleman, J. S. (1988). Social capital in the creation of human capital. *American Journal of Sociology, 94*(1), 95–120. doi:10.1086/228943

Davenport, T. H., & Prusak, L. (1998). *Working knowledge: How organizations manage what they know*. Boston: Harvard Business School Press.

Davis, E., Spekman, R., & Hall, P. (2004). *The Extended Enterprise. Gaining competitive advantage through collaborative supply-chain*. New York: FT Prentice Hall.

Day, G. S. (1994). The capabilities of market-driven organisations. *Journal of Marketing, 58*(4), 37–52. doi:10.2307/1251915

De Clercq, D., & Sapienza, H. J. (2001). The creation of relational rents in venture capitalist-entrepreneur dyads. *Venture Capital, 3*(2), 107–128. doi:10.1080/13691060110045661

Dittrich, K., & Duysters, G. M. (2007). Networking as a means to strategic change: the case of open innovation in mobile telephony. *Journal of Product Innovation Management, 24*(6), 510–521. doi:10.1111/j.1540-5885.2007.00268.x

Drucker, P. (1993). *Post-Capitalist society*. New York: Harper Business.

Dyer, J., & Nobeoka, I. (2000). Creating and Managing a High-Performance knowledge sharing network: the Toyota Case. *Strategic Management Journal, 21*(3), 345–367. doi:10.1002/(SICI)1097-0266(200003)21:3<345::AID-SMJ96>3.0.CO;2-N

Filieri, R. (2009). *Overcoming knowledge sharing barriers through communities of practice*. Unpublished doctoral dissertation, Università degli Studi Roma 3, Italia

Filieri, R., & Al Guezaui, S. (2009). Knowledge Sharing Barriers: a Meta-Analytic Review. In *Proceedings of the 10th European Conference of Knowledge Management (ECKM)*, Università Degli Studi Di Padova, Vicenza, Italy, September 3-4.

Flap, H. D., & De Graaf, N. D. (1986). Social capital and attained occupational status. *Netherlands Journal of Sociology*, *22*(2), 145–161.

Fleming, L., Mingo, S., & Chen, D. (2007). Collaborative Brokerage, Generative Creativity, and Creative Success. *Administrative Science Quarterly*, *52*(3), 443–475.

Foster, R., & Kaplan, S. (2001). *Creative Destruction*. New York: Double Day.

Fukuyama, F. (1995). *Trust: Social Virtues and the Creation of Prosperity*. New York: Free Press.

Gabbay, S. M., & Leenders, R. Th. A. J. (1999). CSC: The structure of advantage and disadvantage. In Leenders, R. Th. A. J., & Gabbay, S. M. (Eds.), *Corporate Social Capital and Liability* (pp. 1–16). Boston: Kluwer Academic.

Gilsing, V. A., & Nooteboom, B. (2005). Density and strength of ties in innovation networks: An analysis of multimedia and biotechnology. *European Management Review*, *2*(3), 179–197. doi:10.1057/palgrave.emr.1500041

Gomes-Casseres, B. (1996). *The Alliance Revolution: The New Shape of Business Rivalry*. Cambridge, MA: Harvard University Press.

Granovetter, M. (1973). The strength of weak ties. *American Journal of Sociology*, *78*(6), 1360–1380. doi:10.1086/225469

Granovetter, M. (1983). The strength of weak ties: A network theory revisited. *Sociological Theory*, *1*(1), 201–233. doi:10.2307/202051

Granovetter, M. (1985). Economic action and social structure: The problem of embeddedness. *American Journal of Sociology*, *91*(3), 481–510. doi:10.1086/228311

Grant, R. M. (1996). Toward a knowledge-based theory of the firm. *Strategic Management Journal*, *17*(10), 109–122.

Gulati, R. (1995). Social Structure and Alliance Formation Patterns: A Longitudinal Analysis. *Administrative Science Quarterly*, *40*(4), 619–652. doi:10.2307/2393756

Hanifan, L. J. (1916). The Rural School Community Center. *The Annals of the American Academy of Political and Social Science*, *67*, 130–138. doi:10.1177/000271621606700118

Hansen, M. T. (1999). The search-transfer problem: the role of weak ties in sharing knowledge across organizational subunits. *Administrative Science Quarterly*, *44*(1), 82–111. doi:10.2307/2667032

Hargadon, A., & Sutton, R. I. (1997). Technology brokering and innovation in a product development firm. *Administrative Science Quarterly*, *42*(4), 716–749. doi:10.2307/2393655

Henderson, R., & Cockburn, I., (1994). Measuring Competence? Exploring Firm Effects in Pharmaceutical Research. *Strategic Management Journal*, *15*(Winter special issue), 63-84.

Holland, J., & Gibbon, J. (1997). SMEs in the Metal Manufacturing, construction and contracting service sectors: environmental awareness and action. *Eco-Management and Auditing*, *4*(1), 4–14. doi:10.1002/(SICI)1099-0925(199703)4:1<7::AID-EMA58>3.0.CO;2-9

Huber, G. (1991). Organizational learning: the contributing process and the literatures. *Organization Science*, *2*(1), 88–115. doi:10.1287/orsc.2.1.88

Huston, L., & Sakkab, N. (2006). Connect and Develop, Inside Procter & Gamble's new model for innovation. *Harvard Business Review*, *84*(3), 58–66.

Inkpen, A., & Tsang, E. (2005). Social Capital Networks and Knowledge Transfer. *Academy of Management Review*, *30*(1), 146–165. doi:10.2307/20159100

Jacobs, J. (1961). *The Death and Life of Great American Cities*. New York: Random House.

Johannisson, B. (2000). Networking and Entrepreneurial Growth. In Sexton, D. L., & Landström, H. (Eds.), *The Blackwell Handbook of Entrepreneurship* (pp. 368–386). Oxford: Blackwell.

Kaufmann, A., & Tödtling, F. (2002). SMEs in Regional Innovation Systems and the Role of Innovation Support: The Case of Upper Austria. *The Journal of Technology Transfer*, *27*(1), 15–26. doi:10.1023/A:1013140318907

Kogut, B., & Zander, U. (1992). Knowledge of the firm, combinative capabilities, and the replication of technology. *Organization Science*, *3*(3), 383–397. doi:10.1287/orsc.3.3.383

Kraatz, M. S. (1998). Learning by association? Interorganizational networks and adaptation to environmental change. *Academy of Management Journal*, *41*(6), 621–643. doi:10.2307/256961

Laursen, K., & Salter, A. (2006). Open For Innovation: The Role Of Openness In Explaining Innovation Performance Among U.K. Manufacturing Firms. *Strategic Management Journal*, *27*(2), 131–150. doi:10.1002/smj.507

Leonard-Barton, D. (1995). *Wellsprings of Knowledge: Building and Sustaining the Sources of Innovation*. Boston: Harvard Business School Press.

Lilien, G. L., Morrison, P. D., Searls, K., Sonnack, M., & von Hippel, E. (2002). Performance Assessment of the Lead User Idea-Generation Process for New Product Development. *Management Science*, *48*(8), 1042–1059. doi:10.1287/mnsc.48.8.1042.171

Lin, N., Cook, K., & Burt, R. (2001). *Social Capital: Theory and Research*. New York: Aldine de Gruyter.

Lind, M. R., & Zmud, R. W. (1995). Improving interorganizational effectiveness through voice mail facilitation of peer-to-peer relationships. *Organization Science*, *6*(4), 445–461. doi:10.1287/orsc.6.4.445

Lorenzoni, G., & Lipparini, A. (1999). The leveraging of interfirm relationships as a distinctive organizational capability: a longitudinal study. *Strategic Management Journal*, *20*(4), 317–338. doi:10.1002/(SICI)1097-0266(199904)20:4<317::AID-SMJ28>3.0.CO;2-3

Loury, G. C. (1977). A dynamic theory of racial income differences. In Wallace, P. A., & La Mond, A. M. (Eds.), *Women, Minorities, and Employment Discrimination* (pp. 153–186). Lexington, MA: Lexington books.

Maier, R. (2007). *Knowledge Management Systems. Information and Communication Technologies for Knowledge Management* (3rd ed.). Berlin: Springer Verlag, Wien.

Matusik, S. F., & Hill, W. L. (1998). The Utilization of Contingent Work, Knowledge Creation, and Competitive Advantage. *Academy of Management Review*, *23*(4), 680–697. doi:10.2307/259057

Moran, P., & Ghoshal, S. (1996). Theories of economic organization: The case for realism and balance. *Academy of Management Review*, *21*(1), 58–72. doi:10.2307/258629

Nonaka, I. (1994). A Dynamic Theory of Organizational Knowledge Creation. *Organization Science*, *5*(1), 14–37. doi:10.1287/orsc.5.1.14

Nonaka, I., & Takeuchi, H. (1995). *The Knowledge-Creating Company: How Japanese Companies Create the Dynamics of Innovation*. Oxford, UK: Oxford University Press.

Obstfeld, D. (2005). Social Networks, the *Tertius Iungens* Orientation, and Involvement in Innovation. *Administrative Science Quarterly*, *50*(1), 100–130.

Orlikowski, W. J., & Yates, J. (1994). Genre repertoire: The structuring of communicative practices in organizations. *Administrative Science Quarterly*, *39*(4), 541–574. doi:10.2307/2393771

Portes, A. (1998). Social Capital: Its Origin and Application in Modern Sociology. *Annual Review of Sociology*, *24*(1), 1–24. doi:10.1146/annurev.soc.24.1.1

Powell, W., Koput, K., & Smith-Doerr, L. (1996). Interorganizational collaboration and the locus of innovation: networks of learning in biotechnology. *Administrative Science Quarterly*, *41*(1), 116–145. doi:10.2307/2393988

Prahalad, C. K., & Hamel, G. (1990). The core competence of the corporation. *Harvard Business Review*, *68*(3), 79–91.

Prahalad, C. K., & Ramaswamy, V. (2004). *The Future of Competition: Co-Creating Unique Value with Customers*. Boston: Harvard Business School Press.

Putnam, R. (1993). The prosperous community: social capital and public life. *The American Prospect*, *13*, 35–42.

Putnam, R. (2000). *Bowling alone: The collapse and Revival of American Community*. New York: Simon & Schuster.

Reagans, R., & McEvily, B. (2003). Network Structure and Knowledge Transfer: The Effects of Cohesion and Range. *Administrative Science Quarterly*, *48*(2), 240–267. doi:10.2307/3556658

Romo, F. P., & Schwartz, M. (1995). The structural embeddedness of business decisions: The migration of manufacturing plants in New York State, 1960 to 1985. *American Sociological Review*, *60*(1), 874–907. doi:10.2307/2096431

Sawhney, M., Verona, G., & Prandelli, E. (2005). Collaborating to create: the internet as a platform for customer engagement in product innovation. *Journal of Interactive Marketing*, *19*(4), 4–17. doi:10.1002/dir.20046

Schiff, M. (1992). Social Capital, Labor, Mobility, and Welfare: The Impact of Uniting States. *Rationality and Society*, *4*(2), 157–175. doi:10.1177/1043463192004002003

Thomke, S., & von Hippel, E. (2002). Customers as Innovators: A New Way to Create Value. *Harvard Business Review*, *80*(2), 5–11.

Toffler, A. (1990). *Powershift*. New York: Bantam.

Utterback, J. M. (1994). *Mastering the Dynamics of Innovation*. Boston: Harvard Business School Press.

Uzzi, B. (1997). Social Structure and Competition in Interfirm Networks: The Paradox of Embeddedness. *Administrative Science Quarterly*, *42*, 35–67. doi:10.2307/2393808

Uzzi, B., & Spiro, J. (2005). Collaboration and Creativity: The Small World Problem. *American Journal of Sociology*, *111*(2), 447–504. doi:10.1086/432782

Van de Ven, A. H. (1986). Central problems in the management of innovation. *Management Science*, *32*(5), 590–607. doi:10.1287/mnsc.32.5.590

Von Hippel, E. (1986). Lead users: a source of novel product concepts. *Management Science*, *32*(7), 791–805. doi:10.1287/mnsc.32.7.791

Welford, R. (1994). Improving Corporate Environmental Performance. *Environmental Management and Health*, *5*, 6–10. doi:10.1108/09566169410057100

Wiggins, R. R., & Ruefli, T. W. (2002). Sustained Competitive Advantage: Temporal Dynamics and the Incidence and Persistence of Superior Economic Performance. *Organization Science*, *13*(1), 82–105.

Williamson, O. E. (1975). *Markets and Hierarchies: Analysis and Antitrust Implications: A Study in the Economy of Internal Organization*. New York: The Free Press.

Woolcock, M. (1998). Social Capital and Economic Development: Toward a Theoretical Synthesis and Policy Framework. *Theory and Society*, *27*(2), 151–208. doi:10.1023/A:1006884930135

Yli-Renko, H., Autio, E., & Sapienza, H. J. (2001). Social capital, knowledge acquisition, and knowledge exploitation in young technology-based firms. *Strategic Management Journal*, *22*, 587–613. doi:10.1002/smj.183

Zack, M. H. (1999). Developing a Knowledge Strategy. *California Management Review*, *41*(3), 125–145.

Chapter 14

Do Top-Down Cluster Policies Succeed in Fostering Social Capital and Innovation?
Some Insights from Italian Cases

Silvia Massa
University of Genoa, Italy

Stefania Testa
University of Genoa, Italy

ABSTRACT

In the last decades, the cluster approach to innovation has become very influential among policymakers. Clusters are recognized as the primary locus of social capital (SC) and civic engagement and are thus able to foster company innovativeness. This chapter investigates the effectiveness of top-down cluster policies in creating new SC and the ability of this SC to create new intellectual capital (IC), hence improving firms' innovativeness. For this purpose, two case studies about top-down clusters in Italy were developed. While in literature the SC-IC-innovation link has been largely investigated as well as the power of traditional clusters to promote SC, to the knowledge of the authors, the same power has not been investigated with reference to top-down clusters. The purpose of this chapter is to investigate such links, thus contributing to fill a gap in the extant literature. Examination of the cases in this study revealed almost opposite results in terms of the links investigated.

INTRODUCTION

Since the seminal work of Kline and Rosenberg (1986) innovation has been increasingly considered as an interactive process. This relatively recent view frames innovation as the result of a process whose success rests upon interactions and exchanges of knowledge involving diverse inter-depending actors. Landry et al. (2002) review the evolution of the concept of innovation from the 1950s, when explanations of innovation were based solely on tangible forms of capital, to present knowledge-based innovation derived from social networks. This more socially embedded concept of innovation has led to new innovation policies aimed at creating a social environment which is able to stimulate en-

DOI: 10.4018/978-1-61520-875-3.ch014

trepreneurship and collective learning, as well as social prestige and acceptance of innovation (Park, 2001). A stream of literature involving social-embedded innovation underlines the importance of regional resources for innovation processes in the more globalised economy. Proximate relations, localised learning and phenomena such as regional clusters and regional innovation systems (cf. Schmitz, 2004, p.2) are increasingly recognized as fundamental drivers of competitiveness of firms and industries. According to Porter (1998b, p.77): "paradoxically, the enduring competitive advantages in a global economy lie increasingly in local things – knowledge, relationships, and motivation that distant rivals cannot match". Two concepts belonging to the family of territorial innovation theory (Moulaert & Sekia, 2003) have demonstrated particular resonance in academic and policy circles: regional innovation systems (RIS) and clusters (Cooke et al., 2004; Acs, 2000; Porter, 2000a). Even though both concepts are closely related, they should not be conflated. Isaksen and Hauge (2002, p. 14) define the latter as "a concentration of 'inter-dependent' firms within the same or adjacent industrial sectors in a small geographic area". A RIS, on the other hand, is defined as "interacting knowledge generation and exploitation subsystems linked to global, national and other regional systems" (Cooke, 2004, p. 3). Several authors compare and contrast the two approaches (see e.g. Isaksen, 2007, Asheim & Coenen, 2005). This chapter takes up the issue of regionalising innovation policy by focusing in particular on the cluster approach. This approach, rooted in the works of Michael Porter (1990, 1998b, 2000), has become very influential among policymakers (see e.g. OECD (Organisation for Economic Co-operation and Development), 2007; LEED (Local Economic and Employment Development), 2009; LEED 2002; European Commission 2002; Ketels et al 2008; Nauwelaers, 2003; Karlsson, 2007). Clusters are recognized as the primary locus of social capital (SC) and civic engagement and are thus able to foster company innovativeness. Moreover, clusters offer the advantage of limiting opportunistic behaviour by encouraging cluster participants to strive for constructive interaction that will positively affect their long term interests (Porter, 1998a). Clusters favour dynamics of learning and knowledge creation based on socially embedded vertical and horizontal linkages of co-locating firms and on their interaction with education/research and development (R&D) (Lagendijk, 2000; Malmberg & Maskell, 2002; Maskell, 2001; Steiner, 1998; Wolfe & Gertler, 2004). Thus, as Porter claims (1998b:89), "Governments should promote cluster formation and upgrading [...]". Indeed, manuals providing rough prescriptions on how to foster and create clusters have been published (Ffowcs-Williams, 2004; Rosenfeld, 2002) and the growing interest of policy makers in "recreating Silicon Valley" is not surprising (see e.g. Feldman & Francis, 2004; Ibrahim, 2009; Science|Business Innovation Board, 2009). Several cluster policies, from the OECD, the World Bank, national governments and regional development agencies, have been undertaken and documented (Boekholt & Thuriaux, 1999; EU Commission, 2002; Raines, 2000, 2001, 2002; Roelandt and den Hertog, 1999; Roelandt et al., 2000). According to the European Cluster Observatory, the EU counts some 2,000 clusters, 70 different national cluster policies, and hundreds of regional programmes (see http://www.clusterobservatory.eu/). The evident emphasis on clusters has been triggered by a general feeling of disappointment with previous supply-side-oriented measures targeted on infrastructure and technology support. In particular, most of the technology parks and technology transfer centres which had been established on a large scale in the 1980s and early 1990s did not match the high expectations their role models had created (Hassink, 1996). Cluster strategies were conceived as a good balance between supply-oriented policies and demand-oriented policies, even though some authors denounce the dominance of supply-sided actions, often towards a few, well known organizations (see e.g. Nauwelaers, 2001).

This chapter investigates the effectiveness of deliberate cluster promotion programs (so called top-down cluster policies) in creating new SC and the ability of this SC to create new intellectual capital (IC) in cluster companies, hence fostering innovation. To this purpose, two cases of top-down clusters in Italy were selected for investigation and comparison. In the literature, the link between SC, IC and innovation as well as the power of traditional clusters to promote SC (Porter, 1990), have been largely investigated (see the next paragraph for references); however, to the knowledge of the authors, the same power has not been investigated with reference to top-down clusters. It is paradoxical that the literature on top-down clusters scarcely mentions SC issues (see e.g. Redzepagic & Stubbs, 2006, among the exceptions) as they are so strictly connected to the concept of cluster itself. The purpose of this chapter is to investigate the link between the concepts of top-down clusters and SC, thus addressing a gap in the extant literature. This gap is most likely due to the fact that top-down clusters and SC are traditionally investigated by separate scientific communities. This chapter's contribution to the literature is the development of a comprehensive framework to link fields that have been traditionally maintained distinct.

This chapter is organized as follows. Firstly, the theoretical background is provided. Secondly, methodology and investigation framework are described. Thirdly, the selected top-down clusters are presented. Fourthly, empirical evidence from the cases is discussed. Finally, some conclusions and future research directions are provided.

THEORETICAL BACKGROUND

SC - IC - Innovation

In recent literature, several theoretical contributions have emphasised the importance of socially embedded collaboration for innovation-driven regional development. The notion of SC focuses on this factor of economic success (Fromhold-Eisebith, 2004) and it has become increasingly popular in a wide range of social sciences, economics and organization studies (Adler & Kwon, 2002). Although literature yields many definitions of SC, a broadly accepted definition is that which defines SC as the sum of the features of social structure that facilitate action (Adler & Kwon, 2002). It is worth noting that SC does not necessarily refer to local, regional, and geographically bounded features (Bellandi, 2000) but for the purposes of this chapter the focus is devoted solely to them. Several dimensions have been introduced to better explain the nature of SC. In line with Nahapiet and Ghoshal (1998), the notion of SC in this chapter comprises three dimensions. The structural dimension concerns properties of social systems and networks of relations as a whole, including: network ties, network configuration and appropriable organization – that is the existence of networks created for a purpose that may be used for another. The cognitive dimension refers to those resources providing shared representations, interpretations and systems of meaning among parties, and therefore includes: shared languages and code and shared narratives. The relational dimension refers to the kind of personal relationships people have developed with each other through a history of interactions, including: trust and trustworthiness, norms and sanctions, obligations and expectations, and identity and identification.

SC has proved to be a powerful explaining factor for several phenomena: entrepreneurship and formation of start-up companies, improved supplier relationships, reduction of transaction costs, as well as inter-firm learning and creation of IC (see special issue of Strategic Management Journal, 21, 3, 2000). In particular, Hargadon and Sutton (1997) and Nahapiet and Ghoshal (1998) investigated the role of SC in facilitating the development of IC. It is worth noting that, in this chapter, SC is an attribute related to local social systems (such as local institutions and networks as well as local norms, values and culture) while IC is an attribute related to individual companies. Many scholars have investigated IC at the company level (for an extended literature review see e.g. Bontis, 1999; Petty & Guthrie, 2000; Marr et al., 2004; Kaufmann & Schneider, 2004; Roos et al., 2006) and this tradition has grown significantly in recent years (Serenko & Bontis, 2004). IC can be broadly defined as the sum of all knowledge firms utilize for competitive advantage (Nahapiet & Ghoshal, 1998; Youndt et al., 2004) but determining its components and measurement approaches has proved to be a difficult task (Schneider & Samkin, 2008). Following Stewart (1999, 2001), Subramanian and Youndt (2005), Youndt, Subramaniam and Snell (2004), Martinez-Torrez (2006) as well as Martin de Castro and Saez (2008) among others, IC is defined as encompassing: i) human capital; ii) structural capital; and iii) relational capital. Human capital represents the combined knowledge, skills, innovativeness and capabilities of a company's individual employees (Youndt & Snell, 2004). Structural capital represents non-human storehouses of knowledge embedded in technology, software, databases, structure and routines (Youndt & Snell, 2004). Schneider and Samkin, (2008) argue that it can be further disaggregated into internal (e.g. organizational structures) and external capital (e.g. trademarks). Relational capital represents knowledge embedded in networks of relations that the organization and its members establish both inside and outside the working environment (Adler & Kwon, 2002). The relation between IC and innovation is so intertwined that the boundaries between the two literatures have been getting thinner and thinner in recent years (Longo & Mura, 2009) and, frequently, the constructs of IC and innovation appear in the same studies where the former is identified as a predictor of the latter (Ahuja, 2000; Subramaniam & Venkatraman, 2001; Tsai & Ghoshal, 1998). Similarly to IC, a plethora of definitions, indicators and measurement methodologies for innovation has been introduced in the literature (for an extended review on this topic see e.g. Garcia & Calantone, 2002). Innovation is measured here in terms of its main components (Bonanno & Haworth, 1998): product changes and process changes.

Top-Down Clusters

The concept of top-down clusters tends to overlap with the concept of hi-tech clusters due to the fact that current policies are oriented towards promoting hi-tech industries, such as biotechnology, new materials, information technology and telecommunications (Enright et al., 2001). While traditional clusters have spontaneously come about in a bottom-up manner, hi-tech clusters are often the result of deliberate and planned public intervention policies (Piccaluga et al., 2003). In Italy, hi-tech clusters were launched in the 2002-2004 National Research Plan as formal institutions with responsibilities to foster cluster activities (Lagendiik, 2000). According to Italian government guidelines, top-down clusters have the responsibility for the following cluster support actions (OECD, 2007):

(G1) Coordinating joint research projects involving higher education institutions and private firms and strengthening R&D infrastructure (buying equipment or building new facilities).

(G2) Attracting and training researchers (e.g., establishing laboratories, scholarships or training courses in the specialised field of the cluster).

(G3) Supporting start-ups and business expansion (including entrepreneurship training, assistance with patenting and marketing business services for small and medium enterprises (SMEs)).

As noted by several authors (Bonaccorsi, 2006; Nauwelaers, 2001) the focus of these policies is essentially set on community building activities among research centres, institutions and firms rather than on financial support to individual firms. Such community building activities include enhancing communication, coordination and commitment among actors. Although the focus of clusters is on social aspects (Garret-Jones, 2004), the "disparate array of indicators and measures" that have been developed (Geisler, 2005) fail to capture these very social features (Gertler, 2004). This is also true for SC that is not reflected at all in these indicators (Davis et al., 2006). As noted by Diez (2001), this is due to the fact that SC and other "soft" benefits of clustering are intangible and difficult to quantify.

Several authors are highly critical of the notion itself of top-down clusters (for a review on the topic see e.g. Fromhold-Eisebith and Eisebith, 2005; 2008). Bresnahan et al. (2001, p.858) argue that "direct, top down policies are most likely to fail." As evidence, Martin and Sunley (2003, p.28) claim that "even cluster enthusiasts find it enormously difficult to point to any examples of deliberate cluster promotion programs that have been unambiguously successful". Other authors (see Malmberg and Power,2003; Atzema and Visser, 2006; Koschatzky and Lo, 2007) note that it is not always necessary to have or promote clusters to stimulate innovation. Other authors are cautious in generalizing about the results of top-down cluster policies claiming that they are determined by various factors, and notably by the personal qualities of individuals involved in initiatives (Fromhold-Eisebith and Eisebith, 2008).

METHODOLOGICAL AND INVESTIGATIONAL FRAMEWORK

Using this background as a starting point, the cases of two hi-tech clusters were chosen for investigation. Both clusters were established in Italy at the beginning of the 2000s. Data collection for each case was based on nine semi-structured interviews with key stakeholders: research centres, institutions and firms. The choice of such stakeholders is rooted in the concept of triple helix. According to the triple helix model, innovation is no longer the realm of companies but, rather, the result of "an expanding network system of interactive spirals" (Leidesdorff & Etzkowitz, 2001, p.1) generated by university, industry and government. It is largely recognized in literature that these three stakeholders have rather different perspectives on topics related to innovation which should be taken into consideration in this kind of investigation (Massa & Testa, 2008a, 2008b, 2009). A list of questions about the links under investigation was developed. These questions, which will be exemplified below, were inspired by relevant literature (see e.g. Grootaert et al., 2004; Bontis, 2001). Due to space limitations, only examples of questions asked to managers and entrepreneurs will be reported. Similar questions were developed and customized for the other stakeholders. All interviews were conducted by the research team, which made audio recordings and took detailed field notes. Subsequently, complete write-ups were carried out as detailed case studies. All interviewees were given various opportunities to add suggestions and comments to drafts of the case studies in an effort to achieve construct validity. Reliability requirements were assured by using the same protocol for each case - which included transcripts of audio recordings from all interviews, transcripts of research team field notes and observations, as well as transcripts of secondary data - and by the development of a complete database in the data collection phase. Data collection was stopped after nine interviews per case because the last interviews conducted provided the research team with narratives similar to those previously heard and the theoretical saturation point seemed to have been reached. Archival documents were collected as well, following the distinct tradition in the literature on social science research methods that advocates different sources of data (and methods) be used to validate one another (Jick, 1979). In particular, data extracted from the clusters' websites and from publications (annual reports, articles, master theses etc.) were used to capture a complete, holistic, and contextual portrayal of the units under study. Evidence chains were established and it was found that the documentary evidence corroborated the stories provided by the various interviewees. The two cases were analyzed in relation to the links introduced in the theoretical background, by coding and classifying the evidence collected by means of content analysis (Bryman & Burgess, 1993). Due to the limited number of cases, no software packages were used for analyzing data.

Three research questions (RQ) were developed.

1st RQ - Are top-down cluster policies effective in creating new SC?

This first RQ was addressed by investigating the possible impacts of cluster support actions (G1, G2, G3) on the increase of each of the three dimensions of SC (structural, cognitive and relational). Therefore, nine links were examined by means of three question sets. Figure 1 represents the links examined, while Figure 2 offers some examples of questions for each of the three sets mentioned.

2nd RQ - Can such newly created SC generate new IC?

This second RQ was addressed by investigating the possible impact that an increase of each of the three dimensions of SC may have on the increase of each of the three dimensions of IC. Accordingly, nine links were explored by means of three question sets. Figure 3 represents the links examined, while Figure 4 offers some examples of questions for each of the three mentioned sets.

3rd RQ - Can such newly created IC improve innovation in firms?

This third RQ was addressed by investigating the possible impact that an increase of each of the three dimensions of IC may have on product and process innovation. Then six links were explored by means of three question sets. Figure 5 represents the examined links, while Figure 6 offers some examples of questions for each of the three sets mentioned.

It is acknowledged that the model proposed is over-simplified, however it may be useful for identifying critical areas when implementing and

Figure 1.

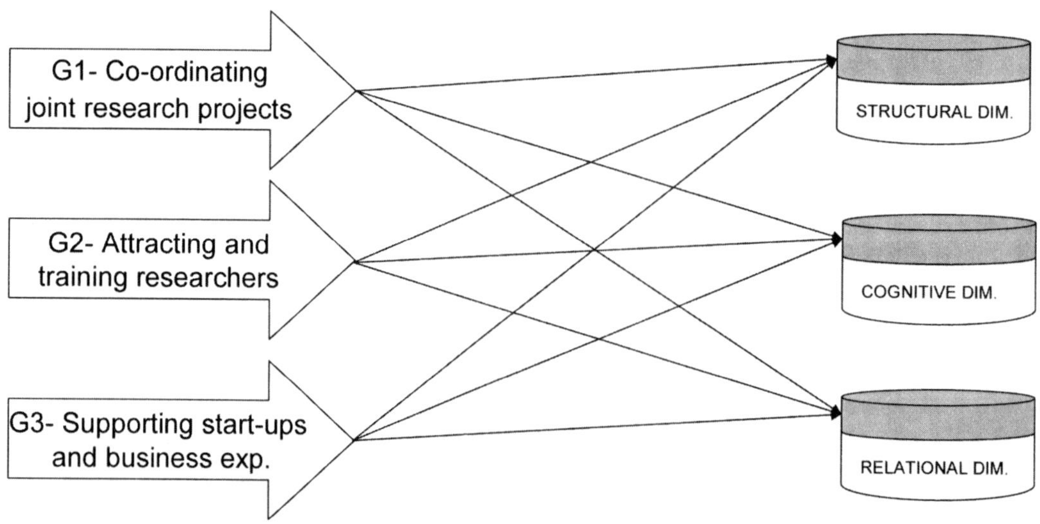

managing top-down clusters or evaluating their performance (see Figure 7). For example, the model does not take into consideration the fact that reciprocal associations exist among the different dimensions of SC as well as the different dimension of IC (see e.g. Tsai & Ghoshal, 1998).

THE TOP-DOWN CLUSTER CASE STUDIES

This chapter compares two case studies of top-down clusters on the basis of the three RQs set forth above. Due to privacy issues, the two clusters will be called: A-CLUST and B-CLUST.

Figure 2.

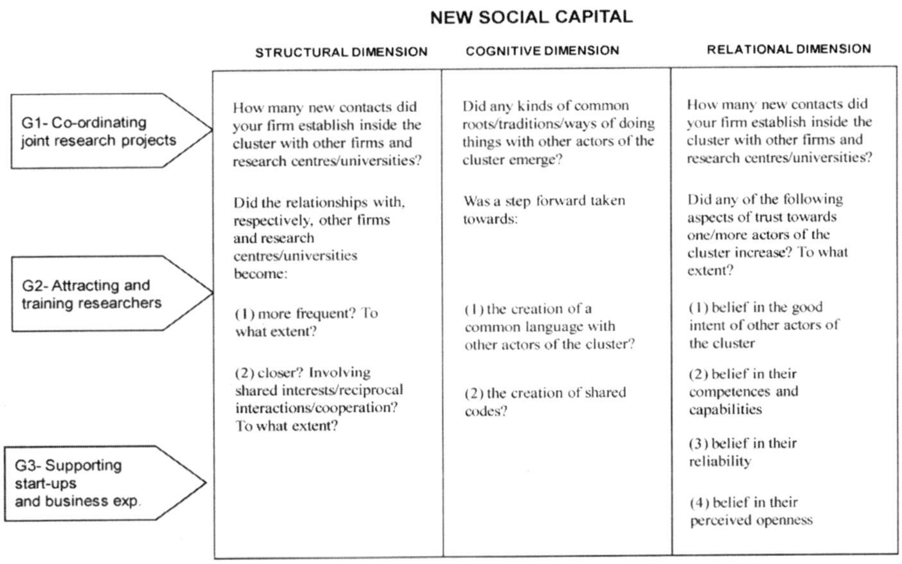

Figure 3.

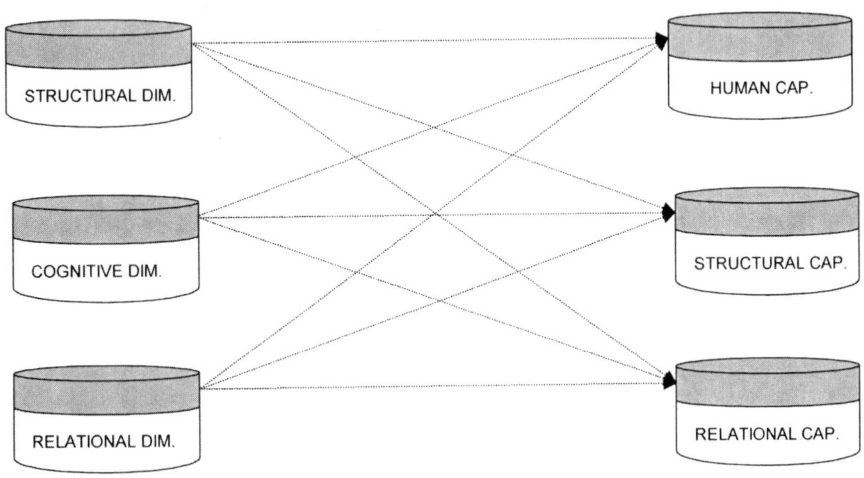

Figure 4.

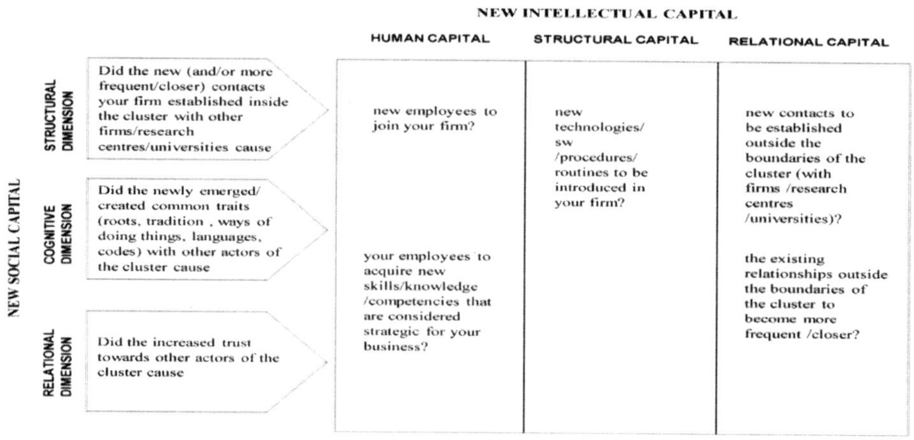

Figure 5.

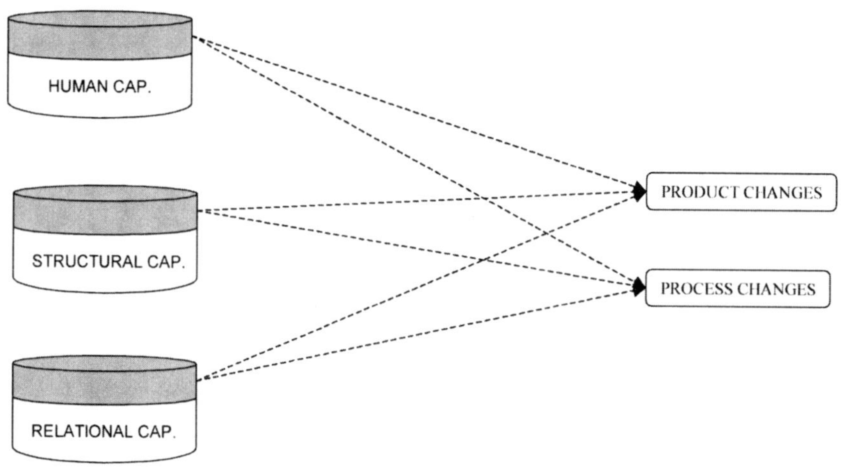

A-CLUST was started in 2003 in an area with a strong cluster vocation in traditional industries. A consultancy firm matched firm needs with university competencies in the local area and selected the best opportunities following an international comparison. The niche sector of nanotechnologies applied to materials was selected as it was considered the most promising in connection with the fabric of the local entrepreneurial system (in the fields of mechanics, mechatronics, textiles, tanning, etc...). The first executive action of the cluster strategy was to establish a consortium composed of the three main local universities and a nanotech-lab. Thus, a link was created between applied research and the industrial system, which is to say between the locus of scientific knowledge and that of technological knowledge. A-CLUST represents the pillar of regional strategy for innovation and local development, as demonstrated by new investments included in the current regional operative program and the Regional Plan for Nanotechnologies 2009-2013.

B-CLUST was started in 2005 in an area without an explicit cluster vocation. From an

Figure 6.

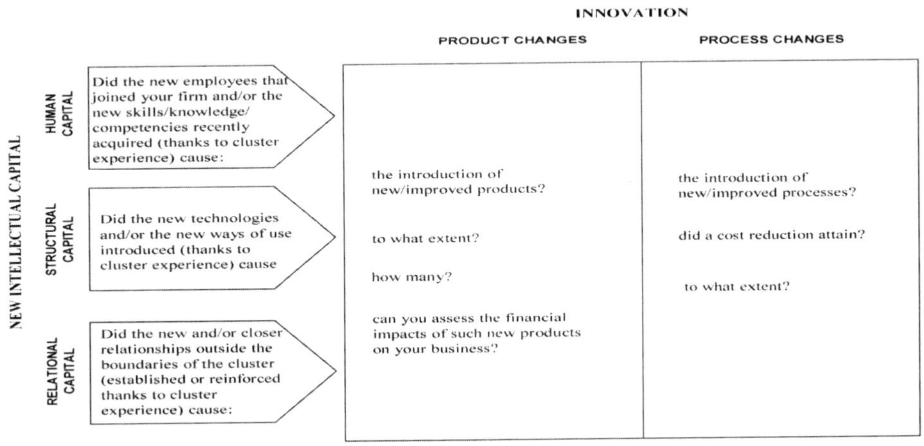

Figure 7.

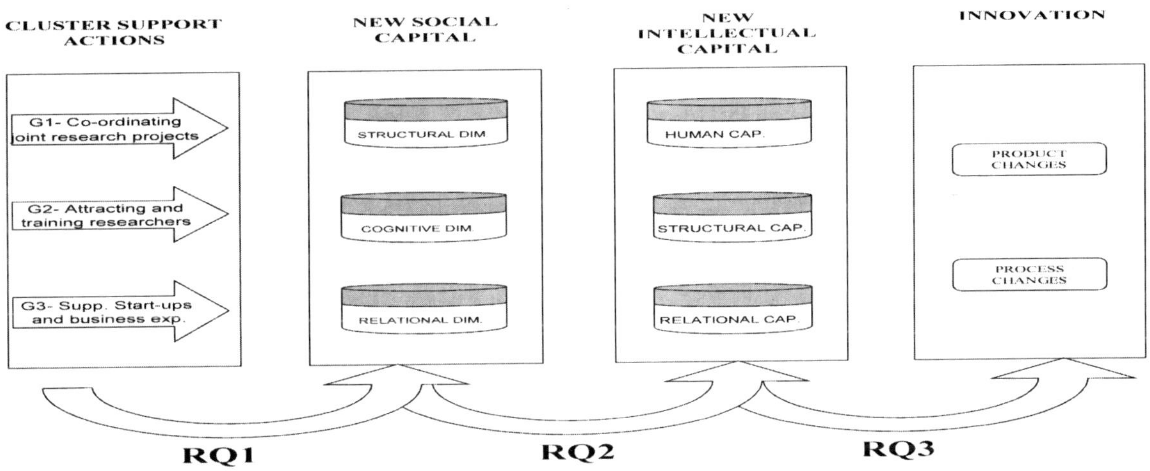

industrial point of view, this area is dominated by a small number of large, originally government-controlled firms in the field of information and communication technologies (ICTs). The territory is also characterized by a large number of scarcely integrated SMEs that generate about 71.4% of the regional value added manufacturing The first step in cluster creation concerned the definition of research topics, which were chosen by considering only fields of interest to large companies and the local university without involving specialised consultants. A group of SMEs had proposed to define cluster research topics based on their ability to maximize potential impact on the local area. Their proposal was rejected as it was considered not to be of interest to the larger players in the cluster. Consequently, the cluster framework that had been settled upon was perceived by SMEs as lacking in any support of their activities. The entrepreneurs interviewed conveyed the perception that large companies simply made use of SMEs since SME participation is mandatory in order to implement top-down cluster policies and receive financial support (see Mussati et al., 2007). According to its detractors, the topics suggested by large firms and the university resulted in cluster research themes that were vague and lacking in focus. B-CLUST is structured as a consortium encompassing local university (51%), regional government, national research council, chamber of commerce, an SME consortium and large firms. The regional government recently announced its exit strategy from cluster experience, offering its shares to consortium partners with pre-emptive rights. None of the partners took advantage of this option and new firms are now expected to enter.

In comparing clusters, the literature investigates several dimensions and demand clearly emerges as an important issue (Gilsing, 2001). At the moment, demand in A-CLUST comes mainly from cluster firms in traditional industries looking for advanced technologies to boost their competitive advantage. Demand in B-CLUST is virtually worldwide and in this respect it has been global from its inception. In fact, cluster firms are mainly on the technology supply side. Local demand could potentially come from public administration.

Another important dimension in comparing clusters is the type of knowledge inside the cluster. A-CLUST is a typical science-based cluster. The locus of knowledge generation is clearly identifiable as cluster laboratories employ researchers from nationwide and international universities and research centres. Cluster firms are involved as testers or customers. In B-CLUST the knowledge type is more engineering-based, the fruit of industrial applied research. The locus of knowledge generation is consequently more complex to identify. Joint laboratories between university and firms are the intended place where different know-how merges to generate new shared knowledge.

DISCUSSION

In this section each RQ is discussed by examining the relevant empirical evidence from both case studies. The data are reported and compared in order to highlight the main differences between the cases. The discussion of the first RQ is subdivided into three sections, one for each cluster support action (G1, G2, G3).

First RQ: Are Top-Down Cluster Policies Effective in Creating New SC?

G1 - Coordinating Joint Research Projects

A-CLUST coordinates both public-funded and firm-commissioned research projects. The regional government has funded the university consortium with €12m to carry out nine R&D projects in the sector of nanobiotechnology and nanotechnology applied to materials. R&D projects are carried out by consortium scientific staff in a purpose built cutting-edge laboratory furnished with over €20m of top-level scientific equipment. Cluster firms participate in such projects as testers. To date privately funded R&D projects totalling approximately €2.8m per year have been commissioned by 68 firms, including 48 large firms and 28 SMEs. Moreover, A-CLUST sets up and leases tools and machinery to universities and private firms.

SC was furthered by these support actions in its structural dimension by creating and reinforcing links between private firms and cluster research structures. This is evidenced by the university consortium manager who declared:

"We make a special effort to persuade firms to share part of the privately-funded research results with other firms of the cluster, provided that they are not direct or potential competitors. Commonly we leverage the fact that research infrastructure and technology platforms are entirely financed with public funds. Currently, most of the customer firms accept to leave public some results".

Sharing knowledge deriving from research projects contributes to reinforcing the network from the structural side, even if relationships among the firms are not direct but mediated by institutions. This seems to be a virtuous circle: the more firms consent to share research results, the more other firms are stimulated to do the same. Such behaviour depends on the trust they have in A-CLUST and contributes to generating new trust, thus affecting SC also from the relational point of view. The relational dimension of the cluster's SC increases in proportion to the trust among its actors and could be compromised by opportunistic behaviours. The role of institutions inside clusters serves to guarantee competition remains fair. The relational dimension of SC in A-CLUST seems to increase year after year, as can be inferred from the increasing number of projects commissioned and the growing number of firms sharing results. In the future, the cognitive dimension may be manifested through the exchange of research results becoming a shared practice, a sort of common tradition among cluster actors.

B-CLUST also activated joint research projects, but in a completely different manner. The main difference regards the projects' proponents. While A-CLUST's projects were commissioned by the firms that fund partially or entirely the research projects, B-CLUST's public research structures (essentially the local university) proposed research projects to cluster firms, following a technology-push approach. This proactive approach of the university, together with the scarce interest demonstrated towards the firms' real needs, is evident from the words of an interviewed academic, who declares:

"hi-tech clusters should become project machines to secure money from the EU".

The prospect of economic support convinces firms to join projects, but subsequently they find it extremely difficult to apply the research results. For example, the entrepreneurs interviewed cite funding programmes requiring SMEs to formulate a research proposal for universities to develop. At least in some cases, the reality was that, the universities themselves suggested to the firms what research products they were to request regardless of their actual needs. Frequently such research products had already been developed and laid unused in laboratories or were already under development as they fell within the areas of interest of university departments. As a result, collaboration between firms and universities was merely simulated. Such behaviour may often be caused by lack of time. Both entrepreneurs and academics noted that calls for proposals, particularly those relevant to regional funding, are generally launched unexpectedly without considering the time required to identify the real needs to be addressed. It is difficult to conceive of an average SME, which is traditionally overwhelmed by daily tasks, having a perfectly updated and ready-to-submit research project on hand. In this situation, SC is scarcely engaged as such a simulated collaboration hardly contributes to enhancing the structural dimension of a network. On the contrary, it seems to play a role in compromising the relationships with universities and public institutions from the point of view of trust, that is to say the relational dimension of SC. Indeed, it appears to be lost opportunity to reduce the never-ending divide between entrepreneurs and academics. It is as if the common belief that research carried out in university laboratories has very little in common with firms' needs was finally proven (Massa & Testa, 2008a, 2008b, 2009).

G2: Attracting and Training Researchers

A-CLUST began its experience by establishing top-level laboratories to perform state-of-the-art research projects on nanobiotechnology and nanotechnology applied to materials. These laboratories are the place where the cluster knowledge base on nanotechnologies is created and constantly updated. Moreover, A-CLUST provides a variety of educational opportunities in the nanotech field, including a postgraduate master's degree (hereafter NIM, now at its seventh edition), workshops, as well as tailored courses to fulfil particular client needs. A-CLUST organizes top-level scientific events on nanotechnologies attracting researchers from all over the world.

This educational support action impacted on the structural dimension of SC by creating and reinforcing links among researchers working inside the cluster's research structures and between such research structures and private firms. In 2007, 360 firms visited the cluster's scientific laboratories and 83 of them requested feasibility studies and eight yearly firm-commissioned educational programmes were started. In this context, reinforcing links between research structures and firms is considered by researchers as a superior goal compared to, for example, publishing scientific papers. As reported by the university consortium manager:

"A high percentage of our researchers' energy and time is devoted to building and enriching strong and stable relationships with local companies".

From the cognitive point of view, the educational activity on nanotechnologies contributes to creating a common language inside the cluster and has a determinant role in stimulating the demand of nanotechnologies inside traditional industries.

B-CLUST activated a postgraduate master degree as one of its first actions as a cluster. However, it experienced extreme difficulty in finding participants who could not easily find placement inside the cluster's firms upon completing the master's programme. This resulted in graduates being dispersed outside the cluster's territory, with a consequent lack of proper return on investment from the point of view of creating new SC. B-CLUST resolved to establish research laboratories to develop joint research projects with cluster firms. However, finding an appropriate location for the laboratories and the subsequent building works have taken a considerable time (nearly two years thus far). Machinery had been already bought in order not to lose the funding and is becoming obsolete in warehouses. This situation contributes to the growing disrepute of cluster governance and to further inhibit any interest in collaboration. As a result, the impact of these actions on SC cannot be considered positive, in particular from the relational point of view of trust and credibility.

G3: Supporting Start-Ups and Business Expansion

In cooperation with another Italian top-down cluster, A-CLUST organizes an annual international business plan competition for the fields of nanotechnology and polymer-based materials, which is now in its fifth edition. The organizers aim at financing start-ups inside their local territories and at soliciting funds from early-stage investors and venture capital firms. The award totals €600,000, the highest amount worldwide for this kind of competition. In efforts to further the most promising technologies, raise more funds and create new ventures, the organizers present top finalists to major Italian and international investors and venture capitalists. The last two editions of the competition received strong support from private enterprise which provided sponsorship and participated in the selection process.

These actions enhanced the structural dimension of SC by enlarging the network through the creation of new nodes – start-ups – and reinforcing the existing local links among research institutions and sponsoring firms. Moreover, the repeated successful editions of the competition contribute to creating a sort of local tradition of excellence in nanotechnologies, thus also impacting on the cognitive dimension.

B-CLUST has not yet activated any actions aimed at supporting start-ups on the cluster's main themes. Local institutions do organize a business plan competition to support the creation of local start-ups, but this initiative is not integrated with the cluster.

Second RQ: Can Such Newly Created SC Generate New IC?

In A-CLUST, newly created SC seems to have a positive impact on the generation of new IC. For the moment this is mainly in its human capital dimension. As recognized by many scholars (see e.g. Nahapiet & Ghoshal, 1998), new IC is the result of prior relationships and trust among individuals and institutions, and not the mere result of financed projects. The development of trust has a major bearing on the success of any collaboration, thus the role of the cluster's actors is mainly to address the need for collaborators to interact socially and trustworthily. Due to new, more frequent and closer contacts between local firms and research centres, as well as of increased trust towards institutions, the number of local companies available to hosting NIM students for internships is constantly increasing. During the first NIM edition, local companies counted for just 23% of the total number of hosting companies, whereas during the latest edition, they counted for 60% of the total. It is worth noting that hosting students in production facilities is demanding for companies, requiring specifically allocated time and resources. This has been facilitated by the fact that most companies hosting NIM students have joint research projects with cluster research centres and thus the involvement of students is framed towards a specific goal. NIM students are often hired by the hosting companies at the end of research projects. This fact guarantees company access to appropriate skills, a crucial condition for successful innovation processes. Furthermore, about 50% of NIM students from outside the cluster are now recruited as new staff in local companies. NIM also seems to be attracting talent into the region through its reputation which is increasing beyond the cluster's boundaries. During the master's seventh edition, 194 applications were received from 28 different countries with foreign students representing about 75% of total students. Coupled with internal awareness, such external recognition is an important indicator of cluster performance (Davis et al., 2006). This is not just the result of top-down cluster policies (namely G1 and G2) but a consequence of the SC the cluster's actors have been able to create.

Any evaluation of the structural dimension of company IC is premature as new technologies based on nanotech have been introduced in a few companies, but they are still in their infancy.

The relational dimension of company IC was enhanced through the establishment of new contacts between cluster research centres and large extra-cluster companies operating in the supply side of nano-materials. It is not yet possible to evaluate whether such collaborations will be durable and sustainable.

As previously noted, no SC has been created in B-CLUST, whether in terms of trust towards the cluster's actors or increased awareness of technology applicability on the demand-side. The cluster's actors have not been able to build extended networks within the business world or the university community. This situation precludes any evaluation of the model's subsequent steps. However, this does not mean that new IC has not been created inside local companies. It simply means that this was the result of individual companies' ties in their knowledge networks and of private investments. This is not surprising given that some scholars (see Atzema & Visser, 2006; Koschatzky & Lo, 2007) claim that promoting clusters is not always necessary to stimulate innovation. Indeed, some of the entrepreneurs interviewed share the same opinion. For example, the entrepreneur of a cluster SME who is a leading figure in the field of Medical Devices for Functional Medicine and Rehabilitation, exemplifies this opinion:

"We certainly don't wait for the cluster help to innovate! We find collaborations by ourselves. We are used to collaborating with university departments, our suppliers, research departments of other firms. Currently we are partner of a university spin-off that is 150 km from here with whom we are studying new applications of medical devices in the veterinary sector."

Third RQ: Can sSuch Newly Created IC Improve Innovation in Firms?

As reported in nanotech literature, time-to-market in the sector can be over ten years. Nevertheless A-CLUST decided to focus solely on technologies which have reached a development phase that may have commercial applications within a five-year period. As an example, a new floor wax was introduced in the product portfolio of an SME as a consequence of the new IC created through interaction with A-CLUST research centres. Other innovations developed in A-CLUST include: anti-scratch coatings and paint, nano-structured coatings with improved mechanical properties to be applied in the bicycle industry, advanced applications in the polymer field and in the development of plastic products and advanced applications in the production of protective motorcycle clothing. Several projects stop at the feasibility study stage due to high implementation costs. Examples include the development of anti-scratch marble kitchen bench tops and ice-resistant pipes for professional ice-cream freezers. However, technology evolves rapidly and implementation costs which are presently prohibitive may drop enough for projects to be re-examined within an acceptable timeframe.

One of the main cluster goals is to foster innovation in local firms, thus creating value for the territory. This derives from an explicit strategic choice of the cluster promoters, as also emerges from an interview with the Regional Councilman for Economy and Innovation:

"*We commit ourselves to make the investment of public money strictly connected to the needs of the local productive system, providing results that are patentable and able to foster a competitive growth. I would expect that our entrepreneurs and also the public administrators understand the importance of these projects: an investment in research is worth more than some public contribution distributed to the firms.*"

CONCLUSION

Discussion of the empirical data and relevant literature (Martin & Sunley, 2003) revealed highly varied experiences going under the name of top-down clusters. Indeed, the same set of norms, policies and instruments set in place to create top-down clusters give birth to deeply different implementations of the cluster concept. In the cases under examination, such different implementations led to nearly opposite results in terms of SC and consequently in IC and innovation generation. The fundamental and seminal difference in the cases is the approach adopted to originate the cluster: A-CLUST was essentially created by demand-driven factors whereas B-CLUST was created by supply–driven factors. The idea at the base of A-CLUST's creation was to identify advanced technologies that could best serve the potential local market of large firms and SMEs in traditional industries, frequently already structured in industrial clusters. Conversely, the idea at the base of B-CLUST's creation was to identify the common denominator between the hi-tech supply of large local firms and the local university's competencies (which, at the end of the process, resulted to be rather a large pool of hi-technologies), as well as to address the global demand. The advantages and disadvantages of the approaches used in both cases have been discussed in the literature. At least at the beginning, the approach A-CLUST employed seems to be less risky and more clearly focused, however it incurs in the limitations that a "protected domestic market" presents. Furthermore, the choice of a "sunrise" industry, such as biotechnology, new materials, information technology and telecommunications

(Enright et al., 2001) increases the risk of destructive competition deriving from the fact that such sectors are frequently targeted (sometimes in neighbouring areas). The approach chosen by B-CLUST may experience slower return on investments but maintains "a strong export-orientation" which has a higher probability of success in the long-term (Bresnahan et al., 2001). Consequently, both approaches can be deemed acceptable and assessing the best one is does not fall within this chapter's objectives.

What can explain such different results in the present top-down cluster policies in terms of SC-IC-innovation generation? Through examining the cases, it emerges that the main strength of A-CLUST is that it sits in the middle, spanning both the supply and demand side of the cluster equation with unique linkages around the world. On the contrary, what seems to fail in the second case is the very core of top-down cluster philosophy. Regional governments should set regional priorities and formulate a view for the future as well as check that the rules of power relationships do not overcome the rules deriving from SC and civic engagement that are supposed to be the essence of clusters. Contradicting this basic rule seems to emerge as the original flaw of B-CLUST. In fact, from the choice of research themes to governance issues, prevailing interests of a minority emerge to the detriment of regional development as a whole. Indeed, in B-CLUST, the thorniest problem is precisely cluster governance, i.e. the definition of who is allowed to define a problem and find a solution. Cluster governance is a challenging task as top-down cluster policies imply complex governance mechanisms – mixtures of markets, firms, alliances, associations, public and private organizations - with no a priori structurally superior solution (Davis et al., 2006). In B-CLUST, large companies and the university were supposed to have the necessary breadth of perspective and capabilities to govern the cluster, yet this was not the case. For example, university members aimed to make the new hi-tech cluster a "project machine", totally neglecting to take on board the views of the other stakeholders, such as the small entrepreneurs. Indeed, the SMEs' mistrust towards the cluster is mainly due to their perception of not having any say in the decisions despite their clear interest in how the cluster is implemented. As noted in literature (Viale & Etzkowitz, 2005), taking "the role of the other" does not mean that universities become firms or governments become businesses. It means that the interests of others are taken into consideration. Furthermore, when lacking an appropriate form of governance, the desire to implement certain support actions may vary since actors involved face conflicting incentives, feelings and interests. As an example, in B-CLUST, large companies do not have any interest in promoting new start-ups (if not controlled by them), thus it is not surprising that action G3 is still far from implementation. This situation exists despite the fact that literature broadly testifies (see e.g. Bresnahan et al, 2001) that the best-performing high-tech clusters are those with a high start-up rate, i.e. with a high rate of new node creation in the structural dimension of SC. Each category of stakeholders seems to act for their own category's interests to prevail.

In conclusion, what distinguishes the successful case from the failure case? In line with the extant literature (see e.g. Nauwelaers, 2001), the answer to this relates more to governance than to support actions themselves. In the success story, the regional government was able to set a governance structure that provided each stakeholder with an active voice in how to implement the initiative. Such involvement contributed to generating trust from the beginning, enhancing the initiative's reputation and attracting new participants year after year, thus positively impacting on SC. In the failure case, the regional government did not take the lead of the initiative, whether by setting regional priorities, formulating a view for the future or checking that the rules of power relationships did not overcome the "good" rules embedded in SC. Moreover, the regional government's exit from the cluster consortium officialises the institution's renunciation of any role in governance. This retreat of the regional government left excessive room to a few large companies and tended to reinforce the status quo. As noted by Roelandt et al. (2000; p. 19), "the notion of clusters is appropriated by (industrial) policy makers and used as an excuse to continue more or less traditional ways of defensive industrial policy making". From this evidence it can be inferred that the most important action to invert the course of B-Cluster should be to improve the governance of the initiative. To this purpose, the keystone could be identifying new committed individuals to lead the cluster. As Fromhold-Eisebith and Eisebith (2008) note about clusters: "new staff can at times 'turn around' the entire setting for the better or worse."

The choice of a case study approach presents advantages and disadvantages. Deeply observing what happens in reality helps to understand certain phenomena better; however, arriving at general implications through the study of a limited number of cases is difficult. Furthermore, it is worth underlining that SC-IC-innovation generation started by top-down cluster policies is modelled as a simple linear sequential process, not taking into account interactive and recursive aspects. These nonlinear aspects can be expected to change the causal relations between input and output. In spite of this limitation, it is believed that this oversimplified model is useful for identifying areas of attention when implementing and managing top-down clusters or evaluating their performance.

FUTURE RESEARCH DIRECTIONS

Further research is needed to delve deeper into RQ2 and RQ3, i.e. the impact of SC on IC and of IC on innovation, as they were scarcely investigable as of now due to the short life of the proposed cases. Such cases should be further monitored, in a longitudinal perspective, in order to add evidence to the study. In fact, this is a field in which governments implementing policies need to be patient as it can clearly take a number of years for programmes to become established and for SC to be generated. In addition, it would be valuable to perform comparisons with top-down clusters in other countries.

REFERENCES

Acs, Z. (Ed.). (2000). *Regional Innovation, Knowledge and Global Change*. New York: Pinter.

Adler, P. S., & Kwon, S. (2002). Social capital: prospects for a new concept. *Academy of Management Review*, *27*(1), 17–40.

Ahuja, G. (2000). Collaborative networks, structural holes, and innovation: a longitudinal study. *Administrative Science Quarterly*, *45*(3), 425–455.

Amable, B., Barré, R., & Boyer, R. (1997). *Les systèmes d'innovation à l'ère de la globalisation*. Paris: Economica.

Asheim, B. T., & Coenen, L. (2005). Knowledge bases and regional innovation systems: Comparing Nordic clusters. *Research Policy*, *34*(8), 1173–1190.

Atzema, O., & Visser, E. (2006 November). *Innovation policy in regions with(out) clusters: a plea for a differentiated and combined network approach*. Paper presented at DIME-LIEE/NTUA Conference, Athens, Greece.

Bellandi, M. (2000, September). *Local Development, Big Firms and Social Capital.* Paper presented at the European Congress of the Regional Science Association, Barcelona, Spain.

Boekholt, P., & Thuriaux, B. (1999). Public policies to facilitate clusters: background, rationale and policy practices in international perspective. In OECD (Ed.), Boosting Innovation: The Cluster Approach (pp. 381-412). Paris: OECD.

Bonaccorsi, A., & Nesci, F. (2006). *Bacini di competenze e processi di agglomerazione – I distretti tecnologici in Europa*. Milano, Italy: Franco Angeli.

Bonanno, G., & Haworth, B. (1998). Intensity of competition and the choice between product and process innovation. *International Journal of Industrial Organization*, *16*(4), 495–510.

Bontis, N. (1998). Intellectual capital: An Exploratory Study that Develops Measures and Models. *Management Decision*, *36*(2), 63–76.

Bontis, N. (2001). Assessing knowledge assets: a review of the models used to measure intellectual capital. *International Journal of Management Reviews*, *3*(1), 41–60.

Bresnahan, T., Gambardella, A., & Saxenian, A. L. (2001). Old economy inputs for new economy outcomes: cluster formation in the new silicon valleys. *Industrial and Corporate Change*, *10*(4), 835–860.

Bryman, A., & Burgess, R. (Eds.). (1993). *Analysing Qualitative Data*. London: Routledge.

Cooke, P. (2004). Evolution of regional innovation systems—emergence, theory, challenge for action. In Cooke, P., Heidenreich, M., & Braczyk, H.-J. (Eds.), *Regional Innovation Systems* (2nd ed., pp. 1–18). London: Routledge.

Cooke, P., Heideneich, M., & Braczyk, H.-J. (Eds.). (2004). *Regional Innovation Systems* (2nd ed.). London: Routledge.

Davis, C., Arthurs, D., Cassidy, E., & Wolfe, D. (2006, September). *What indicators for clusters policies in the 21st century?* Paper presented at Conference Blue Sky II, Ottawa, Canada.

De Castro, G. M., & Saez, P. L. (2008). Intellectual capital in high-tech firms. The case of Spain. *Journal of Intellectual Capital*, *9*(1), 25–36.

Diez, M. (2001). The evaluation of regional innovation and cluster policies: towards a participatory approach. *European Planning Studies*, *9*(7), 907–923.

Enright, M., & Ffowcs-Williams, I. (2001). *Local partnership, clusters and SME globalisation*. OECD Workshop Paper. OECD, Paris (http://www.oecd.org).

European Commission. (2002). *Regional Clusters in Europe. Observatory of European SMEs* (2002/no. 3). Report coordinated by A. Isaksen & E. Hauge. Brussels: European Union. Retrieved July 10th, 2009, from http://europa.eu.int/comm/enterprise/enterprise policy/analysis/doc/smesobervatory 2002 report3 en.pdf

Feldman, M. P., & Francis, J. L. (2004). Homegrown Solutions: Fostering Cluster Formation. *Economic Development Quarterly*, *18*(2), 127–137.

Ffowcs-Williams, I. (2004). *Cluster Navigator*. Wellington, New Zealand: Cluster Navigators Ltd.

Fromhold-Eisebith, M. (2004). Innovative milieu and social capital: complementary or redundant concepts of collaboration-based regional development? *European Planning Studies*, *12*(6), 747–765.

Fromhold-Eisebith, M., & Eisebith, G. (2005). How to institutionalize innovative cluster? Comparing explicit top-down and implicit bottom-up approaches. *Research Policy*, *34*(8), 1250–1268.

Fromhold-Eisebith, M., & Eisebith, G. (2008). Looking behind facades: evaluating effects of (automotive) cluster promotion. *Regional Studies*, *42*(10), 1343–1356.

Garcia, R., & Calantone, R. (2002). A critical look at technological innovation typology and innovativeness terminology: a literature review. *Journal of Product Innovation Management*, *19*(2), 110–132.

Garrett-Jones, S. (2004). From citadels to clusters: the evolution of regional innovation policies in Australia. *R & D Management, 34*(1), 3–16.

Geisler, E. (2005). The measurement of scientific activity: research in linking the philosophy of science and metrics of science and technology outputs. *Scientometrics, 62*(2), 269–284.

Gilsing, V. (2001). Towards second generation cluster policy: the case of the Netherlands. In den Hertog, P., Bergman, E., & Remoe, S. (Eds.), *Innovative Clusters. Drivers of National Innovation Systems* (pp. 361–376). Paris, France: OECD.

Jick, T. (1979). Mixing Qualitative and Quantitative Methods: Triangulation in Action. *Administrative Science Quarterly, 24*(4), 602–611.

Grootaert, C., Narayan, D., Nyan Jones, V., & Woolcock, M. (2004). *Measuring social capital: an integrated questionnaire.* Washington, DC: The World Bank.

Hargadon, A., & Sutton, R. (1997). Technology brokering and innovation in a product development firm. *Administrative Science Quarterly, 42,* 716–749.

Hassink, R. (1996). Technology Transfer Agencies and Regional Economic Development. *European Planning Studies, 4*(2), 167–184.

Ibrahim, D. (2009). *Financing the Next Silicon Valley.* Legal Studies Research Paper Series, Paper No. 1065. Retrieved July 10th, 2009, from http://works.bepress.com/cgi/viewcontent.cgi?article=1003&context=darian_ibrahim

Isaksen, A. (2007) Clusters, innovation and the local learning paradox. *International Journal of Entrepreneurship and Innovation Management,* 7(2/3/4/5), 366-384.

Isaksen, A., & Hauge, E. (2002). *Regional Clusters in Europe. Observatory of European SMEs report 2002 No. 3.* Luxembourg: European Communities.

Karlsson, C. (2007). *Clusters, functional regions and cluster policies.* CESIS Electronic Working Paper Series Retrieved July 10th, 2009, from http://www.infra.kth.se/cesis/documents/WP84.pdf

Kaufman, L., & Schneider, Y. (2004). Intangibles: a synthesis of current research. *Journal of Intellectual Capital, 5*(3), 366–388.

Ketels, C., Lindqvist, G., & Solvell, O. (2008). *Clusters and Cluster Initiatives.* Retrieved July 10th, 2009, from http://www.clusterobservatory.eu/upload/ClustersAndClusterOrganisations.pdf

Kline, J., & Rosenberg, N. (1986). An overview of innovation. In Landau, R., & Rosenberg, N. (Eds.), *The positive sum strategies: harnessing technology for economic growth* (pp. 275–305). Washington, DC: National Academic Press.

Koschatzky, K., & Lo, V. (2007) *Methodological framework for cluster analysis.* Working papers firms and region (R1/2007). Karlsruhe, Germany: Fraunhofer Institute for Systems and Innovation Research, Competence Centre "Policy and Regions."

Lagendijk, A. (2000). Learning in non-core regions: towards intelligent clusters; addressing business and regional needs. In Boekema, F., Morgan, K., Bakkers, S., & Rutten, R. (Eds.), *Knowledge, Innovation and Economic Growth* (pp. 165–191). London: Edward Elgar.

Landry, R., Amara, N., & Lamari, M. (2002). Does social capital determine innovation? To what extent? *Technological Forecasting and Social Change, 69*(7), 681–701.

LEED. (2009). *Clusters, Innovation and Entrepreneurship.* Paris, France: OECD Publishing.

LEED. (2002). *LEED-CEI-EBRD project on Industrial Districts and Local Clusters 2001-2002.* Paris, France: OECD Publishing.

Leydesdorff, L., & Etzkowitz, H. (1996). Emergence of a Triple Helix of University-Industry-Government Relations. *Science & Public Policy, 23*, 279–286.

Leydesdorff, L., & Etzkowitz, H. (2001). The transformation of University-Industry-Government relations. *Electronic Journal of Sociology*. Retrieved July 10th, 2009, from http://sociology.org/content/vol005.004/th.html

Longo, M., & Mura, M. (2009, February). *Intellectual capital and innovative capabilities. Does health care?* Paper presented at the International Forum on Knowledge Assets Dynamics (IFKAD), Glasgow, UK.

Malmberg, A., & Maskell, P. (2002). The elusive concept of localization economies: towards a knowledge-based theory of spatial clustering. *Environment & Planning A, 34*, 429–449.

Marr, B., Gray, D., & Schiuma, G. (2004). Measuring intellectual capital: what, why and how? In M. Bourne, (Ed.), Handbook of performance measurement. London: Gee.

Martin, R., & Sunley, P. (2003). Deconstructing clusters: chaotic concept or policy panacea? *Journal of Economic Geography, 3*(1), 5–35.

Maskell, P. (2001). Towards a knowledge-based theory of the geographical cluster. *Industrial and Corporate Change, 10*, 921–943.

Massa, S., & Testa, S. (2008a). Which roles in innovation processes? A matter of perspectives. In Terziovski, M. (Ed.), *Energizing Management through Innovation and Entrepreneurship* (pp. 146–166). London: Routledge.

Massa, S., & Testa, S. (2008b). Innovation and SMEs: misaligned perspectives and goals among entrepreneurs, academics and policy makers. *Technovation, 28*(7), 393–407.

Massa, S., & Testa, S. (2009). Innovation topics under debate at the EC: the opinions of Italian entrepreneurs, academics and policy makers. *International Journal of Entrepreneurship and Innovation Management, 10*(3/4), 361–376.

Moulaert, F., & Sekia, F. (2003). Territorial Innovation models: a critical survey. *Regional Studies, 37*(3), 289–302.

Mussati, G., Pedrana, M., & Postigliola, M. (2007). *Country Report: Italy*. Retrieved July 7th, 2009, from http://www.clusterobservatory.eu/upload/

Nahapiet, J., & Ghoshal, S. (1998). Social capital, intellectual capital, and the organizational advantage. *Academy of Management Review, 23*(2), 242–267.

Nauwelaers, C. (2003, May). *Innovative Hot Spots in Europe: Policies to promote trans-border clusters of creative activity.* Paper presented at Trend Chart Policy Workshop, Luxemburg.

Nauwelaers, C. (2001). Path-dependency and the role of institutions in cluster policy generation. In Mariussen, A. (Ed.), *Cluster Policies—Cluster Development?* (pp. 93–107). Stockholm, Sweden: Nordregio.

Martinez-Torres, M. R. (2006). A procedure to design a structural and measurement model of Intellectual Capital: An exploratory study. *Information & Management, 43*(5), 617–626.

OECD. (2007). *Reviews of regional innovation. Competitive regional clusters. National policy approaches*. Paris, France: OECD.

Park, S. H. (2001). Regional innovation strategies in the knowledge-based economy. *GeoJournal, 53*(1), 29–38.

Petty, R., & Guthrie, J. (2000). Intellectual capital literature review: measurement, reporting and management. *Journal of Intellectual Capital, 1*(2), 155–176.

Piccaluga, A., & Cesaroni, F. (2003). *Distretti industriali e distretti tecnologici. Modelli possibili per il Mezzogiorno*. Milano, Italy: Franco Angeli.

Porter, M. (1990). *The competitive advantage of nations*. London: Macmillan.

Porter, M. (1998a). *On competition*. Boston: Harvard Business School Press.

Porter, M. (1998b). Clusters and the new economics of competition. *Harvard Business Review*, *76*(6), 77–90.

Porter, M. E. (2000a). Locations, clusters and company strategy. In Clark, G., Feldman, M., & Gertler, M. (Eds.), *The Oxford Handbook of Economic Geography* (pp. 253–274). Oxford, UK: Oxford University Press.

Porter, M. (2000b). Location, competition and economic development: Local clusters in a global economy. *Economic Development Quarterly*, *14*(1), 15–34.

Raines, P. (2000). *Developing Cluster Policies in Seven European Regions* (Regional and Industrial Policy Research Paper No. 42). Glasgow, UK: University of Strathclyde, European Policies Research Centre.

Raines, P. (2001). *The Cluster Approach and the Dynamics of Regional Policy-Making* (Regional and Industrial Policy Research Paper No. 47). Glasgow, UK: University of Strathclyde, European Policies Research Centre.

Raines, P. (2002). Clusters and prisms. In Raines, P. (Ed.), *Cluster Development and Policy* (pp. 159–177). Aldershot, UK: Ashgate.

Redzepagic, D., & Stubbs, P. (2006, September). *Rethinking clusters and social capital in Croatia*. Paper presented at the European Association for Comparative Economics Studies (EACES) 9th Biennial Conference, Brighton, UK.

Roelandt, T. J. A., & den Hertog, P. (1999). Cluster analysis and cluster based policy making in OECD countries. An introduction to the theme. In OECD (Ed.), Boosting Innovation: The Cluster Approach (pp. 9–23). Paris, France: OECD Publishing.

Roelandt, T. J. A., Gilsing, V. A., & van Sinderen, J. (2000, December). *New policies for the new economy. Cluster-based innovation policy: international experiences*. Paper Presented at the Fourth Annual EUNIP Conference, Tilburg, The Netherlands.

Roos, G., Pike, S., & Fernström, L. (2006). Managing Intellectual Capital in Practice. New York: Butterworth-Heinemann, an imprint of Elsevier, USA.

Rosenfeld, S. A. (2002). *Creating smart systems. A guide to cluster strategies in less favoured regions*. Paper Presented at a Seminar on European Union Regional Innovation Strategies, Brussels, Belgium. Retrieved July 7th, 2009, from http://europa.eu.int/comm/regional policy/innovation/pdf/

Serenko, A., & Bontis, N. (2004). Meta-review of knowledge management and intellectual capital literature: citation impact and research productivity rankings. *Knowledge and Process Management*, *11*(3), 185–198.

Science|Business Innovation Board. (2009). *Clustering For Growth. How to build dynamic innovation clusters in Europe*. Retrieved July 7th, 2009, from ww.sciencebusiness.net/documents/

Schmitz, H. (2004). Globalized localities: introduction. In Schmitz, H. (Ed.), *Local Enterprises in the Global Economy* (pp. 1–19). Cheltenham, UK: Edward Elgar.

Schneider, A., & Samkin, G. (2008). Intellectual capital reporting by the New Zealand local government sector. *Journal of Intellectual Capital*, *9*(3), 456–486.

Steiner, M. (Ed.). (1998). *Clusters and Regional Specialisation: On Geography, Technology and Networks*. London: Pion.

Stewart, T. A. (1999). *Intellectual Capital: The New Wealth of Organizations*. New York: Currency/Doubleday.

Stewart, T. A. (2001). *The Wealth of Knowledge: Intellectual Capital and the Twenty-First Century Organization*. London: Nicholas Brealey.

Subramanian, M., & Youndt, M. A. (2005). The influence of intellectual capital on the types of innovative capabilities. *Academy of Management Journal*, *48*(3), 450–463.

Subramaniam, M., & Venkatraman, N. (2001). Determinants of transnational new product development capability: testing the influence of transferring and deploying tacit overseas knowledge. *Strategic Management Journal*, *22*(4), 359–378.

Tsai, W., & Ghoshal, S. (1998). Social capital and value creation: the role of intrafirm networks. *Academy of Management Journal*, *41*(4), 464–478.

Viale, R., & Etzkowitz, H. (2005, May). *Third academic revolution: polyvalent knowledge; the 'DNA' of the triple helix*. Paper presented at the 5th Triple Helix Conference, Turin, Italy.

Wolfe, D. A., & Gertler, M. (2004). Clusters from the inside and out: local dynamics and global linkages. *Urban Studies (Edinburgh, Scotland)*, *41*(5-6), 1071–1093.

Youndt, M. A., Subramaniam, M., & Snell, S. A. (2004). Intellectual capital profiles: An examination of investments and returns. *Journal of Management Studies*, *41*(2), 335–361.

Youndt, M. A., & Snell, S. A. (2004). Human resource configurations, intellectual capital, and organizational performance. *Journal of Managerial Issues*, *16*(3), 337–360.

Compilation of References

Abeysekera, I., & Guthrie, J. (2004). Human capital reporting in a developing nation. *The British Accounting Review*, *36*, 251–268. doi:10.1016/j.bar.2004.03.004

Achilladelis, B., & Antonakis, N. (2001). The Dynamics of Technological Innovation: The Case of the Pharmaceutical Industry. *Research Policy*, *30*(4), 535–588. doi:10.1016/S0048-7333(00)00093-7

Acs, Z. (Ed.). (2000). *Regional Innovation, Knowledge and Global Change*. New York: Pinter.

Adamides, E. D., & Karacapilidis, N. (2006). Information Technology Support for the Knowledge and Social Process of Innovation Management. *Technovation*, *26*, 50–59. doi:10.1016/j.technovation.2004.07.019

Adams, J. L. (1974). *Conceptual blockbusting: A guide to better ideas*. New York: W.H. Freeman & Co Ltd.

Adams, R., Bessant, J., & Phelps, R. (2006). Innovation Management measurement: a review. *International Journal of Management Reviews*, *8*(1), 21–47. doi:10.1111/j.1468-2370.2006.00119.x

Adler, P. S., & Kwon, S. W. (2002). Social capital: Prospects for a new concept. *Academy of Management Review*, *27*(1), 17–40. doi:10.2307/4134367

Adler, P. S., & Seok-Woo, K. (2002). Social Capital: Prospects for a New Concept. *Academy of Management Review*, *27*, 17–40. doi:10.2307/4134367

Adner, R. (2002). When are Technologies Disruptive? A Demand-Based View of the Emergence of Competition. *Strategic Management Journal*, *23*, 667–688. doi:10.1002/smj.246

Ahn, J.-H., & Chang, S.-G. (2004). Assessing the Contribution of Knowledge to Business Performance: The KP3 Methodology. *Decision Support Systems*, *36*, 403–416. doi:10.1016/S0167-9236(03)00029-0

Ahonen, G., & Grojer, J.-E. (2005). Social accounting in the Nordic countries: from social accounting towards accounting in a social context. In Jonsson, S., & Mouritsen, J. (Eds.), *Accounting in Scandinavia –The Northern Lights* (pp. 43–60). Copenhagen: Copenhagen Business School Press.

Ahuja, G. (2000). Collaborative networks, structural holes, and innovation: a longitudinal study. *Administrative Science Quarterly*, *45*(3), 425–455. doi:10.2307/2667105

Aiman-Smith, L., Goodrich, N., Roberts, D., & Scinta, J. (2005). Assessing Your Organization's Potential for Value Innovation. *Research Technology Management*, *48*(2), 37–42.

Akgün, A. E., Keskin, H., & Aren, S. (2007). Emotional and Learning Capabilities and their Impact on Product Innovativeness and Firm Performance. *Technovation*, *27*, 501–513. doi:10.1016/j.technovation.2007.03.001

Akgun, A. E., Keskin, H., Byrne, J. C., & Aren, S. (2007). Emotional and Learning Capability and their Impact on Product Innovativeness and Firm Performance. *Technovation*, *27*, 501–513. doi:10.1016/j.technovation.2007.03.001

Al-Ali, N. (2003). *Comprehensive intellectual capital management: Step-by-step*. Mahwah, NJ: John Wiley & Sons.

Alavi, M., & Leidner, D. E. (2001). Review: knowledge management and knowledge management systems: conceptual foundations and research issues. *Management Information Systems Quarterly, 25*(1), 107–136. doi:10.2307/3250961

Albert, S., & Whetten, D. (1985): Organizational Identity. In L. L. Cummings, B. M. y Staw, (Eds.), Research in Organizational Behavior, (pp. 263-295). Greenwich, CT: JAI Press.

Albert, S., Ashforth, B. E., & Dutton, J. (2000). Organizational identity and Identification: Charting new waters and building new bridges. *Academy of Management Review, 25*(1), 13–17.

Alegre, J. & Lapiedra, R. (2005). Gestión del Conocimiento y Desempeño Innovador: un Estudio del Papel Mediador del Repertorio de Competencias Distintivas. *Cuadernos de Economía y Dirección de la empresa, 23*, 117-138.

Alegre, J., Lapiedra, R., & Chiva, R. (2005). *Propuesta y Validación de una Escala de Medida del Desempeño Innovador de la Empresa.* ACEDE Conference, La Laguna.

Alegre-Vidal, J., Lapiedra-Alcamí, R., & Chiva-Gómez, R. (2004). Linking Operations Strategy and Product Innovation: An Empirical Study of Spanish Ceramic Tile Producers. *Research Policy, 33*, 829–839. doi:10.1016/j.respol.2004.01.003

Alguezaui, S., & Filieri, R. (2008, May). *Innovation Across Tech-Firm's Boundaries: a Knowledge-Based View.* Paper presented at the Conference on Offshoring & Outsourcing: The Organizational and Geographical Relocation of High-Value Company Functions, SDA Bocconi, Milan.

Allee, V. (1998). Creating value in the knowledge economy. *HR Monthly*, (April): 12–17.

Allee, V. (1999). The art and practice of being a revolutionary. *Journal of Knowledge Management, 3*(2), 121–132. doi:10.1108/13673279910275576

Allee, V. (2003). *The Future of Knowledge.* New York: Butterworth-Heinemann.

Allen, J., James, A. D., & Gamlen, P. (2007). Formal versus informal knowledge networks in R&D: a case study using social network analysis. *R & D Management, 37*(3), 179–195. doi:10.1111/j.1467-9310.2007.00468.x

Almeida, P., & Phene, A. (2004). Subsidiaries and Knowledge Creation: The Influence of the MNC and Host Country on Innovation. *Strategic Management Journal, 25*, 847–864. doi:10.1002/smj.388

Amabile, T. M. (1998). How to Kill Creativity. *Harvard Business Review, 76*, 76–87.

Amabile, T. M., Barsade, S., Mueller, J., & Staw, B. (2007). La conexión entre las emociones y la creatividad en el trabajo. *Harvard Deusto Business Review, 159*, 36–44.

Amable, B., Barré, R., & Boyer, R. (1997). *Les systèmes d'innovation à l'ère de la globalisation.* Paris: Economica.

Amit, R., & Schoemaker, P. J. H. (1993). Strategic Assets and Organizacional Rent. *Strategic Management Journal, 14*, 33–46. doi:10.1002/smj.4250140105

Amit, R., & Zott, C. (2001). Value creation in e-business. *Strategic Management Journal, 22*, 493–520. doi:10.1002/smj.187

Anand, B. N., & Khanna, T. (2000). Do firms learn to create value? The case of alliances. *Strategic Management Journal, 21*(3), 295–315. doi:10.1002/(SICI)1097-0266(200003)21:3<295::AID-SMJ91>3.0.CO;2-O

Anderson, N., Dreu, C., & Nijstad, B. (2004). The routinization of innovation research: a constructively critical review of the state-of-the-art-science. *Journal of Organizational Behavior, 25*, 147–173. doi:10.1002/job.236

Anderson-Levitt, K. M. (2003). *Local meanings, global schooling: Anthropology and world culture theory.* New York: Palgrave MacMillan. doi:10.1057/9781403980359

Andriessen, D. (2001). Weightless wealth: four modifications to standard IC theory. *Journal of Intellectual Capital, 2*(3), 204–214. doi:10.1108/14691930110399941

Andriessen, D. (2002). *The financial value of intangibles: searching for the Holy Grail.* Paper presented at the 5th World Congress on the Management of Intellectual Capital, McMaster University, Hamilton, 16-18 December.

Andriessen, D. (2004). *Making sense of intellectual capital: Designing a method for the valuation of intangibles.* Boston: Butterworth-Heinemann.

Andriessen, D. (2004a). *Making Sense of Intellectual Capital: Designing a Method for the Valuation of Intangibles.* Oxford, UK: Elsevier Butterworth-Heinemann.

Andriessen, D. (2004b). IC valuation and measurement: classifying the state-of-the-art. *Journal of Intellectual Capital, 5*(3), 230–242. doi:10.1108/14691930410533669

Andriessen, D. (2006). On the metaphorical nature of intellectual capital: a textual analysis. *Journal of Intellectual Capital, 7*(1), 93–110. doi:10.1108/14691930610639796

Apostolou, D., & Mentzas, G. (1999a). Managing Corporate Knowledge: A Comparative Analysis of Experiences in Consulting Firms: Part 1. *Knowledge and Process Management, 6*(3), 129–138. doi:10.1002/(SICI)1099-1441(199909)6:3<129::AID-KPM64>3.0.CO;2-3

Apostolou, D., & Mentzas, G. (1999b). Managing Corporate Knowledge: A Comparative Analysis of Experiences in Consulting firms: Part 2. *Knowledge and Process Management, 6*(3). doi:10.1002/(SICI)1099-1441(199909)6:3<129::AID-KPM64>3.0.CO;2-3

Apple, M. W., Kenway, J., & Singh, M. (2005). *Globalizing education: policies, pedagogies, & politics.* New York: Peter Lang Publishing, Inc.

Arbussá, A., & Coenders, G. (2007). Innovation Activities, Use of Appropriation Instruments and Absorptive Capacity: Evidence from Spanish Firms. *Research Policy, 36*(10), 1545–1558. doi:10.1016/j.respol.2007.04.013

Argote, L., & Ingram, P. (2000). Knowledge transfer: A basis for competitive advantage in firms. *Organizational Behavior and Human Decision Processes, 82*(1), 150–169. doi:10.1006/obhd.2000.2893

Argote, L., McEvily, B., & Reagans, R. (2003). Managing Knowledge in Organizations: An Integrative Framework and Review of Emerging Themes. *Management Science, 49*(4), 571–582. doi:10.1287/mnsc.49.4.571.14424

Argyris, C. (1992). *On Organizational Learning.* Cambridge, MA: Blackwell.

Argyris, C., & Scho¨n, D. A. (1996). *Organizational learning II: Theory, Method and Practice.* Reading, MA: Addison Wesley.

Argyris, C., & Schön, D. (1978). *Organizational Learning: A Theory of Action Perspective.* Reading, MA: Addison-Wesley.

Arrow, K. J. (1974). *Organization and Information. The Limits of Organization.* New York: W.W. Norton & Co.

Asheim, B. T., & Coenen, L. (2005). Knowledge bases and regional innovation systems: Comparing Nordic clusters. *Research Policy, 34*(8), 1173–1190.

Atzema, O., & Visser, E. (2006 November). *Innovation policy in regions with(out) clusters: a plea for a differentiated and combined network approach.* Paper presented at DIME-LIEE/NTUA Conference, Athens, Greece.

Augier, M., & Teece, D. J. (2007). dynamic capabilities and Multinational Enterprise: Penrosean Insights and Omissions. *Management International Review, 47*(2), 175–192. doi:10.1007/s11575-007-0010-8

Baker, S. (2008). *The Numerati.* New York: Houghton Mifflin Company.

Baldridge, J. V., & Burnham, R. A. (1975). Organizational Innovation - Individual, Organizational and Environmental Impacts. *Administrative Science Quarterly, 20,* 165–176. doi:10.2307/2391692

Baldwin, C., Hienerth, C., & von Hippel, E. (2006). How user innovations become commercial products: A theoretical investigation and case study. *Research Policy, 35,* 1291–1313. doi:10.1016/j.respol.2006.04.012

Barañano, A. M., Bommer, M., & Jalajas, D. S. (2005). Sources of Innovation for High-Tech SMEs: A Comparison of USA, Canada, and Portugal. *International Journal of Technology Management*, *30*(1/2), 205–219. doi:10.1504/IJTM.2005.006351

Barney, J. (1991). Firm Resources and Sustained Competitive Advantage. *Journal of Management*, *17*, 99–120. doi:10.1177/014920639101700108

Barney, J. (2001). Is the Resource-Based View a Useful Perspective for Strategic Management Research? Yes. *Academy of Management Review*, *26*, 41–57. doi:10.2307/259393

Barney, J. B. (1986). Organizational Culture: Can It Be a Source of Sustained Competitive Advantage. *Academy of Management Review*, *11*(3), 656–665. doi:10.2307/258317

Barney, J. B. (1991). Firm Resources and Sustained Competitive Advantage. *Journal of Management*, *17*(1), 99–120. doi:10.1177/014920639101700108

Barney, J. B. (2001). Is the resource-based view a useful prospective for strategic management research? *Yes. Academy of Management Review*, *26*(1), 41–56. doi:10.2307/259393

Bassi, L. J. (1997). Harnessing the power of intellectual capital. *Training & Development*, *51*(12), 25.

Becheikh, N., Landry, R., & Amara, N. (2005). Lessons from Innovation Empirical Studies in the Manufacturing Sector: A Systematic Review of the Literature from 1993-2003. *Technovation*, *26*(5/6), 644–664.

Beckman, S. L., & Barry, M. (2007). Innovation as a learning process: Embedding design thinking. *California Management Review*, *50*(1), 31.

Beldebos, R., Caree, M., & Lokshin, B. (2006). Complementary in R&D cooperation strategies. *Review of Industrial Organization*, *28*, 402–426.

Bellandi, M. (2000, September). *Local Development, Big Firms and Social Capital.* Paper presented at the European Congress of the Regional Science Association, Barcelona, Spain.

Belliveau, M. A., O'Reilly, C. A., & Wade, J. B. (1996). Social capital at the top: effects of social similarity and status on CEO compensation. *Academy of Management Journal*, *39*(6), 1568–1593. doi:10.2307/257069

Beneito, P. (2006). The innovative performance of in-house and contracted R&D in terms of patents and utility models. *Research Policy*, *35*, 502–517. doi:10.1016/j.respol.2006.01.007

Benito Torres, J. L., & Varela González, J. A. (2002). Influencia del Tipo de Proceso y del Grado de Novedad sobre las Actividades Ejecutadas durante el Desarrollo de Nuevos Productos. *Revista Europea de Dirección y Economía de la Empresa*, *11*(2), 173–187.

Benson, G., Finegold, D., & Mohrman, S. (2004). You paid for the skills, now keep them: tuition reimbursement and voluntary turnover. *Academy of Management Journal*, *47*(3), 315–333. doi:10.2307/20159584

Berends, H., Boersma, F. K., & Weggeman, M. P. (2001). *The structuration of organizational learning, Working Paper 01.12.* Eindhoven, The Netherlands: Eindhoven Centre for Innovation Studies.

Berger, P. L., & Luckman, T. (1968). *La construcción social de la realidad.* Buenos Aires, Argentina: Amorrortu Editores.

Bernstein, J. I., & Nadiri, M. I. (1988). Interindustry R&D spillovers, rates of return, and production in high-tech industries. *The American Economic Review*, *78*(2), 429–434.

Bernstein, J. I., & Nadiri, M. I. (1989). Research and development and intra-industry spillovers: An empirical application of dynamic duality. *The Review of Economic Studies*, *56*, 245–269. doi:10.2307/2297460

Bloom, D. E., & Cohen, J. E. (2001). *The unfinished revolution: Universal basic and secondary education.* Paper presented at the American Academy of Arts and Sciences, Cambridge, MA, July.

Bloom, D. E., & Cohen, J. E. (2002). Education for all: An unfinished revolution. *Daedalus*, (summer), 84-95.

Blumentritt, T., & Danis, W. M. (2006). Business Strategy Types and Innovative Practices. *Journal of Managerial Issues*, *18*(2), 274–291.

Blyler, M., & Coff, R. W. (2003). Dynamic Capabilities, Social Capital, and Rent Appropriation: Ties that Split Pies. *Strategic Management Journal*, *24*, 677–686. doi:10.1002/smj.327

Boekholt, P., & Thuriaux, B. (1999). Public policies to facilitate clusters: background, rationale and policy practices in international perspective. In OECD (Ed.), Boosting Innovation: The Cluster Approach (pp. 381-412). Paris: OECD.

Boer, H. & During, W.E. (2001). Innovation, What Innovation? A Comparison between Product, Process and Organizational Innovation. *International Journal of Technology Management, 22* (1/2/3), 83-107.

Bollinger, L. (1983). A Review of Literature and Hypotheses on New Technology-Based Firms. *Research Policy*, *12*(1), 1–14. doi:10.1016/0048-7333(83)90023-9

Bonaccorsi, A., & Nesci, F. (2006). *Bacini di competenze e processi di agglomerazione – I distretti tecnologici in Europa*. Milano, Italy: Franco Angeli.

Bonanno, G., & Haworth, B. (1998). Intensity of competition and the choice between product and process innovation. *International Journal of Industrial Organization*, *16*(4), 495–510.

Bontis, N. (1996). There's a Price on your Head: Managing Intellectual Capital Strategically. *Business Quarterly*, *60*, 41–47.

Bontis, N. (1996). Intellectual Capital: an exploratory study that develops measures and models. *Working paper* 96-11, University of Western Ontario. Published in *Management Decision (1998)*, 36 (2), 67-76.

Bontis, N. (1998). Intellectual capital: An exploratory study that develops measures and models. *Management Decision*, *36*(2), 63–76. doi:10.1108/00251749810204142

Bontis, N. (1999). Managing organizational knowledge by diagnosing intellectual capital: framing and advancing the state of the field. *International Journal of Technology Management, 18* (5/6/7/8), 433-62.

Bontis, N. (2000). Intellectual capital and business performance in Malaysian industries. *Journal of Intellectual Capital*, *1*(1), 85–100. doi:10.1108/14691930010324188

Bontis, N. (2001). Assessing knowledge assets: a review of the models used to measure intellectual capital. *International Journal of Management Reviews*, *3*(1), 41–60. doi:10.1111/1468-2370.00053

Bontis, N. (2004). National Intellectual Capital Index: a united nations initiatives for the Arab region. *Journal of Intellectual Capital*, *5*(1), 13–39. doi:10.1108/14691930410512905

Bontis, N., & Fitz-enz, J. (2002). Intellectual capital ROI: a causal map of human capital antecedents and consequents. *Journal of Intellectual Capital*, *3*(3), 223–247. doi:10.1108/14691930210435589

Bontis, N., Crossan, M., & Hulland, J. (2002). Managing and organizational learning system by aligning stocks and flows. *Journal of Management Studies*, *39*(4), 437–469. doi:10.1111/1467-6486.t01-1-00299

Bontis, N., Keow, W. C., & Richardson, S. (2000). Intellectual capital and business performance in Malaysian industries. *Journal of Intellectual Capital*, *1*(1), 85–100. doi:10.1108/14691930010324188

Borgatti, S. P., & Foster, P. (2003). The network paradigm in organizational research: A review and typology. *Journal of Management*, *29*(6), 991–1013.

Borgatti, S., & Cross, R. (2003). A Social Network View of Organizational Learning: Relational and Structural Dimensions of 'Know Who.'. *Management Science*, *49*(4), 432–445. doi:10.1287/mnsc.49.4.432.14428

Bornemann, M., Knapp, A., Schneider, U., & Sixl, K. I. (1999, June). *Holistic Measurement of Intellectual Capital*. Paper presented at the International Symposium Measuring and Reporting Intellectual Capital: Experiences, Issues, and Prospects, Amsterdam, Netherlands.

Bossink, B. A. G. (2002). The Development of Co-Innovation Strategies: Stages and Interaction Patterns in Interfirm Innovation. *R & D Management*, *32*(4), 311–320. doi:10.1111/1467-9310.00263

Boston Consulting Group. (2005). *ITC Survey of Senior Executives*. Retrieved from http://www.signallake.com/innovation/BCG2005ITCSurveyReport.pdf

Bouchikhi, H., et al. (1998). The Identity of Organizations. In Whetten, D. A. & Godfrey, P. C. (Eds), Identity in Organizations: Building Theory Trough Conversations (pp. 33-80). Thousand Oaks, CA: Sage Publication.

Boulton, R., Libert, B., & Samek, S. (2000). A business model for the new economy. *The Journal of Business Strategy*, *21*, 29–35. doi:10.1108/eb040102

Bounfour, A., & Edvinsson, L. (2005). *Intellectual capital for communities: Nations, regions and cites*. Burlington, MA: Butterworth Heinemenn.

Bourdieu, P. (1986). The forms of capital. In Richardson, J. G. (Ed.), *Handbook of Theory and Research for the Sociology of Education* (pp. 241–258). New York: Greenwood Press.

Bourdieu, P., & Wacquant, L. (1992). *An invitation to reflexive sociology*. Chicago: University of Chicago Press.

Bouty, I. (2000). Interpersonal and Interaction Influences on Informal Resource Exchanges between R&D Researches across Organizational Boundaries. *Academy of Management Journal*, *43*, 50–65. doi:10.2307/1556385

Bowen, D. E., & Ostroff, C. (2004). Understanding HRM-Firm performance linkages: The role of the "strength" of the HRM system. *Academy of Management Review*, *29*(2), 203–221. doi:10.2307/20159029

Bower, J. L., & Christensen, C. M. (1995). Disruptive technologies: catching the wave. *Harvard Business Review*, *17*(1), 43–53.

Bowman, E. H., & Helfat, C. E. (2001). Does Corporate Strategy Matter? *Strategic Management Journal*, *22*, 1–23. doi:10.1002/1097-0266(200101)22:1<1::AID-SMJ143>3.0.CO;2-T

Boyer, M. A., Brown, S. W., Butler, M. J., Niv-Solomon, A., Urlacher, B., & Hudson, N. F. (2007, July). Experimenting with global governance: understanding the potential for generational change. *Globalisation, Societies and Education*, *5*(2), 153–180. doi:10.1080/14767720701425727

Bozbura, F. T. (2004). Measurement and application of intellectual capital in Turkey. *The Learning Organization*, *11*(4/5), 357–367. doi:10.1108/09696470410538251

Bozbura, F. T., & Beskese, A. (2005). Prioritization of organizational capital measurement indicators using fuzzy AHP. In *Proceedings of the 11th international fuzzy systems association world Congress*, (pp. 1756–1760).

Bozbura, F. T., Beskese, A., & Kahraman, C. (2007). Prioritization of human capital measurement indicators using fuzzy AHP. *Expert Systems with Applications*, *32*(4), 1100–1112. doi:10.1016/j.eswa.2006.02.006

Brennan, N. (2001). Reporting intellectual capital in annual reports- evidence from Ireland. *Accounting, Auditing & Accountability Journal*, *14*(4), 423–436. doi:10.1108/09513570110403443

Brennan, N., & Connell, B. (2000). Intellectual capital: current issues and policy implications. *Journal of Intellectual Capital*, *1*(3), 206–240. doi:10.1108/14691930010350792

Bresnahan, T., Gambardella, A., & Saxenian, A. L. (2001). Old economy inputs for new economy outcomes: cluster formation in the new silicon valleys. *Industrial and Corporate Change*, *10*(4), 835–860.

Brooking, A. (1996). *Intellectual Capital. Core Asset for the Third Millennium Enterprise*. London: International Thomson Business Press.

Brooking, A. (1997). The management of intellectual capital. *Long Range Planning*, *30*(3), 364–365. doi:10.1016/S0024-6301(97)80911-9

Brown, A. D., & Humphreys, M. (2006). Organizational Identity and Place: A Discursive Exploration of Hegemony and Resistance. *Journal of Management Studies*, *43*(2), 231–257. doi:10.1111/j.1467-6486.2006.00589.x

Brown, A. D., Humphreys, M., & Gurney, P. M. (2005). Narrative, identity and change: a case study of Laskarina Holidays. *Journal of Organizational Change Management, 18*(4), 312–326. doi:10.1108/09534810510607029

Brown, J., & Duguid, P. (1991). Organizational Learning and Communities of Practice: Towards a Unified View of Working, Learning and Innovation. *Organization Science, 2*, 40–57. doi:10.1287/orsc.2.1.40

Brown, M. G., & Svenson, R. A. (1998). Measuring R&D productivity. *Research Technology Management, 41*(6), 30–35.

Bruggen, A., Vergauwen, P., & Dao, M. (2009). Determinants of intellectual capital disclosure: evidence from Australia. *Management Decision, 47*(2), 233–245. doi:10.1108/00251740910938894

Bryan, A., & Vavrus, F. (2005, July). The promise and peril of education: the teaching of in/tolerance in an era of globalization. *Globalisation, Societies and Education, 3*(2), 183–202. doi:10.1080/14767720500167033

Bryman, A., & Burgess, R. (Eds.). (1993). *Analysing Qualitative Data*. London: Routledge.

Bueno, E. (1998). El capital intangible como clave estratégica en la competencia actual. Boletín de estudios económicos, 3(August), 207-229.

Bueno, E. (1998). El capital intangible como clave estratégica en la competencia actual. *Boletín de Estudios Económicos, 53*(164), 207–229.

Bueno, E. (2002a). Enfoques principales y tendencias en dirección del conocimiento (knowledge management). In Hernández, R. (Ed.), *Gestión del conocimiento: Desarrollos teóricos y aplicaciones* (pp. 21–54). Cáceres, Spain: Ediciones La Coria-Fundación Xavier Salas.

Bueno, E. (2002b). El capital social en el nuevo enfoque del capital intelectual de las organizaciones. *Revista de Psicología del Trabajo y de las Organizaciones, 2/3*, 177–345.

Bueno, E. (2005). Fundamentos epistemológicos de dirección del conocimiento organizativo: desarrollo, medición y gestión de intangibles. *Economía Industrial, 357*, 13–26.

Bueno, E., Morcillo, P., & Salmador, M. P. (2006). *Dirección estratégica: Nuevas perspectivas teóricas.* Madrid: Pirámide.

Bueno, E., Salmador, M. P., & Longo, M. (2008, September). *Marco conceptual y enfoques principales del concepto identidad organizativa: Una aproximación empírica.* Paper presented at the XVIII meeting of Asociación Científica de Economía y Dirección de Empresas, León, Spain.

Bueno, E., Salmador, M. P., & Rodriguez, O. (2004). The role of social capital in today's economy: empirical evidence and proposal of a new model of intellectual capital. *Journal of Intellectual Capital, 5*(4), 556–574. doi:10.1108/14691930410567013

Bukh, P. N. (2002). The relevance of intellectual capital disclosure: a paradox? *Accounting, Auditing & Accountability Journal, 16*(1), 49–56. doi:10.1108/09513570310464273

Bukh, P. N., & Jensen, I. K. (2008). Intellectual capital statements in the Danish utility sector: materialization and enactment. *Journal of Human Resource Costing & Accounting, 12*(3), 148–178. doi:10.1108/14013380810919831

Bukh, P. N., Larsen, H. T., & Mouritsen, J. (1999). *Developing intellectual capital statements: Lessons from 23 Danish Firms.* Paper for Workshop on Accounting for Intangibles and the Virtual Organization, Brussels, February 12-13.

Bukh, P. N., Larsen, H. T., & Mouritsen, J. (2001). Constructing intellectual capital statements. *Scandinavian Journal of Management, 17*(1), 87–108. doi:10.1016/S0956-5221(00)00034-8

Burgelman, R., Maidique, M. A., & Wheelright, S. C. (2004). *Strategic Management of Technology and Innovation* (pp. 8–12). New York: McGraw-Hill.

Bürgi, P., & Oliver, D. (2005). Organizational Identity as a Strategy Practice. *Imagination Lab Foundation.* Retrieved July 26, 2008, from http://imagilab.org/publications_wp.html#61

Bürgi, P., Roos, J., & Oliver, D. (2002). Organizational Identity and Strategy. *Imagination Lab Foundation*. Retrieved July 26, 2008, from http:/ /imagilab.org/publications_wp.html#14

Burt, R. (1992). *Toward a Structural Theory of Action*. New York: Academic Press.

Burt, R. (1997). The contingent value of social capital. *Administrative Science Quarterly*, *42*(2), 339–365. doi:10.2307/2393923

Burt, R. (2000). The network structure of social capital. In Staw, B., & Sutton, R. (Eds.), *Research in organizational behavior* (*Vol. 22*). New York: JAI Press.

Burt, R. (2001). *Social Capital: Theory and Research*. New York: Aldine De Gruyter.

Burt, R. (2004). Structural Holes and Good Ideas. *American Journal of Sociology*, *110*(2), 349–399. doi:10.1086/421787

Burt, R. S. (1992). *Structural holes: The social structure of competition*. Cambridge, MA: Harvard University Press.

Burt, R. S. (1997). The contingent value of social capital. *Administrative Science Quarterly*, *42*(2), 339–365. doi:10.2307/2393923

Butchart, R. (1987). A new UK definition of high technology industries. *Economy Trends*, *400*(February), 82–88.

Buzbera, F. T. (2004). Measurement and application of intellectual capital in Turkey. *The Learning Organization*, *11*(4), 357–367. doi:10.1108/09696470410538251

Cabello-Medina, C., Carmona-Lavado, A., & Valle-Cabrera, R. (2006). Identifying the Variables Associated with Types of Innovation, Radical or Incremental: Strategic Flexibility, Organisation and Context. *International Journal of Technology Management*, *35*, 80–106. doi:10.1504/IJTM.2006.009230

Cacioppo, J. T., & Patrick, W. (2008). *Loneliness*. New York: W. W. Norton and Company.

Calantone, R. J., Benedetto, C. A., & Schmidt, J. B. (1999). Using the analytic hierarchy process in new product screening. *Journal of Product Innovation Management*, *16*(1), 65–76. doi:10.1016/S0737-6782(98)00036-8

Camisón, C., Palacios, D., & Devece, C. (2000). *Un nuevo modelo para la medición del capital intelectual: el modelo Nova*. Paper presented at X meeting of Asociación Científica de Economía y Dirección de Empresas, Oviedo, Spain.

Canadian Management Accountants (CMA). (1999, April 16). *Focus Group Draft: Measuring Knowledge Assets*. Mississauga, Canada: Author.

Capaldo, A. (2006). Network structure and innovation: The leveraging of a dual network as a distinctive relational capability. *Strategic Management Journal*, *28*(6), 585–608. doi:10.1002/smj.621

Carlucci, D., Marr, B., & Schiuma, G. (2004). The knowledge value chain—how intellectual capital impacts business performance. *International Journal of Technology Management*, *27*, 575–590. doi:10.1504/IJTM.2004.004903

Carmeli, A. (2004). The link between organizational elements, perceived external prestige and performance. *Corporate Reputation Review*, *6*, 314–331. doi:10.1057/palgrave.crr.1540002

Carmeli, A., & Tishler, A. (2004). The relationships between intangible organizational elements and organizational performance. *Strategic Management Journal*, *25*, 1257–1278. doi:10.1002/smj.428

Carpenter, M. A., Sanders, G., & Gregersen, H. B. (2001). Bundling Human Capital with Organizational Context: the Impact of International Assignment Experience on Multinational Firm Performance and CEO Pay. *Academy of Management Journal*, *44*, 493–511. doi:10.2307/3069366

Carson, E., Ranzijn, R., Winefield, A., & Marsden, H. (2004). Intellectual capital: Mapping employee and work group attributes. *Journal of Intellectual Capital*, *5*(3), 443–455. doi:10.1108/14691930410550390

Cascio, W. (1998). Valuing and costing intellectual capital. *HR Monthly*, (February): 12–16.

Castro, G. M., & Lopez, P. (2008). Intellectual capital in high-tech firms - The case of Spain. *Journal of Intellectual Capital*, *9*(1), 25–36. doi:10.1108/14691930810845786

CE. (1995). *Libro Verde de la Innovación*. Bruselas.

Cefis, E., & Marsili, O. (2006). Survivor: The role of innovation in firm's survival. *Research Policy*, *35*, 626–641. doi:10.1016/j.respol.2006.02.006

Celemi. (1998). *Growing a knowledge company*. www.celemi.com Site includes Celemi intangible assets monitor, http://www.celemi.se/sbc/sbc2.html

CEPAL. (2005a). *Preliminary overview of the economies of Latin America and the Caribbean 2005*. Santiago, Chile: CEPAL.

CEPAL. (2005b). *La inversión extranjera en América Latina y el Caribe*. Santiago, Chile: CEPAL.

CEPAL. (2006). *América Latina y el Caribe: proyecciones 2006-2007*. Santiago, Chile: CEPAL.

Chan, D. (1996). Cognitive misfit of problem-solving style at work: A facet of person-organization fit. *Organizational Behavior and Human Decision Processes*, *68*(3), 194–207. doi:10.1006/obhd.1996.0099

Chan, K. H. (2009). Impact of intellectual capital on organizational performance, An empirical study of companies in the Hang Seng Index (Part 1). *The Learning Organization*, *16*(1), 4–21. doi:10.1108/09696470910927641

Chang, Y. (2003). Benefits of Co-operation on Innovative Performance: Evidence from Integrated circuits and Biotecnology Firms in the UK and Taiwan. *R & D Management*, *33*, 425–437. doi:10.1111/1467-9310.00308

Chatzkel, J. (2002). *Intellectual Capital*. Mankato, MN: Capstone.

Chen, H. H. (2009). Measuring intellectual capital using fuzzy analytic hierarchy process. *International Journal of Innovation and Learning*, *6*(1), 51–61. doi:10.1504/IJIL.2009.021682

Chen, J., Zhu, Z., & Xie, H. Y. (2004). Measuring Intellectual Capital: a New Model and Empirical Study. *Journal of Intellectual Capital*, *5*, 195–212. doi:10.1108/14691930410513003

Chen, M. H., & Wang, M. C. (2008). Social Networks and New Venture's Innovative Capability: The Role of Trust within Entrepreneurial Teams. *R & D Management*, *38*, 253–264. doi:10.1111/j.1467-9310.2008.00515.x

Chen, M.-C., Cheng, S.-J., & Hwang, Y. (2005). An empirical investigation of the relationship between intellectual capital and firms' market value and financial performance. *Journal of Intellectual Capital*, *6*(2), 159–176. doi:10.1108/14691930510592771

Cheng, M. Y., Lin, J. Y., Hsiao, T. Y., & Lin, T. W. (2008). Censoring model for evaluating intellectual capital value drivers. *Journal of Intellectual Capital*, *9*(4), 639–654. doi:10.1108/14691930810913195

Chesbrough, H. (2003). *Open Innovation: The New Imperative for Creating and Profiting from Technology*. Boston: Harvard Business School Press.

Chesbrough, H. (2007). Why firms should have open business models. *MIT Sloan Management Review*, *48*(2), 21–28.

Chesbrough, H. Vanhaverbeke, W. & West, J. (2006) Open Innovation: Researching a New Paradigm. New York: Oxford University Press, USA.

Chesbrough, H. W. (2003). Environmental influences upon firm entry into new sub-markets Evidence from the worldwide hard disk drive industry. *Research Policy*, *32*(4), 659–678. doi:10.1016/S0048-7333(02)00033-1

Chesbrough, H. W., & Teece, D. J. (2003). Organizarse para Innovar: ¿Cuándo es Virtuoso lo Virtual? *Harvard Deusto Business Review*, *112*, 22–30.

Chiesa, V., Coughlan, P., & Voss, C. A. (1996). Development of a technical innovation audit. *Journal of Product Innovation Management*, *13*(2), 105–136. doi:10.1016/0737-6782(95)00109-3

Christensen, C. M and Overdorf, M. (2000). Meeting the challenge of disruptive innovation. *Harvard Business Review, 78*(2), 66-76.

Christensen, C. M. (1997). *The Innovator's Dilemma*. Cambridge, MA: Harvard Business School Press.

Chunghwa Telecom Co. Ltd. Telecommunication Laboratories, from http://qqq.cht.com.tw/webform/index.htm

CIA. (2006). *The world factbook*. Washington, DC: Central Intelligence Agency.

CIC (2003). *Modelo Intellectus: Medición y Gestión del Capital Intelectual* (Serie Documentos Intellectus No. 5). Madrid: Centro de Investigación sobre la Sociedad del Conocimiento (CIC).

CIC. (2001). Identificación y Medición del Capital Tecnológico de la Empresa, ed. CIC-IADE, Madrid.

CIC. (2002). *Guías y Directrices de Utilización del Modelo Intellectus (Documento Intellectus-abril)*. Madrid: Universidad Autónoma de Madrid, Society Research Centre.

CIC. (2003a). Modelo Intellectus: Medición y Gestión del Capital Intelectual, ed. CIC-IADE, Madrid.

CIC. (2003b). Metodología para Elaboración de Indicadores de Capital Intelectual, ed. CIC-IADE, Madrid.

Clark, K. B., & Fujimoto, T. (1990). The power of product Integrity. *Harvard Business Review, 68*(6), 107–118.

Cohen, D., & Prusak, L. (2001). *In good company: how Social Capital make organizations work*. Boston: Harvard University School Press.

Cohen, M. D., & Levinthal, D. A. (1990). Absorptive Capacity: A New Perspective of Learning and Innovation. *Administrative Science Quarterly, 35*(1), 128–152. doi:10.2307/2393553

Cohen, S. L. & Backer, N. K. (1999). Making and mining intellectual capital: method or madness? *Training and Development, September* (5), 46.

Cohen, W., & Levinthal, D. (1990). Absorptive Capacity: A New Perspective on Learning and Innovation. *Administrative Science Quarterly, 35*, 128–152. doi:10.2307/2393553

Cole, R. E. (1998). Introduction. *California Management Review, 45*(3), 15–21.

Coleman, J. S. (1988). Social capital in the creation of human capital. *American Journal of Sociology, 94*(1), 95–120. doi:10.1086/228943

Coleman, J. S. (1988). Social capital in the creation of human capital. *American Journal of Sociology, 94*(Supplement), S95–S120. doi:10.1086/228943

Coleman, J. S. (1990). *Foundations of social theory*. Cambridge, MA: Harvard University Press.

Collins, D. (1997). Changes in the value relevance of earnings and book values over the past forty years. *Journal of Accounting and Economics, 24*(1), 39–67. doi:10.1016/S0165-4101(97)00015-3

Collins, D. J. (1994). Research Note: How Valuable are Organizational Capabilities? *Strategic Management Journal, 15*(Winter Special Issue), 143-152.

Comptroller General of the Republic. (2005). *Panama in numbers: Years 2000-2004*. Panama: General Comptroller's of the Republic.

Conner, K. R. (1991). A historical comparison of resource-based theory and five schools of thought within industrial organization economics: Do we have a new theory of the firm? *Journal of Management, 17*(1), 121–154. doi:10.1177/014920639101700109

Conner, K. R., & Prahalad, C. K. (1996). A Resource-based Theory of the Firm: Knowledge Versus Opportunism. *Organization Science, 7*(5), 477–501. doi:10.1287/orsc.7.5.477

Cook, S. D. N., & Brown, J. S. (1999). Bridging epistemologies: the generative dance between organizational knowledge and organizational knowing. *Organization Science, 10*(4), 381–400. doi:10.1287/orsc.10.4.381

Cooke, P. (2004). Evolution of regional innovation systems—emergence, theory, challenge for action. In Cooke, P., Heidenreich, M., & Braczyk, H.-J. (Eds.), *Regional Innovation Systems* (2nd ed., pp. 1–18). London: Routledge.

Cooke, P., Heideneich, M., & Braczyk, H.-J. (Eds.). (2004). *Regional Innovation Systems* (2nd ed.). London: Routledge.

Coombs, J. E., & Bierly, P. E. (2006). Measuring Technological Capability and Performance. *R & D Management, 36*, 421–438. doi:10.1111/j.1467-9310.2006.00444.x

Cooper, A. C. (1971). *The Founding of Technology-Based Firms*. Milwaukee, WI: Centre for Venture Management.

Cooper, R. G., & Kleinschmidt, E. J. (1997). Winning businesses in product development: The critical success factors. *Journal of Product Innovation Management, 14*(2), 132–132. doi:10.1016/S0737-6782(97)83946-X

Cooper, R., & Burrel, G. (1988). Modernism, Postmodernism and Organizational Analysis: An Introduction. *Organization Studies, 9*(1), 91–112. doi:10.1177/017084068800900112

Corley, K. G. (2006). Guiding Organizational Identity through Aged Adolescence. *Journal of Management Inquiry, 15*(2), 85–99. doi:10.1177/1056492605285930

Cornelissen, J. P. (2002a). On the "Organizational Identity" Metaphor. *British Journal of Management, 13*(3), 259–288. doi:10.1111/1467-8551.00242

Cornelissen, J. P. (2002b). The Merit and Mischief of Metaphor: A Reply to Gioia, Schultz and Corley. *British Journal of Management, 13*(3), 277–279. doi:10.1111/1467-8551.00244

Cornelissen, J. P. (2006). Metaphor and the Dynamics of Knowlege in Organization Theory: A Case Study of the Organizational Identity Metaphor. *Journal of Management Studies, 43*(4), 683–709. doi:10.1111/j.1467-6486.2006.00607.x

Covaleski, M. A. (1998). The Calculated and the Avowed: Techniques of Discipline and Struggles Over Identity in Big Six Public Accounting Firms. *Administrative Science Quarterly, 43*, 293–327. doi:10.2307/2393854

Cuganesan, S. (2006). Reporting organizational performance in managing human resources. *Journal of Human Resource Costing & Accounting, 10*(3), 164–188. doi:10.1108/14013380610718629

Cyert, R. M., & March, J. G. (1963). *A Behavioral Theory of the Firm*. Englewood Cliffs, NJ: Prentice-Hall.

Dahlman, C., Routti, J., & Ylä-Anttila, P. (2005). *Finland as a knowledge economy: Elements of success and lessons learned*. Washington, DC: World Bank.

Damanpour, F. (1987). The Adoption of Technological, Administrative, and Ancillary Innovations: Impact of Organizational Factors. *Journal of Management, 13*(4), 675–688. doi:10.1177/014920638701300408

Damanpour, F. (1991). Organizational Innovation: A Meta-Analysis of Effects of Determinants and Moderators. *Academy of Management Journal, 34*(3), 555–590. doi:10.2307/256406

Damanpour, F., & Evan, W. M. (1984). Organizational Innovation and Performance: The Problem of 'Organizational Lag. *Administrative Science Quarterly, 29*, 392–409. doi:10.2307/2393031

Damanpour, F., & Gopalakrishnan, S. (1998). Theories of Organizational Structure and Innovation Adoption: the Role of Environmental Change. *Journal of Engineering and Technology Management, 15*, 1–24. doi:10.1016/S0923-4748(97)00029-5

Darroch, J. (2003). Developing a measure of knowledge management behaviors and practices. *Journal of Knowledge Management, 7*(5), 41–54. doi:10.1108/13673270310505377

DATI (Danish Agency for Trade and Industry). (1997). *Intellectual Capital Accounts: Reporting and Managing Intellectual Capital*. Copenhagen: Danish Agency for Trade and Industry.

DATI (Danish Agency for Trade and Industry). (1998). *Intellectual Capital Accounts: New Tool for Companies, (English version)*. Copenhagen: DTI Council.

DATI (Danish Agency for Trade and Industry). (1999). *Developing Intellectual Capital Accounts. Experiences from 19 Companies*. Copenhagen: Ministry of Business and Industry.

DATI (Danish Agency for Trade and Industry). (2000). *A Guideline for Intellectual Capital Statements*. Copenhagen: Danish Agency for Trade and Industry.

Davenport, T. H., & Prusak, L. (1998). *Working knowledge: How organizations manage what they know*. Boston: Harvard Business School Press.

Davis, C., Arthurs, D., Cassidy, E., & Wolfe, D. (2006, September). *What indicators for clusters policies in the 21st century?* Paper presented at Conference Blue Sky II, Ottawa, Canada.

Davis, E., Spekman, R., & Hall, P. (2004). *The Extended Enterprise. Gaining competitive advantage through collaborative supply-chain*. New York: FT Prentice Hall.

Davis, J. P., & Eisenhardt, K. (2007). *Rotating leadership and symbiotic organization: relationship processes in the context of collaborative innovation*. MIT Sloan School of Management Working Papers.

Day, G. S. (1994). The capabilities of market-driven organisations. *Journal of Marketing*, *58*(4), 37 52. doi:10.2307/1251915

De Castro, G. M., & Saez, P. L. (2008). Intellectual capital in high-tech firms. The case of Spain. *Journal of Intellectual Capital*, *9*(1), 25–36.

De Clercq, D., & Sapienza, H. J. (2001). The creation of relational rents in venture capitalist-entrepreneur dyads. *Venture Capital*, *3*(2), 107–128. doi:10.1080/13691060110045661

De Geus, A. (1997). *The Living Company: Habits for survival in a Turbulent Business Enviroment*. Boston: Harvard Business School Press.

De Saá, P., & Díaz, N. L. (2007). Incidencia de los Recursos Humanos de I+D Internos y Contratados en la Innovación. *Cuadernos de Economía y Dirección de la Empresa*, *33*, 7–30.

Dean, T. J., & Mayer, G. D. (1996). Industry environments and new venture formations in U.S. manufacturing: A conceptual and empirical analysis of demand determinations. *Journal of Business Venturing*, *11*, 107–132. doi:10.1016/0883-9026(95)00109-3

Dell, O., C., & Grayson, C. J. (1998). If Only We Knew What We Know: Identification and Transfer of Internal Best Practices. *California Management Review*, *40*(3), 154–174.

Demsetz, H. (1988). The theory of the firm revisited. *Journal of Law Economics and Organization*, *4*(1), 141–162.

Dent, E. B. (2002). The messy history of OB&D: How three strands came to be seen as one rope. *Journal of Management History*, *40*(3), 266–280.

Deward, R. D., & Dutton, J. E. (1986). The Adoption of Radical and Incremental Innovations: an Empirical Analisys. *Management Science*, *32*(11), 1422–1433. doi:10.1287/mnsc.32.11.1422

Dhanaraj, C., & Parkhe, A. (2006). Orchestrating innovation networks. *Academy of Management Review*, *31*(3), 659–669. doi:10.2307/20159234

Díaz, N., Aguiar, I., & De Saá, P. (2006). El Conocimiento Organizativo Tecnológico y la Capacidad de Innovación. Evidencia para la Empresa Industrial Española. *Cuadernos de Economía y Dirección de la Empresa*, *27*, 33–60.

Díaz-Díaz, N. L., & De Saá-Pérez, P. (2007). *El Papel de los Recursos Humanos de I+D en la Absorcio n del Conocimiento Adquirido mediante Alianzas*. ACEDE Conference, Sevilla.

Díaz-Díaz, N. L., Aguiar-Díaz, I., & De Saá-Pérez, P. (2006). Technological Knowledge Assets and Innovation. *International Journal of Technology Management*, *35*(1-4), 29–51. doi:10.1504/IJTM.2006.009228

Dierickx, I., & Cool, K. (1989). Asset Stock Accumulation and Sustainability of Competitive Advantage [Comment; Reply]. *Management Science*, *35*(12), 1504–1514. doi:10.1287/mnsc.35.12.1504

Diez, M. (2001). The evaluation of regional innovation and cluster policies: towards a participatory approach. *European Planning Studies*, *9*(7), 907–923.

Dittrich, K., & Duysters, G. M. (2007). Networking as a means to strategic change: the case of open innovation in mobile telephony. *Journal of Product Innovation Management*, *24*(6), 510–521. doi:10.1111/j.1540-5885.2007.00268.x

do Rosario Cabrita, M., & Vaz, J. L. (2005). IC and value creation: Evidence from the Portuguese banking Industry. *Electronic Journal of Knowledge Management*, *4*(1), 11–20.

Drucker, P. (1965). *The Future of Industrial Man*. London: New American Library.

Drucker, P. (1993). *Post-Capitalist society*. New York: Harper Business.

Dunbar, R. (1996). *Grooming, Gossip, and the Evolution of Language*. Cambridge: Harvard University Press.

Dutton, J., & Dukerich, J. M. (1991). Keeping an Eye on the Mirror: Image and Identity in organizational Adaptation. *Academy of Management Journal*, *34*, 517–554. doi:10.2307/256405

Dyer, J. H. (1997). Effective interfirm collaboration: How firms minimize transaction costs and maximize transaction value. *Strategic Management Journal*, *18*(7), 535–556. doi:10.1002/(SICI)1097-0266(199708)18:7<535::AID-SMJ885>3.0.CO;2-Z

Dyer, J. H., & Nobeoka, K. (2000). Creating and managing a high-performance knowledge-sharing network: the Toyota case. *Strategic Management Journal*, *21*(3), 345–367. doi:10.1002/(SICI)1097-0266(200003)21:3<345::AID-SMJ96>3.0.CO;2-N

Dyer, J. H., & Singh, H. (1998). The Relational View: Cooperative Strategy and Sources of Interorganizational Competitive Advantage. *Academy of Management Review*, *23*, 660–679. doi:10.2307/259056

Dyer, J., & Nobeoka, I. (2000). Creating and Managing a High-Performance knowledge sharing network: the Toyota Case. *Strategic Management Journal*, *21*(3), 345–367. doi:10.1002/(SICI)1097-0266(200003)21:3<345::AID-SMJ96>3.0.CO;2-N

Edvinsson, L. (1997). Developing IC at Skandia. *Long Range Planning*, *30*(3), 366–373. doi:10.1016/S0024-6301(97)90248-X

Edvinsson, L. (1997). Developing intellectual capital at Skandia. *Long Range Planning*, *30*(3), 266–373.

Edvinsson, L., & Malone, M. S. (1997). *Intellectual Capital. Realizing your Companies True Value by Finding its Hidden Brainpower*. New York: Harper Business Publisher.

Edvinsson, L., & Malone, M. S. (1997). *Intellectual Capital*. London: Piatkus.

Edvinsson, L., & Sullivan, P. (1996). Developing a Model for Managing Intellectual Capital. *European Management Journal*, *14*, 356–364. doi:10.1016/0263-2373(96)00022-9

Egbetokun, A., Siyanbola, W., Sanni, M., & Olamade, O. (2009). What drive innovation? inferences from an industry wide survey in Nigeria. *International Journal of Technology Management*, *45*(1/2), 123–140. doi:10.1504/IJTM.2009.021524

Ehin, C. (1995). *The Quest for Empowering Organizations: Some Lessons From our Foraging Past. Organization Science*. November-December.

Ehin, C. (2000). *Unleashing Intellectual Capital*. Boston: Butterworth-Heinemann.

Ehin, C. (2005). *Hidden Assets: Harnessing the Power of Informal Networks*. Boston: Springer.

Ehin, C. (2009). *The Organizational Sweet Spot: Engaging the Innovative Dynamics of Your Social Networks*. New York: Springer.

EIRMA. (2004). *Assessing R&D Effectiveness. Working Group Number 62 Report, Europe Industrial Research Management Association*. Paris: EIRMA.

Eisenhardt, K. M. (1989). Building theories from case study research. *Academy of Management Review, 14*, 532–550. doi:10.2307/258557

Eisenhardt, K. M., & Martin, J. A. (2000). Dynamic Capabilities: What are They? *Strategic Management Journal, 21*(10-11), 1105–1121. doi:10.1002/1097-0266(200010/11)21:10/11<1105::AID-SMJ133>3.0.CO;2-E

Ellickson, M. C. (2002). Determinants of job satisfaction of municipal government employees. *Public Personnel Management, 31*(3), 343–358.

Ellis, A. (1997). *A reason and emotion in psycho-therapy*. New York: Lyle Stuart.

Engstrom, T. E. J., Westnes, P., & Westness, S. F. (2003). Evaluating intellectual capital in the hotel industry. *Journal of Intellectual Capital, 4*(3), 287–303. doi:10.1108/14691930310487761

Enright, M., & Ffowcs-Williams, I. (2001). *Local partnership, clusters and SME globalisation*. OECD Workshop Paper. OECD, Paris (http://www.oecd.org).

Erickson, G., & Jacobson, R. (1992). Gaining comparative advantage through discretionary expenditures: The returns to R&D and advertising. *Management Science, 38*(9), 1264–1279. doi:10.1287/mnsc.38.9.1264

European Commission. (2002). *Regional Clusters in Europe. Observatory of European SMEs* (2002/no. 3). Report coordinated by A. Isaksen & E. Hauge. Brussels: European Union. Retrieved July 10th, 2009, from http://europa.eu.int/comm/enterprise/enterprise policy/analysis/doc/smesobervatory 2002 report3 en.pdf

Evangelista, R., Perani, G., Rapiti, F., & Archibugi, D. (1997). Nature and impact of innovation in manufacturing: some evidence from the Italian innovation survey. *Research Policy, 26*, 521–536. doi:10.1016/S0048-7333(97)00028-0

Farenga, S. J., & Joyce, B. A. (2001). Hardware versus brainware: Where are technology dollars being invested? *Journal of Technology and Teacher Education*, (Autumn): 313.

FASB. (2001). *Improving business reporting: insights into enhancing voluntary disclosures*. Norwalk, CT: Business Reporting Research Project.

Feiwel, G. R. (1975). *The IC of Michael Kaleck: a study in economic theory and policy*. Knoxville, TN: The University of Tennessee Press.

Feldman, M. P., & Francis, J. L. (2004). Homegrown Solutions: Fostering Cluster Formation. *Economic Development Quarterly, 18*(2), 127–137.

Fernandez, E., Montes, J., & Vazquez, C. (1998). Los Recursos Intangibles como Factores de Competitividad en la Empresa. *Dirección y Organización, 20*, 83–98.

Fernström, L. (2005). A marketing perspective on intellectual capital. In Marr, B. (Ed.), *Perspectives on Intellectual capital* (pp. 82–95). Burlington, MA: Elsevier Butterworth Heinemann. doi:10.1016/B978-0-7506-7799-8.50011-5

Ferrier, F. (2001). *Managing, measuring and maximizing intangible assets: an eclectic view of some of the latest international developments*. Paper presented at the Intellectual Property: Managing, measuring and maximizing intangible assets, Sydney, 15 November.

Ffowcs-Williams, I. (2004). *Cluster Navigator*. Wellington, New Zealand: Cluster Navigators Ltd.

Field, A. (2005). *Discovering statistics using SPSS*. London: Sage.

Filieri, R. (2009). *Overcoming knowledge sharing barriers through communities of practice*. Unpublished doctoral dissertation, Università degli Studi Roma 3, Italia

Filieri, R., & Al Guezaui, S. (2009). Knowledge Sharing Barriers: a Meta-Analytic Review. In *Proceedings of the 10th European Conference of Knowledge Management (ECKM)*, Università Degli Studi Di Padova, Vicenza, Italy, September 3-4.

Fiol, C. M. (2002). Capitalizing on paradox: The role of language in transforming organizational identities. *Organization Science, 13*(6), 653–666. doi:10.1287/orsc.13.6.653.502

Fiol, C. M., & Huff, A. S. (1992). Maps for Managers: Where are we? Where do we go from here. *Journal of Management Studies*, *29*(3), 267–285.

Firer, S., & Williams, S. M. (2003). Intellectual capital and traditional measures of corporate performance. *Journal of Intellectual Capital*, *4*(3), 348–360. doi:10.1108/14691930310487806

Fischer, M. M. (2001). Innovation, knowledge creation and systems of innovation. *The Annals of Regional Science*, *35*, 199–216. doi:10.1007/s001680000034

Flamholtz, E. (1985). *Human research accounting*. Los Angeles: Jossey-Bass.

Flamholtz, E. G., Bullen, M. L., & Hua, W. (2002). Human resource accounting: a historical perspective and future implications. *Management Decision*, *40*, 947–954. doi:10.1108/00251740210452818

Flap, H. D., & De Graaf, N. D. (1986). Social capital and attained occupational status. *Netherlands Journal of Sociology*, *22*(2), 145–161.

Fleming, L., Mingo, S., & Chen, D. (2007). Collaborative Brokerage, Generative Creativity, and Creative Success. *Administrative Science Quarterly*, *52*(3), 443–475.

Florin, J., Lubatkin, M., & Schulze, W. (2003). A social capital model of high-growth ventures. *Academy of Management Journal*, *46*(3), 374–384. doi:10.2307/30040630

Flostrand, P. (2006). The sell side – observations on intellectual capital indicators. *Journal of Intellectual Capital*, *7*(4), 457–473. doi:10.1108/14691930610709112

Fosfuri, A., & Ronde, T. (2007). *Leveraging resistance to change and the Skunk Works model of innovation*. Corporate intellectual property (RIPE) Conference, 7-8 December 2007, Copenhagen, Denmark.

Foss, N. (1996). Knowledge-based approaches to the theory of the firm: Some critical comments. *Organization Science*, *7*(5), 470–476. doi:10.1287/orsc.7.5.470

Foster, R., & Kaplan, S. (2001). *Creative Destruction*. New York: Double Day.

Franco, H. (2003). *Private universities*. Panama: UNESCO.

Freeman, C. (1987). *Technology policy and economic performance: lessons from Japan*. London: Pinter.

Fromhold-Eisebith, M. (2004). Innovative milieu and social capital: complementary or redundant concepts of collaboration-based regional development? *European Planning Studies*, [REMOVED HYPERLINK FIELD]*12*(6), 747–765.

Fromhold-Eisebith, M., & Eisebith, G. (2005). How to institutionalize innovative cluster? Comparing explicit top-down and implicit bottom-up approaches. *Research Policy*, *34*(8), 1250–1268.

Fromhold-Eisebith, M., & Eisebith, G. (2008). Looking behind facades: evaluating effects of (automotive) cluster promotion. *Regional Studies*, *42*(10), 1343–1356.

Fukuda, K., & Watanabe, C. (2008, October). *Catalyst role of government R&D inducing hybrid management*. Paper presented at 1st International Conference on Technology Management, Tokyo, Japan.

Fukuyama, F. (1995). *Trust: Social Virtues and the Creation of Prosperity*. New York: Free Press.

Gabbay, S. M., & Leenders, R. Th. A. J. (1999). CSC: The structure of advantage and disadvantage. In Leenders, R. Th. A. J., & Gabbay, S. M. (Eds.), *Corporate Social Capital and Liability* (pp. 1–16). Boston: Kluwer Academic.

Galbraith, J. K. (1969). *A theory of price control*. Cambridge, MA: Harvard University Press.

Galbreath, J. (2005). Which Resources Matter the Most to Firm Success? An Exploratory Study of Resource-Based Theory. *Technovation*, *25*, 979–987. doi:10.1016/j.technovation.2004.02.008

Galende, J. (2006). Analysis of Technological Innovation from Business Economics and Management. *Technovation*, *26*, 300–311. doi:10.1016/j.technovation.2005.04.006

Galende, J., & De la Fuente, J. (2003). El Proceso de Innovación de la Empresa Española: Identificación de Patrones de Innovación. *Cuadernos de Economía y Dirección de la Empresa*, *16*, 145–171.

Gallego, A., & Casanueva, C. (2007). *El Peso de la Cooperación en la Innovación de la Empresa Industrial Española*. ACEDE Conference, Sevilla.

Gallego, I., & Rodríguez, L. (2005). Situation of Intangibles Assets in Spanish Firms: An Empirical Análisis. *Journal of Intellectual Capital*, *6*, 105–126. doi:10.1108/14691930510574690

Garcia, R., & Calantone, R. (2002). A critical look at technological innovation typology and innovativeness terminology: a literature review. *Journal of Product Innovation Management*, *19*(2), 110–132.

García, T., & Mulero, E. (2007). Medida de los Factores Claves del Éxito de la I+D: El Constructo y sus Dimensiones. *Cuadernos de Economía y Dirección de la Empresa*, *32*, 15–48.

Garcia-Vega, M. (2006). Does technological diversification promote innovation? An empirical analysis for European firms. *Research Policy*, *35*(2), 30–246.

Garrett-Jones, S. (2004). From citadels to clusters: the evolution of regional innovation policies in Australia. *R & D Management*, *34*(1), 3–16.

Geisler, E. (2005). The measurement of scientific activity: research in linking the philosophy of science and metrics of science and technology outputs. *Scientometrics*, *62*(2), 269–284.

General Comptroller's of the Republic. (2005). *Panama in numbers: Years 2000-2004*. Panama: General Comptroller's of the Republic.

Gilsing, V. (2001). Towards second generation cluster policy: the case of the Netherlands. In den Hertog, P., Bergman, E., & Remoe, S. (Eds.), *Innovative Clusters. Drivers of National Innovation Systems* (pp. 361–376). Paris, France: OECD.

Gilsing, V. A., & Nooteboom, B. (2005). Density and strength of ties in innovation networks: An analysis of multimedia and biotechnology. *European Management Review*, *2*(3), 179–197. doi:10.1057/palgrave.emr.1500041

Gioia, D. A. (1998). From Individual to Organizational Identity. In Whetten, D. A., & Godfrey, P. C. (Eds), Identity in Organizations: Building Theory Trough Conversations (pp. 17-31). Thousand Oaks, CA: Sage Publication.

Gioia, D. A., & Chittipeddi, K. (1991). Sensemaking and sensegiving in strategic change initiation. *Strategic Management Journal*, *12*(6), 443–458. doi:10.1002/smj.4250120604

Gioia, D. A., & Pitre, E. (1990). Multiparadigm perspectives on theory bulding. *Academy of Management Review*, *15*(4), 584–602. doi:10.2307/258683

Gioia, D. A., & Thomas, J. (1996). Identity, image and issue interpretation: Sensemaking during strategic change in academia. *Administrative Science Quarterly*, *41*, 370–403. doi:10.2307/2393936

Gioia, D. A., Schultz, M., & Corley, K. G. (2002a). On Celebrating the Organizational Identity Metaphor: A Rejoinder to Cornelissen. *British Journal of Management*, *13*(3), 269–275. doi:10.1111/1467-8551.00243

Gioia, D. A., Schultz, M., & Corley, K. G. (2002b). Metaphorical Shadow Boxing: A Response to Cornelissen's Reply to our Rejoinder. *British Journal of Management*, *13*(3), 281–281. doi:10.1111/1467-8551.00245

Glaser, B. G., & Strauss, A. L. (1967). *The Discovery of Grounded Theory: Strategies for Qualitative Research*. New York: Aldine Publishing Company.

Glynn, M. A. (2000). When Cymbals Become Symbols: Conflict Over Organizational Identity Within a Symphony Orchestra. *Organization Science*, *11*(3), 285–298. doi:10.1287/orsc.11.3.285.12496

Glynn, M. A., & Abzug, R. A. (2002). Institutionalizing identity: Symbolic isomorphism and organizational names. *Academy of Management Journal*, *45*(1), 267–280. doi:10.2307/3069296

Gnuschke, J. (2001). Obtaining market value from intellectual capital. *Business Perspectives, 6*(Summer-Fall), 4.

Goh, P. C. (2005). Intellectual capital performance of commercial banks in Malaysia. *Journal of Intellectual Capital*, 385–396. doi:10.1108/14691930510611120

Golden-Biddle, K., & Rao, H. (1997). Breaches in boardroom: Organizational Identity and Conflicts of Commitment in a Nonprofit Organization. *Organization Science*, *8*(6), 593–609. doi:10.1287/orsc.8.6.593

Gomes-Casseres, B. (1996). *The Alliance Revolution: The New Shape of Business Rivalry*. Cambridge, MA: Harvard University Press.

Good, M. L., (1996). The globalization of technology. *Physics Today, August, Part 1, 49*(8), 23-27.

Gorsuch, R. L. (1983). *Factor analysis*. Hillsdale, NJ: L. Erlbaum Associates.

Goto, A., & Suzuki, K. (1989). R&D capital, rate of return on R&D investment and spillover of R&D in Japanese manufacturing industries. *The Review of Economics and Statistics*, *71*(4), 555–564. doi:10.2307/1928096

Granovetter, M. (1973). The strength of weak ties. *American Journal of Sociology*, *78*(6), 1360–1380. doi:10.1086/225469

Granovetter, M. (1983). The strength of weak ties: A network theory revisited. *Sociological Theory*, *1*(1), 201–233. doi:10.2307/202051

Granovetter, M. (1985). Economic action and social structure: The problem of embeddedness. *American Journal of Sociology*, *91*(3), 481–510. doi:10.1086/228311

Granovetter, M. S. (1973). The strength of weak ties. *American Journal of Sociology*, *78*, 1360–1380. doi:10.1086/225469

Granovetter, M. S. (1985). Economic action and social structure: The problem of embeddedness. *American Journal of Sociology*, *91*, 481–510. doi:10.1086/228311

Granovetter, M. S. (1992). Problems of explanation in economy sociology. In Nohria, N., & Eccles, R. (Eds.), *Networks and organizations: Structure, form and action* (pp. 25–56). Boston: Harvard Business School Press.

Grant, R. M. & Baden-Fuller, C. (1995). A Knowledge-Based Theory of Inter.-Firm Collaboration. *Academy of Management Journal*, Best Paper Proceedings, 17-21.

Grant, R. M. (1991). The Resource-Based Theory of Competitive Advantage: Implications for Strategy Formulation. *California Management Review*, *33*(3), 114–135.

Grant, R. M. (1996). Prospering in Dynamically Competitive Environments. *Organization Science*, *7*(4), 375–387. doi:10.1287/orsc.7.4.375

Grant, R. M. (1996). Toward a Knowledge-Based Theory of the Firm. *Strategic Management Journal, 17*(Winter Special Issue), 109-122.

Grant, R. M. (1996). Toward a knowledge-based theory of the firm. *Strategic Management Journal*, *17*(10), 109–122.

Grant, R. M. (1996a). Prospering in dynamically-competitive environments: organizational capability as knowledge integration. *Organization Science*, *7*(4), 375–388. doi:10.1287/orsc.7.4.375

Grant, R. M. (1996b). Toward a knowledge-based theory in the firm. *Strategic Management Journal*, *17*, 109–122.

Grasenick, K., & Low, J. (2004). Shaken, not stirred defining and connecting indicators for the measurement and valuation of intangibles. *Journal of Intellectual Capital*, *5*(2), 268–281. doi:10.1108/14691930410533696

Greenwood, J., & Jovanovic, B. (1999). The information-technology revolution and the stock market. *The American Economic Review*, *89*(2), 116–122.

Griliches, Z. (1979). Issues in assessing the contribution of R&D to productivity growth. *The Bell Journal of Economics*, *10*, 92–116. doi:10.2307/3003321

Griliches, Z., & Lichtenburg, F. (1984). Inter-industry technology flows and productivity growth: A reexamination. *The Review of Economics and Statistics*, *66*(2), 324–329. doi:10.2307/1925836

Grootaert, C., Narayan, D., Nyan Jones, V., & Woolcock, M. (2004). *Measuring social capital: an integrated questionnaire*. Washington, DC: The World Bank.

Groth, J. C. (1994). Human Capital and Europe Future. *European Business Review, 94*(5), 20–27. doi:10.1108/09555349410068858

Guilding, C., & Pike, R. (1990). Intangible marketing assets: a managerial accounting perspective. *Accounting and Business Review, 21*(18), 41–49.

Gulati, R. (1995). Social Structure and Alliance Formation Patterns: A Longitudinal Analysis. *Administrative Science Quarterly, 40*(4), 619–652. doi:10.2307/2393756

Gulati, R., Nohria, N., & Zaheer, A. (2000). Strategic networks. *Strategic Management Journal, 21*(3), 203–215. doi:10.1002/(SICI)1097-0266(200003)21:3<203::AID-SMJ102>3.0.CO;2-K

Gupta, A. K., Brockhoff, K., & Weisenfeld, U. (1992). Making trade-offs in new product development process: A German/US comparison. *Journal of Product Innovation Management, 9*(1), 11–18.

Guthrie, J. (2000). Measuring up to change. *Financial Management*, (Dec): 11.

Guthrie, J. (2001). The management, measurement and reporting of intellectual capital. *Journal of Intellectual Capital, 2*(1), 27–41. doi:10.1108/14691930110380473

Guthrie, J., & Petty, R. (2000). Intellectual Capital: Australian annual reporting practices. *Journal of Intellectual Capital, 1*(3), 241–251. doi:10.1108/14691930010350800

Guthrie, J., Petty, R., & Yongvanich, K. (2004). Using content analysis as a research method to inquire into intellectual capital reporting. *Journal of Intellectual Capital, 5*(2), 282–293. doi:10.1108/14691930410533704

Guthrie, J., Petty, R., Ferrier, F., & Wells, R. (1999). *There is no accounting for intellectual capital in Australia: a review of annual reporting practices and the internal measurement of Intangibles*. Amsterdam: OECD.

Haanes, K., & Lowendhal, B. (1997). The Unit of Activity: Towards an Alternative to the Theories of the Firm. In *Strategy, Structure and Style*. London: John Wiley and Sons.

Hagedoorn, J., & Duysters, G. (2002). Learning in Dynamic Inter-Firm Networks: The Efficacy of Multiple Contracts. *Organization Studies, 23*(4), 525–548. doi:10.1177/0170840602234002

Hair, J. F., Black, W. C., Babin, B. J., & Anderson, R. E. (2009). *Multivariate data analysis* (7th ed.). New York: Prentice Hall.

Hall, R. (1992). The strategic analysis of intangible resources. *Strategic Management Journal, 13*, 135–144. doi:10.1002/smj.4250130205

Hall, R. (1993). A Framework Linking Intangible Resources and Capabilities to Sustainable Competitive Advantage. *Strategic Management Journal, 14*, 607–618. doi:10.1002/smj.4250140804

Halstead, J. (2008). U.S. Employee Engagement Deteriorates as Financial Crisis Worsens. *Modern Survey Website*, November 12. Retrieved March 2, 2009 from http://modernsurvey.com/news/?p=124

Hammerer, G. (1996). *Intangible investments in Austria*. Paper presented at workshop, OECD New Indicators for the Knowledge-based Economy, OECD, Paris.

Han, D., & Han, I. (2004). Prioritization and selection of intellectual capital measurement indicators using analytic hierarchy process for the mobile telecommunications industry. *Expert Systems with Applications, 26*(4), 519–527. doi:10.1016/j.eswa.2003.10.010

Han, J. K., Kim, N., & Srivastava, K. (1998). Market Orientation and Organizational Performance: Is Innovation a Missing Link? *Journal of Marketing, 62*(4), 30–45. doi:10.2307/1252285

Hanifan, L. J. (1916). The Rural School Community Center. *The Annals of the American Academy of Political and Social Science, 67*, 130–138. doi:10.1177/000271621606700118

Hanks, K., Belliston, L., & Edwards, D. (1977). *Design Yourself. Los Altos, CA.* W.: Kauffman Inc.

Hannah, D. R. (2005). Should I keep a secret? The effects of trade secret protection procedures on employees' obligations to protect trade secrets. *Organization Science, 16*(1), 71–84. doi:10.1287/orsc.1040.0113

Hansen, M. T. (1999). The search-transfer problem: the role of weak ties in sharing knowledge across organizational subunits. *Administrative Science Quarterly, 44*(1), 82–111. doi:10.2307/2667032

Hargadon, A., & Sutton, R. I. (1997). Technology brokering and innovation in a product development firm. *Administrative Science Quarterly, 42*(4), 716–749. doi:10.2307/2393655

Hassink, R. (1996). Technology Transfer Agencies and Regional Economic Development. *European Planning Studies, 4*(2), 167–184.

Hatck, N., & Dyer, J. (2004). Human Capital and Learning as a Source of Sustainable Competitive Advantage. *Strategic Management Journal, 25*, 1155–1178. doi:10.1002/smj.421

Hayashi, F., & Prescott, E. (2002). The 1990s in Japan: A lost decade. *Review of Economic Dynamics, 5*(1), 206–235. doi:10.1006/redy.2001.0149

Hayek, F. A. (1945). The use of knowledge in society. *The American Economic Review, 35*(4), 519–530.

Hayton, J. C. (2005). Competing in the New Economy: the Effect of Intellectual Capital on Corporate Entrepreneurship in High-Technology New Ventures. *R & D Management, 35*, 137–155. doi:10.1111/j.1467-9310.2005.00379.x

Hedlund, G. (1994). A Model of Knowledge Management and the N-Form Corporation. *Strategic Management Journal, 15*, 73–90.

Hegde, D., & Shapira, P. (2007). Knowledge, Technology Trajectories, and Innovation in a Developing Country Context: Evidence from a Survey of Malaysian Firms. *International Journal of Technology Management, 40*, 349–370. doi:10.1504/IJTM.2007.015757

Heiman, B. A., & Burnett, W. R. (2007). *The Isolation of Design Thinking in Firms and Management Education: Definitions, Impediments, and Ways Forward. Society for Global Business and Economic Development.* Kyoto, Japan: SGBED.

Heiman, B. A., & Nickerson, J. A. (2004). Empirical evidence regarding the tension between knowledge sharing and knowledge expropriation in collaborations. *Managerial and Decision Economics, 25*(6-7), 401–420. doi:10.1002/mde.1198

Heiman, B. A., Nickerson, J. A., & Zenger, T. (2008). *Knowledge governance: Multidisciplinary perspectives* (Michailova, S., & Foss, N., Eds.). Oxford, UK: Oxford University Press.

Henderson, R. & Cockburn, I. (1994). Measuring Competence? Exploring Firm Effects in Pharmaceutical Research. *Strategic Management Journal, 15*(Winter Special Issue), 63-84.

Henderson, R. M., & Clark, K. B. (1990). Architectural Innovation: the Reconfiguration of Existing Product Technologies and the Failure of Established Firms. *Administrative Science Quarterly, 35*, 9–30. doi:10.2307/2393549

Henderson, R., & Cockburn, I., (1994). Measuring Competence? Exploring Firm Effects in Pharmaceutical Research. *Strategic Management Journal, 15*(Winter special issue), 63-84.

Hendriksen, E. S., & van Breda, M. F. (1992). *Accounting Theory* (5th ed.). Burr Ridge, IL: Irwin.

Hermans, R., & Kauranen, I. (2005). Value Creation Potential of Intellectual Capital in Biotechnology – Empirical Evidence from Finland. *R & D Management, 35*, 171–185. doi:10.1111/j.1467-9310.2005.00381.x

Hersey, P., & Blanchard, K. (1988). *Management of Organizational Behavior.* Englewood Cliffs, NJ: Prentice Hall, Inc.

Hill, C., & Rothaermel, F. (2003). The Performance of Incumbent Firms in the Face of Radical Technological Innovation. *Academy of Management Review, 28*, 257–274. doi:10.2307/30040712

Hirota, T. (1994). R&D system in Sumitomo Electric Industries, Ltd. *Kansai Daigaku Shogakuronshu, 38*(6), 917–941.

Hirota, T. (1995). New business creation by corporate venture system: A case of compound semiconductor in Sumitomo Electric Industries, Ltd. *Kansai Daigaku Shogakuronshu, 40*(4-5), 589–610.

Hitt, M., Bierman, L., Shimizu, K., & Kochhar, R. (2001). Direct and Moderating Effects of Human Capital on Strategy and Performance in Professional Service Firms: A Resource-Based Perspective. *Academy of Management Journal, 44*, 13–28. doi:10.2307/3069334

Hobjin, B., & Jovanovic, B. (2001). The information-technology revolution and the stock market: Evidence. *The American Economic Review, 91*(5), 1203–1220.

Hofstede, G. (1991). *Cultures and Organizations: Software in the Mind.* Maidenhead, UK: McGraw Hill.

Holland, J., & Gibbon, J. (1997). SMEs in the Metal Manufacturing, construction and contracting service sectors: environmental awareness and action. *Eco-Management and Auditing, 4*(1), 4–14. doi:10.1002/(SICI)1099-0925(199703)4:1<7::AID-EMA58>3.0.CO;2-9

Huang, C. C., Luther, R., & Tayles, M. (2007). An evidence-based taxonomy of intellectual capital. *Journal of Intellectual Capital, 8*(3), 386–403. doi:10.1108/14691930710774830

Huang, K. T. (1998). Capitalizing on intellectual assets. *IBM Systems Journal, 37*(4), 570–583. doi:10.1147/sj.374.0570

Huber, G. (1991). Organizational learning: the contributing process and the literatures. *Organization Science, 2*(1), 88–115. doi:10.1287/orsc.2.1.88

Huergo, E. (2006). The Role of Technological Management as a Source of Innovation: Evidence from Spanish Manufacturing Firms. *Research Policy, 35*, 1377–1388. doi:10.1016/j.respol.2006.07.005

Hung, Y. C., & Chang, E. L. (2006). Measurement scale of intellectual capital for the information service industry. *International Journal of Services and Standards, 2*(2), 154–175. doi:10.1504/IJSS.2006.008730

Huq, M. (2004). Building technological capability in the context of globalization: opportunities and challenges facing developing countries. *International Journal of Technology Management and Sustainable Development, 3*(3). doi:10.1386/ijtm.3.3.155/0

Hur, K. I., & Watanabe, C. (2001). Unintentional technology spillover between two sectors: Kinetic approach. *Technovation, 21*, 227–235. doi:10.1016/S0166-4972(00)00037-7

Hurmelinna-Laukkanen, P., & Puumalainen, K. (2007). The nature and dynamics of appropriability – Strategies for appropriating returns on innovation. *R & D Management, 37*(2), 95–112. doi:10.1111/j.1467-9310.2007.00460.x

Hurwitz, J., Lines, S., Montgomery, B., & Schmidt, J. (2002). The linkage between management practices, intangibles performance and stock returns. *Journal of Intellectual Capital, 3*(1), 51–61. doi:10.1108/14691930210412845

Huselid, M. (1995). The Impact of Human Resource Management Practices on Turnover, Productivity, and Corporate Financial Performance. *Academy of Management Journal, 38*, 635–672. doi:10.2307/256741

Hussi, T., & Ahonen, G. (2002). Managing intangible assets – a question of integration and delicate balance. *Journal of Intellectual Capital, 3*(3), 277–286. doi:10.1108/14691930210435615

Huston, L., & Sakkab, N. (2006). Connect and Develop, Inside Procter & Gamble's new model for innovation. *Harvard Business Review, 84*(3), 58–66.

Ibrahim, D. (2009). *Financing the Next Silicon Valley.* Legal Studies Research Paper Series, Paper No. 1065. Retrieved July 10th, 2009, from http://works.bepress.com/cgi/viewcontent.cgi?article=1003&context=darian_ibrahim

IFAC. (1998). *The Measurement and Management of Intellectual Capital: An Introduction.* Financial and Management Accounting Committee IFAC 1998 Study 7.

Inkpen, A., & Tsang, E. (2005). Social Capital Networks and Knowledge Transfer. *Academy of Management Review, 30*(1), 146–165. doi:10.2307/20159100

Isaksen, A. (2007) Clusters, innovation and the local learning paradox. *International Journal of Entrepreneurship and Innovation Management, 7*(2/3/4/5), 366-384.

Isaksen, A., & Hauge, E. (2002). *Regional Clusters in Europe. Observatory of European SMEs report 2002 No. 3.* Luxembourg: European Communities.

Itami, H., & Roehl, T. (1987). *Mobilizing Invisible Assets.* Cambridge, MA: Harvard University Press.

Itami, H., & Roehl, T. W. (1987). *Mobilizing Invisible Assets.* Boston: Harvard University Press.

Jacobs, J. (1961). *The Death and Life of Great American Cities.* New York: Random House.

Jacobsen, K., & Hofman-Bang, P. (2005). The IC Rating model by Intellectual Capital Sweden. *Journal of Intellectual Capital, 6*(4), 570–587. doi:10.1108/14691930510628834

Jaffe, A. B. (1986). Technological opportunity and spillovers of R&D: Evidence from firm's patents, profits and market value. *The American Economic Review, 76*(5), 984–1001.

Jashapara, A. (2003). *Knowledge Management – An integrated approach (pp. 269).* New York: Prentice Hall.

Jensen, B., & Harmsen, H. (2001). Implementation of success factors in new product development – the missing links? *European Journal of Innovation Management, 4*(1), 37–52. doi:10.1108/14601060110365565

Jick, T. (1979). Mixing Qualitative and Quantitative Methods: Triangulation in Action. *Administrative Science Quarterly, 24*(4), 602–611.

Jin, J. Y., Perote-Pena, J., & Troege, M. (2004). Learning by doing, spillovers and shakeouts. *Journal of Evolutionary Economics, 14*, 85–98. doi:10.1007/s00191-003-0177-5

Johannessen, J.-A., Olsen, B., & Olaisen, J. (1999). Aspects of innovation theory based on knowledge-management. *International Journal of Information Management, 19*, 121–139. doi:10.1016/S0268-4012(99)00004-3

Johannisson, B. (2000). Networking and Entrepreneurial Growth. In Sexton, D. L., & Landström, H. (Eds.), *The Blackwell Handbook of Entrepreneurship* (pp. 368–386). Oxford: Blackwell.

Johanson, U. (1999). Why the concept of human resource costing and accounting do not work. *Personnel Review, 28*(1-2), 91–107. doi:10.1108/00483489910249018

Johanson, U. MaÊrtensson, M. & Skoog, M. (1999a). *Measuring and managing intangibles: eleven Swedish exploratory case studies.* Paper presented at the International Symposium Measuring Reporting Intellectual Capital: Experiences, Issues, and Prospects, OECD, Amsterdam, June.

Johanson, U., Eklov, G., Holmgren, M. & MaÊrtensson, M. (1999b). *Human resource costing and accounting versus the Balanced Scorecard: a literature survey of experience with the concepts.* Paper presented at the International Symposium Measuring Reporting Intellectual Capital: Experiences, Issues, and Prospects, OECD, Amsterdam, June.

Johanson, U., Koga, C., Skoog, M., & Henningsson, J. (2006). The Japanese Government's intellectual capital reporting guideline. *Journal of Intellectual Capital, 7*(4), 474–491. doi:10.1108/14691930610709121

Johanson, U., Martensson, M., & Skoog, M. (1999, June). *Measuring and Managing Intangibles: Eleven Swedish Qualitative Exploratory Case Studies.* Paper presented at the International Symposium on Measuring and Reporting Intellectual, Amsterdam, Netherlands.

Johanson, U., Martensson, M., & Skoog, M. (2001). Measuring to understand intangible performance drivers. *European Accounting Review, 10*(3), 407–437. doi:10.1080/09638180120081562

Johnson, L. D., Neave, E. H., & Pazderka, B. (2002). Knowledge, Innovation and Share Value. *International Journal of Management Reviews*, *4*, 101–134. doi:10.1111/1468-2370.00080

Joia, A. (2004). Are Frequent Customer Always a Company's Intangible Asset? Some Findings drawn from an Exploratory case Study. *Journal of Intellectual Capital*, *5*, 586–601. doi:10.1108/14691930410567031

Jones, C., Hesterly, W. S., & Borgatti, S. P. (1997). A general theory of network governance: Exchange conditions and social mechanisms. *Academy of Management Review*, *22*(4), 911–945. doi:10.2307/259249

Jørgensen, K. M., & Boje, D. M. (2006). Time, place and mind of intellectual capital: on Jørgensen's 'Conceptualising intellectual capital as language game and power.'. *Journal of Intellectual Capital*, *7*(1), 78–88.

Kaiser, H. F. (1970). A second generation little jiffy. *Psychometrika*, *35*(4), 401–415. doi:10.1007/BF02291817

Kaiser, H. F., & Rice, J. (1974). Little jiffy, Mark IV. *Educational and Psychological Measurement*, *34*(1), 111–117. doi:10.1177/001316447403400115

Kale, P., Singh, H., & Perlmutter, H. (2000). Learning and protection of proprietary assets in strategic alliances: building relational capital. *Strategic Management Journal*, *21*(3), 217–237. doi:10.1002/(SICI)1097-0266(200003)21:3<217::AID-SMJ95>3.0.CO;2-Y

Kamath, G. B. (2008). Intellectual capital and corporate performance in Indian pharmaceutical industry. *Journal of Intellectual Capital*, *9*(4), 684–704. doi:10.1108/14691930810913221

Kannan, G., & Aulbur, G., W. (2004). Intellectual Capital: Measurement effectiveness. *Journal of Intellectual Capital*, *5*(3), 389–403. doi:10.1108/14691930410550363

Kaplan, R. S., & Norton, D. P. (1992). The Balanced Scorecard-measures that drive performance. *Harvard Business Review*, (January-February): 71–79.

Kaplan, R. S., & Norton, D. P. (1996). *The Balanced Scorecard*. Cambridge, MA: Harvard Business School Press.

Kaplan, R. S., & Norton, D. P. (2004). Measuring the strategic readiness of intangibles assets. *Harvard Business Review*, *82*(2), 52–63.

Kaplan, R., & Norton, D. (1992). The balanced scorecard: measures that drive performance. *Harvard Business Review*, (January-February): 71–79.

Karlsson, C. (2007). *Clusters, functional regions and cluster policies*. CESIS Electronic Working Paper Series Retrieved July 10th, 2009, from http://www.infra.kth.se/cesis/documents/WP84.pdf

Kaufman, L., & Schneider, Y. (2004). Intangibles: a synthesis of current research. *Journal of Intellectual Capital*, *5*(3), 366–388.

Kaufmann, A., & Tödtling, F. (2002). SMEs in Regional Innovation Systems and the Role of Innovation Support: The Case of Upper Austria. *The Journal of Technology Transfer*, *27*(1), 15–26. doi:10.1023/A:1013140318907

Kaufmann, L., & Schneider, Y. (2004). Intangibles: A synthesis of current research. *Journal of Intellectual Capital*, *5*(3), 366–388. doi:10.1108/14691930410550354

Kauser, S., Oakey, R., & During, W. (2002). *New Technology-Based Firms in the New Millenium* (*Vol. II*). London: Pergamon/Elsevier Science Ltd.

Kauser, S., Oakey, R., & During, W. (2004). *New Technology-Based Firms in the New Millenium* (*Vol. III*). London: Pergamon/Elsevier Science Ltd.

Kauser, S., Oakey, R., & During, W. (2005). *New Technology-Based Firms in the New Millenium* (*Vol. IV*). London: Pergamon/Elsevier Science Ltd.

Kennedy, C. (1964). Induced bias in innovation and the theory of distribution. *The Economic Journal*, *74*(295), 541–547. doi:10.2307/2228295

Kerssens-van Drongelen, I. C., & Cooke, A. (1997). Design principles for the development of measurement systems for research and development processes. *R & D Management*, *27*(4), 345–357. doi:10.1111/1467-9310.00070

Ketels, C., Lindqvist, G., & Solvell, O. (2008). *Clusters and Cluster Initiatives*. Retrieved July 10th, 2009, from http://www.clusterobservatory.eu/upload/ClustersAnd-ClusterOrganisations.pdf

Kim, D. Y., & Kumar, V. (2009). A framework for prioritization of intellectual capital indicators in R&D. *Journal of Intellectual Capital, 10*(2), 277–293. doi:10.1108/14691930910952669

King, D. R., Covin, J. G., & Hegarty, W. H. (2003). Complementary Resources and the Exploitation of Technological Innovations. *Journal of Management, 29*(4), 589–606.

Kirton, M. J. (1980). Adaptors and Innovators in Organizations. *Human Relations, 3*, 213–224. doi:10.1177/001872678003300401

Kirton, M. J. (2003). *Adaption-Innovation in the Context of Diversity and Change*. New York: Routledge.

Kline, J., & Rosenberg, N. (1986). An overview of innovation. In Landau, R., & Rosenberg, N. (Eds.), *The positive sum strategies: harnessing technology for economic growth* (pp. 275–305). Washington, DC: National Academic Press.

Klock, M., & Megna, P. (2000). Measuring and valuing intangible capital in the mobile communications industry. *The Quarterly Review of Economics and Finance, 40*(4), 519–532. doi:10.1016/S1062-9769(00)00055-7

Knight, F. H. (1921). *Risk, Uncertainty and Profit*. New York: Hart, Schaffner & Mark.

Knorr-Cetina, K. (1981). *The Manufacture of Knowledge: An Essay on the Constructivist and Contextual Nature of the Science*. New York: Pergamon Press.

Koberg, C. S., Detienne, D. R., & Heppard, K. A. (2003). An Empirical Test of Environmental, Organizational, and Process Factors Affecting Incremental and Radical Innovation. *The Journal of High Technology Management Research, 14*, 21–45. doi:10.1016/S1047-8310(03)00003-8

Koberg, D., & Bagnall, J. (1974). *The Universal Traveler*. Los Altos, CA: William Kaufmann, Inc.

Kogut, B. (2000). The Network as Knowledge: Generative Rules and Emergence of Structure. *Strategic Management Journal, 21*, 405–425. doi:10.1002/(SICI)1097-0266(200003)21:3<405::AID-SMJ103>3.0.CO;2-5

Kogut, B., & Zander, U. (1992). Knowledge of the firm, combinative capabilities, and the replication of technology. *Organization Science, 3*(3), 383–397. doi:10.1287/orsc.3.3.383

Kogut, B., & Zander, U. (1993). Knowledge of the Firm and the Evolutionary Theory of the Multinacional Corporation. *Journal of International Business Studies, 24*, 625–645. doi:10.1057/palgrave.jibs.8490248

Kogut, B., & Zander, U. (1996). What firms do? Coordination, identity and learning. *Organization Science, 7*(5), 502–518. doi:10.1287/orsc.7.5.502

Konrad Group (The). (1988). *The invisible balance sheet*. Retrieved from www.sveiby.com/articles/IntangAss/DenOsynliga.pdf

Kor, Y., & Mahoney, J. T. (2005). How Dynamics, Management, and Governance of Resource Deployments influence Firm-level Performance. *Strategic Management Journal, 26*(5), 489–496. doi:10.1002/smj.459

Koschatzky, K., & Lo, V. (2007) *Methodological framework for cluster analysis*. Working papers firms and region (R1/2007). Karlsruhe, Germany: Fraunhofer Institute for Systems and Innovation Research, Competence Centre "Policy and Regions."

Kostova, T., & Roth, K. (2003). Social capital in multinational corporations and a micro-macro model of its formation. *Academy of Management Review, 28*(2), 297–317. doi:10.2307/30040714

KPMG Management Consulting. Case Study: Building a Platform for Corporate Knowledge, *(1998a)*.

Kraatz, M. S. (1998). Learning by association? Interorganizational networks and adaptation to environmental change. *Academy of Management Journal, 41*(6), 621–643. doi:10.2307/256961

Kreps, D. (1990). A Course in Microeconomic Theory. In *Modeling competitive situations*. Princeton, NJ: Princeton University Press.

Kuhn, T. (1970). *The Structure of Scientific Revolutions*. Chicago: Chicago University Press.

Kujansivu, P., & Lonnqvist, A. (2007). Investigating the value and efficiency of intellectual capital. *Journal of Intellectual Capital*, *8*(2), 272–287. doi:10.1108/14691930710742844

Kyriakopoulos, K., & De Ruyter, K. (2004). Knowledge Stocks and Information Flows in New Product Development. *Journal of Management Studies*, *41*, 1469–1498. doi:10.1111/j.1467-6486.2004.00482.x

Labianca, G., Brass, D. J., & Gray, B. (1998). Social networks and perceptions of intergroup conflict: The role of negative relationships and third parties. *Academy of Management Journal*, *41*(1), 55–67. doi:10.2307/256897

Lagendijk, A. (2000). Learning in non-core regions: towards intelligent clusters; addressing business and regional needs. In Boekema, F., Morgan, K., Bakkers, S., & Rutten, R. (Eds.), *Knowledge, Innovation and Economic Growth* (pp. 165–191). London: Edward Elgar.

Lai, L. F. (2007). A knowledge engineering approach to knowledge management. *Journal of Information Sciences*. doi:10.1016/j.ins.2007.02.028 (available online at www.sciencedirect.com)

Landry, R., Amara, N., & Lamari, M. (2002). Does social capital determine innovation? To what extent? *Technological Forecasting and Social Change*, *69*(7), 681–701.

Lane, P. J., & Lubatkin, M. (1998). Relative absorptive capacity and interorganizational learning. *Strategic Management Journal*, *19*(5), 461–477. doi:10.1002/(SICI)1097-0266(199805)19:5<461::AID-SMJ953>3.0.CO;2-L

Larsen, E., & Lomi, A. (2002). Representing change: A system model of organizational inertia and capabilities as dynamic accumulation process. *Simulation Modelling Practice and Theory*, *10*, 271–296. doi:10.1016/S1569-190X(02)00085-0

Lauder, H., Brown, P., Dillabough, J., & Halsey, A. H. (2006). *Education, globalization, and social change*. New York: Oxford University Press.

Laursen, K., & Salter, A. (2006). Open For Innovation: The Role Of Openness In Explaining Innovation Performance Among U.K. Manufacturing Firms. *Strategic Management Journal*, *27*(2), 131–150. doi:10.1002/smj.507

Leana, C. R., & Pil, F. K. (2006). Social capital and organizational performance: Evidence from urban public schools. *Organization Science*, *17*(3), 353–368. doi:10.1287/orsc.1060.0191

Leana, C. R., & Van Buren, H. J. III. (1999). Organizational social capital and employment practices. *Academy of Management Review*, *24*(3), 538–545. doi:10.2307/259141

Lee, M., Son, B., & Lee, H. (1996). Measuring R&D effectiveness in Korean companies. *Research Technology Management*, *39*(6), 28–31.

LEED. (2002). *LEED-CEI-EBRD project on Industrial Districts and Local Clusters 2001-2002*. Paris, France: OECD Publishing.

LEED. (2009). *Clusters, Innovation and Entrepreneurship*. Paris, France: OECD Publishing.

Lehrer, J. (2008). The Eureka Moment. *New Yorker (New York, N.Y.)*, (July): 28, 40–45.

Leiponen, A. (2006). Managing Knowledge for Innovation: The Case of Business-to-Business Services. *Journal of Product Innovation Management*, *23*, 238–258. doi:10.1111/j.1540-5885.2006.00196.x

Leitner, K. (2005). Managing and Reporting Intangible Assets in Research Technology Organisations. *R & D Management*, *35*, 125–136. doi:10.1111/j.1467-9310.2005.00378.x

Leonard-Barton, D. (1992). Core capabilities and core rigidities: a paradox in managing new product development. *Strategic Management Journal*, *13*, S111–S125. doi:10.1002/smj.4250131009

Leonard-Barton, D. (1995). *Wellsprings of Knowledge: Building and Sustaining the Sources of Innovation*. Boston: Harvard Business School Press.

Leskovar-Spacapan, G., & Bastic, M. (2007). Differences in Innovation Capability in Transition Economy: Internal Aspect of the Organization's Strategic Orientation. *Technovation, 27*(9), 533–546. doi:10.1016/j.technovation.2007.05.012

Lev, B. (1997). *The boundaries of financial reporting and how to extend them. Industrial Competitiveness in the Knowledge-Based Economy*. Stockholm: OECD.

Lev, B. (1998). *Intangibles and gains from insider trading*. New York: New York University.

Lev, B. (2001). *Intangibles: Management, Measurement, and Reporting*. Washington, DC: Brookings Institute Press.

Lev, B., & Sougiannis, T. (1996). The capitalization, amortization, and value-relevance of R&D. *Journal of Accounting and Economics, 21*(1), 107–138. doi:10.1016/0165-4101(95)00410-6

Levinthal, D. A., & March, J. G. (1993). The Myopia of Learning. *Strategic Management Journal, 14*, 95–112. doi:10.1002/smj.4250141009

Lew, B., & Zarowin, P. (1998). The boundaries of financial reporting and how to extend them. *Working Paper*. New York: New York University

Leydesdorff, L., & Etzkowitz, H. (1996). Emergence of a Triple Helix of University-Industry-Government Relations. *Science & Public Policy, 23*, 279–286.

Leydesdorff, L., & Etzkowitz, H. (2001). The transformation of University-Industry-Government relations. *Electronic Journal of Sociology*. Retrieved July 10th, 2009, from http://sociology.org/content/vol005.004/th.html

Li, H., & Atuahene-Gima, K. (2002). The Adoption of Agency Business Activity, Product Innovation, and Performance in Chinese Technology Ventures. *Strategic Management Journal, 23*, 469–490. doi:10.1002/smj.233

Lichtenthaler, U. (2008). Open innovation in practice: An analysis of strategic approaches to technology transactions. *IEEE Transactions on Engineering Management, 55*(1), 148–157. doi:10.1109/TEM.2007.912932

Liebowitz, J., & Suen, C. Y. (2000). Developing Knowledge management metrics for measuring intellectual capital. *Journal of Intellectual Capital, 1*(1), 54–67. doi:10.1108/14691930010324160

Liebowitz, J., & Wright, K. (1999). Does measuring knowledge make "cents"? *Expert Systems with Applications, 17*(5), 99–103. doi:10.1016/S0957-4174(99)00027-5

Lilien, G. L., Morrison, P. D., Searls, K., Sonnack, M., & von Hippel, E. (2002). Performance Assessment of the Lead User Idea-Generation Process for New Product Development. *Management Science, 48*(8), 1042–1059. doi:10.1287/mnsc.48.8.1042.171

Lim, L. L. K., & Dallimore, P. (2004). Intellectual capital: management attitudes in service industries. *Journal of Intellectual Capital, 5*(1), 181–194. doi:10.1108/14691930410512996

Lin, N. (2001). Building a network theory of social capital. In Lin, N., Cook, K., & Burt, R. (Eds.), *Social capital: Theory and research* (pp. 3–29). Piscataway, NJ: Aldine Transaction.

Lin, N., Cook, K., & Burt, R. (2001). *Social Capital: Theory and Research*. New York: Aldine de Gruyter.

Lincoln, Y. S. (Ed.). (1985). *Organizational Theory and inquiry: The paradigm revolution*. Beverly Hills, CA: Sage.

Lind, M. R., & Zmud, R. W. (1995). Improving interorganizational effectiveness through voice mail facilitation of peer-to-peer relationships. *Organization Science, 6*(4), 445–461. doi:10.1287/orsc.6.4.445

Litschka, M., Andreas, M., & Susanne, S. (2006). Measuring and analyzing intellectual assets: An integrative approach. *Journal of Intellectual Capital, 7*(2), 160–173. doi:10.1108/14691930610661836

Litschka, M., Markom, A., & Schunder, S. (2006). Measuring and analyzing intellectual assets: an integrative approach. *Journal of Intellectual Capital*, *7*(2), 160–171. doi:10.1108/14691930610661836

Little, A. D. (1997). *New Technology-Based Firms in the United Kingdom and the Federal Republic of Germany*. London: Wilton House.

Litwin, G. H., & Stringer, R. A. (1968). *Motivation and Organizational Climate*. Boston: Harvard University.

Liu, C.-C. (2006). Developing measurements of intellectual capital in the e-learning platform industry by the analytic hierarchy process. *International Journal of Innovation and Learning*, *3*(4), 374–386.

Liu, P. L., Chen, W. C., & Tsai, C. H. (2005). An Empirical Study on the Correlation Between the Knowledge Management Method and the New Product Development Strategy on Product Performance in Taiwan's Industries. *Technovation*, *25*, 637–644. doi:10.1016/j.technovation.2003.11.001

Longo, M., & Mura, M. (2009, February). *Intellectual capital and innovative capabilities. Does health care?* Paper presented at the International Forum on Knowledge Assets Dynamics (IFKAD), Glasgow, UK.

Lorenzoni, G., & Lipparini, A. (1999). The leveraging of interfirm relationships as a distinctive organizational capability: a longitudinal study. *Strategic Management Journal*, *20*(4), 317–338. doi:10.1002/(SICI)1097-0266(199904)20:4<317::AID-SMJ28>3.0.CO;2-3

Loury, G. C. (1977). A dynamic theory of racial income differences. In Wallace, P. A., & La Mond, A. M. (Eds.), *Women, Minorities, and Employment Discrimination* (pp. 153–186). Lexington, MA: Lexington books.

Low, J. (2000). The value creation index. *Journal of Intellectual Capital*, *1*, 252–262. doi:10.1108/14691930010377919

Lundvall, B.-Å. (Ed.). (1992). *National systems of innovation: Towards a theory of innovation and interactive learning*. London: Pinter.

Lynn, B. E. (2000). Intellectual capital: unearthing hidden value by managing intellectual assets. *Ivey Business Journal*, *64*(3), 48–52.

Macher, J. T. (2006). Technological Development and the Boundaries of the Firm: A Knowledge-Based Examination in Semiconductor Manufacturing. *Management Science*, *52*(6), 826–843. doi:10.1287/mnsc.1060.0511

Machlup, F. (1980). *Knowledge: Its Creation, Distribution and Economic Significance*. Princeton, NJ: Princeton University Press.

Madhavan, R., & Grover, R. (1998). From Embedded Knowledge to Embodied Knowledge: New Product Development as Knowledge Management. *Journal of Marketing*, *62*, 1–12. doi:10.2307/1252283

Mahoney, J. T., & Pandian, R. J. (1992). The Resource-based view within the conversation of strategic management. *Strategic Management Journal*, *13*(5), 363–380. doi:10.1002/smj.4250130505

Maier, R. (2007). *Knowledge Management Systems. Information and Communication Technologies for Knowledge Management* (3rd ed.). Berlin: Springer Verlag, Wien.

Makadok, R. (2001). Toward a Synthesis of Resource-Based and Dynamic-Capability Views of Rent Creation. *Strategic Management Journal*, *22*(5), 387–401. doi:10.1002/smj.158

Malhotra, A., & Majchrzak, A. (2004). Enabling Knowledge Creation In Far-Flung Teams: Best Practices For IT Support And Knowledge Sharing. *Journal of Knowledge Management*, *8*(4), 75–88. doi:10.1108/13673270410548496

Malhotra, Y. (2000). Knowledge assets in the global economy: Assessment of national intellectual capital. *Journal of Global Information Management*, (July-September): 5.

Malmberg, A., & Maskell, P. (2002). The elusive concept of localization economies: towards a knowledge-based theory of spatial clustering. *Environment & Planning A*, *34*, 429–449.

Mansury, M. A., & Love, J. H. (2008). Innovation, Productivity and Growth in US Business Services: A Firm-Level Analysis. *Technovation, 28*, 52–62. doi:10.1016/j.technovation.2007.06.002

March, J. (1991). Exploration and Exploitation in Organizational Learning. *Organization Science, 2*(1), 71–87. doi:10.1287/orsc.2.1.71

Marginson, S., & Sawir, E. (2005, November). Interrogating global flows in higher education. *Globalisation, Societies and Education, 3*(3), 281–309. doi:10.1080/14767720500166878

Marr, B. (2005). *Perspectives on intellectual capital: Multidisciplinary insights into management, measurement and reporting.* Boston: Elsevier.

Marr, B., Gray, D., & Neely, A. (2003). Why do firms measure their intellectual capital? *Journal of Intellectual Capital, 4*(4), 441–464. doi:10.1108/14691930310504509

Marr, B., Gray, D., & Schiuma, G. (2004). Measuring intellectual capital: what, why and how? In M. Bourne, (Ed.), Handbook of performance measurement. London: Gee.

Marr, B., Schiuma, G., & Neely, A. (2002). Assessing strategic knowledge assets in e-business. *International Journal of Business Performance Management, 4*, 279–295. doi:10.1504/IJBPM.2002.000119

Marr, B., Schiuma, G., & Neely, A. (2004). Intellectual capital – defining key performance indicators for organizational knowledge assets. *Business Process Management Journal, 10*(5), 551–569. doi:10.1108/14637150410559225

Marsili, O. (2000). *Technological regimes and sources of entrepreneurship.* Working Paper 00.10, Eindhoven Centre for Innovation Studies, Eindhoven.

Martin, R., & Sunley, P. (2003). Deconstructing clusters: chaotic concept or policy panacea? *Journal of Economic Geography, 3*(1), 5–35.

Martínez, A., Vela, M. J., Pérez, M., & De Luis, P. (2007). Flexibilidad e Innovación: El Efecto Moderador de la Cooperación. *Revista Europea de Dirección y Economía de la Empresa, 16*, 69–88.

Martinez-Torres, M. R. (2006). A procedure to design a structural and measurement model of Intellectual Capital: An exploratory study. *Information & Management, 43*(5), 617–626.

Maskell, P. (2001). Towards a knowledge-based theory of the geographical cluster. *Industrial and Corporate Change, 10*, 921–943.

Massa, S., & Testa, S. (2008a). Which roles in innovation processes? A matter of perspectives. In Terziovski, M. (Ed.), *Energizing Management through Innovation and Entrepreneurship* (pp. 146–166). London: Routledge.

Massa, S., & Testa, S. (2008b). Innovation and SMEs: misaligned perspectives and goals among entrepreneurs, academics and policy makers. *Technovation, 28*(7), 393–407.

Massa, S., & Testa, S. (2009). Innovation topics under debate at the EC: the opinions of Italian entrepreneurs, academics and policy makers. *International Journal of Entrepreneurship and Innovation Management, 10*(3/4), 361–376.

Matsumoto, K., Ouchi, N., Watanabe, C., & Griffy-Brown, C. (2002). Optimal timing of the development of innovative goods with generation – an empirical analysis focusing on Canon's printer series. *Technovation, 22*, 175–185. doi:10.1016/S0166-4972(01)00003-7

Mattessich, R. (2003). Accounting representation and the onion model of reality: a comparison with Baudrillard's orders of simulacra and his hyperreality. *Accounting, Organizations and Society, 28*(5), 443. doi:10.1016/S0361-3682(02)00024-7

Maturana, H., & Valera, F. (1984). *El árbol del conocimiento.* Santiago de Chile, Chile: Universitaria.

Matusik, S. F., & Hill, W. L. (1998). The Utilization of Contingent Work, Knowledge Creation, and Competitive Advantage. *Academy of Management Review, 23*(4), 680–697. doi:10.2307/259057

Mavridis, D. (2004). The intellectual capital performance of the Japanese banking sector. *Journal of Intellectual Capital, 5*(1), 92–115. doi:10.1108/14691930410512941

Mayo, A. (2001). *The Human Value of the Enterprise: Valuing People as Assets: Monitoring, Measuring, Managing*. London: Nicholas Brealey Publishing.

McClintock, C., Brannon, D., & Maynard-Moody, S. (1979). Applying the logic of sample surveys to qualitative case studies: The case cluster method. *Administrative Science Quarterly*, *24*(4), 612–629. doi:10.2307/2392367

McDermott, R. (1999). Why Information Technology Inspired But Cannot Deliver Knowledge Management. *California Management Review*, *41*(4), 103–117.

McElroy, M. V. (2001). Social innovation capital draft. *Macroinnovation Associates*, July, 1-14.

McElroy, M. W. (2002). Social innovation capital. *Journal of Intellectual Capital*, *3*(1), 30–39. doi:10.1108/14691930210412827

McEvily, S. K., & Chakravarthy, B. (2002). The Persistence of Knowledge-Based Advantage: An Empirical Test for Product Performance and Technological Knowledge. *Strategic Management Journal*, *23*, 285–305. doi:10.1002/smj.223

McEvily, S. K., Eisenhardt, K. M., & Prescott, J. E. (2004). The Global Acquisition, Leverage, and Protection of Technological Competencies. *Strategic Management Journal*, *25*(8/9), 713–722. doi:10.1002/smj.425

McGahan, A. M., & Silverman, B. S. (2006). Profiting from technological innovation by others: The effect of competitor patenting on firm value. *Research Policy*, *31*, 1222–1242. doi:10.1016/j.respol.2006.09.006

Meyer, A. D. (1991). What is strategy's distinctive competence? *Journal of Management*, *17*, 821–833. doi:10.1177/014920639101700413

Miles, M. B., & Huberman, A. M. (1984). *Analyzing qualitative data: A source book for new methods*. Beverly Hill, CA: Sage.

Miller, M., DuPont, B. D., Fera, V., Jeffrey, R., Mahon, B., Payer, B. M., & Starr, A. (1999). *Measuring and reporting intellectual capital from a diverse Canadian industry perspective: experiences, issues and prospects*. Paper presented at the International Symposium – Measuring and Reporting Intellectual Capital: Experience, Issues, and Prospects, 9-11 June, Amsterdam.

Milliron, M. D. (2007). Transcendence and globalization: our education and workforce development challenge. *New Directions for Community Colleges, Summer* (138).

Ministry of Economy and Finance. (2005). *Economic statistics*. Panama: Ministry of Economy and Finance.

Mintzberg, H. (1979). *The Structure of Organizations*. Englewood Cliffs, NJ: Prentice-Hall.

Mizuhara, N. (2007). Meeting the challenge of creating a new material for producing environmentally friendly, small, and high-efficiency devices. *Sumitomo Electric Industries, R&D Professionals,* from http://www.sei.co.jp/RandD_e/researcher/012/index.html

Montes, J. L., Moreno, A. R., & Morales, V. G. (2005). Influence of Support Leadership and Teamwork Cohesion on Organizational Learning, Innovation and Performance: An Empirical Examination. *Technovation*, *25*(10), 1159–1172. doi:10.1016/j.technovation.2004.05.002

Moran, P. (2005). Structural vs. relational embeddedness: Social capital and managerial performance. *Strategic Management Journal*, *26*(12), 1129–1151. doi:10.1002/smj.486

Moran, P., & Ghoshal, S. (1996). Theories of economic organization: The case for realism and balance. *Academy of Management Review*, *21*(1), 58–72. doi:10.2307/258629

Mortensen, J., Eustace, C., & Lannoo, K. (1997). *Intangibles in the European economy: a review essay.* Paper presented at the Workshop on Intangibles in the European Economy, CEPS, Brussels.

Motohashi, K. (2005). University-industry collaborations in Japan: The role of new technology-based firms in transforming the national innovation system. *Research Policy*, *34*, 583–594. doi:10.1016/j.respol.2005.03.001

Moulaert, F., & Sekia, F. (2003). Territorial Innovation models: a critical survey. *Regional Studies, 37*(3), 289–302.

Mouritsen, J. (2002). Developing and managing knowledge through IC statements. *Journal of Intellectual Capital, 3*(1), 10–29. doi:10.1108/14691930210412818

Mouritsen, J. (2004). Measuring and intervening: how do we theorize intellectual capital management? *Journal of Intellectual Capital, 5*(2), 257–267. doi:10.1108/14691930410533687

Mouritsen, J., & Larsen, H. T. (2005). The 2nd wave of knowledge management: the management control of knowledge resources through intellectual capital information. *Journal of Management Accounting Research, 16*, 371–394. doi:10.1016/j.mar.2005.06.006

Mouritsen, J., Bukh, P. N., Larsen, H. T., & Johansen, M. R. (2002). Developing and managing knowledge through intellectual capital statements. *Journal of Intellectual Capital, 3*(1), 10–29. doi:10.1108/14691930210412818

Mouritsen, J., Larsen, H. T., & Bukh, P. N. (2005). Dealing with the knowledge economy: intellectual capital versus balanced scorecard. *Journal of Intellectual Capital, 6*(1), 8–27. doi:10.1108/14691930510574636

Mouritsen, J., Larsen, H. T., & Bukh, P. N. D. (2001). *Intellectual capital and the capable firm: narrating, visualizing and numbering for managing knowledge* (*Vol. 26*). Accounting, Organizations and Society.

Mouritsen, J., Larsen, H., & Bukh, P. (2002). Understanding intellectual capital statements: designing and communicating knowledge management strategies. In N. Bontis (Ed.), *World Congress on intellectual capital readings,* (pp. 179-202). Boston: Butterworth-Heinemann.

Muñoz, A., & Cordón, E. (2002). Tamaño, Estructura e Innovación Organizacional. *Revista Europea de Dirección y Economía de la Empresa, 11*, 103–120.

Mussati, G., Pedrana, M., & Postigliola, M. (2007). *Country Report: Italy*. Retrieved July 7th, 2009, from http://www.clusterobservatory.eu/upload/

Mustafa, M. (2003). Internet Access. *World Bank Public Policy for the Private Sector,* 256.

Myers, S., & Marquis, D. G. (1969). *Successful Industrial Innovation.* Washington, DC: National Science Foundation.

Nabil, A., Awerbuch, B., Slonim, J., Wegner, P., & Yesha, Y. (1997). Computer database: globalizing business, education, culture through the Internet. *Communications of the ACM, 40*(2).

Nag, J., Corley, K. G., & Gioia, D. A. (2007). The intersection of Organizational Identity, Knowledge, and Practice: Attempting strategic change via knowledge grafting. *Academy of Management Journal, 50*(4), 821–847. doi:10.2307/20159892

Nahapiet, J., & Goshal, S. (1998). Social Capital, Intellectual Capital and the organisational advantage. *Academy of Management Review, 23*, 242–266. doi:10.2307/259373

Nakagawa, M. (2008). *An empirical analysis in nonferrous metal technology: Change in technology spillover structure in Japan's nonferrous metal industry.* Unpublished doctoral dissertation, Tokyo Institute of Technology, Tokyo, Japan.

Nakagawa, M., & Watanabe, C. (2007). Moving beyond organizational inertia as a survival strategy for resource-based industry in a service-oriented economy: Lessons from cross-sector technology spillover in the nonferrous metal industry. *Journal of Service Research, 7*(1), 7–35.

Nakagawa, M., & Watanabe, C. (in press). A revival of compound semiconductor material innovation: Strategic technology spillovers in Japan's nonferrous metal industry in the 2000s. *Journal of Service Research.*

Nakagawa, M., Watanabe, C., & Griffy-Brown, C. (2009). Changes in the technology spillover structure due to economic paradigm shifts: A driver of the economic revival in Japan's material industry beyond the year 2000. *Technovation, 29*, 5–22. doi:10.1016/j.technovation.2008.08.003

Nakanishi, Y. (2002). Empirical evidence of inter-industry R&D spillover in Japan. *Journal of Economic Research*, *7*, 91–104.

National Academy of Sciences. (2005). *Rising above the gathering storm: Engineering and employing America for a brighter economic future*. Washington, DC: National Academy Press.

Nauwelaers, C. (2001). Path-dependency and the role of institutions in cluster policy generation. In Mariussen, A. (Ed.), *Cluster Policies—Cluster Development?* (pp. 93–107). Stockholm, Sweden: Nordregio.

Nauwelaers, C. (2003, May). *Innovative Hot Spots in Europe: Policies to promote trans-border clusters of creative activity.* Paper presented at Trend Chart Policy Workshop, Luxemburg.

Nelson, R. R. (1991). Why do firms differ, and how does it matter? *Strategic Management Journal*, *12*(8), 61–74. doi:10.1002/smj.4250121006

Nelson, R. R., & Winter, S. G. (1982). *An Evolutionary Theory of Economic Change*. Cambridge, UK: Belknap Press.

Nelson, R.R. (1991). Why Do Firms Differ, and How Does it Matter? *Strategic Management Journal, 12*(Winter Special Issue), 61-74.

Nerkar, A., & Roberts, P. W. (2004). Technological and Product-Market Experience and the Success of New Product Introductions in the Pharmaceutical Industry. *Strategic Management Journal*, *25*, 779–800. doi:10.1002/smj.417

Nickerson, J. A., & Zenger, T. R. (2004). A knowledge-based theory of governance choice—A problem-solving approach. *Organization Science*, *15*(6), 617–632. doi:10.1287/orsc.1040.0093

Nickerson, J. A., Silverman, B. S., & Zenger, T. R. (2007). The 'problem' of creating and capturing value. *Strategic Organization*, *5*(3), 211–225. doi:10.1177/1476127007079969

Nieto, M. (2001). *Bases para el Estudio del Proceso de Innovación Tecnológica en la Empresa*. León, Spain: Universidad de León.

Nieto, M. J., & Santamaria, L. (2007). The Importance of Diverse Collaborative Networks for the Novelty of Product Innovation. *Technovation*, *27*, 367–377. doi:10.1016/j.technovation.2006.10.001

Nonaka, I. (1991). The Knowledge-Creating Company. *Harvard Business Review*, *69*, 96–104.

Nonaka, I. (1994). A Dynamic Theory of Organizational Knowledge Creation. *Organization Science*, *5*(1), 14–37. doi:10.1287/orsc.5.1.14

Nonaka, I. (2002). A Dynamic Theory of Organizational Knowledge Creation. In Choo, C. W., & Bondis, N. (Eds.), *The Strategic Management of Intellectual capital and Organizational Knowledge* (pp. 437–462). New York: Oxford University Press.

Nonaka, I., & Takeuchi, H. (1995). *The Knowledge-Creating Company: How Japanese Companies Create the Dynamics of Innovation*. Oxford, UK: Oxford University Press.

Nonaka, I., Reinmoller, P., & Toyama, R. (2001). Integrated information technology systems for knowledge creation. In Dierkes, M., Berthoin-Antal, A., Child, J., & Nonaka, I. (Eds.), *Handbook of Organizational Learning and Knowledge*. Oxford, UK: Oxford University Press.

Nordhaug, O. (1998). Competence specificities in organizations. *International Studies of Management and Organization*, *28*(1), 8–27.

O'Donnell, D. (2003). Human interaction: The critical source of intangible value. *Journal of Intellectual Capital*, *4*(1), 82–99. doi:10.1108/14691930310455405

O'Donnell, D., & O'Regan, P. (2000). The structural dimensions of intellectual capital: emerging challenges for management and accounting. *Southern African Business Review*, *4*(2), 14–20.

O'Meara, P., & Mehlinger, H. D. D.H & Newman R. M., (2001). Changing perspectives on international education. Bloomington, IN: Indiana University Press.

Obstfeld, D. (2005). Social Networks, the *Tertius Iungens* Orientation, and Involvement in Innovation. *Administrative Science Quarterly*, *50*(1), 100–130.

OECD. (1981). *Proposed standard, practice for surveys of research and experimental development- Frascati Manual*. Paris: OECD.

OECD. (1992). *OECD proposed guidelines for collecting and interpreting technological innovation data- Oslo manual*. Paris: OECD.

OECD. (1997). *Oslo Manual - proposed guidelines for collecting and interpreting technological innovation data*. Paris: OECD.

OECD. (2006). *Creating Value from Intellectual Assets*. Paris: Author.

OECD. (2006). *Manual de Oslo. Guía para la recogida e interpretación de datos sobre innovación* (3rd ed.). París: OECD and Eurostat.

OECD. (2007). *Reviews of regional innovation. Competitive regional clusters. National policy approaches*. Paris, France: OECD.

OECD. (2008a). *OECD ICT (Information and Communication Technology) outlook*. Paris: OECD.

OECD. (2008b). *OECD STI (Science, Technology and Innovation) outlook*. Paris: OECD.

Oh, H., Labianca, G., & Chung, M. H. (2006). A multilevel model of group social capital. *Academy of Management Review*, *31*(3), 569–582. doi:10.2307/20159229

Ojanen, V., & Vuola, O. (2006). Coping with the multiple dimensions of R&D performance analysis. *International Journal of Technology Management*, *33*(2), 279–290. doi:10.1504/IJTM.2006.008315

Ordonez de Pablos, P. (2002). Evidence of intellectual capital measurement from Asia, Europe and the middle-East. *Journal of Intellectual Capital*, *3*(3), 287–302. doi:10.1108/14691930210435624

Ordóñez de Pablos, P. (2003). Intellectual capital reporting in Spain: a comparative view. *Journal of Intellectual Capital*, *4*(1), 61–81. doi:10.1108/14691930310455397

Ordonez de Pablos, P. (2004). Measuring and reporting structural capital - Lessons from European learning firms. *Journal of Intellectual Capital*, *5*(4), 629–647. doi:10.1108/14691930410567059

Ordonez de Pablos, P. (2005). Intellectual capital reports in India: lessons from a case study. *Journal of Intellectual Capital*, *6*(1), 141–149. doi:10.1108/14691930510574717

Organization for Economic Co-operation and Development (OECD). (1999). Guidelines and instructions for OECD Symposium. In *International Symposium Measuring Reporting Intellectual Capital: Experiences, Issues, and Prospects*, June, Amsterdam, OECD, Paris.

Orlikowski, W. J., & Yates, J. (1994). Genre repertoire: The structuring of communicative practices in organizations. *Administrative Science Quarterly*, *39*(4), 541–574. doi:10.2307/2393771

Ornaghi, C. (2006). Spillovers in product and process innovation: Evidence from manufacturing firms. *International Journal of Industrial Organization*, *24*, 349–380. doi:10.1016/j.ijindorg.2005.07.002

Oxley, J. (1997). Appropriability hazards and governance in strategic alliances: A transaction cost approach. *Journal of Law Economics and Organization*, *13*(2), 387–409.

Palacios, T. M. B., & Galvan, R. S. (2007). Intangible measurement guidelines: a comparative study in Europe. *Journal of Intellectual Capital*, *8*(2), 192–204. doi:10.1108/14691930710742790

Pang, N. S. (2006). *Globalization: Educational research, change and reform. SHA TIN, N.T.* Hong Kong: The Chinese University Press.

Park, S. H. (2001). Regional innovation strategies in the knowledge-based economy. *GeoJournal*, *53*(1), 29–38.

Park, Y., & Kim, S. (2006). Knowledge Management System for Fourth Generation R&D. *Technovation*, *26*, 595–602. doi:10.1016/j.technovation.2004.10.008

Pearse, N. J. (2009). The role of experiences in creating and developing intellectual capital. *Management Research News*, *32*(4), 371–382. doi:10.1108/01409170910944317

Pedrini, M. (2007). Human capital convergence in intellectual capital and sustainability reports. *Journal of Intellectual Capital*, *8*(2), 346–354. doi:10.1108/14691930710742880

Peng, T. A., Pike, S., & Roos, G. (2005). Intellectual capital and performance indicators- Taiwanese healthcare sector. *Journal of Intellectual Capital*, *8*(3), 538–556. doi:10.1108/14691930710774902

Penrose, E. T. (1959). *The Theory of the Growth of the Firm*. Oxford, UK: Basil Blackwell.

Peppard, J., & Rylander, A. (2001). Using an intellectual capital perspective to design and implement a growth strategy: the case of Apion. *European Management Journal*, *19*(5), 510–525. doi:10.1016/S0263-2373(01)00065-2

Peteraf, M. A. (1993). The cornerstones of competitive advantage – a Resource-based view. *Strategic Management Journal*, *14*(3), 179–191. doi:10.1002/smj.4250140303

Peteraf, M., & Barney, J. B. (2003). Unrevailing the Resource-Based Tangle. *Managerial and Decision Economics*, *24*, 309–323. doi:10.1002/mde.1126

Petrash, G. (1996). Dow's journey to a knowledge value management culture. *European Management Journal*, *14*(4), 365–373. doi:10.1016/0263-2373(96)00023-0

Petty, R., & Guthrie, J. (2000). Intellectual Capital: literature review – measurement, reporting and management. *Journal of Intellectual Capital*, *1*(2), 155–169. doi:10.1108/14691930010348731

Petty, R., & Guthrie, J. (2000b). Knowledge management and the measurement of intangibles. *Management Today*, March.

Phene, A., Fladmoe-Lindquist, K., & Marsh, L. (2006). Breakthrough Innovations in the U.S. Biotechnology Industry: The Effects of Technological Space and Geographic Origin. *Strategic Management Journal*, *27*, 369–388. doi:10.1002/smj.522

Piccaluga, A., & Cesaroni, F. (2003). *Distretti industriali e distretti tecnologici. Modelli possibili per il Mezzogiorno*. Milano, Italy: Franco Angeli.

Pike, S., Fernstrom, L., & Roos, G. (2005a). Intellectual capital management approach in ICS Ltd. *Journal of Intellectual Capital*, *6*(4), 489–509. doi:10.1108/14691930510628780

Pike, S., Roos, G., & Marr, B. (2005). Strategic Management of Intangible Assets and Value Drivers in R&D Organizations. *R & D Management*, *35*, 111–124. doi:10.1111/j.1467-9310.2005.00377.x

Pike, S., Roos, G., & Rylander, A. (2002). Intellectual capital management and disclosure. In Bontis, N., & Choo, C. W. (Eds.), *The Strategic Management of Intellectual Capital and Organizational Knowledge (pp. 657-73)*. New York: Oxford University Press.

Pisano, G.P. (1994). Knowledge, Integration, and the Locus of Learning: An Empirical Analysis of Process Development. *Strategic Management Journal*, *15*(Winter Special Issue), 85-100.

Pizarro, I., Real, J. C., & De la Rosa, M. D. (2007). *El Papel del Capital Humano y la Cultura Emprendedora en la Innovación*. ACEDE Conference, Sevilla.

Polanyi, M. (1958). *Personal Knowledge: Towards a Post-Critical Philosophy*. Chicago: University of Chicago Press.

Polanyi, M. (1969). *Knowing and being*. Chicago: University of Chicago Press.

Porter, L. W. (1996). Forty years of organizational studies: Reflections from a micro perspective. *Administrative Science Quarterly*, *41*, 262–269. doi:10.2307/2393718

Porter, M. (1990). *The competitive advantage of nations*. London: Macmillan.

Porter, M. (1998a). *On competition*. Boston: Harvard Business School Press.

Porter, M. (1998b). Clusters and the new economics of competition. *Harvard Business Review*, *76*(6), 77–90.

Porter, M. (2000b). Location, competition and economic development: Local clusters in a global economy. *Economic Development Quarterly*, *14*(1), 15–34.

Porter, M. E. (1980). *Competitive Strategy*. New York: The Free press.

Porter, M. E. (1981). The Contribution of Industrial Organization to Strategic Management. *Academy of Management Review*, *6*(4), 609–620. doi:10.2307/257639

Porter, M. E. (1996). What is Strategy? *Harvard Business Review*, *74*(6), 61–78.

Porter, M. E. (2000a). Locations, clusters and company strategy. In Clark, G., Feldman, M., & Gertler, M. (Eds.), *The Oxford Handbook of Economic Geography* (pp. 253–274). Oxford, UK: Oxford University Press.

Porter, T. B. (2001, August). *Theorizing Organizational Identity*. Paper presented at the meeting of Academy of Management Meeting, Washington, D. C.

Portes, A. (1998). Social Capital: Its Origin and Application in Modern Sociology. *Annual Review of Sociology*, *24*(1), 1–24. doi:10.1146/annurev.soc.24.1.1

Powell, W., Koput, K., & Smith-Doerr, L. (1996). Interorganizational collaboration and the locus of innovation: networks of learning in biotechnology. *Administrative Science Quarterly*, *41*(1), 116–145. doi:10.2307/2393988

Prahalad, C. K., & Hamel, G. (1990). The core competence of the corporation. *Harvard Business Review*, *68*(3), 79–91.

Prahalad, C. K., & Ramaswamy, V. (2004). *The Future of Competition: Co-Creating Unique Value with Customers*. Boston: Harvard Business School Press.

Prajogo, D. I., & Ahmed, P. K. (2006). Relationships between Innovation Stimulus, Innovation Capacity, and Innovation Performance. *R & D Management*, *36*, 499–515. doi:10.1111/j.1467-9310.2006.00450.x

Priem, R. L., & Butler, J. E. (2001). Is the Resource-Based "View" a useful perspective for Strategic Management Research? *Academy of Management Review*, *26*, 22–40. doi:10.2307/259392

Priem, R. L., & Butler, J. E. (2001a). Is the resource-based 'view' a useful prospective for strategic management research? *Academy of Management Review*, *26*(1), 22–40. doi:10.2307/259392

Priem, R. L., & Butler, J. E. (2001b). Tautology in the resource-based view and the implications of externality determined resource value: further comments. *Academy of Management Review*, *26*(1), 57–66. doi:10.2307/259394

Priem, R., & Butler, J. (2001). Is the Resource-Based View a Useful Perspective for Strategic Management Research? *Academy of Management Review*, *26*, 22–40. doi:10.2307/259392

Prusak, L. (1998). *Working Knowledge: How Organizations Manage What They Know*. Cambridge, MA: Harvard Business School Press.

Publishers, Inc.

Pulic, A. (2000). *MVA and VAICTM Analysis of Randomly Selected Companies from FTSE 250*. Graz, London: Austrian Intellectual Capital Research Center.

Pulic, A. (2004). Intellectual capital: does it create or destroy value? *Measuring Business Excellence*, *8*(1), 62–68. doi:10.1108/13683040410524757

Putnam, R. (1993). The prosperous community: social capital and public life. *The American Prospect*, *13*, 35–42.

Putnam, R. D. (1993). *Making Democracy Work: Civic Traditions in Modern Italy*. Princeton, NJ: Princeton University Press.

Putnam, R. D. (1995). Bowling alone: America's declining social capital. *Journal of Democracy*, *6*(1), 65–78. doi:10.1353/jod.1995.0002

Putnam, R. D. (2000). *Bowling Alone: The Collapse and Revival of American Community*. New York: Simon & Schuster.

Quinn, J. B. (1992). *Intelligent Enterprise: A Knowledge and Service Based Paradigm for Industry*. New York: The Free Press.

Rabinow, P., & Sullivan, W. M. (1979). *Interpretive Social Science: A reader*. Berkeley, CA: University of California Press.

Raines, P. (2000). *Developing Cluster Policies in Seven European Regions* (Regional and Industrial Policy Research Paper No. 42). Glasgow, UK: University of Strathclyde, European Policies Research Centre.

Raines, P. (2001). *The Cluster Approach and the Dynamics of Regional Policy-Making* (Regional and Industrial Policy Research Paper No. 47). Glasgow, UK: University of Strathclyde, European Policies Research Centre.

Raines, P. (2002). Clusters and prisms. In Raines, P. (Ed.), *Cluster Development and Policy* (pp. 159–177). Aldershot, UK: Ashgate.

Rao, H., Monin, P., & Durand, R. (2003). Institutional change in Toque Ville: Nouvelle cuisine as an identity movement in French gastronomy. *American Journal of Sociology*, *108*(4), 795–843. doi:10.1086/367917

Rao, L., & Osei-Bryson, K. M. (2007). Towards defining dimensions of knowledge systems quality. *Expert Systems with Applications*, *33*, 368–378. doi:10.1016/j.eswa.2006.05.003

Rastogi, P. (2000b). Sustaining enterprise competitiveness – is human capital the answer? *Human Systems Management*, *19*(3), 193–203.

Reagans, R., & McEvily, B. (2003). Network Structure and Knowledge Transfer: The Effects of Cohesion and Range. *Administrative Science Quarterly*, *48*(2), 240–267. doi:10.2307/3556658

Redzepagic, D., & Stubbs, P. (2006, September). *Rethinking clusters and social capital in Croatia*. Paper presented at the European Association for Comparative Economics Studies (EACES) 9th Biennial Conference, Brighton, UK.

Reed, K., Lubatkin, M., & Srinivasan, N. (2006). Proposing and Testing an Intellectual Capital-Based View of the Firm. *Journal of Management Studies*, *43*, 867–893. doi:10.1111/j.1467-6486.2006.00614.x

Reger, (1994). Reframing the organization: Why implementing total quality is easier said than done. *Academy of Management Review*, *19*(3), 565–584. doi:10.2307/258939

RICYT. (2005). The source for this adaptation is on: http://www.ricyt.org/interior/interior.asp?Nivel1=1&Nivel2=1&Idioma

Riegler, C., & Hollerschmid, C. (2006). Voluntary disclosure on project intangibles from R&D in business reporting. *Journal of Intellectual Capital*, *7*(4), 511–528. doi:10.1108/14691930610709149

Ring, P. S., & Van de Ven, A. H. (1992). Structuring cooperative relationships between organizations. *Strategic Management Journal*, *13*(7), 483–498. doi:10.1002/smj.4250130702

Risher, H., & Stopper, W. G. (2002). Attracting and retaining intellectual capital. Human Resource Planning, (March), 9.

Ritzer, G. (2003). *Teoría sociológica moderna: Quinta edición*. Madrid: McGrawHill.

Rizova, P. (2006). Are you networked for successful innovation? *MIT Sloan Management Review*, *47*(3), 49–55.

Robinson, H. S., Anumba, C. J., Carrillo, P. M., & Al-Ghassani, A. M. (2006). STEPS: a knowledge management maturity roadmap for corporate sustainability. *Business Process Management Journal*, *12*(6), 793–808. doi:10.1108/14637150610710936

Rock, D. & Schwartz, J. (2006). The Neuroscience of Leadership. *Strategy+Business*, (43), 06207.

Roelandt, T. J. A., & den Hertog, P. (1999). Cluster analysis and cluster based policy making in OECD countries. An introduction to the theme. In OECD (Ed.), Boosting Innovation: The Cluster Approach (pp. 9–23). Paris, France: OECD Publishing.

Roelandt, T. J. A., Gilsing, V. A., & van Sinderen, J. (2000, December). *New policies for the new economy. Cluster-based innovation policy: international experiences*. Paper Presented at the Fourth Annual EUNIP Conference, Tilburg, The Netherlands.

Roller, L.-H., & Sinclair-Desgagne, B. (1996). Industrial organization and business strategy on the heterogeneity of firms. *European Economic Review*, *40*, 531–539. doi:10.1016/0014-2921(95)00066-6

Romer, P. (1994). The origins of endogenous growth. *The Journal of Economic Perspectives*, *8*, 3–22.

Romer, P. (1998). Bank of America roundtable on the soft revolution. *Journal of Applied Corporate Finance*, *11*(2), 9–14.

Romijn, H., & Albu, M. (2001). *Explaining innovativeness in small high-technology firms in the United Kingdom*. Working Paper 01.01, Eindhoven Centre for Innovation Studies, Eindhoven University of Technology, Eindhoven.

Romo, F. P., & Schwartz, M. (1995). The structural embeddedness of business decisions: The migration of manufacturing plants in New York State, 1960 to 1985. *American Sociological Review*, *60*(1), 874–907. doi:10.2307/2096431

Roos, G., Pike, S., & Fernström, L. (2006). Managing Intellectual Capital in Practice. New York: Butterworth-Heinemann, an imprint of Elsevier, USA.

Roos, J. (1998). *Intellectual capital: Navigating in the New Business Landscape*. New York: University Press.

Roos, J., & Roos, G. (1997). *Intellectual Capital: Navigating the New Business Landscape*. London: Macmillan.

Roos, J., Roos, G., & Edvinsson, L. (1997). *Intellectual Capital – Navigating the New Business Landscape*. Basingstoke, UK: Macmillan.

Roos, J., Roos, G., Dragonetti, N., & Edvinsson, L. (1998). *Intellectual capital: Navigation in the new business landscape*. New York: New York University Press.

Rosenfeld, S. A. (2002). *Creating smart systems. A guide to cluster strategies in less favoured regions*. Paper Presented at a Seminar on European Union Regional Innovation Strategies, Brussels, Belgium. Retrieved July 7th, 2009, from http://europa.eu.int/comm/regional policy/innovation/pdf/

Rothwell, R. (1992). Successful industrial innovation: critical factors for the 1990's. *R & D Management*, *22*(3), 221–239. doi:10.1111/j.1467-9310.1992.tb00812.x

Rothwell, R. (1994). Towards the fifth-generation innovation process. *International Marketing Review*, *11*(1), 7–31. doi:10.1108/02651339410057491

Rouse, M. J., & Daellenbach, U. S. (1999). Rethinking Research Methods for the Resource-Based Perspective: Isolating Sources of Sustainable Competitive Advantage. *Strategic Management Journal*, *20*, 487–494. doi:10.1002/(SICI)1097-0266(199905)20:5<487::AID-SMJ26>3.0.CO;2-K

Rousseau, D. (1995). *Psychological Contracts in Organizations*. Thousand Oaks, CA: Sage.

Rousseau, D. M., & Shperling, Z. (2003). Pieces of the action: Ownership and the changing employment relationship. *Academy of Management Review*, *28*(4), 553–570. doi:10.2307/30040747

Rowley, T., Behrens, D., & Krackhardt, D. (2000). Redundant governance structures: An analysis of structural and relational embeddedness in the steel and semiconductor industries. *Strategic Management Journal*, *21*(3), 369–386. doi:10.1002/(SICI)1097-0266(200003)21:3<369::AID-SMJ93>3.0.CO;2-M

Rumelt, R. (1991). How Much Does Industry Matter? *Strategic Management Journal*, *12*, 167–185. doi:10.1002/smj.4250120302

Rumelt, R. P. (1984). Towards a Strategic Theory of the Firm. In Lamb, R. (Ed.), *Competitive Strategic Management* (pp. 556–570). Englewood Cliffs, NJ: Prentice-Hall.

Rumelt, R. P. (1987). Theory, Strategy, and Entrepreneurship. In Teece, D. (Ed.), *The Competitive Challenge: Strategies for Industrial Innovation and Renewal* (pp. 137–158). Cambridge, MA: Ballinger.

Rumelt, R. P. (1991). How much does Industry Matter? *Strategic Management Journal*, *12*(3), 167–185. doi:10.1002/smj.4250120302

Saaty, T. L. (1980). *The analytical hierarchy process*. New York: McGraw Hill.

Saaty, T. L. (1995). *Decision making for leaders*. Pittsburgh, PA: RWS Publications.

Saint Onge, H. (1996). Tacit knowledge: the key to the strategic alignment of Intellectual Capital. *Strategy and Leadership*, *24*(2), 10–14. doi:10.1108/eb054547

Saint-Onge, H. (1996). Tacit knowledge: the key to the strategic alignment of intellectual capital. *Strategy and Leadership*, *24*(2), 10–15. doi:10.1108/eb054547

Salman, N., & Saives, A. (2005). Indirect Networks: an Intangible Resource for Biotechnology Innovation. *R & D Management*, *35*, 203–215. doi:10.1111/j.1467-9310.2005.00383.x

Sampson, R. C. (2007). R & D Alliances and Firm Performance: The Impact of Technological Diversity and Alliance Organization on Innovation. *Academy of Management Journal*, *50*, 364–386. doi:10.2307/20159859

Sanchez, P., Chaminade, C., & Olea, M. (2000). Management of intangibles: An attempt to build a theory. *Journal of Intellectual Capital*, *1*(4), 312–327. doi:10.1108/14691930010359225

Sanchez-Canizares, S. M., Munoz, M. A. A., & Lopez-Guzman, T. (2007). Organizational culture and intellectual capital: a new model. *Journal of Intellectual Capital*, *8*(3), 409–430. doi:10.1108/14691930710774849

Sawhney, M., Verona, G., & Prandelli, E. (2005). Collaborating to create: the internet as a platform for customer engagement in product innovation. *Journal of Interactive Marketing*, *19*(4), 4–17. doi:10.1002/dir.20046

Schiff, M. (1992). Social Capital, Labor, Mobility, and Welfare: The Impact of Uniting States. *Rationality and Society*, *4*(2), 157–175. doi:10.1177/1043463192004002003

Schiuma, G., Lerro, A., & Carlucci, D. (2008). The Knoware Tree and the Regional Intellectual Capital Index: An assessment within Italy. *Journal of Intellectual Capital*, *9*(2), 283–300. doi:10.1108/14691930810870346

Schmitz, H. (2004). Globalized localities: introduction. In Schmitz, H. (Ed.), *Local Enterprises in the Global Economy* (pp. 1–19). Cheltenham, UK: Edward Elgar.

Schneider, A., & Samkin, G. (2008). Intellectual capital reporting by the New Zealand local government sector. *Journal of Intellectual Capital*, *9*(3), 456–486.

Schultz, M., & Hatch, M. J. (1996). Living with multiple paradigms: The case of paradigm interplay in organizational culture studies. *Academy of Management Review*, *21*(2), 529–527. doi:10.2307/258671

Schumpeter, J. (1934). *The Theory of Economic Development*. Cambridge, MA: Harvard University Press.

Schumpeter, J. A. (1912). Teoría del Desenvolvimiento Económico, ed. Fondo de Cultura Económica, México, tercera edición.

Schumpeter, J. A. (1942). *Capitalism, Socialism and Democracy*. London: G. Allen & Unwin Ltd.

Schwandt, T. A. (1998). Constructivist, Interpretivist Approaches to Human Inquiry. In Denzin, N. K., & Lincoln, I. S. (Eds.), *Handbook of Qualitative Research* (pp. 221–259). Thousand Oaks, CA: Sage Publication.

Schwandt, T. A. (2000). Three epistemological stances for qualitative inquiry. In Denzin, N. K., & Lincoln, I. S. (Eds.), *Handbook of Qualitative Research* (2nd ed., pp. 189–213). Thousand Oaks, CA: Sage Publication.

Science|Business Innovation Board. (2009). *Clustering For Growth. How to build dynamic innovation clusters in Europe*. Retrieved July 7th, 2009, from ww.sciencebusiness.net/documents/

Scott, S. G., & Lane, V. R. (2000). A stakeholder approach to organizational identity. *Academy of Management Review*, *25*(1), 43–62. doi:10.2307/259262

Searls, D. (2007). It's the Relationship, smarty. *Weblog*, December 26. Retrieved January 3, 2008.

Seetharaman, A., Low, K. L., & Saravanan, A. S. (2004). Comparative justification on intellectual capital. *Journal of Intellectual Capital*, *5*, 522–539. doi:10.1108/14691930410566997

Seleim, A., Ashour, A., & Bontis, N. (2004). Intellectual capital in Egyptian software firms. *The Learning Organization*, *11*(4/5), 332–346. doi:10.1108/09696470410538233

SENACYT. (2002). *National Strategic Plan for the Development of Science, Technology and Innovation. Unpublished preliminary proposal*. Panama: SENACYT.

SENACYT. (2004). *Science and Technology Activities Report. Unpublished preliminary report*. Panama: SENACYT.

Senge, P. M. (1990). *The Fifth Discipline: the Art and Practice of the Learning Organization*. New York: Doubleday.

Serenko, A., & Bontis, N. (2004). Meta-review of knowledge management and intellectual capital literature: citation impact and research productivity rankings. *Knowledge and Process Management, 11*(3), 185–198.

Shapiro, C., & Varian, H. R. (1999). *Information Rules: A Strategic Guide to the Network Economy*. Boston: Harvard Business School Press.

Shearman, C., & Burrell, G. (1988). New technology-based firms and the emergence of new industries: some employment implications. *New Technology, Work and Employment, 3*(2), 87–99. doi:10.1111/j.1468-005X.1988.tb00092.x

Simon, E. S., McKeough, D. T., Ayers, A. D., Rinehart, E., & Alexia, B. (2003). How do You Best Organize for Radical Innovation? *Research Technology Management, 46*(5), 17–20.

Simon, H. (1945). *Administrative Behavior*. New York: The Free Press.

Sirilli, G., & Evangelista, R. (1998). Technological innovation in services and manufacturing: results from Italian surveys. *Research Policy, 27*, 881–899. doi:10.1016/S0048-7333(98)00084-5

Skandia. (1994). *Visualizing intellectual capital in Skandia*. Supplement to Skandia's 1994 Annual Report, Skandia, Stockholm.

Skandia. (1995a). *Renewal and development: intellectual capital*. Supplement to Skandia's 1995 Interim Annual Report, Skandia, Stockholm.

Skandia. (1995b). *Value-creating processes: intellectual capital*. Supplement to Skandia's 1995 Annual Report, Skandia, Stockholm.

Skandia. (1996a). *Power of innovation: intellectual capital*. Supplement to Skandia's 1996 Interim Annual Report, Skandia, Stockholm.

Skandia. (1996b). *Customer value*. Supplement to Skandia's 1996 Annual Report, Skandia, Stockholm.

Skandia. (1997). *Intelligent enterprising*. Supplement to Skandia's 1997 Annual Report, Skandia, Stockholm.

Skandia. (1998). *Human capital in transformation*. Supplement to Skandia's 1998 Annual Report, Skandia, Stockholm.

Smilor, R. W. (1990). University spin-out companies.- Technology Starts ups from university of Texas at Austin. *Journal of Business Venturing, 5*(1), 63–76. doi:10.1016/0883-9026(90)90027-Q

Smircich, L. (1983). Concepts of Culture and Organizational Analysis. *Administrative Science Quarterly, 28*(3), 339–358. doi:10.2307/2392246

Smith, R., & Sharif, N. (2007). Understanding and Acquiring Technology Assets for Global Competition. *Technovation, 27*, 643–649. doi:10.1016/j.technovation.2007.04.001

Society of Management Accountants of Canada (SMAC). (1998). *The management of intellectual capital: the issues and the practice, (Paper No. 16)*. Hamilton, Canada: The Society of Management Accountants of Canada.

Solomons, D. (1991). Accounting and social change: a neutralist view. *Accounting, Organizations and Society, 16*(3), 287–295. doi:10.1016/0361-3682(91)90005-Y

Solow, R. M. (1957). Technical change and the aggregate production function. *The Review of Economics and Statistics, 39*(3), 312–320. doi:10.2307/1926047

Spender, J. C. (1996). Making Knowledge the Basic of a Dynamic Theory of the Firm. *Strategic Management Journal*, (17): 45–62.

Spender, J. C. (1996a). Making Knowledge the Basis of a Dynamic Theory of the Firm. *Strategic Management Journal, 17*, 45–62.

Spender, J. C. (1996b). Organizational Knowledge, Learning, and Memory: Three Concepts in Search of a Theory. *Journal of Organizational Change Management, 9*, 63–78. doi:10.1108/09534819610156813

Stacey, R. D., Griffin, D., & Shaw, P. (2000). *Complexity and management: Fad or radical challenge to systems thinking?* London: Rutledge.

Steiner, M. (Ed.). (1998). *Clusters and Regional Specialisation: On Geography, Technology and Networks.* London: Pion.

Stevens, A., & Price, T. (1996). *Evolutionary Psychiatry: A New Beginning.* New York: Routledge.

Stewart, D. W. (1981). The application and misapplication of factor analysis in marketing research. *JMR, Journal of Marketing Research, 18*(1), 51–62. doi:10.2307/3151313

Stewart, T. A. (1991). Brainpower. *Fortune, 123*(11), 44–60.

Stewart, T. A. (1997). *Intellectual Capital: The New Wealth of Organizations.* New York: Doubleday.

Stewart, T. A. (1999). *Intellectual Capital: The New Wealth of Organizations.* New York: Currency/Doubleday.

Stewart, T. A. (2001). *The Wealth of Knowledge: Intellectual Capital and the Twenty-First Century Organization.* London: Nicholas Brealey.

Stieglitz, N., & Heine, K. (2007). Innovations and the Role of Complementarities in a Strategic Theory of the Firm. *Strategic Management Journal, 28*, 1–15. doi:10.1002/smj.565

Stromquist, N. P. (2002). *Education in a globalized world: The connectivity of economic power, technology, and knowledge.* Lanham, MD: Rowman & Littlefield.

Stuart, T. (2000). Interorganizational Alliances and the Performance of Firms: A Study of Growth and Innovation Rates in a High-Technology Industry. *Strategic Management Journal, 21*, 791–811. doi:10.1002/1097-0266(200008)21:8<791::AID-SMJ121>3.0.CO;2-K

Suarez-Orozco, M. (2007). *Learning in the global era: international perspectives on globalization and education.* Los Angeles, CA: University of California Press.

Suarez-Orozco, M. M., & Quin-Hillard, D. B. (2004). *Globalization culture and education in the new millennium.* London: University of California Press, Ltd.

Suarez-Orozco, M., & Sattin, C. (2007). *Learning in the global era: International perspectives on globalization and education.* Los Angeles, CA: University of California Press.

Subramaniam, M., & Venkatraman, N. (2001). Determinants of transnational new product development capability: testing the influence of transferring and deploying tacit overseas knowledge. *Strategic Management Journal, 22*(4), 359–378.

Subramaniam, M., & Youndt, M. A. (2005). The Influence of Intellectual Capital on the Types of Innovative Capabilities. *Academy of Management Journal, 48*, 450–463. doi:10.2307/20159670

Sullivan, P. H. (2000). *Value-driven Intellectual Capital. How to Convert Intangible Corporate Assets into Market Value.* London: John Wiley and Sons.

Sullivan, P. H. (2001). *Rentabilizar el Capital Intelectual. Técnicas para Optimizar el Valor de la Innovación.* Buenos Aires: Paidós.

Sutton, R., & Callahan, A. (1987). The Stigma of Bankruptcy: Spoiled Organizational Image and Its Management. *Academy of Management Journal, 30*, 405–436. doi:10.2307/256007

Sveiby, K. (1997). *The New Organizational Wealth.* New York: Berret-Koehler.

Sveiby, K. E. (1988). *Den nya AÊ rsredovisningen [The New Annual Report]* (in Swedish). Stockholm, Pub. details, see www.sveiby.com.au for English translation.

Sveiby, K. E. (1997). *The new organizational wealth: Managing and measuring knowledge-based assets*. San Francisco: Berrett-Koehler.

Sveiby, K. E. (1997b). The intangible assets monitor. *Journal of Human Resource Costing & Accounting*, *1*, 73–97. doi:10.1108/eb029036

Sveiby, K. E. (1998). Intellectual capital: thinking ahead. Australian CPA, June (pp. 18-22).

Sveiby, K. E. (2005). *Method of measuring intangible assets* (pp. 1-8). Retrieved January 2, 2006 from www.sveiby.com/articles/IntangibleMethods.htm

Sveiby, K.-E. (1997). *The new organizational wealth: Managing and measuring knowledge-based assets*. San Francisco: Berrett-Koehler.

Sveiby, K.E., (2001). A knowledge based theory of the firm to guide strategy formulation. *Journal of Intellectual Capital*, (2/4).

Swanson, W. H. (2009, Jan. 5). A plea to upgrade education. [Letter to the editor]. *Aviation Week & Space Technology*, 58.

Swart, J. (2006). Intellectual Capital: disentangling an enigmatic concept. *Journal of Intellectual Capital*, *7*(2), 136–159. doi:10.1108/14691930610661827

Swart, J. (2006). Intellectual Capital: Disentangling an Enigmatic Concept. *Journal of Intellectual Capital*, *7*, 136–159. doi:10.1108/14691930610661827

Taylor, F. W. (1911). *The Principles of Scientific Management*. New York: Harper & Brothers.

Teece, D. (1998). Capturing Value from Knowledge Assets: The New Economy, Markets for Know-How and Intangible Assets. *California Management Review*, *40*, 55–79.

Teece, D. J. (1986). Profiting from technological innovation: Implications for integration, collaboration, licensing and public policy. *Research Policy*, *15*(6), 285–305. doi:10.1016/0048-7333(86)90027-2

Teece, D. J. (1996). Firm Organization, Industrial Structure, and Technological Innovation. *Journal of Economic Behavior & Organization*, *31*(2), 193–224. doi:10.1016/S0167-2681(96)00895-5

Teece, D. J. (2000). *Managing Intellectual Capital: Organizational, Strategic, and Policy Dimensions*. Oxford, UK: Oxford University Press.

Teece, D. J. (2000). Strategies for managing knowledge assets: the role of firm structure and industrial context. *Long Range Planning*, *33*, 35–54. doi:10.1016/S0024-6301(99)00117-X

Teece, D. J. (2006). Reflections on "Profiting from innovation.". *Research Policy*, *35*(8), 1131–1146. doi:10.1016/j.respol.2006.09.009

Teece, D. J. (2007). Explicating dynamic capabilities. The Nature and Microfoundations of (Sustainable) Enterprise Performance. *Strategic Management Journal*, *28*(13), 1319–1350. doi:10.1002/smj.640

Teece, D. J., Pisano, G., & Shuen, A. (1997). Dynamic capabilities and strategic management. *Strategic Management Journal*, *18*(7), 509–533. doi:10.1002/(SICI)1097-0266(199708)18:7<509::AID-SMJ882>3.0.CO;2-Z

Teece, D.J. (2004). *The emerging science of Strategic Management*. Prepared for a panel discussion at Lappeenranta University of Technology May 13, 2004.

The Swedish Coalition of Service Industries (SCSI). (1995). *Valuation of Service Companies*. Stockholm: Author.

Thiétart, R. A. (2001). *Doing Management Research. A Comprehensive Guide*. London: SAGE Publications.

Thomke, S., & von Hippel, E. (2002). Customers as Innovators: A New Way to Create Value. *Harvard Business Review*, *80*(2), 5–11.

Tidd, J. (2001). Innovation Management in Context: Environment, Organiztion and Performance. *International Journal of Management Reviews*, *3*, 169–183. doi:10.1111/1468-2370.00062

Tidd, J., Bessant, J., & Pavitt, K. (2005). *Managing Innovation: Integrating Technological, Market and Organizational Change*. Chichester, UK: John Wiley & Sons.

Tippins, M., & Sohi, R. (2003). IT Competency and Firm Performance: Is Organizational Learning a Missing Link? *Strategic Management Journal, 24*, 745–761. doi:10.1002/smj.337

Todtling, F., Lehner, P., & Kaufmann, A. (2009). Do different types of innovaiton rely on specific kinds of knowledge interactions? *Technovation*, 59–71. doi:10.1016/j.technovation.2008.05.002

Toffler, A. (1990). *Powershift*. New York: Bantam.

Tovstiga, G., & Tulugurova, E. (2007). Intellectual capital practices and performance in Russian enterprises. *Journal of Intellectual Capital, 8*(4), 695–707. doi:10.1108/14691930710830846

Tovstiga, G., & Tulugurova, E. (2009). Intellectual capital practices: a four-region comparative study. *Journal of Intellectual Capital, 10*(1), 70–80. doi:10.1108/14691930910922905

Tsai, W. (2000). Social capital, strategic relatedness and the formation of intraorganizational linkages. *Strategic Management Journal, 21*(9), 925–939. doi:10.1002/1097-0266(200009)21:9<925::AID-SMJ129>3.0.CO;2-I

Tsai, W. (2001). Knowledge transfer in intraorganizational networks: Effects of network position and absorptive capacity on business unit innovation and performance. *Academy of Management Journal, 44*(5), 996–1004. doi:10.2307/3069443

Tsai, W., & Ghosal, S. (1998). Social Capital and Value Creation: The Role of Intrafirm Networks. *Academy of Management Journal, 41*, 464–476. doi:10.2307/257085

Tseng, C., & Goo, Y. J. (2005). Intellectual capital and corporate value in an emerging economy: empirical study of Taiwanese manufacturers. *R & D Management, 35*(2), 187–201. doi:10.1111/j.1467-9310.2005.00382.x

Tseng, C., & Goo, Y. J. (2005). Intellectual Capital and Corporate Value in an Emerging Economy: Empirical Study of Taiwanese Manufacturers. *R & D Management, 35*, 187–201. doi:10.1111/j.1467-9310.2005.00382.x

Tushman, M., & Nadler, D. (1986). Organizing for Innovation. *California Management Review, 28*(3), 74–92.

Tushman, M., & Nadler, D. (1986). Organizing for Innovation. *California Management Review, 28*, 74–92.

Un, C. A., & Cuervo-Cazurra, A. (2004). Strategies for Knowledge Creation in Firms. *British Journal of Management, 15*, S27–S41. doi:10.1111/j.1467-8551.2004.00404.x

UNDP United Nations Development Program. (2005). *World development report 2005: International cooperation at a crossroads, aid, trade, and security in an unequal world. United Nations Development Program.* New York: Hoechstetter Printing Co.

Utterback, J. M. (1994). *Mastering the Dynamics of Innovation*. Boston: Harvard Business School Press.

Uzzi, B. (1996). The sources and consequences of embeddedness for the economic performance of organizations: The network effect. *American Sociological Review, 61*(4), 674–698. doi:10.2307/2096399

Uzzi, B. (1997). Social Structure and Competition in Interfirm Networks: The Paradox of Embeddedness. *Administrative Science Quarterly, 42*, 35–67. doi:10.2307/2393808

Uzzi, B., & Spiro, J. (2005). Collaboration and Creativity: The Small World Problem. *American Journal of Sociology, 111*(2), 447–504. doi:10.1086/432782

Van Aken, J. E. (2001). *Mode 2 knowledge production in the field of management*. Working Paper 01.13, Eindhoven Centre for Innovation Studies, Eindhoven University of Technology, Eindhoven.

Van Buren, M. E. (1999). A yardstick for knowledge management. *Training & Development, 53*(5).

Van Buren, M. E. (1999). *Making knowledge counts: knowledge management systems and the human element.* Unpublished manuscript.

Van de Ven, A. H. (1986). Central problems in the management of innovation. *Management Science*, *32*(5), 590–607. doi:10.1287/mnsc.32.5.590

Van der Zahn, M. J.-L. W., Tower, G., & Neilson, J. (2004). *Intellectual Capital and the Efficiency of Value Added: Trends in the Singapore Capital Market 2000-2002*. Burleigh, UK: Poseidon Books.

Viale, R., & Etzkowitz, H. (2005, May). *Third academic revolution: polyvalent knowledge; the 'DNA' of the triple helix*. Paper presented at the 5th Triple Helix Conference, Turin, Italy.

Von Hippel, E. (1986). Lead users: a source of novel product concepts. *Management Science*, *32*(7), 791–805. doi:10.1287/mnsc.32.7.791

Von Hippel, E. (1994). 'Sticky information' and the locus of problem solving: Implications for Innovation. *Management Science*, *40*(4), 429–438. doi:10.1287/mnsc.40.4.429

Von Hippel, E., & Tyre, M. (1995). How Learning by Doing is Done: Problem Identification in Novel Process Equipment. *Research Policy*, *24*(1), 1–12. doi:10.1016/0048-7333(93)00747-H

Von Krogh, G., & Ross, J. (1995). *Organizational epistemology*. New York: MacMillan and St Martin's Press.

Vuontisjarvi, T. (2006). Corporate social reporting in the European context and human resource disclosures: an analysis of Finnish companies. *Journal of Business Ethics*, *68*, 331–354. doi:10.1007/s10551-006-9094-5

Wang, C. L., & Ahmed, P. K. (2007). dynamic capabilities: A Review and Research Agenda. *International Journal of Management Reviews*, *9*(1), 31–51. doi:10.1111/j.1468-2370.2007.00201.x

Wang, W., & Chang, C. (2005). Intellectual Capital and Performance in Causal Models: Evidence from the Information Technology Industry in Taiwan. *Journal of Intellectual Capital*, *6*(2), 222–236. doi:10.1108/14691930510592816

Wasko, M., & Faraj, S. (2005). Why Should I Share? Examining Social Capital and Knowledge Contribution in Electronic Networks of Practice. *Management Information Systems Quarterly*, *29*(1), 35–37.

Watanabe, C. (1999). Systems option for sustainable development. *Research Policy*, *28*(7), 719–749. doi:10.1016/S0048-7333(99)00018-9

Watanabe, C., & Ane, B. K. (2003). Coevolution of manufacturing and service industry functions. *Journal of Service Research*, *3*(1), 101–118.

Watanabe, C., & Tokumasu, S. (2003). Optimal timing of R&D for effective utilization of potential resources in innovation. *Journal of Advances in Management Research*, *1*(1), 11–27.

Watanabe, C., & Wakabayashi, K. (1996). The perspective of techno-metabolism and its insight into national strategies. *Research Evaluation*, *6*(2), 69–76.

Watanabe, C., Takayama, M., Nagamatsu, A., & Tagami, T. (2002). Technology spillover as a complement for high-level R&D intensity in the pharmaceutical industry. *Technovation*, *22*, 245–258. doi:10.1016/S0166-4972(01)00004-9

Watanabe, C., Zhu, B., Griffy-Brown, C., & Asgari, B. (2001). Global technology spillover and its impact on industry's R&D strategies. *Technovation*, *21*, 281–291. doi:10.1016/S0166-4972(00)00048-1

Webb, E., Campbell, D. T., Schwartz, R. D., & Serchrest, L. (1996). *Unobtrusive measures: Non-reactive research in the social sciences*. Chicago: Rand McNally.

Webster, J., & Watson, R. T. (2002). Analyzing the past to prepare for the future: Writing a literature review. *Management Information Systems Quarterly*, *26*(2), xiii–xxiii.

WEF. (2009). *The global competitiveness report 2009-2010*. Geneva: World Economic Forum.

Welford, R. (1994). Improving Corporate Environmental Performance. *Environmental Management and Health*, *5*, 6–10. doi:10.1108/09566169410057100

Wells, R. (2006). Education's effect on income inequality: an economic globalization perspective. *Globalisation, Societies and Education*, *4*(3), 371–391. doi:10.1080/14767720600955428

Wenerfelt, B. (1989). From critical resources to corporate strategy. *Journal of General Management*, *14*(3), 4–12.

Wernelfelt, B. (1984). A Resource-Based View of the Firm. *Strategic Management Journal*, *5*, 171–180. doi:10.1002/smj.4250050207

Werner, B. M., & Souder, W. E. (1997). Measuring R&D performance: State of the art. *Research Technology Management*, *40*(2), 34–42.

Wernerfelt, B. (1984). A Resource-Based View of the Firm. *Strategic Management Journal*, *5*, 171–180. doi:10.1002/smj.4250050207

Wernerfelt, B. (1995). The resource-based view of the firm: ten years after. *Strategic Management Journal*, *5*(2), 171–174. doi:10.1002/smj.4250050207

Whetten, D. A., & Mackey, A. (2002). A social actor conception of organizational identity and its implications for the study of organizational reputation. *Business & Society*, *41*(4), 393–414. doi:10.1177/0007650302238775

White, H. C. (2008). *Identity and Control: How Social Formations Emerge*. Princeton, NJ: Princeton University Press.

Whittington, R. (2001). *What Is Strategy and Does It Matter?*London: Thomson Learning Business Press.

Wiggins, R. R., & Ruefli, T. W. (2002). Sustained Competitive Advantage: Temporal Dynamics and the Incidence and Persistence of Superior Economic Performance. *Organization Science*, *13*(1), 82–105.

Wiig, K. M. (2004). *People-focused knowledge management: how effective decision making leads to corporate success* (pp. 213–237). New York: Elsevier Inc.

Wilcox, A., Fowler, S., & Zeithaml, C. (2001). Managing Orgnizational Competencies for Competitive Advantage: The Middle-Management Edge. *The Academy of Management Executive*, *15*, 95–106.

Williams, S. M. (2000). *Is a company's intellectual capital performance and intellectual capital disclosure practices related?: Evidence from publicly listed companies from the FTSE 100* (pp. 1-41). Retrieved March 29, 2006 from www.vaic-on.net/start.htm

Williamson, O. E. (1975). *Markets and Hierarchies: Analysis and Antitrust Implications: A Study in the Economy of Internal Organization*. New York: The Free Press.

Williamson, O. E. (1985). *The Economic Institutions of Capitalism*. New York: Macmillan Inc.

Williamson, O. E. (1991). Strategizing, economizing, and economic organization. *Strategic Management Journal*, *12*(8), 75–94. doi:10.1002/smj.4250121007

Williamson, O. E. (1999). Strategy research: governance and competence perspectives. *Strategic Management Journal*, *20*(12), 1087–1108. doi:10.1002/(SICI)1097-0266(199912)20:12<1087::AID-SMJ71>3.0.CO;2-Z

Winter, S. G. (2003). Understanding dynamic capabilities. *Strategic Management Journal*, *24*(10), 991–995. doi:10.1002/smj.318

Winter, S. G. (2006). The logic of appropriability: From Schumpeter to Arrow to Teece. *Research Policy*, *35*(8), 1100–1106. doi:10.1016/j.respol.2006.09.010

WIPO. (2004). *WIPO intellectual property handbook: Policy, law and use*. Geneva: World Intellectual Property Organization.

Witt, U. (2000). Changing cognitive frames - changing organizational forms: an entrepreneurial theory of organizational development. *Industrial and Corporate Change*, *9*(4), 733–755. doi:10.1093/icc/9.4.733

Wolfe, D. A., & Gertler, M. (2004). Clusters from the inside and out: local dynamics and global linkages. *Urban Studies (Edinburgh, Scotland)*, *41*(5-6), 1071–1093.

Woolcock, M. (1998). Social Capital and Economic Development: Toward a Theoretical Synthesis and Policy Framework. *Theory and Society*, *27*(2), 151–208. doi:10.1023/A:1006884930135

World Bank. (1999). *Knowledge for development*. New York: Oxford University Press.

World Bank. (2004). *A better investment climate for everyone. World Development Report*. Washington, DC: World Bank and Oxford University Press.

Wright, P. M., Dunford, B. B., & Snell, S. A. (2001). Human resources and the resource based view of the firm. *Journal of Management, 27*, 701–721. doi:10.1177/014920630102700607

Wu, S., Lin, L., & Hsu, M. (2007). Intellectual Capital, Dynamic Capabilities and Innovative Performance of Organisations. *International Journal of Technology Management, 39*, 279–296. doi:10.1504/IJTM.2007.013496

Wu, W., Tsai, H., Cheng, K., & Lai, M. (2006). Assessment of Intellectual Capital Management in Taiwanese IC Design Companies: Using DEA and the Malmquist Productivity Index. *R & D Management, 36*, 531–545. doi:10.1111/j.1467-9310.2006.00452.x

Wu, W.-Y., Chang, M.-L., & Chen, C.-W. (2008). Promoting innovation through the accumulation of intellectual capital, social capital and entrepreneurial orientation. *R & D Management, 38*(3), 265–277. doi:10.1111/j.1467-9310.2008.00512.x

Yam, R. C. M., Cheng, J., Fai, K., & Tang, E. P. Y. (2004). An Audit of Technological Innovation Capabilities in Chinese Firms: Some Empirical Findings in Beijing, China. *Research Policy, 33*, 1123–1140. doi:10.1016/j.respol.2004.05.004

Yam, R., Guan, J. C., Pun, K. F., & Tang, E. (2004). An audit of technological innovation capabilities in Chinese firms: some empirical findings in Beijing, China. *Research Policy, 33*, 1123–1140. doi:10.1016/j.respol.2004.05.004

Yin, R. K. (1984). *Case Study Research: Design and Methods*. San Francisco: Sage Publications.

Yli-Renko, H., Autio, E., & Sapienza, H. J. (2001). Social capital, knowledge acquisition, and knowledge exploitation in young technology-based firms. *Strategic Management Journal, 22*, 587–613. doi:10.1002/smj.183

Youndt, M., & Snell, S. (2004). Human Resource Configurations, Intellectual Capital and Organization Performance. *Journal of Managerial Issues, 16*, 337–360.

Youndt, M., Subramanian, M., & Snell, S. (2004). Intellectual Capital Profiles: An Examination of Investments and Returns. *Journal of Management Studies, 42*, 335–361. doi:10.1111/j.1467-6486.2004.00435.x

Zack, M. H. (1999). Developing a Knowledge Strategy. *California Management Review, 41*(3), 125–145.

Zahra, S. A., & George, G. (2002). Absorptive capacity: A Review, reconceptualization, and extension. *Academy of Management Review, 27*(2), 185–203. doi:10.2307/4134351

Zahra, S. A., & George, G. (2002). Absorptive Capacity: A Review, Reconceptualization, and Extension. *Academy of Management Review, 27*(2), 185–203. doi:10.2307/4134351

Zollo, M., & Winter, S. G. (2002). Deliberate Learning and the Evolution of dynamic capabilities. *Organization Science, 13*(3), 339–351. doi:10.1287/orsc.13.3.339.2780

Zuckerman, E. W. (2003). Robust identity or nonentities? Typecasting in the feature-firm market. *American Journal of Sociology, 108*(5), 1018–1074. doi:10.1086/377518

About the Contributors

Miriam Delgado Verde is Assistant Professor of Business Administration at Universidad Complutense de Madrid, Spain. She was a Fellow at Manchester Institute of Innovation Research (Manchester Business School-University of Manchester) from 2008 to 2009. Also, Professor Delgado Verde has worked in several research projects and she is author and co-author of several papers, books, and book chapters on the Resource-Based View, Intellectual Capital, Technological Innovation and Knowledge Management, which have been published in international and national journals as well as editorial groups.

* * *

Salma Alguezaui has got a BA in business engineering from 'Al Akhawayn University' and a Master in e-business from the 'e-business management School', Scuola Superiore ISUFI, Universita' del Salento. Actually, she is a PhD student and she is ending her research at 'e-business management School', Scuola Superiore ISUFI, Universita' del Salento. She has published several papers in peer reviewed conferences and her research interests include social capital, knowledge management and innovation at industrial districts, regional and country level.

Pablo Blasco Bocigas is a professor of Business Economy at the Complutense University of Madrid. He received his Bachelor in Bussiness Administration and Economic Science from the University of Zaragoza and his Master in Human Resource Management from ESIC Business School. Currently he is doing his PhD. in Business Adminitration at the Complutense University of Madrid and his research interest is based on the analysis of the social capital impact on firms' outcomes from the strategic management perspective. In addition he is combining this academic work with the domain of consultancy. He is working as a consultant in different organizational projects related to the business management.

Eduardo Bueno is Professor of business administration and the Chair of the Knowledge Management and Business Innovation Research Institut (IADE) at Universidad Autónoma de Madrid. He received his Ph.D from Universidad Complutense de Madrid in economy. His current research focuses on knowledge and strategic management, intellectual capital and innovation management.

Yuan-Chieh Chang is associate professor, Institute of Technology Management, National Tsing Hua University, Taiwan. He earned a Ph.D. in Science and Technology Policy and Management, University of Manchester (2001), England and a MBA in management of technology at the Asian Institute of Technology (1996), Thailand. He was visiting scholar in Copenhagen Business School, Denmark (2006), Tsing

Hua University, China (2008), and Stanford University, the US (2008). Dr. Chang is Deputy General Secretary, Chinese Society for Management of Technology (CSMOT) in Taiwan. His current research interests are measuring intellectual capital, science and technology policy and R&D management. He serves editorial advisory boards of R&D Management and International Journal of Innovation Systems Management. He published papers in the journals of Research Policy, Regional Studies, Scientometrics, R&D Management, Research Technology Management, Technology in Society, and International Journal of Technology Management.

Hui-Ru Chi is a Ph.D. student in the Graduate institute of Human Resource Management at National Changhua University of Education in Taiwan. Her major is in the areas of technology management and organizational innovation management. She is also a lecturer in the Department of Finance at National Taichung Institute of Technology in Taiwan.

Jorge Cruz González got a degree in business management and is PhD candidate in business management at Universidad Complutense de Madrid (Spain). The author has participated in several research projects developed in the Department of Business Administration and funded by various public and private institutions. He is member of the research group about "Strategy, Knowledge and Innovation" in the firm (E.C.I.) at Universidad Complutense de Madrid. His main research topics are focused on the analysis of Dynamic Capabilities, Absorptive Capacity, Innovation Management and Organizational Adaptation through Learning. At these topics he has presented theoretical and empirical studies at national and international scientific conferences.

Charles (Kalev) Ehin is an Emeritus Professor of Management and former Dean of the Gore School of Business at Westminster College in Salt Lake City, Utah USA. He is a recognized authority on innovation dynamics having published three groundbreaking books on the subject. They include Unleashing Intellectual Capital (Butterworth-Heinemann, 2000), Hidden Assets: Harnessing the Power of Informal Networks (Springer, 2005) and his latest work The Organizational Sweet Spot: Engaging the Innovative Dynamics of Your Social Networks (Springer, 2009). Ehin is also the author of Aftermath (Publish America, 2004), which is a personal account of the origins and after effects of World War II in Europe. He received his undergraduate degree from Colgate University, his MBA from Syracuse University and his Ph.D. from the University of Oklahoma. His e-mail address is kalev1@msn.com and his website is www.UnManagement.com.

Raffaele Filieri is the Research Fellow at Dipartimento di Sociologia e Comunicazione, Universita' degli Studi di Roma 'La Sapienza', Italia; and Kemmy Business School, University of Limerick, Limerick, Ireland. Her research interests include knowledge sharing, innovation, social networks, process reengineering, marketing intelligence, tourism management. Her previous studies and academic experience include being a visiting PhD at e-Business Management School - Scuola Superiore ISUFI, Universita' del Salento 92005- 2008), as well as being a PhD student at Industrial Complex Systems PhD School (SICS) within the University of Roma 'Roma 3' (2005-2008), her PhD thesis defence, 'Overcoming knowledge sharing barriers through communities of practice' (January 2009) and a BA/MSC in Communication Studies at Faculty of Social Sciences. Previous (non academic) positions include consultant at Focus Management, Upper Saddle River, NJ: at Clio S.r.l., Lecce, Instructor at Lattanzio & Associati, Milano, and Project Manager, Cittadinanzattiva (n.g.o.).

Bruce Heiman's education includes an A.B. in psychology from Stanford (1981), an M.S. from Stanford (General Engineering, Product Design (1983), an M.B.A. from Universiteit Nijenrode (1990) and a PhD from UC Berkeley, Haas School of Business (2002). He has published eclectically in a substantial number of journals and conference proceedings. Research interests include transaction cost economics, problem finding and solving, and design thinking, He has worked previously as a faculty member at Universiteit Maastricht, NL, Faculty of Economics. Prof. Heiman has over 12 years experience working full time or consulting in strategic management, with an emphasis on innovation processes.

Janet Holland completed a Ph.D. in Teaching and Leadership, Instructional Design and Technology, with a minor in Communications from the University of Kansas. Dr. Holland currently serves as an Assistant Professor at Emporia State University, teaching pre-service teachers and master degree students of Instructional Design and Technology. Research interests include improving curriculum pedagogy issues including, service learning, affective learning communities, peer mentoring, and the globalization of instruction. As an instructional designer, new technologies are continually examined in an effort to inspire innovative teaching and learning practices. Dr. Holland's goal is to create authentic, relevant, meaningful, engaging learning experiences.

Joe Holland has a Ph.D. degree in Nuclear Engineering (N.E.) from Texas A&M University and a M.S. (N.E.) and a B.S. (N.E.) from Kansas State University. Dr. Holland currently is the Chief Engineer and Scientist of Advatech Pacific, Inc. which is an aerospace oriented research, development and engineering services organization. Advatech Pacific's current research areas range from advanced system engineering for space vehicles, launch systems, tactical communications systems and advanced structural analysis of non-isotropic materials. As Chief Engineer/Scientist, Dr. Holland is continually investigating new technologies, processes and tools, and more importantly finding and training personnel to use them.

Pia Hurmelinna-Laukkanen is a Professor of International Marketing (acting) at the Faculty of Economics and Business Administration at University of Oulu, a Senior Researcher in Business Law at School of Business, and a project manager at the Technology Business Research Center (TBRC) at Lappeenranta University of Technology (Finland). She has published her work in over 25 refereed articles in journals such as R&D Management, Technovation, and International Journal of Production Economics, over 70 conference papers, and in other scientific and managerial publications.

Carlos Kan is a Panamanian. He received his first degree in Electronic Engineering, Universidad Santa Maria La Antigua, Panama. Recently, he obtained MBA degree in Technology Management, College of Technology Management, National Tsing Hua University, Taiwan.

Amir Khanlari is PhD student of Marketing Management in the management faculty of University of Tehran. He received a master degree in information technology management. His research interests include customer relationship management, knowledge management, and business intelligence. He has published his papers in the Journal of Computers and Industrial Engineering, Iranian Journal of Accounting & Auditing Review and several local journals.

Mónica Longo is an adjunct professor of financial economy and accounting at Universidad Complutense de Madrid and researcher of IADE at Universidad Autónoma de Madrid. She is studying for a PhD in Business Administration at Universidad Autónoma de Madrid. Her current research interests include organizational identity, knowledge management and strategic management.

Masahiro Nakagawa graduated from Kyoto University with a bachelor's degree in science. He received his Ph. D. (industrial engineering and management) from Tokyo Institute of Technology. Nakagawa joined Sumitomo Electric Industries, Ltd. (SEI) and spent a quarter century there. He conducted research and development in manufacturing process on compound semiconductor materials; engaged in management of technology in research and development unit, particularly evaluated profitability and marketability of R&D outcomes. He also started up a new business department from an R&D group. Nakagawa is currently a senior research analyst in Japan Science and Technology Agency (JST); conducting evaluation of outputs and outcomes of government funded basic research programs.His research fields are management of technology, technological innovation, and evaluation of technology.

Iman Raeesi is PhD student of Systems Management in the management faculty of University of Tehran. He received a master degree from University of Tehran in Information Technology management. His research interests include knowledge management, business intelligence, enterprise resource planning and information resources management. He has presented and published his papers in several international and Iranian conferences and journals on knowledge management, e-learning, knowledge sharing and related fields.

Elsa Alama Salazar is a Professor at the Faculty of Economics and Business at the Universidad de Piura, Peru. She obtained the Degree of Doctor in Management from the Universidad Complutense de Madrid in 2008. Her research area is Knowledge Management and Intellectual Capital. She has published several articles in scholar journals, such as, "The role of intellectual capital in technological innovation" which is an application to professional services firms in Spain, in Notebooks of Economics and Business, Issue 40, September 2009, "Intellectual Capital. A proposal to classify and measure", in Latin American Journal of Management, Volume 37, 2007. Besides, she has participated as a lecturer at different congresses, inside an outside the country, like ACEDE Congress, held in León - Spain (2008), and soon in the Congress of the Iberoamerican Academy of Management to be held in Buenos Aires - Argentina (December 2009).

M. Paz Salmador is Assistant Professor of Strategic Management at Universidad Autónoma de Madrid (Spain) and Senior Researcher in the Knowledge Society Research Centre in the Science Park of Madrid. Previously, she was Visiting Researcher at the Japan Advanced Institute of Science and Technology (Ishikawa, Japan), and at the National North-Western University (Resistencia, Argentina), and Senior Fulbright Scholar at Texas A&M University.

Babak Sohrabi is Associate Professor in Department of Information Technology Management, Faculty of Management at University of Tehran. He received his PhD degree in Management Science, from Lancaster University, England in 2000. He is the Editor-in-Chief of IT management Journal at University of Tehran. His research interests include Artificial Intelligence, Knowledge management, electronic commerce, decision support systems and organizational impact of information technology.

The results of his research activities have appeared in Journal of Interdisciplinary Mathematics, International Journal of Engineering, Journal of operational Research Society, Journal of production Research, Computers & Industrial Engineering, Review of business information systems and several other local and international journals and conferences.

Chihiro Watanabe graduated from Tokyo University with a bachelor's degree in engineering (urban planning) and received his Ph. D. (arts and science), also from Tokyo University. Professor Watanabe joined Japan's Ministry of International Trade and Industry (MITI) and spent a quarter century there. He is former Deputy Director-General of Technology Development.. Professor Watanabe is currently a professor of the Department of Industrial Management of Tokyo Seitoku University while a professor emeritus of Tokyo Institute of Technology. He is also a visiting professor of National University of Singapore, an adjunct professor of University of Jyvaskyla, and a guest scholar of the International Institute for Applied Systems Analysis (IIASA). His research fields are technology innovation management, institutional innovation, techno-economics, technology innovation, and technology policy. Professor Watanabe has published 140 refereed papers in the above fields, and awarded Minister of Education, Science & Technology Award, JSSPRM (Japan Society for Science Policy & Research Management) Award, and IAMOT (Intl. Association of Management of Technology) Research Award in 2008.

Pei-Ju Yu is an assistant researcher in the Division of International Economy at Chung-Hua Institution for Economic Research in Taiwan. Her research interests are in the areas of technology management, science, technology and innovation policy, service innovation and service science. Her current projects focus on the transformation of manufacturing as well as advanced manufacturing system.

Index

J

K

Q

R

S